Exploring the Solar System

COMPANION WEBSITE

This book has a companion website:

www.wiley.com/go/bond/solarsystem

with Figures and Tables from the book for downloading

Exploring the Solar System

Peter Bond

WILEY-BLACKWELL

A John Wiley & Sons, Ltd., Publication

Library of Congress Cataloging-in-Publication Data
Bond, Peter, 1948–
    Exploring the solar system / Peter Bond.
       p. cm.
    Includes index.
    ISBN 978-1-4443-5108-8 (cloth) – ISBN 978-1-4051-3499-6 (pbk.)
 1. Solar system.   2. Space astronomy.   I. Title.
    QB501.B68 2011
 523.2–dc23
                                            2011030179

A catalogue record for this book is available from the British Library.

Wiley also publishes its books in a variety of electronic formats. Some content that appears in print may not be available in electronic books.

Set in 9/11.5 pt Minion by Toppan Best-set Premedia Limited
Printed and bound in Malaysia by Vivar Printing Sdn Bhd

1   2012

Dedication

To my wife Edna, with whom I have travelled the rocky road of life.

Conversion Table

The metric units used in this book can be converted to English units by using the approximate conversions given below:

Length

1 kilometer = 0.62 of a mile
1 meter = 39.37 inches
1 centimeter = 0.39 inches
1 millimeter = 0.039 inches

Area

1 sq. kilometer (km^2) = 0.04 sq. miles
 1 sq. meter (m^2) = 1.2 sq. yards
 1 sq. centimeter (cm^2) = 0.155 sq. inches

Temperature

To convert °Celsius to °Fahrenheit, multiply °C by 1.8 and add 32

Speed

1000 km/h = 277.8 m/s = 621.37 mph

Volume

1 cubic cm (cm^3) = 0.061 cubic inches

Contents

COMPANION WEBSITE

This book has a companion website:

www.wiley.com/go/bond/solarsystem

with Figures and Tables from the book for downloading

Introduction

We shall not cease from exploration and the end of all our exploring will be to arrive where we started... and know the place for the first time. (T.S. Elliot)

This book is about a unique corner of the Universe, a small expanse of largely empty space that surrounds an ordinary star in the suburbs of the Milky Way galaxy. Known as the Solar System, this region is populated by the Sun, eight planets, dozens of satellites and dwarf planets, and a multitude of smaller objects.

Why is it important to explore and understand the Solar System? Because the third planet from the Sun is our home: Earth is the only place yet discovered where living organisms and intelligent life exist, or have ever existed. This unique, "Goldilocks" world is the cradle of humankind, a fragile oasis in the vastness of space.

However, spaceship Earth is subject to many threats and stresses. Some are man-made, such as deforestation, atmospheric pollution or emissions of ozone-destroying chemicals. Some are natural planetary processes, such as crustal movement and changing sea level. Others are external, including solar flares and marauding asteroids.

As news reports of natural disasters constantly remind us, Earth is an ever-changing world, subject to ice ages, mountain building, earthquakes, volcanic eruptions and devastating cosmic impacts. Since its birth some 4.5 billion years ago, the planet has endured all of these natural forces to evolve into the largely benign place we see today. If we can understand how this evolution occurred, we will then have a better chance of predicting how it will change in the future.

This is where studies of the Sun, planets and other inhabitants of the Solar System come to the fore. Only by comparing and contrasting the evolution of these very different objects can we hope to understand the past, present and future of our Earth.

This scientific endeavour has been made possible by the advent of the Space Age. During this great age of discovery, modern technology has enabled us to construct automated spacecraft and robots that can act as surrogate explorers, venturing forth into the vast, hostile ocean of space to seek out and study new worlds.

Over more than half a century, hundreds of robotic spacecraft have been sent from Earth to examine at close quarters all of the planets, and many other objects, in our Solar System. This book is based on the flood of data sent back by these probes, which has enabled scientists to assemble, piece by piece, a realistic picture of our Solar System. For the first time, human eyes have been able to see towering cliffs, dust devils, erupting volcanoes, dry river beds and ice formations on dozens of distant worlds, most of them totally alien to our experience here on Earth.

Many years ago, my imagination was captured by books that described the family of alien worlds that circle our Sun, although, at that time, most of the information available was pure speculation. I have been fascinated by the many and varied members of the Solar System ever since. It is my hope that readers of this book will be similarly fascinated and inspired.

Exploring the Solar System has been written as an introductory text book for undergraduate students with a modest background in science. However, it is also intended to inform and inspire anyone who looks up at the night sky and wishes to know more about the alien worlds that inhabit our corner of the Universe.

After an introductory chapter which provides an overview of the Solar System, the book sets out to systematically describe the main characteristics of each major planet and its retinue of satellites, as well as the smaller members of the Sun's retinue. The final chapter enables the reader to compare and contrast our Solar System with systems around distant stars, where huge numbers of strange and exotic exoplanets are now being discovered.

Questions at the end of each chapter have been added to help students to recognize and comprehend the main points of each chapter, and to compare each planetary system. Useful reference material is provided in the form of numerous appendices, an extensive reading list, and a comprehensive glossary.

Exploring the Solar System, First Edition. Peter Bond.
© 2012 by Peter Bond. Published 2012 by Blackwell Publishing Ltd.

This book would not have been possible without the support and encouragement of Ian Francis, Senior Commissioning Editor for Wiley-Blackwell, and Delia Sandford, the Managing Editor for this project. I am most grateful for their patience and forbearance as the book has edged towards completion.

My sincere thanks also go to Kelvin Matthews of Wiley-Blackwell, who has checked all of the illustrations, to the production team, especially Kathy Syplywczak, and to the various reviewers whose helpful comments and criticisms played such an important role in shaping the final text.

Much of the information this book is based on original scientific papers, many of which are listed in the final pages. Numerous other scientific sources – many now available on the Internet – were also used, including magazine articles, press releases, and other information provided by space agencies – particularly NASA - and universities. I am also very grateful to everyone who helped me to obtain, or provided me with, the spectacular images that illuminate this story of outreach and discovery.

Finally, I would like to thank my wife, Edna, who first encouraged me to describe and explain the wonders of our Solar System.

Peter Bond

ONE
Beginnings

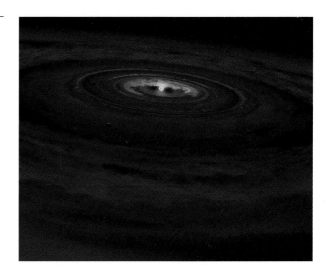

For millennia, people have studied the heavens and wondered about the nature and origins of the Sun, Moon and planets. Indeed, Solar System studies dominated the field of astronomy until the introduction of powerful telescopes and advanced instruments in the 19th century. In the last 50 years, spacecraft have flown past or orbited all of the major planets, landed on the Moon, Mars, Titan and an asteroid, and brought back samples of Moon rock, the solar wind, asteroid and comet dust. This era of robotic and human exploration has revolutionized scientists' knowledge of our corner of the Galaxy, and further astounding revelations are expected in the decades to come.

Wandering Stars

Since time immemorial, people have stared in wonderment at the night sky. In previous millennia, when the darkness of the sky was not degraded by artificial lighting, it was easy to recognize how the stellar patterns drifted from horizon to horizon as the night progressed, and how they changed as the seasons passed.

However, in addition to the familiar, twinkling stars, observers noted seven objects that moved with varying speeds against the background of "fixed" stars.[1] In order of greatest apparent brightness, they were the Sun, Moon, Venus, Jupiter, Mars, Mercury and Saturn. The ancient Greeks called them "planetes" ("wandering stars"), a designation we still use for all but the Sun and Moon.

For ancient astrologers and astronomers (the two disciplines were inextricably intertwined for many centuries) the most important of the wanderers were the Sun, which was responsible for daylight, and the Moon, which dominated the night. Both of these objects displayed visible disks and moved quite rapidly across the sky. Careful study of their regular motions and apparitions enabled people to devise calendars and introduce convenient ways of measuring time. Thus, a year was the period of time before the Sun returned to the same place in the sky, while the month was the period that elapsed between each new or full Moon.

The other five planets were rather less noticeable, though each had its own peculiar characteristics. For example, Mercury and Venus never strayed far from the Sun in the twilight skies of morning or evening. The other three moved more slowly from constellation to constellation, sometimes describing loops in the sky as they appeared to temporarily reverse direction.

It was also evident that the seven planets often came together in the sky or even passed behind the Moon during occultations. They always remained within a narrow band on the sky, known as the zodiac (after the Greek word for "animal"). The Sun's annual path across the sky, called the ecliptic, ran along the center of this celestial highway. Clearly, the planes of the planets' orbits were closely aligned with each other.

The Earth-Centered Universe

Until the mid-16th century, it was accepted as an established fact by most civilizations that Earth lay at the center of the Universe.[2] Like the axle of a wheel, everything else rotated around it.

[1] For a time, the ancient Greeks thought there were nine planets. Venus was named both as the Evening Star (Hesperus) and the Morning Star (Phosphorus). Similarly, Mercury was thought to be two different planets – Lucifer and Hermes.

[2] A Sun-centered, heliocentric model of the Universe was proposed by the Greek astronomer Aristarchus in the 3rd century BC, but it was not widely accepted.

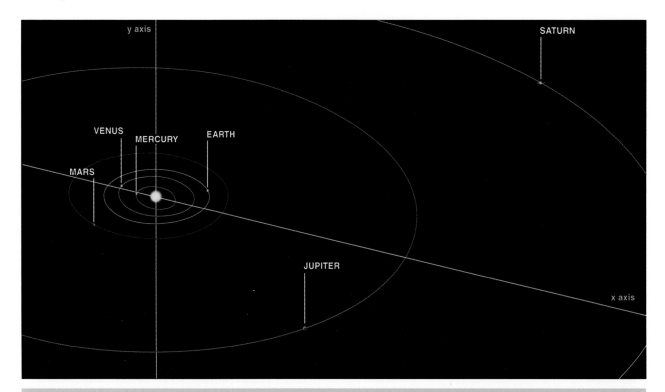

Figure 1.1   The relative sizes of the orbits of the seven "planets" (including the Sun and Moon) visible to the naked eye and recognized by ancient astronomers. All of the orbits are slightly elliptical and nearly in the same plane as Earth's orbit (the ecliptic). The diagram is from a view above the ecliptic plane and away from the perpendicular axis that goes through the Sun. (Lunar and Planetary Institute)

The reasons for this thinking seemed self evident. All of the celestial objects, including the Sun, moved across the sky from east to west (with the occasional exception of a comet or a shooting star). However, since no one experienced any of the sensations that would be expected if Earth was continually spinning, it seemed logical to believe that it was the heavens which were in motion around Earth.

According to this geocentric theory, the Sun, Moon and planets were carried by invisible, crystalline spheres which were centered on the Earth. A much larger celestial sphere carried the fixed stars around the central Earth once every day.

Although early civilizations accepted the visual evidence that Earth is (more or less) flat, this idea was contradicted by several lines of evidence (see Chapter 3). For example, different star patterns or constellations are visible from different places. However, if Earth is flat, then the same constellations should be visible everywhere at a certain time.

One key piece of evidence was the curved outline of Earth's shadow as it drifted across the face of the full Moon during a total lunar eclipse. This was the case no matter where the observation was made or at what time it took place. Since only a spherical body can cast a round shadow in all orientations, it seemed clear that Earth was round.

Similarly, observations of a sailing ship disappearing over the horizon showed that, instead of simply becoming smaller and smaller, its hull disappeared from view before the sails and mast. This could only be explained on a curved ocean.

Measuring Distances and Sizes

One of the most fundamental problems facing early astronomers was the scale of the Universe. How big were the Earth, Sun and Moon, and how far away were they? It seemed evident that Earth was huge compared with every other object, and since it was the home of humanity, it was assumed that Earth was pre-eminent.

The question of the size of the spherical Earth was solved in the 3rd century BC by Eratosthenes, who compared the length of shadows made at different locations at the time of the spring equinox (see Chapter 3). Some facts were also known about the relative sizes and distances of other objects.

Since its shadow easily covered the entire Moon during lunar eclipses, Earth had to be substantially larger than its satellite. During a solar eclipse, the Moon passed in front of the Sun, so the latter had to be further away. However, since their apparent sizes were identical, the Sun must be considerably larger than the Moon. Similarly, the Moon sometimes occulted or passed in front of stars and planets, so these, too, had to be much more remote.

Calculations by the Greek astronomers Aristarchus (c.310–c.230 BC) and Hipparchus (c.190–120 BC), based on the size of Earth's shadow, suggested that the Moon's diameter is about one third that of Earth and that its distance is nearly 59 times Earth's radius. This established the scale of the Earth-Moon system with a fair degree of accuracy. However, their simple geometric methods grossly underestimated the Sun's distance.

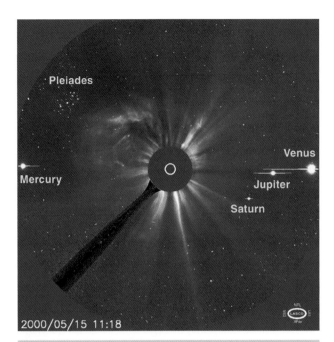

Figure 1.2 All of the major planets follow orbits that lie within 8° of the Sun's path across the sky – the ecliptic. This narrow celestial belt is known as the zodiac. In this image from the SOHO spacecraft, four planets appear close to the Sun (whose light is blocked by an occulting disk). Also in view are some background "fixed" stars, including the Pleiades cluster. (NASA-ESA)

Determination of the planetary distances remained problematical for a long time. It soon became clear to observers in the classical world that some planets move more slowly through the constellations of the night sky. Since a slow-moving planet such as Saturn was also fainter than the faster moving objects, Mars and Jupiter, it seemed logical that Saturn is further away from Earth.

It was also clear that the Sun, Moon and planets did not move at uniform speeds or follow simple curved paths across the sky. One of the most difficult observations to explain was an occasional "loop" in the motions of the more distant planets. This occurred when Mars, Jupiter and Saturn were shining brightly around midnight. At such times, the planet's nightly eastward ("prograde") motion would gradually come to a stop. It would then reverse direction toward the west, becoming "retrograde", before resuming its general movement toward the east.

The explanation for this motion had to wait until astronomers realized that the Sun was at the center of the planetary system, and that Earth orbited the Sun (see "The Central Sun" below). The loops could then be accounted for by Earth traveling along a smaller orbit so that it would catch up with, then overtake, the outer planets (see Fig. 1.3) – like an athlete on an inside track.

Accurate calculations of planetary distances also had to wait until the 17th century, when observers were able to measure angular distances with reasonable accuracy. The basic geometrical method they used was called parallax.

This involved measurement of the apparent shift in position of an object when viewed from two different locations. To illustrate

this, hold one finger upright in front of your nose and close first one eye and then the other. The finger seems to shift position against the background, although it is, of course, stationary. When the finger is moved closer, the shift appears larger, and vice versa.

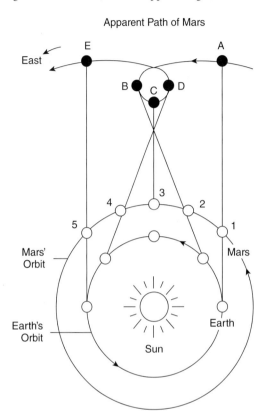

Figure 1.3 The apparent retrograde loops in the motions of Mars, Jupiter and Saturn are now known to be caused by the relative orbital movement of the planets and Earth. Since Earth moves faster along its orbit than the more distant planets, it overtakes them on the inside track. As Earth approaches and passes Mars, the slower moving outer planet (points 2 to 4) appears to move backward (points B to D) for a few months against the backcloth of "fixed" stars. (Kenneth R. Lang / Tufts University, NASA)

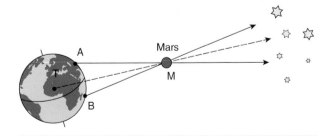

Figure 1.4 The distance of a planet such as Mars can be calculated by measuring its angle of sight – its location against the background of fixed stars – from two or more places on Earth. If the length of the baseline (e.g. the distance between two viewing sites, A–B) is known, the distance can be found by using simple trigonometry. (ESO)

Astronomers realized that, if a parallax shift in a planet's position could be measured from two widely separated locations, then its distance could be calculated. This method was first used by a French astronomer, Jean Richer, working in Cayenne (French Guiana) together with Giovanni Domenico Cassini and Jean Picard in Paris. They made simultaneous parallax observations of Mars during its closest approach in 1671, using the recently invented pendulum clocks to ensure that the measurements were made at precisely the same moment.[3]

Cassini's calculations led to a value of about 140 million km for the astronomical unit (AU) – the mean Sun-Earth distance. Now that this distance was known with reasonable accuracy, Kepler's third law (see Box 1.1) could be used to calculate the distances of the Sun and planets for the first time.

During the 18th century a great deal of time, money and effort was spent in attempting to refine these figures. One method was to observe rare transits of Venus across the face of the Sun from many different locations. The most famous transit observations took place in 1761 and 1769 when the British explorer, Captain James Cook, sailed to the Pacific as part of an army of 150 observers scattered across the globe, but these gave very inaccurate results (see Chapter 6).

More successful was the world-wide effort to determine the parallax of the asteroid Eros when it passed close to Earth in 1931. Highly accurate measurements were possible since Eros has no atmosphere and appears as a mere point of light in even the largest telescopes. The value of the astronomical unit turned out to be 149.6 million km.

Since then, more sophisticated techniques have been introduced to refine the scale of the Solar System. One of the most successful is radar, when radio signals are reflected from the surfaces of distant objects (see Chapters 5, 6 and 13). Since the velocity of these microwaves is known and the time taken between emission and reception can be measured to a fraction of a second, the distance can be readily calculated. (Radar has also revealed the sizes and shapes of hundreds of asteroids.) A similar technique used to calculate changes in the Earth-Moon distance involves the use of laser pulses bounced off special reflectors left on the lunar surface.

Once an object's distance is accurately known, the diameter can be determined from its apparent angular size, as seen in a telescope. Unfortunately, this is very difficult for the smaller or more distant members of the Solar System, particularly if their albedo, or surface reflectivity, is uncertain.

In general, the larger an object, the more light its surface reflects. However, some objects are much better mirrors than others. A small, reflective object can have the same apparent brightness as a large, dark object. New observations of some Kuiper Belt objects, beyond the orbit of Pluto, indicate that their albedos are greater than previously believed. Since they are more reflective than anticipated, astronomers have revised their diameters downwards.

Another method, involving the occultation of a star by a planet or other object, is especially valuable in relation to bodies which are normally difficult to observe. The object's diameter is calculated from the length of time during which it hides the star from view. Unfortunately, if it possesses a dense, cloudy atmosphere, the occultation only gives the diameter at the cloud tops.

The Central Sun

The difficult task of breaking with tradition and accepting the Sun as the center of the Universe began with a Polish priest and astronomer named Nicolaus Copernicus (1473–1543). He decided that the only way to make sense of the planetary orbits was to relegate Earth to the status of a planet that orbited the Sun. The movement of the stars across the sky was then explained by the rotation of the spherical Earth, while the calendar of seasons and changing constellations in the heavens were accounted for by its year-long journey around the Sun.

Copernicus' most significant work, called *De Revolutionibus Orbium Celestium* (Concerning the Revolutions of the Celestial Spheres), was published shortly before his death. Curiously, this did not provoke a violent reaction by the establishment of the day, nor did it immediately lead to any major upheaval in scientific thought. Lacking enough evidence to swing the argument one way or the other, the great minds of the day were faced with an impasse.

Half a century passed before the interventions of two great scholars swung the argument in favor of Copernicus' heliocentric theory. The first breakthrough was made in 1609 by a young German named Johannes Kepler. By one of those strange twists of irony, Kepler was a pupil of Tycho Brahe, one of the leading opponents of the Copernican order. Given the unenviable task of finding an explanation for the retrograde motion of Mars (see Fig. 1.3), Kepler was able to draw upon the excellent observational data recorded by his employer.

Brahe died in 1601, but Kepler continued to laboriously examine the problem before finally arriving at his eureka moment. The planetary orbits, he declared, were not circles but ellipses.[4] Within a short time, Kepler was able to draw up the first two laws of planetary motion (see Box 1.2). His third, and probably most important law, followed in 1619.

As a result, the relative distance of each planet from the Sun could be calculated accurately. Saturn, the most remote planet known at the time, turned out to be nearly ten times further from the Sun than Earth. Since the actual distances remained unknown, the standard unit of measurement became the astronomical unit.

In the same year that Kepler discovered elliptical orbits, an Italian scientist named Galileo Galilei made a simple refracting telescope, comprising two lenses at either end of a narrow tube, and began to study the heavens. Within a short time, he had obtained visual evidence to support the theories of Copernicus and Kepler. Galileo became the first person in history to see the phases of Venus caused by its movement around the Sun. He also observed mountains and craters on the Moon, and saw the planets as disks, rather than points of light.

Most significant of all was his discovery of four star-like objects close to Jupiter. By watching their daily motions, he was able to

---

[3]A by-product of this experiment was the discovery that a pendulum swung more slowly at Cayenne than at Paris, showing that gravity is slightly weaker at the equator. Newton later used this result to show that Earth's diameter is greatest at the equator.
[4]Kepler's task was made slightly easier by the fact that, of the five known planets, only Mercury followed a more elliptical path than Mars.

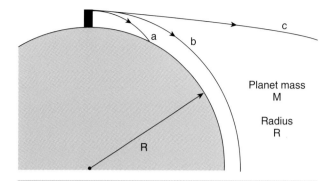

Figure 1.6 (a) If a spacecraft does not accelerate to orbital velocity, it will fall back to the planet's surface. (b) If it reaches orbital velocity, it will remain in a closed path (orbit) around the planet under free fall conditions. (c) If the spacecraft reaches escape velocity, it will be able to break free from the planet's gravitational pull and travel to another planet. The same rules apply to planets and spacecraft in orbit around the Sun.

Figure 1.5 In January 1610, Galileo Galilei used his simple refracting telescope to discover three "stars" aligned on either side of Jupiter. Over a period of several weeks, a fourth "star" appeared. As they shifted positions, Galileo correctly deduced that these were satellites. *Occ.* is the Latin abbreviation for "west" and *Ori.* Stands for "east."

calculate their orbital periods and show that they were Jovian moons (see Chapter 7). The discovery of the first planetary satellites (other than the Moon) supported theories that Earth was not at the center of the Universe and that everything did not revolve around our world.

Galileo's discoveries caused a sensation, although the leaders of the Roman Catholic Church obstinately continued to support a geocentric Universe. In 1633, Galileo was brought before the Inquisition and forced to recant under threat of torture.

Newton and Gravity

The next challenge was to find an explanation for Kepler's laws. Although Galileo conducted numerous experiments into the effects of gravity, he did not realize the full significance of his discoveries. This was left to an Englishman, Isaac Newton, who was born in 1642, the year that Galileo died.

One anecdote attributes Newton's discovery of universal gravitation to him observing an apple falling from a tree. Whatever the truth, by 1684 Newton was able to explain planetary motions. His law of gravitation stated that all objects attract each other, and that the strength of this gravitational attraction is proportional to their mass (see Chapter 8).

Clearly, since the Sun has nearly all of the mass in the Solar System, it should pull all of the other bodies into it. Newton explained that this did not happen because their orbital velocities are just sufficient to counteract the Sun's gravity. The result is that the planets fall towards the Sun in such a way that the curve of their fall takes them completely around it. This is sometimes known as free fall. (The same explanation, of course, applies to artificial satellites.)

Newton's law also stated that the strength of gravitational attraction decreases with distance. For example, if planet A is twice as far from the Sun as planet B, then the gravitational force exerted by the Sun on planet A is one quarter that exerted on planet B. In practical terms, this means that a satellite in low Earth orbit must travel at 8 km/s, whereas the Moon only has to circle the Earth at 1 km/s in order to avoid crashing into our planet. Similarly, planets further from the Sun are able to move more slowly around their orbits than those in the inner Solar System.

Newton's law also explained why a planet's orbital speed increased as it approached perihelion (closest point to the Sun) and slowed near aphelion (furthest point from the Sun).

From this time on the orbital mechanics of the Solar System were very well understood. With the exception of Mercury, whose orbital motion refused to obey Newton's law (see Chapter 5), the only significant problems involved minor variations in orbits caused by gravitational interactions between the planets, particularly those involving massive Jupiter. Careful study of unexpected changes in the orbital velocity of Uranus even enabled the position of an unknown planet, Neptune, to be successfully calculated (see Chapter 11) – although there are those who consider the discovery to be pure chance.

What is a Planet?

In the ancient world, astronomers counted eight planets. When the Sun, Earth and Moon are removed from their list, the number of planets visible to the naked eye is reduced to five: Mercury, Venus, Mars, Jupiter and Saturn.

With the invention of the telescope, the possibility arose of finding fainter, more remote planets. The first newcomer, Uranus, was discovered far beyond the orbit of Saturn by William Herschel in 1781. The list was further increased in 1801, when Giuseppe Piazzi found Ceres in the gap between the orbits of Jupiter and Mars. Pallas, Juno and Vesta – objects in similar orbits to Ceres – were discovered between 1802 and 1807. Since they were clearly much smaller and less substantial than the other planets, they were soon downgraded to "minor planets" or "asteroids" (star-like objects).

Almost 40 years passed before the eighth planet, Neptune, was discovered by Johann Galle and Heinrich D'Arrest. However, neither Uranus nor Neptune seemed to be following its expected path, suggesting that an even more distant planet might be

Box 1.1   Orbits

The direction a spacecraft or other body travels in orbit can be prograde, when a satellite moves in the same direction as the planet (or star) rotates, or retrograde, when it goes in a direction opposite to the planet's (or star's) rotation. All of the planets orbit the Sun in a prograde direction – west to east or counterclockwise as observed from above the Sun's north pole. However, many comets move in a retrograde (clockwise) direction.

Various technical terms are used to describe the characteristics of these orbits. The time an object takes to complete one orbit is known as the **orbital period**. The closest point of an orbit has the prefix "peri" – hence perigee for a satellite of the Earth and perihelion for an object orbiting the Sun. (Helios = Sun) The furthest point in an orbit has the prefix "ap" – as in apogee and aphelion.

The plane of Earth's orbit around the Sun is called the ecliptic. The orbits of the other planets, comets and asteroids are tilted to this plane. The angle of the tilt is the **orbital** inclination. The inclination of a satellite's orbit is measured with respect to the planet's equator. Hence, an orbit directly above the equator has an inclination of 0°, while one passing over a planet's poles has an inclination of 90°.

A planet, asteroid or comet crosses the ecliptic twice during each orbit of the Sun. The points where an orbit crosses a plane are known as nodes. When an orbiting body crosses the ecliptic plane going north, the node is referred to as the **ascending node**. Going south, it is the **descending node**. The line that joins the ascending node and the descending node of an orbit is called the **line of nodes**.

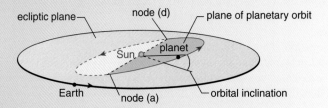

Figure. 1.7   Some important characteristics of a planet's orbit. Here the planet is inferior, that is, closer to the Sun than Earth. Its orbit is inclined to the ecliptic – the plane of Earth's orbit. The planet's orbit crosses the ecliptic at two nodes – the ascending node (a) and the descending node (d). (After Open University)

One of the most important orbital, or Keplerian, elements is the semimajor axis, the average distance of an object from its primary (planet or Sun). The shape of the orbit is described by its eccentricity, measured as a number between zero and 1. An eccentricity of zero indicates a circular orbit. A parabola has an eccentricity of 1.

influencing the movements of its neighbors. The search for this world concluded in 1930 when Clyde Tombaugh recorded the tiny image of Pluto.

For many years, it was generally accepted that there were nine planets, despite growing concerns that Pluto seemed to be too small and lacking in mass to deserve this title. The crunch came in 2003, when Mike Brown discovered 2003 UB313 (now named Eris), an object that is comparable in size to Pluto. With the introduction of ever more sensitive detectors, it seemed likely that there would soon be dozens of Pluto-sized planets.

Aware that there was no generally accepted definition of the term "planet," and faced with a fierce debate over whether Pluto should be demoted, members of the International Astronomical Union gathered in Prague for the 2006 General Assembly. After a lengthy discussion, they agreed to define a planet as a celestial body that: (a) is in orbit around the Sun, (b) has sufficient mass

for its self-gravity to overcome rigid body forces so that it assumes a hydrostatic equilibrium (nearly round) shape, and (c) has cleared the neighborhood around its orbit.

Based on these criteria, the Solar System now consists of eight planets: Mercury, Venus, Earth, Mars, Jupiter, Saturn, Uranus and Neptune. A new distinct class of objects called "dwarf planets" was also introduced. To be classified as a dwarf planet, an object must orbit the Sun and have a nearly round shape. The first dwarf planets to be announced were Ceres (the largest asteroid), Pluto and Eris, although many more are expected to be discovered in the future.

This decision has not met with universal approval. One common criticism relates to what exactly is meant by a planet "clearing its neighborhood." For example, critics argue that Neptune is accepted as a planet, even though many Kuiper Belt objects (including Pluto) cross its orbit. Perhaps, they suggest, it would be more appropriate to use size as a criterion, particularly

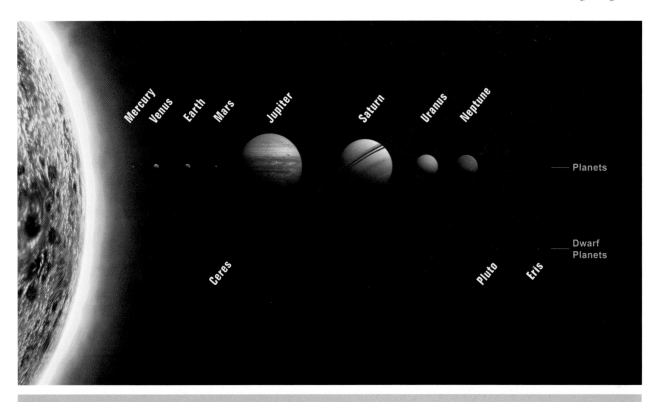

Figure 1.8 In the "new" Solar System, as defined by the International Astronomical Union in 2006, there are eight planets: Mercury, Venus, Earth, Mars, Jupiter, Saturn, Uranus and Neptune (see here in order of their distance from the Sun). A new, distinct class of objects called "dwarf planets" includes the largest asteroid, Ceres, and the two largest known Kuiper Belt objects, Pluto and Eris. The relative sizes of the planets and the Sun are shown. Jupiter's diameter is about eleven times that of Earth, and the Sun's diameter is about ten times that of Jupiter. The distances of the planets are not shown to scale. (IAU)

bearing in mind the diameters of objects that are large enough for gravity to dominate structural strength. There is also some discomfiture with defining Ceres – the largest of the asteroids – as a dwarf planet.

Another complication arises when the current definition is extended to extrasolar planets, that is, planets orbiting other stars (see Chapter 14). Size is not a useful factor, since many of these planets are similar in size and mass to small, cool "failed stars" known as brown dwarfs. Instead, astronomers attempt to distinguish between a giant extrasolar planet and a brown dwarf by determining how they were born. A star is formed during the gravitational collapse of a gaseous nebula, whereas a planet is the product of collisions and accretion (snowball-like growth) between particles in a disk of gas and dust around a central star. Even so, this method of differentiation is difficult to apply, especially in the case of planet-sized objects that have been flung into interstellar space and no longer orbit any star.

The Solar System

Fifty years ago, the population of the Solar System included one central star, nine planets, 31 satellites and thousands of comets and asteroids. However, since the arrival of the Space Age and the development of ever more sensitive ground-based instruments, the inventory of objects has swollen remarkably. Today, the astronomical community recognizes eight planets and five dwarf planets, the tally of planetary satellites has passed 150 and the number of identified small objects is climbing rapidly as increasingly sensitive searches discover thousands of Sun-grazing comets and icy Kuiper Belt objects that orbit beyond Neptune.

In terms of numbers, the Solar System is dominated by debris, in the form of comets, asteroids, meteorites and dust. These are the left-overs from the formation of the planets, four and a half billion years ago. The main asteroid belt, between Mars and Jupiter, is populated by millions of rocky objects that are shepherded by the powerful gravity of the nearby gas giant. They are thought to represent planetesimals – small planetary building blocks – that were unable to accrete due to the gravitational interference of Jupiter.

Beyond the orbit of Neptune are two more swarms of small objects, this time largely made of ice. The inner region, known as the Kuiper Belt, is where short-period comets originate. Pluto and Eris are the largest known inhabitants. The orbital periods of Kuiper Belt objects range from 200–400 years for objects such as Pluto to 1 000 years or longer for those that follow very elliptical orbits that take them far from the Sun.

The Kuiper Belt poses a serious challenge for theories of planet formation, since it contains less than one percent of the mass expected from the protosolar nebula theory. If the Kuiper

Table 1.1 The Planets: Relationship between solar distance and mean density.		
Planet	Distance from Sun (AU)	Mean Density (g/cm³)
Mercury	0.3871	5.43
Venus	0.7233	5.24
Earth	1.0	5.52
Mars	1.5237	3.91
Jupiter	5.2028	1.33
Saturn	9.5388	0.69
Uranus	19.1914	1.29
Neptune	30.0611	1.64

Belt objects formed like the terrestrial planets, growing by accumulating smaller objects as they orbit the Sun, it would take longer than the age of the Solar System to make one KBO!

Even further out – indeed, so far that none of the objects have ever been observed – is the postulated Oort Cloud, the home of most long-period comets.

The basic characteristics of the Solar System are straightforward to describe. Close to the Sun, where temperatures are higher, there are four quite small, but dense, "terrestrial" planets that are composed of rock. Beyond Mars, where temperatures are always well below zero, is the realm of the gas giants, Jupiter and Saturn, and the ice giants, Uranus and Neptune.

The orbits of the major planets are approximately circular, and close to the ecliptic plane. All of the planets and main belt asteroids circle the Sun in the same direction – counter-clockwise as seen from above the Sun's north pole. This is also the direction of the Sun's rotation. However, the beautiful symmetry breaks down when it comes to the smaller, more icy members of the Solar System. Comets can arrive from any direction, and the orbits of the Kuiper Belt objects have no particular orientation, suggesting that there is a spherical swarm of these objects surrounding the Sun and major planets.

Of the four inner planets, Venus and Earth both possess dense atmospheres – though they are very different in nature – while Mercury is too lightweight to have retained a substantial gaseous envelope. Whereas the most common gas on both Venus and Mars is carbon dioxide, Earth is something of an oddball, with an atmosphere dominated by nitrogen and oxygen. This latter gas can be accounted for by the fact that Earth is – as far as we know – the only abode of life in our Solar System. Satellites are rare: Earth is orbited by the Moon, while Mars has two small companions that are generally considered to be captured asteroids.

As their name suggests, the giants are characterized by their large size – tens to thousands of times bigger than Earth – and low bulk densities which can be accounted for by the dominance of hydrogen and helium in their interiors. All four of the giants have ring systems composed of dust, ice and rocky debris, and their gravitational influence is such that they retain dozens of satellites – most of them captured billions of years ago.

Since they are relatively close to the Sun, all of the terrestrial planets have high orbital velocities with periods of less than two Earth years (see Box 1.2). In contrast, their axial rotations are slow and their axial inclinations are very different.

Mercury's axis is almost at right angles to its orbit. It takes 58 days to rotate once, or about two thirds of the time it takes to orbit the Sun. Venus resembles a top that has been knocked completely upside down. As a result, it rotates in a retrograde direction that takes 243 Earth days, longer than its orbital period. Earth and Mars have very similar days and seasons – at least in the present epoch – since their sidereal periods of axial rotation are both around 24 hours and both axes are inclined about 24–25° to their orbits.

The motions of the outer planets are very different. Their large distances from the Sun require modest velocities to maintain their orbits. Orbital periods range from almost 12 years for Jupiter to about 165 years for Neptune. However, despite their swollen spheres, they all spin much faster on their axes than their terrestrial siblings, with sidereal periods in the range 10–20 hours.[5] However, there is considerable variation in their axial tilts. Jupiter is almost upright, Saturn and Neptune are inclined more than Earth and Mars, while Uranus spins on its side so that the polar regions alternately point toward or away from the Sun.

The Birth of the Solar System – Early Ideas

The Sun, which contains over 99% of the system's mass, completes one rotation in about 24 days. In contrast, the largest planets, Jupiter and Saturn, rotate once in about ten hours. When combined with their orbital motion, it turns out that Jupiter accounts for some 60% of the Solar System's angular momentum, with another 25% accounted for by Saturn. This compares with about 2% for the sluggardly Sun.

Any theory of cosmogony that attempts to account for the formation of the Solar System must take into account the angular momentum of the Solar System objects, as well as the facts that all of the planets travel in the same direction and more or less in the same plane. The obvious conclusion is that they all formed in the same manner and at about the same time.

There seem to be two main possibilities: the planets were either created by material derived from the Sun or a nearby companion star, or they formed from a cloud of diffuse matter that surrounded the Sun. However, theorists have struggled for centuries to match the hypotheses to the known facts, in order to choose between them.

One of the earliest, and most successful, attempts to explain how the Solar System came about was the nebular hypothesis – the idea that the Sun and planets formed from a vast, slowly rotating disk of gas and dust. A modified version of this hypothesis is the generally accepted explanation today.

---

[5] The sidereal period is the time a planet takes to orbit the Sun, with respect to a particular background star.

Figure 1.9 These four panels show the scale of the Solar System as we know it today. At top left are the orbits of the inner planets and the main asteroid belt. Top right shows the orbits of the outer planets and the Kuiper Belt. Lower right shows the orbit and current location of Sedna, which travels further from the Sun than any other known object in the Solar System. Lower left shows that even Sedna's highly elliptical orbit lies well inside the inner edge of the proposed Oort Cloud (shown in blue). This spherical cloud contains millions of icy bodies orbiting at the limits of the Sun's gravitational pull. (NASA/JPL/R. Hurt, SSC-Caltech)

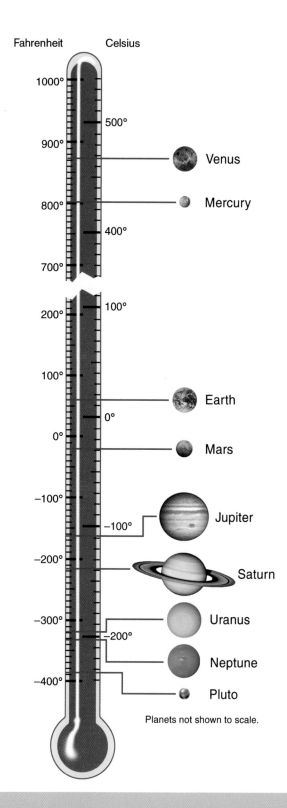

Figure 1.10   In general, a planet's surface temperature decreases with its distance from the Sun. Venus is the exception, since its dense carbon dioxide atmosphere traps infrared radiation. The runaway greenhouse effect raises its surface temperature to 467°C. Mercury's slow rotation and thin atmosphere result in the night-side temperature being more than 500°C colder than the day-side temperature shown above. Temperatures for Jupiter, Saturn, Uranus, and Neptune are shown for an altitude in the atmosphere where pressure is equal to that at sea level on Earth. Earth lies in the center of the "habitable zone," where water can exist as a liquid and conditions are favorable to life. (Lunar and Planetary Institute)

Some of the key evidence comes from modern observations of distant star systems. Today, spaceborne telescopes can peer into the hearts of giant molecular clouds, such as the Orion Nebula, and search for young, Sun-like stars that replicate the conditions that prevailed in our Solar System some 4.6 billion years ago.

These observations show that so-called protoplanetary disks, or proplyds, exist around most very young stars – those less than ten million years old. Many of the disks are larger than our Solar System. Observations of slightly older stars show how these disks evolve as time goes by, with the formation of swarms of rocky and icy debris and gaps in the clouds created by fledgling planets.

As currently envisaged, the Solar System began with the collapse of a cloud of interstellar gas. The trigger for this collapse may have been the passage of an externally generated shock wave – perhaps from a supernova explosion, density waves passing through the galaxy, or a major reduction in the cloud's magnetic field or temperature. The first of these explanations is a prime candidate, since many stars form in clusters within clouds containing thousands of solar masses of material. When the giant stars of the cluster run through their short life spans, they are likely to produce a series of supernovas, preceded by powerful stellar winds.

Over millions of years, the original cloud may be broken up into smaller fragments, each mixed with heavier elements from the dying stars, as well as the ubiquitous hydrogen and helium gas. Once a fragment reaches a critical density, it is able to overcome the forces associated with gas pressure and begins to collapse under its own gravity.

The contracting cloud begins to rotate, slowly at first, then faster and faster – rather like when an ice skater pulls in his arms. Since material falling from above and below the plane of rotation collides at the mid-plane of the collapsing cloud, its motion is cancelled out. The cloud begins to flatten into a disk, with a bulge at the center where the protostar is forming. The disk was probably thicker at a greater distance from the protostar, where gas pressure was lower.

Such a nebula would almost certainly rotate slowly in the early stages, but as it contracts, conservation of angular momentum causes the cloud to spin faster. If this process continues, the core forming at the center of the nebula will spin up so fast that it flies apart before it has a chance to form a star. Somehow, that angular momentum must be removed before a star can form.

Studies of other young stars and their surrounding disks provide evidence that, as the interstellar gas collapses, it also winds up the magnetic field which permeates the nebula. Gas which is rotating too fast to collapse is expelled and dispersed along the magnetic field.

This process naturally forms a spiral-shaped magnetic field that helps to generate polar jets and outflows associated with very young stars. At the same time, the jets remove angular momentum, allowing other material to accrete and collapse. Gravitational instability, turbulence and tidal forces within the "lumpy" disk may also play a part, helping to transfer much of the angular momentum to the outer regions of the forming disk.

The protoplanetary disk is heated by the infall of material. The inner regions, where the cloud is most massive, become hot enough to vaporize dust and ionize gas. As contraction continues and the cloud becomes increasingly dense, the temperature at its core reaches the point where nuclear fusion commences. The merging protostar begins to emit copious amounts of ultraviolet radiation. Radiation pressure drives away much of the nearby dust, causing the star to decouple from its nebula.

The youngster may remain in this T Tauri stage for perhaps 10 million years, after which most of the residual nebula has evaporated or been driven into interstellar space. All that remains of the original cloud is a rarefied disk of dust particles, mainly silicates and ice crystals.

Meanwhile, the seeds of the planets have begun to appear. More refractory elements condense in the warm, inner regions of the nebula, while icy grains condense in the cold outer regions. Individual grains collide and stick together, growing into centimeter-sized particles. These swirl around at different rates within the flared disk, partly due to turbulence and partly as the result of differences in the drag exerted by the gas. After a few million years, these dusty or icy golf balls accrete into kilometer-sized planetesimals and gravity becomes the dominant force.

The Solar System now resembles a shooting gallery, with objects moving at high speed in chaotic fashion and enduring frequent collisions with each other. Some of these impacts are destructive, causing the objects to shatter and generate large amounts of dust or debris. Other collisions are constructive, resulting in a snowballing process. Over time, the energy loss resulting from collisions means that construction eventually dominates.

Eventually, the system contains a relatively small number of large bodies or protoplanets. Millions of years pass as they continue to mop up material from the remnants of the solar nebula and to collide with each other, finally resulting in a population of widely separated worlds occupying stable orbits and traveling in the same direction around the young central star.

It is worth noting here that computer simulations of the early Solar System show that even the slightest differences in initial conditions can produce different planetary systems. Depending on exactly where each embryo started out, the orbital positions of new planets vary randomly from simulation to simulation. The total number of planets – and hence, their final masses – may also vary greatly. It seems that planet formation is a very chaotic process.

Rocky Planets

Modeling suggests that collisions between planetesimals initially occur at low velocities, allowing them to merge and grow. At the Earth's distance from the Sun, it takes only about 1 000 years for 1 km sized objects to grow into 100 km objects. Another 10,000 years produces 1 000 km diameter protoplanets, which double in diameter over the next 10,000 years. Such models indicate that Moon-sized objects can form in a little over 20,000 years.

As planetesimals within the protosolar disk grow larger and more massive, their gravity increases, and once a few of the objects reach a size of 1 000 km, they begin to stir up the remaining smaller objects. Near encounters accelerate the smaller, asteroid-sized chunks of rock to higher and higher speeds. Eventually, they

Box 1.2 Kepler's Laws of Planetary Motion

Johannes Kepler (1571–1630) was one of the most important characters in the story of unraveling how the Solar System works. The German-born mathematician was appointed assistant to Tycho Brahe (1546–1601), the most famous observer of the day. Granted access to Brahe's catalog of positional data, Kepler was given the task of explaining the orbit of Mars. After four years of calculations, Kepler finally realized in 1605 that the orbits of the planets were not perfect circles, but elongated circles known as ellipses.

Whereas a circle has one central point, an ellipse has two key interior points called foci (singular: focus). **The sum of the distances from the foci to any point on the ellipse is a constant.** For Solar System objects, the Sun always lies at one focus (see below).

In order to draw an ellipse, place two drawing pins some distance apart and loop a piece of string around them. Place a pencil inside the string, draw the string tight and move the pencil around the pins. Now move one of the pins and repeat the process. Note how the shape of the ellipse has changed.

The amount of "stretching" or "flattening" of the ellipse is termed its eccentricity. All ellipses have eccentricities lying between zero and one. A circle may be regarded as an ellipse with zero eccentricity. As the ellipse becomes more stretched, its eccentricity approaches one.

Figure 1.11   A circle has an eccentricity of zero. As the ellipse becomes more stretched (i.e. the foci move further apart) the eccentricity approaches one. Half of the major axis is termed the semimajor axis. The average distance of a planet from the Sun as it follows its elliptical orbit is equal to the length of the semimajor axis. The eccentricity is calculated by dividing the distance between the two foci by the length of the major axis. (Open University)

In reality, most of the planets follow orbits that are only slightly elliptical. Their eccentricities are so small that they look circular at first glance. Pluto and Mercury are the main exceptions, with eccentricities exceeding 0.2.

Another key characteristic of an ellipse is its maximum width, known as the **major axis**. Half of the major axis is termed the semimajor axis. The average distance of a planet from the Sun as it goes around its elliptical orbit is equal to the length of the semimajor axis.

After intensive work on the implications of his discovery, Kepler eventually formulated his Three Laws of Planetary Motion.

- Kepler's First Law: The orbits of the planets are ellipses, with the Sun at one focus of the ellipse. (Generally, there is nothing at the other focus).
- Kepler's Second Law: The line joining the planet to the Sun sweeps out equal areas in equal times as the planet travels around the ellipse. In order to do so, a planet must move faster along its orbit near the Sun and more slowly when it is far away. A planet's point of nearest approach to the Sun is termed perihelion; the furthest point from the Sun on its orbit is termed aphelion. Hence, a planet moves fastest when it is near perihelion and slowest when it is near aphelion.
- Kepler's Third Law: The square of a planet's sidereal (orbital) period is proportional to the cube of its mean distance (semimajor axis) from the Sun. This means that the period, or length of time a planet takes to complete one orbit around the Sun, increases rapidly with its distance from the Sun. Thus, Mercury, the innermost planet, takes only 88 days to orbit the Sun, whereas remote Pluto takes 248 years to do the same.

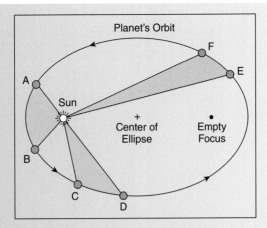

Figure 1.12 Kepler's first law states that the orbit of a planet about the Sun is an ellipse with the Sun at one focus. The other focus of the ellipse is empty. According to Kepler's second law, the line joining a planet to the Sun sweeps out equal areas in equal times. In this diagram, the three shaded areas ABS, CDS and EFS all have equal areas. A planet takes as long to travel from A to B as from C to D and E to F. It moves most rapidly when it is nearest the Sun (at perihelion) and slowest when it is farthest from the Sun (at aphelion). (Kenneth R. Lang, Tufts University)

This law can be used to make some useful, but fairly simple, calculations. For example, if the period is measured in Earth years and the distance is measured in astronomical units (AU), the law may be written in the simple form: $P(\text{years})^2 = R(\text{AU})^3$.

This equation may also be written as: $P(\text{years}) = R(\text{AU})^{3/2}$. Thus, if we know that Pluto's average distance from the Sun (semimajor axis) is 39.44 AU, we can calculate that the orbital period is: $P = (39.44)^{3/2} = 247.69$ years. Similarly, if we know that Mars takes 1.88 Earth years to orbit the Sun, we can calculate that its semimajor axis is: $R = (1.88)^{2/3} = 1.52$ AU.

Figure 1.13 A graph showing the orbital periods of the planets plotted against their semimajor axes, using a logarithmic scale. The straight line that connects the planets has a slope of 3/2, verifying Kepler's third law which states that the squares of the orbital periods increase with the cubes of the planetary distances. This law applies to any bodies in elliptical orbits, including Jupiter's four largest satellites (inset). (Kenneth R. Lang, Tufts University)

Figure 1.14   The planets formed about 4.5 billion years ago from a huge nebula – cloud of gas and dust – that surrounded the young Sun. Within a few million years, colliding particles in the nebula accreted until golf-ball-sized pebbles appeared. Further collisions caused these to increase in mass, eventually growing into the planets we see today. Some of these, further from the Sun, were able to pull in huge atmospheres of hydrogen and helium. Those in the warmer, inner regions were made of rock rather than ices and light gases. The remnants formed comets and asteroids. (NASA/JPL-Caltech/T. Pyle, SSC)

Figure 1.15   The early stages of star and planet formation: (**a**) A Hubble Space Telescope (HST) view of five young stars in the Orion Nebula. Four are surrounded by gas and dust trapped in orbit as the stars formed. These are possibly protoplanetary disks, or "proplyds," that might eventually produce planets. The bright proplyds are closest to the hottest stars of the parent star cluster, while the object farthest from the hottest stars appears dark. (C.R. O'Dell/Rice University; NASA) (**b**) This HST image shows Herbig-Haro 30, a young star surrounded by a thin, dark disk. The disk extends 64 billion km, dividing the nebula in two. The central star is hidden from direct view, but its light reflects off the upper and lower surfaces of the disk to produce the pair of reddish nebulas. Gaseous jets (green) remove material from above and below the disk and transfer angular momentum outwards. (**c**) A computer simulation showing how a protoplanetary disk surrounding a young star begins to fragment and form gas giant planets with stable orbits. (Mayer, Quinn, Wadsley, Stadel).

Box 1.3 Mass and Density

Two of the basic properties of Solar System objects are mass and density. Mass is a measure of the amount of matter in a particle or object. The standard unit of mass in the International System (SI) is the kilogram (kg). This is usually determined by measuring the object's gravitational influence on other objects, for example, natural or artificial satellites.

Once the volume of an object is known, its bulk density can be calculated. In this book, density is usually expressed in grams per cubic centimeter (g/cm^3). As a guide, the density of water is $1.0\,g/cm^3$. Objects which have a density lower than water are able to float (assuming enough water is available!).

If a planet has a high density, it means that it is largely made of dense, rocky or metallic materials. Objects often have low densities because they contain a lot of gases or ices, but few rocky materials. This is why all of the giant planets in the Solar System have low densities, despite their huge size.

The planet with the lowest density ($0.7\,g/cm^3$) is Saturn. The reason that Saturn has such a low density is that it is mainly composed of gas, particularly hydrogen and helium. There is only a small rocky core at its center.

Other objects, including many small satellites and asteroids, have low densities because they are piles of loosely consolidated rubble or highly porous, that is, they contain numerous empty spaces.

The densities of planets are also a reflection of their size and the layering of their interiors. Earth has the highest density of all the planets in the Solar System because it is made of dense, rocky materials. At the surface, crustal rocks have densities between 2.5 and $3.5\,g/cm^3$. However, Earth's average density is much higher ($5.5\,g/cm^3$).

This is partly because the denser elements, such as iron and nickel, have sunk to the center of the planet, while the less dense materials have risen to the surface. Many planets were internally differentiated in this way early in their lives.

The centers of the planets are also more compressed by the weight of the overlying material. In the case of Earth, for example, the normal, uncompressed density of its rocks is about $4.4\,g/cm^3$, but the central core is compressed to greater than normal density by the overlying layers.

More massive planets should experience greater compression at their centers, and hence higher average densities, if they are made of the same rocky and metallic materials as Earth. The opposite should apply to smaller planets. However, the smallest of the rocky planets, Mercury, actually has an average density of $5.4\,g/cm^3$, only slightly lower than Earth's.

Mercury's density rises to a remarkable $5.3\,g/cm^3$ after it has been corrected for the effects of internal compression – much higher than Earth's. The only way to explain this is to assume that the little planet has a huge core of iron and nickel that takes up almost half of its interior (see Chapter 5).

are traveling so quickly that when they collide, they pulverize each other instead of merging.

While the largest protoplanets continue to grow, the remaining rocky planetesimals grind each other into dust. Some of this dust is gathered in by the surviving planets, while much of the remainder is swept out of the Solar System when the Sun evolves into a hydrogen-burning star. (A cloud of micron-sized dust particles still exists in the ecliptic plane of the Solar System. Known as the zodiacal cloud, it is composed of silicate particles that are largely derived from collisions between main belt asteroids.)

One of the problems that has to be solved by Solar System theorists is an explanation for the silicate and metal–rich nature of the terrestrial planets and the dominance of hydrogen and helium in the outer planets. Clearly, the marked difference in composition between the inner and outer planets must be related to the materials that made up different regions of the disk.

The dense, rocky nature of the Earth and its neighbors suggests that they simply formed through the accretion of dust grains in the solar nebula. However, studies of primitive chondritic meteorites show the presence of millimeter-sized droplets (chondrules) that were once liquid.

It seems that, before they amalgamated to form the meteorites, these existed for a brief period as independent spheroids at temperatures above 1 500°C. Some chondrules seem to include other

chondrules, indicative of being exposed to high temperatures on more than one occasion (see Chapter 13). The source of the heating is uncertain, although shock waves, solar heating and collisions between planetesimals have been suggested.

Laboratory experiments indicate that these molten globules were cooled very rapidly within ten million years of the collapse of the molecular cloud. The cause of such sudden cooling events remains unclear. What does seem certain is that the chondrules and dust began to stick together and grow in size, creating chunks of chondritic material. Drag from gas in the nebula encouraged the pebble-sized objects to creep inward, all the time gathering in more material.

Once a population of large planetesimals evolved, their destiny was determined largely by chance. A fast, head-on collision caused the objects to break apart. A slow, gentle encounter enabled the participants to merge into an even larger object. In this way, the terrestrial planets grew to more or less their current size over a period of some ten million years.

The huge amounts of kinetic energy dumped in the planets by frequent, massive impacts caused partial or total melting and the creation of magma oceans. This led to internal differentiation, with the denser elements, such as iron, sinking to the core and the lighter ones rising to the surface to create silicate crusts. Early atmospheres were generated by outgassing of volatile molecules such as water, methane, ammonia, hydrogen, nitrogen and carbon

Box 1.4   Key Steps in the formation of rocky planets (after Kenyon and Bromley)

1. A molecular cloud made up of gas and dust begins to collapse.
2. A protostar begins to form at the core of the collapsing nebula.
3. A disk-shaped nebula of orbiting dust and gas develops in the protostar's equatorial plane.
4. Dust grains in the disk collide and merge.
5. Large (1 mm) dust grains fall into a thin, dusty sheet.
6. Collisions produce planetesimals 1 meters to 1 km across.
7. More collisions between planetesimals produce planetary embryos.
8. Planetary embryos stir up the leftover planetesimals.
9. Planetesimals then collide and fragment.
10. A cascade of collisions reduces fragments to dust.
11. Planets sweep up some of the dust.
12. Radiation and a "wind" of charged particles from the central star remove the remaining gas and dust.

dioxide. A final heavy bombardment, which ended about 3.8 billion years ago, is clearly marked in the crater record of the Moon, and this has been applied to other planets and satellites.

Occasionally a satellite was created as the by-product of a major impact. Such is thought to be the case with Earth and its Moon. Debris from an ancient collision between the young Earth and a Mars-sized planetesimal created a ring of debris that eventually came together to form the Moon. A similar explanation has been put forward for the Pluto-Charon system (see Chapters 3 and 12).

Gas Giants and Ice Giants

In the outer reaches of the solar nebula, temperatures were low enough for ices to form. Indeed, it seems that ice particles were much more abundant than silicate dust particles. This being the case, any planetesimals born in the frigid outer zone would have resembled icy dirt balls, much like the comets we see today. Unfortunately, the main constituents of Jupiter and Saturn are hydrogen and helium, rather than water. Since temperatures in the nebula would have been too warm for these gases to condense, accretion of hydrogen and helium snowflakes cannot have occurred. Another explanation must be found.

There seem to be two possibilities. Studies of gas giant interiors suggest that Jupiter and Saturn may possess rocky cores at least as large as the Earth. It may be, therefore, that the early stages of growth of these planets resembled the accretion taking place in the inner Solar System, with the growth of massive nuclei of ice and dust. Once these became sufficiently large, about 5 to 15 times the mass of Earth, they were able to attract and hold onto even the lightest gases in the surrounding solar nebula. As their mass and gravitational grasp grew, their spheres became ever more bloated.

Alternatively, they could simply have developed as the result of large-scale gravitational instabilities in the solar nebula. Since the disk in the outer reaches contained both dust and condensed ices, there was plenty of raw material for large planets to develop and grow.

Any theory must also account for the fact that Jupiter and Saturn are huge hydrogen-helium planets, whereas Uranus and Neptune are notably smaller and composed mostly of elements that form ices: oxygen, carbon and nitrogen. If the latter pair began as icy nuclei, they must have grown quite slowly in the more rarefied conditions of the presolar nebula beyond about 15 AU. By the time they were massive enough to draw in large amounts of gas, the nebula had dissipated and the supply was cut off.

Migrating Planets

Our picture of the early Solar System is complicated by the likelihood that the giant planets migrated considerable distances before they ended up in their present positions. Such large-scale movement is supported by the discovery of numerous large, extrasolar planets that orbit within a fraction of an astronomical unit of a star.

In the case of our Solar System, this migration can be explained by the exchange of orbital momentum between giant planets and innumerable planetesimals. One current model (the Nice model) envisages a chaotic early Solar System occupied by the major planets out to a distance of about 15 AU (closer than the present orbit of Uranus). Jupiter may have been born a little farther out in the Solar System than it is today, whereas the other giants were closer to the infant Sun than at present. Beyond the planets was a region swarming with leftover planetesimals.

Whereas Jupiter was massive enough to eject large numbers of planetesimals to the outer reaches of the Solar System or out of the system altogether, the three, smaller, giants were unable to do this. Instead, they flung similar numbers of planetesimals toward the Sun and away from it. Whenever Uranus or Neptune decelerated a nearby planetesimal, causing the object to move closer to the Sun, the planet gained a tiny amount of momentum. The resultant acceleration caused it to drift away from the Sun.

Over time, after billions of such gravitational interactions, Jupiter spiraled inward a modest distance, while Saturn drifted outward. When Jupiter reached a distance of 5.3 AU and Saturn arrived at 8.3 AU, the two planets were in a 2:1 orbital resonance, so that one orbit of Saturn lasted precisely two Jupiter orbits. The repeated gravitational pull of Jupiter caused Saturn's orbit to become much more elongated.

Saturn began to create havoc with the orbits of Uranus and Neptune, causing them to become more elliptical. They began to plow through the outer swarm of icy planetesimals, scattering billions of them in all directions. By the time the planets had cleared most of the intruders from their vicinities and the system had settled down again, Saturn had migrated out to about 9.5 AU. The effect on the outer planetary pair was even more extreme. Uranus had moved from about 13 to 19 AU, while Neptune had been catapulted from 15 to 30 AU.

Another consequence of this 500 million year long planetary reshuffle was that the remaining planetesimals, perhaps 0.1% of

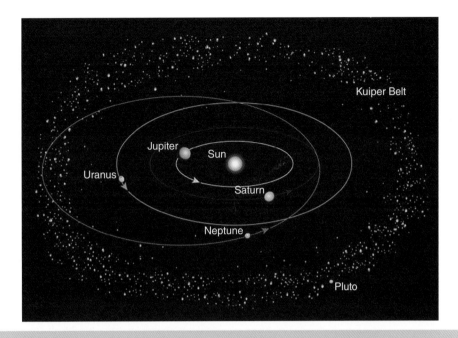

Figure 1.16   According to a recent model, Saturn scattered Neptune outwards beyond Uranus into the Kuiper Belt during the epoch of planet migration. The gravitational interaction of Neptune with icy planetesimals sent billions of Kuiper Belt objects inward, towards the Sun. As a result, Neptune and Uranus migrated outward to their present orbits. Some KBOs, such as Pluto, were locked into orbital resonances with Neptune. (Nature).

the original planet-building population, were relocated beyond 30 AU, where they now reside as Kuiper Belt objects.

Furthermore, the asteroid belt was also strongly perturbed during this burst of migration, adding to the sudden, massive delivery of planetesimals to the inner Solar System. As their pock-marked surfaces show, the Moon and terrestrial planets suffered heavily during this Late Heavy Bombardment, around four billion years ago.

Planetary Satellites

The Solar System contains well over 100 planetary satellites, but, as might be expected from the wide range of sizes and composi-tions, these seem to have arisen in several different ways.

As mentioned above, Earth's Moon is thought to have been born during a massive, grazing collision between the young Earth and a Mars-sized planetary embryo. The mixture of debris from both objects formed a ring around the scarred Earth, eventually accreting into a large satellite.

Other satellites may also have been created by sizeable impacts early in the Solar System's history. For example, the Pluto-Charon system may have originated during a collision between two large, icy planetesimals over four billion years ago. Simulations show that some of the debris from the collision would be blasted into orbit around the surviving protoplanet, eventually coalescing to form Charon and several smaller satellites.

Most of the major satellites seem to have followed a less trau-matic path, gradually accreting from a protoplanetary disk, much like the planets. The most obvious example is the Jovian system,

with its family of four Galilean moons. The inner pair, Io and Europa, are smaller but more dense (with a higher proportion of rock) than the outer pair, ice-rich Ganymede and Callisto. All of them orbit Jupiter in the same direction and in more or less the same plane.

These characteristics can be explained if the moons were born from a spherical cloud of dust and gas being drawn inward from the solar nebula by a fledgling planet. As time went by, the cloud flattened into a disk around the protoplanetary core. This disk was hotter and denser near the center, allowing condensation and accretion of the less volatile materials. Further out, the icy volatiles could also condense and accrete to form Ganymede and Callisto.

Although the Saturnian family of satellites is dominated by planet-sized Titan, none of the members are particularly rocky, with many only slightly denser than water. Titan itself is similar in density to Ganymede and Callisto. If the proto-Saturn was surrounded by a collapsing cloud, it seems to have been only about one quarter the mass of Jupiter's. This suggests that the cloud contained less silicate (rocky) material and more ice than its counterpart in the warmer environs of Jupiter.

Certainly, there is a general increase in size and mass moving outward from Saturn toward Titan, with a marked decrease in both properties beyond Titan. This has led theorists to suggest that Titan grew sufficiently quickly to collect much of the solid material in the disk around Saturn, leaving only a modest amount for the medium-sized satellites to accumulate.

A modified version of the accretion scenario has recently been proposed by Robin Canup and William Ward of the Southwest Research Institute. They suggested that a growing satellite's gravity induces spiral waves in a surrounding disk of gas, primarily

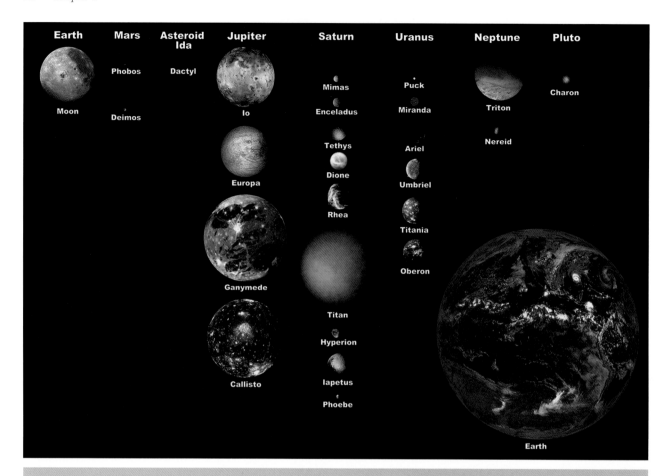

Figure 1.17   The most significant satellites in our Solar System are shown beside the Earth, with their correct relative sizes and colors. Ganymede and Titan are larger than Mercury and eight satellites are larger than Pluto. Earth's Moon is the fifth largest, with a diameter of 3 476 km. Most of them are thought to have formed from a disk of gas and dust in orbit around their home planet. However Triton and many of the smallest satellites (including the moons of Mars) are thought to be captured asteroids or Kuiper Belt objects that formed elsewhere in the Solar System. Earth's Moon and Charon (and possibly the moons of Uranus) are thought to have formed as the result of major impacts. (NASA)

hydrogen. Gravitational interactions between these waves and the satellite cause the moon's orbit to contract. This effect becomes stronger as a satellite grows, so that the bigger a satellite gets, the faster its orbit spirals inward toward the planet. They proposed that the balance between the inflow of material to the satellites and the loss of satellites through collision with the planet implies a maximum size for a satellite of a gas giant.

Numerical simulations and analytical estimates of the growth and loss of satellites showed that multiple generations of satellites were likely, with today's satellites being the last surviving generation to form as the planet's growth ceased and the gas disk dissipated.

The origin of the Uranian (and Neptunian) satellites is open to debate. Models suggest that the ice giants grew more slowly than their larger cousins. By the time they were large enough to gather a disk of material, most of the gas and dust had been dispersed, probably after the young Sun entered its active T Tauri phase. If the regular satellites of Uranus and Neptune could not have formed through large-scale accretion from a circumplanetary disk, how did they come about?

One idea is that the planets were larger and hotter during their accretion phase. As they subsequently cooled and contracted, they left behind a "spinout disk" from which small satellites could grow by accretion.

One complication is the fact that the Uranian satellites orbit in circles close to the planet's equator, even though it spins on its side. Neptune's rotation axis is also tilted quite markedly, aligned at about 30° to its orbital plane, while the orbits of its small satellites are circular and near-equatorial. This suggests that the planets were involved in major impacts early in their histories, and that the satellites were born during or after these collisions.

It may be that impacts with planet-sized objects blasted out clouds of hot material that formed orbiting disks around the ice giants. When the material cooled and condensed, the ice-rock ingredients were available for medium-sized satellites to form.

The major exception is Triton, the largest satellite of Neptune. One clue to its origin is that most of its bulk properties are very similar to those of Pluto, one of the largest known members of the Kuiper Belt. Furthermore, unlike the other Neptunian moons, it follows a retrograde path which is quite steeply inclined to the planet's equator. This unusual orbit has led to speculation that Triton was a Kuiper Belt object that ventured too close to Neptune and was somehow captured.

The Heliosphere

The motion of superhot plasma inside the Sun generates a powerful magnetic field. The Sun's atmosphere extends into interplanetary space through the motion of the electrically charged particles (mainly electrons and protons) of the solar wind, which streams outward in all directions at typical speeds of between 400 and 7 500 km/s. As the particles spiral around the Sun, they carve out an invisible bubble which extends outward for many billions of kilometers. Although electrically neutral atoms, cosmic rays and dust particles from interstellar space can penetrate this bubble, virtually all of the atomic particles in the heliosphere originate in the Sun itself.

The region of space in which the Sun's magnetic field and the wind of charged particles (solar wind) dominate the interstellar medium is known as the heliosphere. The shape of the heliosphere and the distance of the heliopause are determined by three main factors: the motion of the Sun as it plows through the interstellar medium, the density of the interstellar plasma and the pressure exerted on its surroundings by the solar wind.

From theoretical studies and spacecraft observations of planetary magnetospheres and the solar wind, it is known that the density of the solar wind decreases as the inverse square of its distance from the Sun. In other words, solar wind density at 4 AU is only one quarter its density at 2 AU. The strength of the Sun's magnetic field also weakens with distance, although at a slower rate. Eventually, the density and magnetic influence of the solar wind decrease so much that its outward motion is impeded by the sparse plasma of the interstellar medium.

The heliosphere acts like an island in a stream, causing interstellar plasma to be diverted around it. At first it was thought that the heliosphere really was spherical, but the two Voyager spacecraft, which are currently heading out of the Solar System on different paths, observed what seemed to be a "squashed" heliosheath. In this new model, the heliosphere resembled a huge wind sock or tadpole – much like a comet's elongated tail – that is shaped by the motion of the Sun as it plows through a hot, tenuous cloud of interstellar gas and dust. Studies of the motion of nearby stars show that the Sun is traversing the cloud at a velocity of 25.5 km/s. The interstellar medium forces the solar wind to turn back and confines it within the heliosphere.

This picture had to be revised in 2009, when data from the IBEX spacecraft and the Cassini spacecraft in orbit around Saturn showed that the heliosphere is roughly spherical – perhaps like an elongated balloon – after all. Instruments on the spacecraft were used to map the intensity of the energetic neutral atoms ejected from the heliosheath as the solar wind interacts with the interstellar medium. The data showed a belt of hot, high-pressure particles where the interstellar wind flows by the heliosphere. Their distribution indicates that the heliosphere resembles a huge bubble which expands and contracts under the influence of the local interstellar magnetic field as it sweeps past.

The interaction of the heliosphere with the interstellar medium takes place in several stages. For a spacecraft traveling out of the Solar System, the first boundary to be reached is the termination shock. This is a standing shock wave where the supersonic solar wind slows dramatically from more than 100 km/s to about half that speed.

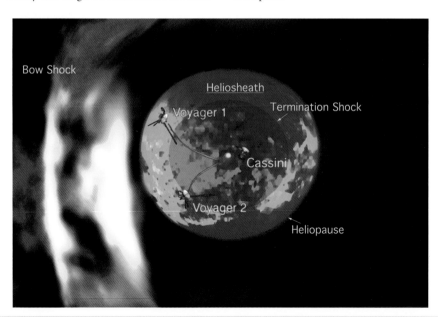

Figure 1.18   For many years the shape of the heliosphere was thought to resemble a comet, with a blunt "head" and an elongated tail. Recent spacecraft observations suggest that it is more like an elongated balloon or bubble. In this artwork the multicolored (blue and green) bubble represents Cassini measurements of the emission of particles known as energetic neutral atoms. The first crossing into interstellar space by Voyager 1 may occur around 2017. (NASA-JPL/JHU-APL).

Beyond the termination shock is a region known as the heliosheath, where particles of the solar wind and interstellar gas mix. In December 2004, Voyager 1 became the first spacecraft to cross into the heliosheath. NASA's two Voyager spacecraft are heading out of the Solar System in different directions. Voyager 1 crossed the termination shock on December 17, 2004, becoming the first spacecraft to enter the heliosheath. Voyager 2 crossed the termination shock on August 30, 2007 – 30 years after it was launched from Florida. The Voyager 2 crossing took place almost 1.6 billion km closer to the Sun than Voyager 1's, confirming that the outer boundary of the Solar System is curved.

Observations by the Magnetospheric Imaging Instrument (MIMI) on board Cassini show that the heliosheath is about 40 to 50 AU (6 billion to 7.5 billion km) thick. Further out is the heliopause, the boundary between the interstellar medium and the heliosphere. Beyond the heliopause, the interstellar ions flow around the heliosphere, modifying its size and shape.

Still further out, there is probably a bow shock, another shock surface where the supersonic flow of the interstellar medium is suddenly slowed as it approaches the heliosphere. All of these boundaries are thought to be moving back and forth at speeds of up to 100 km/s as the heliosphere is squeezed and released due to gusts in the solar wind and variations in the interstellar magnetic field.

The Future

The Solar System is continually evolving and changing. The collision of comet Shoemaker-Levy 9 with Jupiter in July 1994 illustrated that impacts and planetary evolution are continuing today. More significantly, the Sun is also evolving, as nuclear fusion continues to create helium from hydrogen in its core.

Since its birth, 4.54 billion years ago, the Sun has grown 30% brighter and this change will continue. Over the next 1.2 billion years, its surface temperature will increase by about 150°C and its luminosity will increase by another 10%. By this time, the oceans will have boiled away. Over the next 2 billion years, even the water vapor is lost, turning Earth into an arid planet comparable to Venus today.

Models suggest that, about 7 billion years into the future, the Sun will swell into a red giant with a diameter perhaps 200 times larger than today's value – large enough to reach almost to Earth's present orbit. However, an increase in the solar wind will cause up to 25% of the Sun's mass to be blown away. This drop in mass will cause the orbits of the planets to expand outwards, so that Venus may recede to Earth's current orbit, while Earth may lie near the present orbit of Mars. However, this outward retreat will probably be partially balanced by solar tidal drag, which will cause our planet to spiral slowly inward. The Earth's fate will hang in the balance.

Further out, Mars will briefly become warm enough to melt its icy volatiles, leading to a temporary spell of warmth with a dense atmosphere. However, the planet's gravity is not strong enough to maintain the situation for very long. Jupiter's ice-rich Galilean moons will also develop thick atmospheres of water vapor, but again, these wet greenhouse conditions will only represent a short-lived Eden. On Saturn's giant moon Titan, an ocean of liquid ammonia may survive for several hundred million years, perhaps providing a brief interlude when primitive life may evolve.

With its hydrogen now exhausted, the Sun will shrink and become 100 times less luminous as it switches to helium for its energy source. However, the process of helium to carbon fusion will only prolong its active life for a few hundred million years. As the helium becomes exhausted, the Sun will expand once more into a red giant. Riven by sudden pulsations in size, it may well consume Earth – if it still exists. One hundred million years after the second red giant phase, the Sun will eject its outer layers, forming a beautiful (from the outside!) planetary nebula. All that will be left is a tiny, extremely hot, superdense core known as a white dwarf.

The final layout of the Solar System is hard to predict, but it may be that the scorched remnants of Earth and Mars, along with the outer giants, will continue to orbit the fading dwarf star, largely undisturbed, for hundreds of billions of years.

Meanwhile, our galactic environment will also have changed dramatically. About five billion years from now, the Andromeda galaxy and our Milky Way will collide, combining to form a single, football-shaped elliptical galaxy. By then, the Sun will be an aging star nearing the red giant phase and the end of its life.

Models suggest that the Solar System likely will reside 100,000 light-years from the center of the new galaxy – four times further than the current distance. Any human descendants observing the future sky will experience a very different view. The band of the Milky Way will be gone, replaced by a huge bulge of billions of stars.

Questions

- What did the word "planet" originally mean? Why were objects given this name?
- How many planets were recognized before the invention of the telescope?
- How many planets are recognized today? What is the current definition of a planet?
- List six characteristics of the present Solar System that any theory of Solar System formation must explain.
- Explain the main features of the nebular hypothesis that is favored by most scientists today.
- Explain three possible origins of planetary satellites. Give likely examples of each type.
- Describe the main features of the heliosphere.

TWO

Sun

The Sun is our nearest star, some 270,000 times closer than the next closest stars of the Alpha Centauri system. When compared with other stars, there is nothing out of the ordinary about the Sun. However, since it accounts for more than 99% of the mass of the Solar System, the Sun is undoubtedly the dominant influence over the local galactic neighborhood. It is the Sun's gravitational influence which holds the planets in their orbits and controls a region of space that extends in all directions for several light years.

The Sun influences every aspect of our lives, most obviously through its tremendous output of light and heat. Plants grow upward toward the Sun and soon die if the supply of light is cut off. Humans require sunlight to produce vitamin D. If sunlight is blocked by thick clouds or disappears at night, the temperature drops noticeably. In winter, when the Sun is above the horizon for shorter periods, the lower temperatures often result in frost and prolonged snow cover. If the Sun was extinguished tomorrow, Earth would soon turn into a frozen, lifeless planet, blanketed in ice.

In recent decades, space observatories have examined our nearest star in incredible detail and at many different wavelengths. Nevertheless, there are still many mysteries that remain to be unraveled, including the causes of solar flares, the mechanism that causes such extreme heating of the corona, and the origin of its eleven-year periodic cycle of surface activity.

The Birth of the Sun

The structure and evolution of Sun-like stars are quite well understood (see Chapter 1). The Sun was formed some five billion years ago in a giant cloud of dust and gas, known as a molecular cloud, located in one of the spiral arms of the Milky Way galaxy. The cloud began to contract, possibly triggered by a shock wave from a nearby supernova explosion. Over tens of millions of years, the cloud continued to collapse under its own gravitation while pulling in material from its neighborhood.

As the gas cloud shrank to a fraction of its original size, and as the central density and pressure increased, a protostar began to form, surrounded by a rotating disk, or nebula. Jets and radiation from the growing star cleared away most of the gas in the surrounding cloud. As it shrank in size, it also began to rotate much faster. If this spin up process had continued for long enough, the fledgling star would have flown apart, but it was able to shed much of its angular momentum by accumulating (accreting) slow-spinning material from a surrounding disk of gas and dust, and by ejecting fast-spinning material through two powerful bipolar jets that were shaped by the local magnetic field and formed perpendicular to the disk.

When the core of the protostar reached a temperature of 10 million degrees Celsius, hydrogen fusion caused the Sun to emit copious amounts of radiation. The remains of the accretion disk around the star eventually clumped together to form a planetary system. Today our star is approximately halfway through its life.

The Sun as a Star

The Sun is one of more than 100 billion stars in the barred spiral galaxy that we call the Milky Way. By mapping the distribution of stars in the sky, astronomers have found that it is located in the Orion Arm, a small, partial arm located between the Sagittarius and Perseus arms of our Galaxy.

The Sun lies in the main disk of the Galaxy, about 28,000 light years from its center. This is quite a desirable location for a planetary system that supports life, far from the excessive gravitational disturbances and intense radiation that exist nearer the densely populated center. As the Galaxy rotates, it takes the Sun about

Exploring the Solar System, First Edition. Peter Bond.
© 2012 by Peter Bond. Published 2012 by Blackwell Publishing Ltd.

Figure 2.1 The Milky Way is a spiral galaxy with a bar-shaped mass of stars at the center. The Sun lies near a small, partial arm called the Orion Arm, or Orion Spur, about 28 000 light years from the center – a little over half way towards the outer edge. (NASA/JPL, R. Hurt/Caltech)

Figure 2.2 Solar rotation deduced from measurements by the Michelson Doppler Imager (MDI) instrument on SOHO. The left side shows differences in rotation speed. Red-yellow is faster than average and blue is slower than average. The light orange belts, which are moving slightly faster than their surroundings, extend to a depth of about 20 000 km. Sunspots, caused by disturbances in the solar magnetic field, tend to form at the edge of these bands. The cutaway (right) reveals rotation speed inside the Sun. The large dark red band is a massive, fast flow of hot, electrically charged gas (plasma) beneath the equator. The blue lines represent plasma flow within two huge circulation cells in the convective zone, extending to a depth of at least 26 000 km. The return flow indicated at the bottom of the convection cells has not yet been observed. (NASA-ESA)

Table 2.1
The Sun

Diameter	1 392 000 km = 109 Earth diameters
Mass	1.9891×10^{30} kg = 333,000 Earth masses
Volume	1.412×10^{33} cm^3 = 1.3 million Earths
Density	1.409 g/cm^3 (water = 1)
Temperature at the surface	5 785 K
Temperature in the corona	>2 000 000 K
Temperature at the centre	15 500 000 K
Luminosity (energy output)	= 3.846×10^{33} erg/s
Absolute magnitude	4.79
Apparent magnitude	−26.78
Rotation time at equator	25 days
Rotation time at 60° latitude	29 days

225 million years to complete one circuit, traveling at a velocity of 220 km/s. This amounts to about 20 trips over the course of its 4.5 billion year lifetime.

The Sun is the central star of our Solar System and by far the largest single object in the System. With a diameter of approximately 1.4 million km, it is more than 100 times wider than Earth. 1.3 million Earths would fit inside the Sun. Indeed, all of the planets combined could easily be swallowed by its enormous sphere. However, its size is modest compared with giant stars such as Betelgeuse in Orion, whose vast bulk would extend all the way to the orbit of Jupiter if it swapped places with the Sun.

The Sun's rotation was first detected by observing the motion of sunspots in the photosphere. Evidence for its gaseous composition is provided by different rotation rates at different latitudes – on solid bodies, such as planets, all parts of the surface have the same period of rotation. At the equator, the Sun spins on its axis once every 25 days, but this increases to more than 30 days at the poles. The cause of this differential rotation is an important area of current research.

The Sun's rotation axis is tilted about 7.25° to the pole of the ecliptic, so we see more of the solar north pole each September and more of its south pole in March. Its gaseous composition and low density result in a mass 'only' 333,000 times that of Earth.

Figure 2.3 A SOHO image of a huge prominence rising from the solar disk, with Earth shown to scale. 330 000 Earths would fit inside the Sun. (NASA/ESA)

Figure 2.4 On July 14, 2000, high energy particles from a solar flare left their mark on the detectors of the SOHO spacecraft for several hours. SOHO is located about 1.5 million km from Earth, so that it can provide advance warning of solar storms heading for the planet. (ESA/NASA)

The temperature of the Sun's visible surface (the photosphere) is a little over 6 000°C, but about 2 000 km above this region is the much hotter corona – which is usually only visible during a total solar eclipse. The temperature of the gas in the corona exceeds 2 million degrees Celsius. However, even this figure is dwarfed by solar flares, sudden releases of energy whose temperature typically reaches 10 or 20 million degrees Celsius, occasionally soaring to 100 million degrees.

The Solar Spectrum

The Sun is yellow in color, which correlates with a surface temperature of about 6 000°C, so it is classified as a yellow dwarf of spectral type G2. The visible white light of the solar spectrum can be split into many wavelengths (colors) by passing it through a raindrop or a prism. The red end of the spectrum is at a slightly longer wavelength (lower frequency) than the violet end.

However, most of the Sun's output of electromagnetic radiation is invisible to our eyes. The first to demonstrate the existence of this invisible light was William Herschel. 19 years after he discovered the planet Uranus, Herschel made another notable breakthrough when he used a prism to split sunlight into its colors and then placed a thermometer just beyond the red end of the spectrum. Noting a higher temperature at this location, he concluded that it must be caused by the existence of "calorific rays" – now known as infrared (literally "below red") radiation.

Subsequent studies have shown that the Sun also emits invisible electromagnetic radiation at much longer and shorter wavelengths. In 1801, Johann Ritter proved the existence of ultraviolet radiation by using a prism to create a solar spectrum and then noting how paper soaked in sodium chloride (salt solution) became darker.

In 1942, James Hey was assigned to find a solution to what seemed to be severe jamming of anti-aircraft radars by the Germans. Instead, he found that the direction of maximum interference seemed to follow the Sun. On checking with the Royal Observatory, he learned that a very active sunspot was traversing the solar disc. Hey had discovered the first radio emissions from the Sun.

Seven years later, solar X-rays were discovered using instruments on board US suborbital rockets. The earliest detectors were simply pieces of photographic film shielded from visible and ultraviolet light by foil made of beryllium or aluminum. Later flights carried more sophisticated Geiger counters. By the 1950s it was possible to confirm that the solar corona consisted of million degree plasma and that active regions on the Sun's surface were also a source of X-rays.[1]

Occasionally, major outbursts on the Sun, involving the interaction of high-energy electrons, protons, and atomic nuclei, have been found to produce gamma rays – a form of electromagnetic

[1] Plasma – often described as the fourth state of matter (after solids, liquids and gases) – is an almost completely ionized gas, in which atoms or molecules are converted into ions by the addition or removal of one or more electrons. The hot mixture comprises negatively charged electrons and positively charged protons derived from high speed collisions between ions. The electrons and protons neutralize each other, so there is no net charge.

(a)

(b)

Figure 2.5 **(a):** The electromagnetic spectrum ranges from extremely short wavelength (high frequency) gamma rays to extremely long wavelength (low frequency) radio waves. The shorter wavelength radiation (gamma rays, X-rays and ultraviolet) is associated with high temperatures and high energy processes. (NASA) **(b)** The solar spectrum in the visible, or white light, region. Spectrograms like this split the light up into different wavelengths (colors). Dark bands superimposed on the colors are absorption lines, created when atoms in the Sun's outer regions absorb light at certain wavelengths. These lines (sometimes called Fraunhofer lines) indicate specific ionized elements in the Sun. (NSO/AURA/NSF)

radiation with extremely low wavelengths (high frequencies) and energies up to one million electron volts.[2]

Since most of the high frequency radiation is screened out by Earth's atmosphere (fortunately for our health), studies of the high energy end of the spectrum generally require instruments to be lofted above the blanket of air, on balloons, sounding rockets or satellites.

Considerable information on the Sun's composition and temperature can be obtained by spreading out the visible spectrum in a spectrograph. By passing light through a series of fine slits or a diffraction grating, it is possible to see hundreds of black lines in the spectrum. These are called absorption lines because they are created when different ionized atoms in the solar atmosphere absorb light at certain wavelengths.[3] The wavelength of each line indicates a specific ionized element in the Sun, while the darkness of each line shows its relative abundance.

This spectral fingerprinting makes it possible to unravel the chemical composition of the solar atmosphere. It turns out that the Sun is mainly composed of the two lightest elements, hydrogen and helium – both of which are rare on Earth. 71% of the Sun (by mass) consists of hydrogen, although this element actually accounts for 92% of the atoms in the Sun. Helium, the second most abundant element in the Sun is so rare on Earth that it was first discovered by studying the solar spectrum during the eclipse of 1868. Helium accounts for 27% of the Sun's mass and 7.8% of its atoms. This means that heavier elements make up only 0.1% of the atoms in the Sun.

Oxygen, carbon, and nitrogen are the three most abundant 'metals', that is, elements heavier than helium. The Sun also has traces of neon, sodium, magnesium, aluminum, silicon, phosphorus, sulfur, potassium, and iron. Altogether some 67 elements have been detected in the solar spectrum.

Nuclear Fusion

Until the early 20th century, the source of the Sun's heat and light was unknown. Although the extreme conditions in the Sun's core were recognized by 1870, there was still no known physical process that could account for the solar furnace. For example, heat from

[2] An electron volt is a measure of energy. It is defined as the kinetic energy gained by an electron passing through a potential difference of one volt.
[3] The black lines are sometimes called Fraunhofer lines after Joseph von Fraunhofer who, in 1814, invented the diffraction grating. Many years passed before Gustav Kirchhoff discovered the link between the spectral lines and the chemical composition of the Sun.

the gravitational collapse of the Sun would only be sufficient to keep it shining for some fifteen million years.

The explanation eventually came after a series of breakthroughs at the Cavendish Laboratory in Cambridge, England, and elsewhere. In 1905, Albert Einstein published his famous equation $E=mc^2$, which basically states that a small quantity of mass is equivalent to a very large amount of energy. The release of energy is always associated with a reduction in mass.

By 1911, Ernest Rutherford had shown that the atom comprised a tiny nucleus which was surrounded by particles called electrons. The nucleus itself was eventually found to be composed of other particles – protons and neutrons. Nuclear reactions not only involved shifts in proton-neutron combinations, which released much greater amounts of energy than any chemical reactions, but they were also the cause of radioactivity – particles which are emitted from nuclei as a result of nuclear instability.

By the 1920s, Francis Aston was able to demonstrate that the mass of the helium atom is slightly smaller than the mass of the four hydrogen atoms (protons) from which it is formed. By now, it was known from laboratory experiments that some elements could be transformed into other elements. Arthur Eddington put together the pieces of the puzzle by proposing that stars such as the Sun were crucibles inside which hydrogen was transformed into helium. This transformation, he suggested, was the source of their energy.

The subsequent confirmation that hydrogen is the most abundant element in the Sun supported the idea that hydrogen nuclei (protons) are fused together deep inside the Sun. With the development of first the atomic (fission) bomb (1945), and then the hydrogen (fusion) bomb (1952), there could be no doubt that nuclear reactions produce vast amounts of radiation and energy.

According to the most generally accepted current theory, the Sun's energy output is the result of a series of nuclear fusion reactions, the most dominant of which is known as the proton-proton or pp chain. These reactions take place in the central core.

In the first stage of the pp chain reaction, two positively charged protons collide with enough energy to overcome the repulsive electrical force between them, uniting to form a deuteron. However, since a deuteron comprises a single proton and a single neutron, one of the protons must be transformed into a neutron by emitting a positively charged particle, known as a positron, along with a low energy neutrino. Some of the energy liberated during this reaction is converted into radiation when the positron collides with an electron.

The second stage occurs less than a second later, when the deuteron collides with another proton to form a nucleus of light helium (helium-3), while releasing a gamma ray. Finally, two helium-3 nuclei meet and fuse to form a nucleus of normal helium. When this occurs, two protons are returned to the surrounding gas. This leaves a helium nucleus comprising two protons and two neutrons.

The entire process may take more than a million years, on average, but the sheer number of reactions that occur each second inside the Sun has a remarkable multiplier effect. Roughly 10 trillion trillion trillion helium nuclei are created every second, resulting in a reduction in the Sun's mass of five million tons per second. This seems an incredible amount, but the loss is

Figure 2.6 The proton-proton chain reaction has three stages. During this nuclear process, four hydrogen nuclei (protons) are united to create a helium nucleus. However, two of the protons have to be changed into neutrons. This is because a helium nucleus comprises two protons and two neutrons. This proton transformation is made possible by the release of a positively charged anti-particle called a positron, leaving behind the neutral (chargeless) neutron. Gamma radiation is also released during this collision process. (UCAR-Randy Russell)

insignificant compared with the Sun's total mass of two thousand trillion trillion tons. Since the helium nucleus is only 0.7% lighter than its original components, the Sun has actually lost only a few hundredths of one percent of its mass during its five billion year history.

More serious is the rate of conversion of hydrogen into helium. About 600 million tons of hydrogen are transformed into helium each second in the Sun's core. Over its lifetime, approximately 37% of the hydrogen has already been converted into helium. After another five billion years of chain reactions, our star will run out of hydrogen and require a new source of fuel.

The Structure of the Sun

Although the Sun is a giant ball made almost entirely of plasma, it is not uniform throughout. Models of the solar interior, based on a combination of theoretical studies and observations, show a layered structure, with two main spherical shells surrounding a high density core.

Each of the regions has different physical characteristics. The core, the innermost 25%, is where the Sun's energy is generated. This energy diffuses outward in the form of radiation (mostly gamma rays and X-rays) through the radiative zone, then continues upward by convective fluid motion through the convection zone, the outermost 30%. A thin interface layer (the tachocline) between the radiative zone and the convection zone is where the Sun's magnetic field is thought to be generated.

Table 2.2
Internal Zones of the Sun

Zone	Radial Distance from Center (%)	Temperature (K)	Density (g/cm³)	Energy Transport
Core	0–25	15–7 million	~150–20	Radiative
Radiative zone	25–70	7–2 million	20–0.2	Radiative
Convective zone	70–100	2 million–5 700	0.2–0.0000002	Convective

Radiation takes over again in the thin, visible surface layer – the photosphere. Above the photosphere are the chromosphere and the corona. They make up the Sun's sparse, but extremely hot, atmosphere, which makes such spectacular viewing during total solar eclipses. This atmosphere is molded and modified by ever-changing magnetic fields and electrical currents. The moving currents and the rotation of the Sun result in a complex magnetic structure, whose strength varies and whose polarity changes in roughly eleven-year cycles.

Relatively empty zones in the corona, known as coronal holes, often occur near the Sun's poles. Sometimes they spread towards the equator or even stretch all the way from pole to pole. They are particularly prominent at times of solar minimum. Since the magnetic fields in these holes are radially aligned to surface, rather than curving back to the Sun, they offer an escape route for the charged particles in the high speed solar wind.

The Core and the Radiative Zone

In the core, gas is squashed by the pressure of the overlying material, so that it occupies a much smaller volume. At the very center, the Sun's density is about 150 g/cm³, 10 times greater than that of solid lead, and the temperature soars to more than 15,000,000°C. As described above, nuclear reactions in the core consume hydrogen to form helium.

Both the temperature and the density decrease with distance from the center, so that nuclear fusion more or less ceases beyond the outer edge of the core (about 25% of the distance to the surface or 175,000 km from the center). At that point the temperature is only half its central value and the density drops to about 20 g/cm³.

The incredibly high temperature is a major factor in preventing the Sun from collapsing under its own gravity. Hydrogen nuclei near the centre collide more frequently and at higher speeds than anywhere else in the Sun, so they exert a greater outward push against the overlying material. In addition to this gas pressure, the colliding particles also emit electromagnetic radiation which slowly escapes toward the surface, further helping to inflate the solar sphere. This latter force is called radiation pressure.

The main mechanism of energy transport in the inner 70% of the Sun is electromagnetic radiation in the form of photons which exhibit both wave-like and particle-like properties.

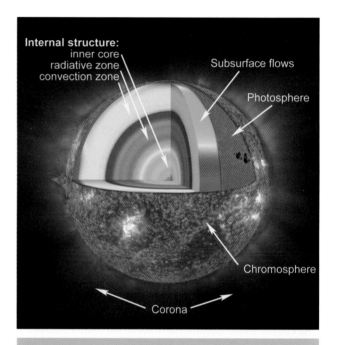

Figure 2.7   At the center of the Sun is an extremely dense core, where the temperature reaches fifteen million degrees Celsius. The core accounts for 2% of the Sun's volume but 60% of its mass. Energy generated during nuclear reactions in the core travels to the surface through the radiative and convective zones. The radiative zone lies between about 25% and 70% of the distance from the center. The convective zone extends to just below the surface. In this region, the reduction in temperature toward the surface is so rapid that the gas becomes unstable, allowing convection currents of rising and falling plasma to dominate. (NASA)

Although gamma ray photons are produced in the core, the photons that emerge from the Sun's surface are mainly in the form of visible light.

Each photon generated in the core travels only a very short distance before it encounters another particle. The photon is then scattered or absorbed, and then re-emitted. The photons produced by these encounters may emerge in any direction, so their zig-zag progress toward the surface via innumerable collisions may take hundreds of thousands of years.

As the photons diffuse outward, they move into regions where the plasma is cooler. As a result, they collide with electrons or ions that are generally moving at slower speeds, and the overall energy of the photon population is gradually reduced.

Above the core is the radiative zone, the Sun's largest internal "shell", which reaches out to 70% of the star's diameter. In this zone, energy from the core is transported by radiation, From the bottom to the top of the radiative zone, the density drops from $20\,g/cm^3$ (about the density of gold) to only $0.2\,g/cm^3$ (much less than the density of water). Meanwhile, the temperature falls from $7,000,000°$ C to about $2,000,000°$ C over the same distance.

The Interface Layer (Tachocline)

Between the radiative zone and the outer convective zone is a relatively thin transition region, known as the interface layer, or tachocline. There, the stable conditions of the radiative zone gradually disappear, to be replaced by the more turbulent fluid motions found in the convection zone.

Although it is extremely difficult to investigate conditions at such depth within the Sun, recent investigations indicate that the Sun's magnetic field is generated by a dynamo effect in this layer. Shear flows – changes in fluid flow velocities across the layer – stretch and increase the strength of magnetic field lines of force. There also seem to be sudden changes in chemical composition across the tachocline. Variations in temperature and pressure, originating in the tachocline and spreading into the convection zone through the transport of heat, may also be the key to differences in the rate of rotation of the Sun's surface.

The Convective Zone

Above the tachocline is the convective zone, which makes up the outer 30% of the solar sphere. The convective zone extends from a depth of about 200,000 km all the way to the visible surface. At the base of the convection zone the temperature is about 2,000,000°C, but this decreases to about 6 000°C at the surface, where the density is only $0.0000002\,gm/cm^3$ (about 1/10,000th the density of air at sea level).

As its name suggests, energy is transported within this zone by convection – the turbulent motion resulting from hot gas rising toward the surface and cooler gas sinking back toward the center.

Although, to us, the temperature in the lower convective zone seems incredibly high, gas in the convective zone is actually cool enough for the heavier ions (such as iron, carbon, nitrogen, calcium and oxygen) to hold on to some of their electrons. This means that the material is more opaque, making it harder for radiation to travel through.

Heat trapped in the lower part of the convective zone makes the plasma unstable, so that it starts to boil or convect. The hotter, less dense, material starts to bubble up from below, and continues to rise as long as its temperature exceeds that of its surroundings. These convective motions carry heat quite rapidly to the surface,

although the gas expands and cools as it rises. On the surface the convection can be seen in the form of granulation and supergranulation, creating a carpet of gaseous cells that rise and fall like bubbles in boiling liquid.

Helioseismic observations by the MDI instrument on the SOHO spacecraft have also revealed two huge circulation cells within the convective zone, probably extending to a depth of at least 26,000 km (4% of the solar radius). Plasma near the surface flows from the equator to the poles at a speed of 10–20 m/s, very slow compared to random motions at the surface of around 1 km/s. Models suggest that there is a return flow at the bottom of the convection zone, though this has not yet been observed. Since the plasma density is much higher at depth, this flow toward the equator must be much slower, perhaps 1–2 m/s. At this rate, it would take about 20 years for material to travel from the polar regions to the equator (see Fig. 2.29).

Helioseismology

Although direct study of its interior is impossible, insights into the conditions – temperature, composition and motions of gas – within the Sun may be gained by observing oscillating waves, rhythmic inward and outward motions of its visible surface. The study of these solar oscillations is called helioseismology. In many ways, it resembles the study of seismic waves generated by earthquakes to learn about the Earth's interior.

The complex pattern of periodic throbbing motions appears on the surface due to acoustic (sound) waves that are trapped inside the Sun. Although they cannot be observed with the naked eye, the

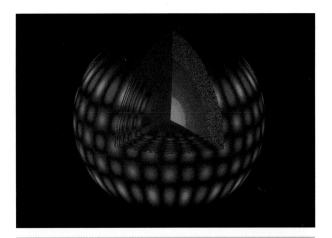

Figure 2.8 A computer representation of one of nearly ten million modes of sound wave oscillations of the Sun, showing receding regions in red and approaching regions in blue. By measuring the frequencies of many such modes and using theoretical models – a technique known as helioseismology – it is possible to infer a great deal about the Sun's internal structure and dynamics. Such oscillations are measured by the ground-based Global Oscillation Network Group (GONG), as well as orbiting observatories such as SOHO. (NSO/AURA/NSF)

Box 2.1   Solar Eclipses

Observation of the Sun usually requires specialized equipment and proper eye protection. However, there are rare occasions when anyone can stare at the Sun (or at least its outer regions) for several minutes without any protection. These spectacular apparitions are total solar eclipses.

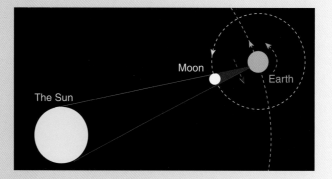

Figure 2.9   Once every four weeks, the Moon moves between the Sun and Earth. This is the lunar phase known as new Moon. Occasionally, the Moon passes precisely in front of the Sun, so that its shadow (the umbra) just reaches Earth. This results in a total solar eclipse for anyone located within the umbra. As the shadow travels across the surface of the rotating Earth, it traces a narrow path of totality, where the Sun's disk is hidden from view for up to 7½ minutes. (ESA)

Solar eclipses occur when the Moon passes directly in front of the Sun, covering its disk. This is possible because of a mathematical coincidence: the diameter of the Sun is some 400 times greater than the Moon's diameter, but, at this particular epoch of time, the Moon is approximately 400 times closer to Earth than the Sun.

Figure 2.10   The total solar eclipse of July 11, 1991 – photographed in red light from Mauna Loa, Hawaii. It shows an active Sun shortly after the solar maximum of late 1989. In addition to the bright coronal streamers, a red, hook-shaped prominence is visible (upper right), with a much smaller prominence at lower right. The bright inner corona has been suppressed to show the faint streamers of the outer corona. (High Altitude Observatory / Rhodes College)

Solar eclipses can only happen at the time of new Moon. If the Moon travels directly between the Sun and Earth, it is clear that our planet will pass through the lunar shadow and an eclipse will occur. However, such alignments do not take place every four weeks, partly because the plane of the Moon's orbit is inclined around 5° to the plane of Earth's orbit (see Chapter 4). Moreover, although the Moon crosses the plane of Earth's orbit twice every four weeks, one of these intersection points (nodes) must intersect a line joining the Sun and Earth for a solar eclipse to occur.

This means that, although the Moon passes between the Sun and Earth during each new Moon phase, the objects are usually not precisely aligned. As a result, there are only two to five total eclipses per year, when the Moon completely covers the Sun's disk.

At such times, the Moon's cone-shaped shadow (the umbra) stretches across space and grazes the Earth. For an observer situated inside the umbra, the eclipse is total. For someone located in the outer part of the shadow (the penumbra), only part of the Sun is masked, so the eclipse is partial.

As the planet rotates, the tip of the umbra travels over the surface at about 1 600 km/h, tracing a path no more than 270 km wide. Within the path of totality the sky goes dark, enabling the stars and planets to appear for up to 7½ minutes.[4]

Just before the Moon completely covers the photosphere, the Sun's light shines through gaps in lunar mountains to produce a glittering "diamond ring" effect. As totality sets in, the normally invisible solar atmosphere – the corona – appears as a pearl-white ring around the black Moon. (The corona is not usually visible in daylight because its luminosity is only about one millionth that of the photosphere.) A number of reddish, flame-like prominences may rise above the lunar limb, held in place by powerful magnetic fields.

The visible structure of the corona is related to the density of electrons in the solar atmosphere that are available to reflect light from the photosphere. Its appearance varies considerably. Near times of solar maximum, when sunspots are most numerous, the corona displays numerous bright "helmet" streamers that emanate all around the solar disk. When the Sun is less active, these streamers are fewer in number, and often missing altogether from the polar regions.

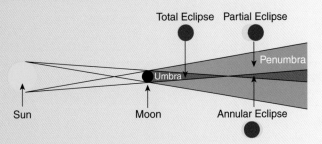

Figure 2.11 The Moon's shadow consists of two cone-shaped areas, known as the umbra and the penumbra. For an observer standing within the umbra, the eclipse is total. If Earth is slightly further away from the Moon, the outer part of the solar disk is not covered and the eclipse is annular (ring-like). For an observer standing in the penumbra, only part of the Sun is masked: the eclipse is partial. (ESA)

Sometimes, when the Moon is near its apogee, the lunar disk does not completely cover the Sun. During such an annular eclipse, the Moon's shadow does not reach Earth, so the dark circle of the Moon is surrounded by a bright ring of the Sun's surface. (Annular comes from annulus, the Latin word for ring.) Annular eclipses are slightly more frequent than total eclipses.

Partial eclipses are seen over a much larger area than total eclipses, so they are much more frequent in any given location on Earth. At these times, only part of the Sun is covered by the Moon, resembling a bite taken out of its disk.

tiny motions can be detected as subtle shifts in the wavelength of the spectral absorption lines. A region that is rising displays a shortening in wavelength which causes a small blue shift. Sinking columns are slightly redshifted. Images showing the overall pattern of these spectral shifts are known as Dopplergrams.

The movements are also visible as minuscule variations in the Sun's light output. Based on shifts in specific spectral lines, the helioseismic images reveal millions of vertical, gaseous motions generated by sound waves all over the photosphere. The most intense of these are low frequency waves that oscillate on a time scale of about 5 minutes, coinciding with velocities of 0.5 km/s. However, the overall pattern is extremely complex – the result of

millions of oscillations, both large and small – that simultaneously resonate with periods ranging from a few minutes to one hour. Motions as slow as a few millimeters per second have been detected, but they may also be remarkably long-lived, persisting for up to one year.

Although observations of the Sun from any single location on Earth (except the poles) are generally limited in duration due to the planet's rotation, a worldwide network of observatories known as the Global Oscillation Network Group (GONG) has been set up to enable 24 hour helioseismic studies. Meanwhile, spacecraft such as SOHO have an uninterrupted view of the Sun from orbit.

[4]The longest total eclipse of the 21st century, which took place over Asia on July 22, 2009, lasted a maximum of 6 minutes 39 seconds. This duration will not be exceeded until 2132.

It turns out that the entire Sun is ringing like a bell, with global oscillations that may continue for weeks. Each of the 10 million sound waves reverberates around the interior before it reaches the surface. Waves of different frequencies descend to different depths. On their return journey, they are influenced by changes in temperature, density and composition, just like seismic waves inside Earth.

The lower-pitched waves, with a frequency of about 3 MHz (a five minute period), have been used to probe the solar interior and even to make images of the far side of the Sun, when they give advance warning of flares and active regions before they appear around the limb and start to impact Earth. One of the most significant results has been the recognition of a small change in sound speed at a radial distance from the center of 71.3%, marking the lower boundary of the convection zone. Measurements of the speed of the waves indicate a helium composition of 23–26% in the convection zone – consistent with other calculations of helium abundance.

The Photosphere

The photosphere is the Sun's visible surface, the region with which we are most familiar. This gaseous layer is only about 100 km deep – extremely thin compared to the 700,000 km radius of the Sun. It appears darker towards the limb, or edge of the visible disk. This limb darkening is the result of looking at the cooler and dimmer regions of the solar disk, whereas the observer looks straight down toward the hotter, brighter regions in the center of the disk.

With a temperature of only 5 700°C, the photosphere is cool enough for molecules to form, so it mostly consists of neutral (not ionized) gas. Most of the spectral Fraunhofer lines are formed here.

The most easily recognized features in the photosphere are dark sunspots, which occur singly or in groups. They grow and then fade over days or weeks. However, for reasons not yet understood, the sunspots increase in overall number, decrease and then return to a peak every eleven years or so. This is called the sunspot cycle.

Convectional cells, found all over the photosphere, are marked by small, cell-like granules. About 1 000 km across, they show where gas is rising, cooling and then sinking. Like a pan of boiling water, the granules are short-lived, lasting for only about 20 minutes before they are replaced by new, upwelling cells. The flow of gas within the granules can reach speeds exceeding 7 km/s, producing sonic "booms" and other noise that generates surface waves.

By measuring the Doppler effect in the solar spectrum, it is possible to measure the motion of material in the photosphere. The measurements show that much larger scale convectional motion occurs over the entire photosphere, creating supergranules about 35,000 km across. Individual supergranules survive for a day or two and have flow speeds of about 0.5 km/s.

Also visible are small, bright regions, known as faculae, which are often associated with sunspots. Faculae occur where strong magnetic fields greatly reduce the local density of the gas. The low

Figure 2.12   A detailed view of convection cells, or granulation, in the photosphere, taken by Japan's Hinode spacecraft. The lighter areas show where gas is rising from below, while the darker, intergranular lanes reveal where cooler gases are sinking. (JAXA/NASA/PPARC)

density makes it nearly transparent, so the lower levels of granules are more easily visible. At these deeper layers, the gas is hotter and radiates more strongly, explaining the brightening.

The Chromosphere

The chromosphere (literally "color-sphere") is an irregular layer, about 5 000 km thick, which lies above the photosphere. It is named after the reddish prominences which can be seen arcing above the Sun during a solar eclipse. In this region, the temperature rises from 6 000°C to about 20,000°C, possibly due to turbulence and the action of shock waves.[5] At these higher temperatures hydrogen emits reddish light known as H-alpha (Hα) emission.

When the Sun is viewed through a spectrograph or a filter that isolates the hydrogen-alpha light, many new features become visible. Huge prominences which tower above the limb are shaped by local magnetic fields, often forming loops and arcs many times bigger than Earth. When they occur above the main solar disk, these dense, gaseous tongues look very different. Instead of red eruptions, they appear as dark, thread-like features, so they are known as filaments. Both filaments and prominences can erupt to considerable heights over the course of a few minutes or hours.

Locally concentrated magnetic fields are also associated with plage (French for "beach"), bright patches surrounding sunspots that are best seen in H-alpha light. Plage also form

---

[5] **The coolest part of the Sun is in the lower chromosphere, about 500 km above the photosphere, where the temperature is about 4370°C.**

Figure 2.15 The Sun's magnetic field allows the release of wave energy from its interior. Magnetic windows, associated with strong knots in the magnetic field, allow these sound waves to leak out, so that they travel upward into the chromosphere through thin fountains of hot gas, or "spicules." (Zina Deretsky, National Science Foundation)

Figure 2.13 Two giant prominences rising from the chromosphere on March 18, 2003. The shape of the plasma structures is molded by the local magnetic field lines. The loops rose to an altitude equal to twenty Earth diameters in a few hours, and may have been associated with a flare and coronal mass ejection. The SOHO image was taken in extreme ultraviolet light. (EIT consortium, ESA/NASA)

Figure 2.14 Taken by Hinode's Solar Optical Telescope on November 11, 2006 – this image reveals fine scale structure in the chromosphere that extends outward from the top of the convection cells, or granulation, of the photosphere. The structure results from the interaction of hot, ionized gas with the magnetic field. (JAXA/NASA)

Spicules

Recent images taken by Japan's Hinode spacecraft have revealed a fine structure comprising hot, ionized gas which extends up from the surface granulation toward the corona. This escaping gas can also be seen in numerous jets known as spicules.

Spicules are small jets of plasma that are propelled upward from the photosphere at speeds of up to 90,000 km/h. They are less than 600 km in diameter, but soar to heights of 5000 km, extending to the chromosphere's outer boundary in only a few minutes. Although each jet is short-lived, usually plunging back towards the surface, they often reform in the same spot every five minutes or so. This five minute periodicity is not a coincidence, since the surface of the Sun oscillates with the same period. There are over 100,000 spicules at any one time in the chromosphere.

Spicules almost always occur very close to swarms of small magnetic flux tubes, where the magnetic field is highly concentrated. If the flux tube is inclined, rather than vertical, the dominant five minute acoustic waves (see helioseismology) can leak all the way up into the atmosphere. When they encounter lower densities of plasma, the waves grow exponentially with height and develop into shock waves. These shock waves then drive mass and energy upwards to create a supersonic jet of plasma: a spicule is formed.

The spicules may reach temperatures of more than 11,000°C, and deliver huge amounts of energy and material into the corona, some of which is likely ejected into space in the form of the solar wind. They have been proposed as one of the mechanisms for superheating the corona (see below).

part of the network of brighter regions that make up the chromospheric network. Once again, concentrated magnetic fields produced by fluid motions in the supergranules result in a patchwork that outlines the giant convectional cells. The network is best seen in H-alpha and the ultraviolet spectral line of calcium (Ca II K).

Figure 2.16 This view of the chromosphere – taken with the Swedish 1 m Solar Telescope on June 16, 2003 – shows an active region (left). The right half of the image is dominated by short, dark features, called spicules. These jets of plasma are shooting upwards at speeds of around 48 000 km/h. The spicules are about 480 km in diameter, and 3 200 to 8 000 km in length. The image was taken in H-alpha light. (Swedish Solar Telescope/Lockheed Martin Solar and Astrophysics Laboratory)

Figure 2.17 A complex series of coronal loops seen in ultraviolet light by the TRACE spacecraft. Shaped by magnetic field lines, the loops extend 120 000 km above the photosphere. (NASA)

The Transition Region

The transition region is a thin, very irregular layer that separates the chromosphere from the much hotter corona. In this region, the temperature increases rapidly from about 20,000°C to 1,000,000°C. At such high temperatures, hydrogen atoms are ionized (stripped of one or more electrons), making them difficult to see. Instead of hydrogen, the light emitted by this region is dominated by carbon, oxygen, and silicon ions. These charged particles emit ultraviolet light, so they can only be observed by spacecraft such as SOHO and TRACE.

The Corona

The corona is the Sun's outer atmosphere. It is extremely hot, with a temperature of about two million degrees Celsius, but the density of its plasma is quite low, so the corona is not very bright. Until the introduction of the coronagraph in 1930, it could only be seen during total eclipses of the Sun.[6]

The corona displays a variety of transient features, including streamers, plumes, and loops, as well as huge eruptions known as coronal mass ejections. The overall shape of the corona also changes during the 11-year sunspot cycle. Dark coronal holes are visible in UV and X-ray images of the polar regions, often extending toward the Sun's equator during times of solar minimum (see Fig. 2.19). These holes are associated with "open" magnetic field lines which allow the high speed solar wind to escape into space.

Long, thin streamers known as polar plumes also project outward along the open magnetic field lines at the poles.

Coronal loops are another common feature, sometimes rising over a million kilometers above the photosphere. These are flows of trapped plasma moving along "alleys" in the arch-shaped magnetic fields of the corona at speeds up to 320,000 km/h. One explanation is that the flows are caused by uneven heating at either extremity of a loop, with plasma racing from the hotter end to the cooler end. The reason for such heating remains uncertain (see below).

Observations of the corona's visible spectrum in the late 19th century revealed mysterious, bright emission lines at wavelengths that did not correspond to any known elements. The true nature of the corona was revealed in 1939, when it was shown that the lines were produced by highly ionized iron and calcium. Such strong ionization could only happen in an environment where temperatures exceed 1,000,000°C. At these temperatures, both hydrogen and helium (the two dominant elements) are completely stripped of their electrons. Even the less common solar constituents, such as carbon, oxygen and nitrogen, are reduced to bare nuclei. Only the heavier elements are able to retain a few of their electrons.

The corona's high temperature means that it emits energy mainly at ultraviolet and X-ray wavelengths. Many active regions, flares and other coronal features are clearly visible in X-ray images taken by satellites such as Yohkoh, SOHO, TRACE and Hinode.

---

[6] A coronagraph is an instrument that uses a disk to blot out the glare of the light from the photosphere.

Coronal Heating

Why is the corona hundreds of times hotter than the photosphere? Since it is physically impossible to transfer thermal energy from the cooler surface to the much hotter corona, the mechanism for such intense heating has intrigued scientists for many years. However, there is now a general consensus that the Sun's localized, highly variable and intense magnetic fields are the cause.

Observations show that coronal heating is more intense in regions with stronger magnetic fields. It is also generally accepted that the origin of coronal heating and activity probably lies in the photosphere and the underlying convective zone. Since convection causes rapid, turbulent motion of the gases in the photosphere, the magnetic field lines emerging from the surface are bent or mixed. This may result in the creation of waves along the magnetic field lines, or the formation of magnetic discontinuities and reconnections in the corona.

Several physical processes have been proposed to explain coronal heating. One is "wave-heating", involving waves in the gas that propagate along the magnetic field lines and then dissipate in the corona.[7] Another is "microflare heating", when a great number of extremely small-scale flares dissipate magnetic energy in the corona through a process called magnetic reconnection. This occurs when magnetic field lines behave rather like rubber bands, snapping and then reconnecting with lines of opposite polarity.

A third candidate is spicules – fountain-like jets of plasma that emanate from the chromosphere. Some fast-moving spicules reach temperatures of more than 1 million degrees, and cause a brightening of the corona. If even some of that super hot plasma stays aloft it would make a contribution to coronal heating.

High resolution X-ray imaging has revealed the presence of numerous, relatively small explosions, called microflares or nanoflares. These sudden bursts of energy occur within coronal loops – thin magnetic tubes filled with very hot plasma that arch high above the surface. Although they are so small that they cannot be studied individually, so many erupt at the same time that their combined effect is quite dramatic.

Nanoflares are associated with the build-up of considerable magnetic stress due to random shaking of the magnetic field lines that are rooted in the solar interior. In this scenario, energy is released through the reconnection of neighboring magnetic field lines with locally opposite polarity. This violent release of energy creates two jets of material at the reconnection site which are accelerated and repelled by the reconnected and highly curved magnetic field lines.

The Hinode spacecraft has measured plasma in active regions with temperatures so high that they can only be produced by impulsive energy bursts associated with storms of nanoflares.

Models based on the nanoflare theory suggest that the plasma strands within coronal loops are confined by magnetic field lines. When a nanoflare occurs, a low-temperature, low-density strand is rapidly heated to around 10 million degrees Celsius. Heat flows from the hot, upper part of the strand toward the base of the coronal loop, where conditions are cooler. Since the plasma at the base is denser, the heat input is only sufficient to raise its temperature to about one million degrees. This dense plasma then expands upward along the strand. Each coronal loop is, therefore, a collection of faint, 5 to 10 million degree strands and bright, one million degree strands.

Although it seems that these processes play an important, and perhaps dominant, role in coronal heating, there is still some doubt over whether the amount of energy released by the nanoflares is sufficient to heat the corona.

Data from Hinode and NASA's Solar Dynamics Observatory indicate that a special kind of magnetic wave, known as Alfvén waves, may account for much of the coronal heating.[8] These waves travel at very high speeds along the magnetic field lines that extend from the photosphere and into the corona. However, observations have shown that short-lived spicules which shoot upward from the chromosphere wiggle sideways up to 1 000 km while they form. Advanced computer simulations suggest that these movements are caused by lateral motion of the magnetic field in the Alfvén waves.

Hinode images have also shown oscillations within solar prominences – large structures of relatively cool plasma that rise through the corona. These oscillations are widely thought to be

Figure 2.18 A false color image of the corona, made from three exposures taken by SOHO. Blue shows plasma at a temperature of 1–1.5 million degrees Celsius, green 1.5–2 million degrees, and red hotter than 2.5 million degrees. A dark coronal hole is visible at upper right. (NASA-ESA)

---

[7] These are generally known as magnetohydrodynamic waves. The most important types involved in coronal heating are magneto-acoustic waves (sound waves that have been modified by the presence of a magnetic field) and Alfvén waves (similar to ULF radio waves that have been modified by interaction with plasma).

[8] The waves are named after Hannes Alfvén, who received a Nobel Prize in 1942 for his work in this field.

caused by Alfvén waves propagating along the threads at speeds of about 20 km/s. Like waves crashing on a beach, they dump their energy in the corona, heating the plasma to millions of degrees.

Figure 2.19 The twist applied to magnetic field lines in the corona induces an upward-moving Alfvén wave which carries enough energy to heat the sparse plasma in the corona. (Queen's University Belfast)

The Magnetic Sun

Magnetism is the key to understanding the Sun. This is because it is mostly composed of plasma, an electrically conducting gas in which the atomic nuclei have been almost entirely stripped of their electrons. The flow of electrically charged ions and electrons in the plasma is readily deflected by local magnetic fields. This is particularly noticeable in the corona, where the gases are extremely rarefied and thus easily shaped by magnetism, rather than gravity.

Observations of the photosphere's magnetic field are made by measuring the splitting of spectral absorption lines (known as the Zeeman effect) and the polarization of light. Various techniques are then used to determine both the strength and the direction of the magnetic field. These magnetic field observations can then be compared with the observed structures in the Sun's outer regions.

The Sun's magnetic field, like Earth's, resembles that of a bar magnet surrounded by a dipole field (i.e. it has two magnetic poles). The field flows out of the Sun at the north pole, and re-enters at the south pole. Usually, the magnetic axis roughly coincides with the rotation axis. During times of high activity, the Sun also features numerous dipoles with multiple bar magnets, each represented by an active region.

The complex "magnetic carpet" seen in the photosphere reaches the surface via the granular network of convection cells that are created as plasma wells up from below. Intense stirring causes magnetic dipoles to grow continually within the cells before being shed into the corona through magnetic reconnection.

Figure 2.20 Based on SOHO data, this image shows irregular magnetic fields (the "magnetic carpet") on the photosphere. The carpet comprises a sprinkling of tens of thousands of magnetic concentrations which have both north and south magnetic poles. These are the bases of magnetic loops extending into the corona. Whiter areas represent more material at a temperature exceeding one million degrees Celsius, darker areas represent less. The black and white spots represent magnetic field concentrations with opposite polarities. Each spot is roughly 8 000 km across. (Stanford-Lockheed Institute for Space Research / NASA-GSFC)

The Sun's magnetic variability waxes and wanes over the eleven-year sunspot cycle, changing in parallel with the number of active regions (see below). The active regions arise in unpredictable locations, emerging from the deep interior and breaking through the photosphere into the corona. Features frequently (but not always) associated with magnetically active regions include sunspots, coronal loops, flares and coronal mass ejections. All of these are most common around the time of solar maximum.

Sunspots occur where very intense magnetic lines of force break through the surface. The sunspot cycle results from the recycling of magnetic fields by the flow of material in the interior. Prominences that rise above the surface are supported by, and threaded through, with magnetic field lines.

Streamers and loops seen in the corona, sometimes reaching heights of several hundred thousand kilometers, are also shaped by magnetic fields. The arches comprise bright strands of plasma that connect two areas in the photosphere with opposite magnetic polarities. Bright blobs of hot plasma race up and down coronal loops at tremendous speeds along threads that follow the magnetic field lines.

The polarity of the Sun's magnetic dipole switches at the end of each eleven-year cycle, as the internal magnetic dynamo reorganizes itself. At solar minimum, the magnetic field resembles a simple dipole, with magnetic field lines running north-south. However, because the equator rotates much faster than the poles, the magnetic field becomes increasingly twisted as time goes by. Over a period of eleven years, the twisted field lines wrap around the Sun, generating areas of intense magnetic fields that appear at the surface as sunspots. The process of tangling ceases when the dynamo readjusts, recreating a dipole field, but with a reversal in polarity, i.e. the north magnetic pole switches to the south magnetic pole, and vice versa.

The change-over is not always smooth. In March 2000, for example, the south magnetic pole faded and was replaced by a north pole. For nearly a month, the Sun had two north poles. The original south pole migrated north and, for a while, became a

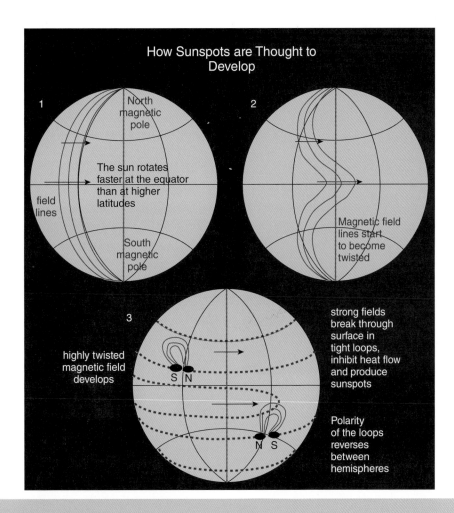

Figure 2.21 At the start of the eleven-year sunspot cycle, the Sun's magnetic field resembles a large bar magnet, with two magnetic poles. However, because the equatorial region rotates much faster than the poles, the magnetic field gets progressively wrapped around the Sun, becoming stretched as it nears the equator. After many rotations the magnetic field becomes extremely complex. The twisted field lines wrap around the Sun, generating areas of intense magnetic field that are visible at the surface as sunspots. The tangling process ends when the dynamo readjusts, recreating a dipole field. (UCAR)

band of south magnetic flux smeared around the equator. By May 2000, it had returned to its usual location near the Sun's southern spin axis. Then, in 2001, the magnetic field completely flipped, so that the south and north magnetic poles swapped positions, which they retained throughout the rest of the cycle.

The Sun's magnetic field is carried out into the Solar System by the charged particles (electrons and protons) of the solar wind (see below).

Sunspots

Sunspots are dark areas of irregular shape on the surface of the Sun. They are often large enough to be seen with the naked eye, and Chinese records of major sunspots go back to at least 28 BC.[9]

Sunspots are much easier to observe than any other solar phenomena because of their dark appearance against much brighter surroundings. This is because they are typically about 1 500°C cooler than the rest of the photosphere, although sunspots are actually fairly luminous.

The spots vary in size from a mere 1 000 km to one million km. Many of them are larger than Earth, and easily visible to the naked eye under the right conditions. In most sunspots, the darkest region is the central core (the umbra), which is surrounded by a lighter penumbra, although either of these may appear without the other. The penumbra displays light and dark linear features, known as filaments, which radiate away from the umbra. These filaments are the result of radial outward flows of gas along channels of nearly horizontal magnetic field.

Spectroscopic studies show that sunspots are associated with strong, concentrated magnetic fields. In close proximity to sunspots, the strength of the magnetic field may increase more than 3 000 times. Field strengths are directly related to their physical size, ranging from 1 000 gauss in smaller examples to 4 000 gauss in giant sunspots. (This compares with an average 0.5 gauss for Earth's surface magnetic field.) These powerful fields partially block the flow of plasma and heat from below, causing the sunspot's temperature to drop more than a thousand degrees compared with its surroundings.

Sunspots are the most visible features associated with active regions, often seen alongside faculae and plage (see above). They almost always occur in groups or pairs of opposite polarity, where magnetic fields project through the photosphere from below: in the center of a sunspot the magnetic field lines are vertical. Isolated sunspot pairs tend to be aligned in an east-west direction.

Magnetograms show that the polarities of sunspot pairs located in the northern and southern hemispheres are reversed. In one hemisphere, the sunspot with a negative magnetic polarity almost

LARGE UMBRA

Penumbral filaments

SMALL UMBRA

Umbral dots

Pores

Faculae

Penumbral filaments

Figure 2.22 Sunspots have a darker, central region (the umbra) surrounded by a lighter outer section (the penumbra) with a linear filamentary structure. Around the sunspot are thousands of granules, the tops of convectional cells 1 000–2 000 km in diameter. Sunspots appear dark because they are cooler than their surroundings. A large sunspot might have a temperature of about 4 000°C compared with about 5 500°C for the nearby bright photosphere. This sunspot was observed with the Swedish 1-m Solar Telescope on La Palma. (Royal Swedish Academy of Sciences)

Figure 2.23 Active regions on the Sun are made up of many relatively small magnetic structures emerging at adjacent locations. In this image, the rising magnetic structures are represented as white arches emerging from the surface. The dark areas at the bases of the arches are sunspots that form when solar magnetic fields become concentrated. The base of each arch has a different polarity. (NASA-GSFC)

---

[9] Direct, naked eye observation of the Sun is very dangerous and proper eye protection is essential.

Figure 2.25 An image of the chromosphere obtained by Hinode on November 20, 2006. Plasma aligned along the solar magnetic field lines is rising vertically from a sunspot (an area of strong magnetic field) toward the corona. (Hinode, JAXA/NASA/PPARC)

Figure 2.24 Flows within and around a sunspot, derived from SOHO data. Red is hot gas, blue is cooler gas. Outflowing plasma at the surface is underlain by material rushing inward, like a giant whirlpool. This inflow is strong enough to pull the magnetic fields together and reduce the amount of heat that normally flows from the interior. The cooled material sinks to a depth of only a few thousand kilometers and then spreads out. (ESA-NASA)

always leads the sunspot which has a positive polarity (with respect to the westward apparent motion due to solar rotation). The behavior is similar in the other hemisphere, except for reversed magnetic polarities. This pattern is a direct result of the internal dynamo that generates the overall magnetic field.

Some sunspots survive only a matter of hours, but many persist for days, weeks, or even months in the case of the very biggest. Each spot resembles a whirlpool, where hot gas near the surface converges and then dives into the interior at speeds of up to 4 000 km/h. The sinking gas in the immediate vicinity of the sunspot reaches depths of only a few thousand kilometers. The descending flow removes the heat that accumulates beneath the spot and eventually brings it to the surface, far from the sunspot, where it is radiated into space.

The first evidence that sunspots are likely to form comes when sound waves deep in the convective zone begin to accelerate through a particular region. Within half a day, intense magnetic fields shoot upwards like a fountain, traveling at 4 500 km/h. A 20,000 km wide column of hot gas, confined by an intense, rope-like magnetic field, reaches almost to the visible surface.

At a depth of 4 000 km, it may separate into strands that make their own way towards the surface, eventually forming smaller sunspots around the main spot complex. The intense magnetic

fields prevent the normal upward flow of energy from the interior, leaving the sunspot much cooler than its surroundings. Immediately below the main spot is a cushion of cooler, less intensely magnetized gas. Each major sunspot complex is associated with a separate magnetic column in which the polarity is often aligned in the opposite direction.

Since more sunspots appear at solar maximum, the solar irradiance reaching Earth during that time might be expected to decrease. However, satellite radiometer measurements show that, while sunspots cause a decrease in the solar irradiance on time scales of days to weeks, the long-term solar irradiance actually increases by about 0.1% as sunspot (magnetic) activity increases. The source of this additional irradiance has been traced to the bright faculae near the limb of the Sun (see above).

Sunspot Cycles

The rise in sunspot activity every eleven years or so is associated with greater magnetic activity, with an increase in the number of active regions, flares and coronal mass ejections. This cycle was discovered in 1843 by German astronomer Heinrich Schwabe as a by-product of his search for a planet closer to the Sun than Mercury.

Astronomers give each eleven-year solar cycle a number. For obscure historical reasons, solar cycle 1 was a fairly ordinary cycle which peaked in 1760. The most recent complete cycle, number 23, peaked in 2001 and ended in 2007. However, cycle 24 was extremely reluctant to start, with the Sun remaining largely blank well into the autumn of 2009, interrupted by an occasional sunspot with reversed magnetic polarity (indicative of the start of a new cycle). In addition to the deepest solar minimum in nearly a century, there was clear evidence of a decline in sunspot magnetism of about 50 gauss per year since 1992.

Figure 2.26   Snapshots of the changing solar magnetic field (left) and the soft X-ray corona (right) from 1991 to 2000 – almost an entire solar cycle. Obtained one year apart between one solar maximum (lower right) and the next, they show the evolution of coronal structure due to changes in the magnetic fields. Note the few magnetic features and lack of X-ray bright loops in the middle, at solar minimum. The strongest magnetic fields (shown in dark blue and white) occur in the active regions and coincide with the brightest coronal X-ray emissions. White shows an upward pointing magnetic field and dark blue a downward pointing field. Pale blue shows a weak field of "mixed" polarity. (Lockheed Martin Solar & Astrophysics Lab, Yohkoh SXT and Kitt Peak National Solar Observatory)

This was not the first time that sunspots largely disappeared. The solar minima of 1901 and 1913, for instance, were even longer than the 2008–2009 hiatus. In the 17th century, the Sun experienced a 70-year sunspot drought, known as the Maunder Minimum, that still baffles scientists. Between 1645 and 1715, the number of observed sunspots plummeted from thousands per year to a few dozen.

The sunspot number is calculated by first counting the number of sunspot groups and then the number of individual sunspots. The final sunspot number represents the sum of the number of individual sunspots and 10 times the number of groups. Since most sunspot groups contain, on average, about ten spots, this formula has provided reliable numbers even when observing conditions were less than ideal and small spots were difficult to see.

Each cycle tends to start with a few small spots close to latitudes 40° north and south. As it progresses, the spots and associated active regions move closer to the equator, and larger, more long-lived spots appear. After approximately eleven years, the final spots of that particular cycle form at latitudes of about 5°, before disappearing altogether.

Why do the bands where sunspots form drift equatorward over time and then disappear? For a long time, it was believed that this was due to a wave-like process involving changing magnetic fields deep inside the Sun. However, recent studies suggest that it may be caused by a giant circulation system in which compressed plasma, 200,000 km below the surface, moves from the poles toward the equator at a speed of about 5 km/h – equivalent to a leisurely walking pace.

Near the equator, the gases rise and head back toward the poles. Flowing through the surface layers, where the plasma is less compressed, they are able to move faster, reaching 32 to 64 km/h. The variable speed means that plasma can take anywhere from 30 to 50 years to complete the full circuit.

Since the speed of this meridional circulation system changes slightly from one sunspot cycle to the next, it may act like an internal clock that sets the period of the sunspot cycle. The circulation is faster in cycles shorter than the average eleven-year period and slower in longer-than-average cycles.

The absence of sunspots in 2008–2009 has been attributed to the fact that, from 1996 onwards, the deep, internal plasma flow associated with the next solar cycle was moving more slowly than usual, while the top of the conveyor belt was moving at record-high speed. This seems to have affected the internal dynamo, delaying the rearrangement of the solar magnetic field. This, in turn, prolonged the period of switchover between cycles, leading to an extended solar minimum.

Near-surface jet streams, which flow from the poles toward the equator at depths of 1 000 to 7 000 km, have also been associated with the sunspot cycle. By using helioseismology to keep track of gas moving below the surface, scientists from the National Solar Observatory, Arizona, found that the Sun generates new jet streams near its poles every eleven years. The streams migrate to the equator and are apparently associated with the production of sunspots once they reach a critical latitude of 22°.

Despite important progress in recent years, predictions of future sunspot cycles are not very accurate. Time will tell if the slow start to cycle 24 marks the start of an extended quiet period of solar activity. However, exactly how the jet streams are generated, and the precise mechanism that enables them to trigger sunspot production remains a mystery.

DAILY SUNSPOT AREA AVERAGED OVER INDIVIDUAL SOLAR ROTATIONS

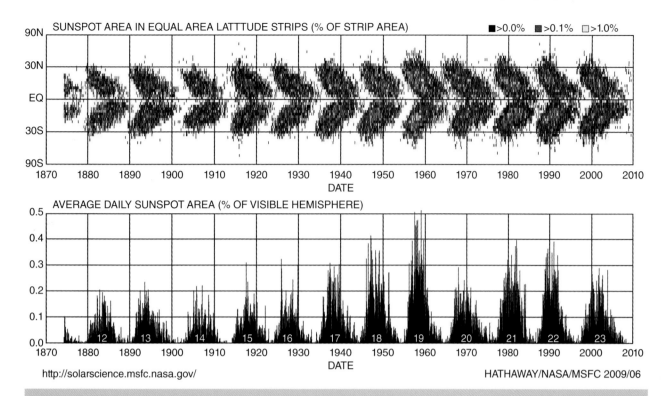

SUNSPOT AREA IN EQUAL AREA LATTTUDE STRIPS (% OF STRIP AREA)   ■>0.0% ■>0.1% □>1.0%

AVERAGE DAILY SUNSPOT AREA (% OF VISIBLE HEMISPHERE)

DATE

http://solarscience.msfc.nasa.gov/   HATHAWAY/NASA/MSFC 2009/06

Figure 2.27 This butterfly diagram (top), named for its characteristic appearance, shows the average positions of sunspots for each rotation of the Sun, based on observations obtained by the Royal Greenwich Observatory, London, since May 1874. The bands first form at mid-latitudes, widen, and then move toward the equator as each cycle progresses. The smallest spots are shown in black, the largest ones in yellow. Below is a plot of the average area covered by sunspots over the same period. (NASA-MSFC)

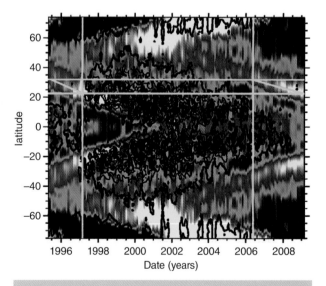

Figure 2.28 The Sun's deep circulation takes the form of a great "conveyor belt" of plasma that flows slowly toward the equator at a depth of 200 000 km. Near the equator it rises toward the surface and then returns toward the poles at a faster rate. The entire circuit takes thirty to fifty years. Changes in the rate of flow may be linked with the duration of the sunspot cycle and number of sunspots. (NASA)

Figure 2.29 A diagram showing roughly east-to-west motion at a depth of about 7 000 km in the solar interior, as derived from helioseismology observations. The diagonal yellow bands are jet streams that are associated with the solar cycle. Flow cycle 23 is shown in yellow on the left side, while the flow for the next cycle (#24), which is notably slower, is shown in yellow on the right side. Overlain on the jet streams are black contours showing the pattern of sunspots. When the streams reach 22° latitude, new sunspots begin to appear. (National Solar Observatory)

Box 2.2   SOHO

The Solar and Heliospheric Observatory (SOHO) is the most important solar observatory ever sent into space. The European-built spacecraft was launched on December 2, 1995, and transferred to the L1 Lagrangian point, approximately 1.5 million km from Earth on its sunward side. At this location, it is relatively easy to keep SOHO in a stable halo orbit from which its instruments can observe the Sun continuously, without being interrupted by eclipses.

SOHO carried twelve instruments that were designed to probe all aspects of the Sun. The spacecraft was designed to operate for two years, but it was still operating in 2011, despite some major technical problems which almost ended the mission.

SOHO observed the Sun throughout an entire eleven-year solar cycle, and its observations led to numerous discoveries. These include:

- Helioseismic data from SOHO and the GONG network of ground stations detected currents of gas beneath the visible surface, giving new insights into the layers of the Sun's interior, the behavior of the magnetic field and the change in sunspot numbers during the solar cycle.
- A 0.1% increase in the Sun's luminosity as the count of sunspots increased 1996–2000. Scientists estimate that high-energy ultraviolet rays from the Sun have become 3% stronger over the past 300 years.
- Until the launch of the Stereo mission, SOHO provided the only reliable way to identify coronal mass ejections that were heading towards Earth. This was done by linking expanding haloes around the Sun to shocks seen in the Earth-facing atmosphere. This gave 2–3 days' warning of these potentially damaging storms.
- Thousands of nanoflares occur every day, due to continual rearrangement of tangled magnetic fields. This helps to explain why the corona is far hotter than its visible surface.
- SOHO helped to locate the sources of the fast and slow solar winds.
- Charged atoms that feed the fast solar wind gain speed very rapidly, apparently driven by strong magnetic waves in the corona. Similar magnetic waves may accelerate the slow wind.
- SOHO found many elements in the solar wind, including the first detections of phosphorus, chlorine, potassium, titanium, chromium and nickel. These give clues to conditions on the Sun and to the history of the Solar System.
- After a solar flare, SOHO observed waves spreading outward across the Sun's visible surface.
- SOHO discovered large tornadoes, where hot gas was spiraling outwards from the Sun's polar regions.
- A wind of particles from distant stars blows through the Solar System, partially counteracting the solar wind. SOHO fixed its direction (from the Ophiuchus constellation) and speed (21 km/s) more accurately.
- Two instruments, SWAN and MDI, detect sound waves reflected from far-side sunspots. This makes it possible to "see" what is happening on the far side of the Sun, giving advance warning of active regions that have yet to appear on the Earth-facing hemisphere.
- More than 2 000 sungrazing comets have been discovered in LASCO images. This makes SOHO by far the most successful comet discoverer in history.

Solar Wind

The corona emits a continuous stream of charged particles, known as the solar wind. This solar wind flows radially outward in all directions at an average speed of about 400 km/s. Almost a million tonnes of solar material is ejected into interplanetary space each second. This outward flow of particles creates an invisible bubble in space, known as the heliosphere, and is responsible for the anti-sunward ion tails of comets and the shape of the magnetospheres around the planets.

The solar wind comprises an equal number of protons and electrons, plus a smaller amount of heavier ionized atoms. These particles originate in the corona where the temperature is so high that not even the Sun's powerful gravity can hold on to them.

Like winds on Earth, the solar wind is highly variable, changing with the state of the corona. During solar minimum, there is usually a fast, steady wind over the poles and a slow variable wind at lower latitudes. At solar maximum, it is highly chaotic, with fast and slow streams of particles interrupted by more frequent coronal mass ejections at all solar latitudes.

The high speed wind (750–800 km/s) is derived from plasma leaking through the corners of a magnetic honeycomb of plasma cells, mainly in large coronal holes. These regions are most commonly located near the poles, where the solar atmosphere is less dense and cooler than surrounding areas. Their open magnetic field lines allow a constant flow of high-density plasma to stream into space. This is particularly noticeable during solar minima, when coronal holes tend be larger and longer-lasting.

In contrast, the slow solar wind (300 km/s) originates over lower latitudes, typically over active regions and streamers, where closed magnetic field lines trap the electrically charged coronal gases, except at the edges of bright, wedge-shaped features called helmets. Many small mass ejections, driven by magnetic explosions, also contribute to the solar wind.

Massive compression and shock waves can result if a fast stream collides with a slow stream. As the Sun rotates, the various streams in the solar wind also rotate, producing a pattern similar to a rotating lawn sprinkler. If a slow stream is followed by a fast stream, the high speed material will catch up and plow into it.

In addition to dramatic changes in speed, spacecraft have recorded the presence of magnetic clouds – clumps of solar particles with embedded magnetic fields – and variations in the composition of the particle population which reflect conditions in their coronal source regions. For example, instruments have shown a higher abundance of magnesium ions compared to oxygen ions in the slow solar wind than in high speed streams.

Despite many years of observation, the precise mechanism of solar wind formation is still not fully understood. Although it is recognized that the fast solar wind originates from coronal holes, images of the outflowing material are still rare. Various source regions have been proposed for the slow solar wind. These include the boundary of coronal holes, helmet streamers located above closed loop structures in the corona, and the edges of active regions.

The suggestion that one of the sources of the slow solar wind is the boundary between coronal holes and active regions has been supported by high resolution images from Hinode. These show a continuous outflow of plasma along apparently open field lines that rise from the edge of an active region adjacent to a coronal hole.

The mechanism that drives the solar wind is also poorly understood. However, recent observations have indicated that Alfvén waves in X-ray jets – fast-moving eruptions of hot plasma that occur near the poles – may accelerate the solar wind to hundreds of kilometers per second. These magnetic oscillations, possibly associated with short-lived spicules in the chromosphere (see above), appear to travel at very high speeds along open field lines that extend from the photosphere and into the corona. The wave energy that leaks into the corona may be sufficient to power the solar wind.

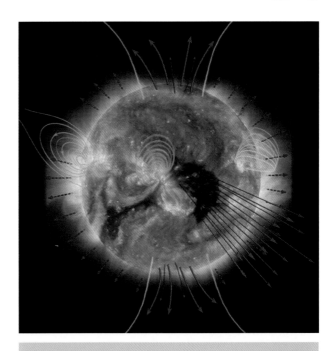

Figure 2.30 The corona is threaded with magnetic fields (yellow lines). Areas with closed magnetic fields give rise to a slow, dense solar wind (short dashed red arrows), while coronal holes with open magnetic fields yield fast, less dense solar wind streams (longer, solid, red arrows). In addition to the permanent coronal holes at the poles, coronal holes sometimes occur closer to the equator, (right of center). The image was taken by SOHO's Extreme ultraviolet Imaging Telescope. (ESA-NASA)

Flares

Flares are tremendous explosions on the surface of the Sun which can last from several seconds to a few hours. The largest flares are also the longest in duration, but they are usually quite rare, occurring only a few times a year until solar maximum approaches. Many smaller flares occur down to the limits of detection of modern instruments at about 10^{27} ergs. These generally last for a short time, down to a few seconds, and they, too, are most frequent near solar maximum, when there may be several dozen flares per day.

As flares heat material to a temperature of up to 100 million degrees Celsius, they release energy equivalent to that released by millions of 100-megaton hydrogen bombs exploding simultaneously. The energy is released in many forms: electromagnetic radiation across virtually the entire spectrum, from radio waves to X-rays and gamma rays; energetic particles (protons and electrons); and mass flows.

The high energy electromagnetic radiation travels at light speed across interplanetary space, sometimes heading towards Earth. Such increases in X-ray flux enhance the ionosphere, causing it to decrease in altitude, but, fortunately, they are unable to penetrate our planet's lower atmosphere.

Not far behind are the particles that are accelerated to near light speed by the flare. Streaming outward along magnetic field lines, they can reach the Earth-Moon system within 20 to 30 minutes, interfering with short wave radio communications, causing short circuits and computer reboots on satellites, and threatening the health of astronauts who are outside their spacecraft.

Flares are usually difficult to see in visible light against the bright background of the photosphere. Most are detected from the invisible radiation they emit. Radio and optical emissions can be observed with telescopes on the Earth, while orbiting observatories are needed to detect energetic emissions such as X-rays and gamma rays.

There are typically three stages to a solar flare, each lasting anything from a few seconds to more than an hour. First is the precursor stage, when the release of magnetic energy is triggered. The soft X-ray emission gradually increases but few, if any, hard X-rays or gamma rays are detected.

In the second, impulsive stage, protons and electrons are accelerated to energies exceeding 1 MeV, whilst hard X-rays and gamma rays are emitted, often rising in many short but intense "spikes", each lasting a few seconds to tens of seconds. The soft X-ray flux also rises more rapidly during this phase, often synchronizing with the hard X-ray profile.

In the third, decay stage, hard X-ray and gamma ray fluxes start to decay rapidly in a matter of minutes, whereas the soft X-ray flux continues to rise, reaching a peak before declining again, sometimes over a period of several hours.

Flares (and coronal mass ejections) occur near sunspots, usually along the dividing line (neutral line) between areas of

Box 2.3   Ulysses

Nearly all scientific spacecraft that observe the Sun or are dispatched across the Solar System follow paths that lie close to the ecliptic. However, the European Space Agency's Ulysses spacecraft followed a highly elliptical orbit that carried it above and below the solar poles, regions that are usually extremely difficult to observe.

Launched on the Space Shuttle Discovery in October 1990, Ulysses received a gravity assist from Jupiter in 1992. This bent its path southward, so that it would fly over the solar poles. Before its mission ended in 2009, the nuclear-powered spacecraft made nearly three complete orbits of the Sun, enabling it to observe the cyclical changes in solar activity.

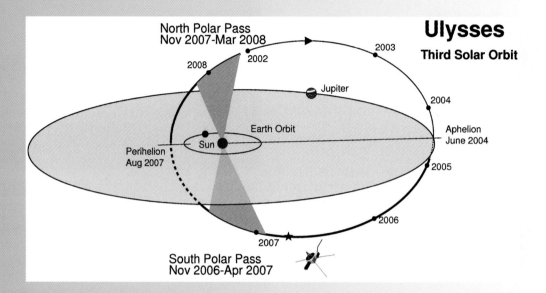

Figure 2.31   The Ulysses spacecraft was the first to leave the ecliptic and explore the Sun's polar regions. Launched in 1990, it was sent to Jupiter for a gravity assist that bent its orbit southward so that it could fly over the star's poles. Ulysses completed almost three orbits before it was shut down in 2009. The final orbit is shown here. (ESA)

Prior to Ulysses, it had only been possible to measure the solar wind near the ecliptic, giving the impression that it typically swept across the Solar System at about 400 km/s, with occasional faster gusts. Ulysses showed that this perception was incorrect. For much of the sunspot cycle, the dominant component is a fast wind from the cooler regions close to the poles that fans out to fill two thirds of the heliosphere. Blowing at a fairly uniform speed of 750 km/s, this far outstrips the slower wind that emerges from the equatorial zone. Rather than being typical, the slow wind is a minor player.

When Ulysses first flew over the polar regions in 1994 and 1995, solar activity was close to minimum, providing a view of the three-dimensional heliosphere at its simplest. The fast solar wind escaping from high latitudes flowed uniformly to fill a large fraction of the heliosphere, and solar wind variability was confined to a narrow region around the equator.

When Ulysses returned to high latitudes in 2000 and 2001, during solar maximum, the Sun displayed many active regions and solar storms were common. Solar wind flows from the poles appeared indistinguishable from flows at lower latitudes.

In 2007, Ulysses made its third polar passage. Compared with observations from the previous solar minimum, the strength of the solar wind pressure had decreased by 20%, and the field strength had decreased by 36%. Although the wind speed was almost the same, the density and pressure were significantly lower.

Of particular interest were Ulysses' observations of the Sun's magnetic field reversal at the change-over between solar cycles. Despite the apparent chaos and complexity of the magnetic field at the surface, the Sun's magnetic equator was quite well defined and stable, clearly separating the negative and positive magnetic hemispheres. However, during the solar minimum of 1994–1995, the magnetic equator was pushed 10° southward with respect to the Sun's rotational equator. The reason for this offset is still not fully understood.

Ulysses' measurements also showed that energetic particles originating in storms close to the Sun's equator are much more mobile than previously thought, and are even able to reach the polar regions. Since these particles tend to move along the magnetic field lines in the solar wind (much like beads on a wire), this means that the structure of the magnetic field is more complex than previously thought.

In addition to studying the Sun, instruments on Ulysses detected small dust particles, hydrogen ions, cosmic rays and neutral helium atoms that entered the heliosphere from interstellar space. The dust flowing into our Solar System was 30 times more abundant than predicted.

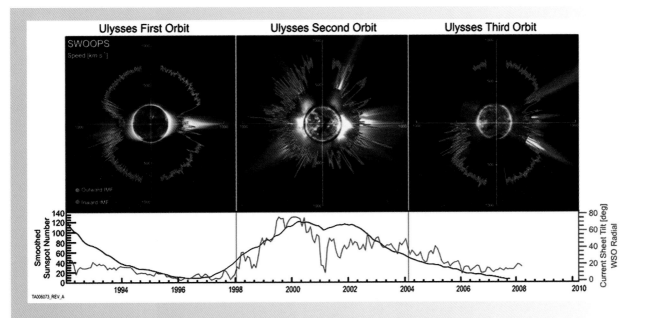

Figure 2.32 Polar plots of solar wind speed over each of Ulysses' orbits. The first orbit occurred during solar minimum, when there was a slow wind over the equator and a fast wind over the poles. The second orbit showed fast and slow winds at all latitudes, consistent with solar maximum. Data from its third orbit, during another solar minimum, indicated that the solar wind speed was similar, even though the density and pressure were significantly lower than in the previous solar minimum. The plot below shows the number of sunspots. (Southwest Research Institute; from McComas et al., 2008)

2003/11/04 19:48

Figure 2.33 SOHO's Extreme Ultraviolet Imager took this image of the most powerful flare ever recorded by spacecraft, which erupted on November 4, 2003. Since the X-ray detector on board NOAA's GOES satellite was saturated by the radiation outburst, there is some uncertainty over its magnitude, with estimates ranging from X28 to X45. The record-breaking series of solar storms in October and November 2003 affected radio communications across the globe, and caused temporary power outages in Europe due to fluctuations in Earth's magnetic field. (NASA-ESA)

Figure 2.34 A flare is created when a magnetic loop in the corona rises to great height, becoming stretched and distorted. When the two sides of the loop get close enough, magnetic reconnection takes place. The loop splits in two, forming a smaller arch at the surface and a separate loop in the corona. The excess energy is released in explosive events like flares and coronal mass ejections (CMEs). The white flash represents a flare generated in the small arch as electrons are accelerated down its magnetic field and slam into the denser plasma near the solar surface, releasing high energy radiation. Plasma trapped in the coronal loop quickly rises and expands, propelling the CME plasma away from the Sun. (NASA)

oppositely directed magnetic fields. A flare occurs when magnetic energy that has built up in the corona is suddenly released as the result of magnetic reconnection.

When a coronal loop rises to great height, it may become stretched and distorted, so that the two sides of the loop move closer together. When magnetic field lines snap and then reconnect, the original loop splits in two, forming a smaller arch at the surface and a separate loop in the corona. The excess energy is released in an explosive flare, often coinciding with a coronal mass ejection which blasts huge amounts of material into space.

The flare generated in the small arch accelerates electrons down the magnetic field lines, causing them to slam into the denser plasma below, producing X-rays, microwaves and a shock wave that heats the surface. The result is a major seismic wave in the

Figure 2.35 Seismic waves ripple away from the site of a moderate-sized solar flare. The image was taken by SOHO's Michelson Doppler Imager on July 9, 1996. Over the course of an hour, the waves traveled for a distance equal to ten Earth diameters before fading into the photosphere. Unlike water ripples that travel outward at a constant velocity, the solar waves accelerated from an initial speed of 35,200 km/h to a maximum of 250,000 km/h before disappearing. (Alexander Kosovichev, Valentina Zharkova, ESA/NASA)

Figure 2.36 On December 13, 2006 – Hinode's Solar Optical Telescope imaged at different wavelengths a new, developing sunspot colliding with an existing spot and then exploding into an X-class solar flare. The bright area in the upper image marks the footprint of a magnetic field that channeled suddenly-released energy from the inner corona down to the surface. The lower image shows that the two sunspots had opposite magnetic polarities. The flare produced high energy protons that reached Earth. (Hinode, JAXA-NASA)

Sun's interior, which can be observed as a series of outward moving ripples – much like the ripples that spread from a rock dropped into a pool of water.

Over the course of an hour, the seismic waves may travel more than 100,000 km before they fade into the fiery background of the Sun's photosphere. The waves can accelerate from an initial speed of 35,200 km/h to a peak of 400,000 km/h before they disappear.

Flares may often erupt one after the other when a particularly active sunspot region appears. One of the most extraordinary sequences of solar storms took place between October 18 and November 5, 2003 – when more than 140 flares were observed – primarily associated with two large sunspot groups. Among them were eleven major X-class flares, including an X17 event on October 28, and an even bigger one on November 4.

The strength of the latter event was difficult to determine, since it saturated the spacecraft detectors, but, based on radio wave-based measurements of the X-rays' effects on Earth's upper atmosphere, it was later uprated to X45, making it by far the largest flare detected since the GOES satellites began their solar X-ray measurements in 1976.

Classification of Solar Flares

Scientists classify solar flares according to their brightness at X-ray wavelengths, that is, their X-ray flux. There are five classes: A (the weakest), B, C, M, and X (the strongest). Each category has nine subdivisions, for example; C1 to C9, M1 to M9, and X1 to X9. Very occasionally, extremely powerful flares are given a much higher designation. For example, one storm of November 4, 2003 has variously been classified as X28 or X45.

These scales are logarithmic, much like the seismic Richter scale, so an M flare is 10 times as strong as a C flare. The most important categories are:

- X-class flares: The most powerful. These major events can trigger radio blackouts around the world and long-lasting radiation storms in Earth's upper atmosphere which can damage or destroy satellites.
- M-class flares: Medium-sized, releasing 10% of the energy of X-class flares. They generally cause brief radio blackouts that affect Earth's polar regions. Minor radiation storms sometimes follow an M-class flare.
- C-class flares: Small, releasing 10% of the energy of M-class flares, with few noticeable consequences on Earth.

Coronal Mass Ejections

Flares are often associated with huge eruptions of ionized material, known as coronal mass ejections (CMEs), that can propel up to 10 billion tons of matter into space These massive bubbles of plasma, threaded with magnetic field lines, are ejected from the Sun over the course of several hours.

We now know that CMEs are very common. During solar maximum, the Sun averages five of these outbursts per day, with about 100 of them heading Earthward per year. During solar minimum, the annual total may drop to about 180, with a strong CME every two days and about 10–15 per year directed at Earth.

Like all solar activity, CMEs are associated with releases of magnetic energy in active regions, particularly the concentrated magnetic fields close to sunspots. However, although they are often associated with flares and prominence eruptions, they can also occur without either of these events taking place.

Forecasting CMEs is not easy. One possible sign of something stirring is the emergence of an inverse-S shape (sigmoid structure) in the X-ray corona. This happens when a magnetic flux tube begins to twist, due to the rotation of one or both of the

Box 2.4   Japanese Solar Observatories

Japan has been playing a leading role in solar studies for several decades. The first of its pioneering orbital observatories was Yohkoh, which was launched August 30, 1991. It carried four main instruments: a Hard X-ray Telescope, the first instrument ever to image high energy X-ray flares; a US-Japanese Soft X-ray Telescope with a field of view that covered the full solar disk but could also obtain a series of small-scale, high-resolution images of flares; a Wide-Band Spectrometer to observe solar radiation in soft X-rays, hard X-rays, and gamma rays, with a fourth detector monitoring Earth's radiation belts; and a UK-US-Japanese spectrometer to study specific spectral regions in soft X-rays.

Yohkoh was the first spacecraft to continuously observe the Sun in X-rays during an entire sunspot cycle. When it was launched, the Sun was near the peak of its eleven-year cycle, so many active regions and flares were imaged. It then observed the subsequent decline and the start of sunspot cycle 23 in the late 1990s. The spacecraft sent back over six million X-ray images before it failed in late 2000.

Yohkoh provided important new data about the corona, including information about how and where this multi-million degree layer is heated to temperatures hundreds of times greater than the solar surface. By tracking the evolution of the corona, it improved understanding of how the Sun's magnetic fields are deformed, twisted, broken and reconnected during flares; and how the coronal plasma is heated to millions of degrees by flares. Various structures, known as sigmoids (S-shaped regions in the corona) and trans-equatorial interconnecting loops (TILs), were shown to be more likely to be the sites of solar eruptions.

Its successor, Hinode, was also a US-UK-Japan collaboration. It was launched on September 23, 2006 – into a Sun-synchronous, near-polar orbit around Earth that allows it to remain in continuous sunlight for nine months each year. It carries an optical telescope that includes a high-resolution imager, a magnetograph that makes rapid observations of the Sun's magnetic and velocity fields, and a spectropolarimeter that makes extremely precise observations of the solar magnetic field. Together these instruments are able, for the first time from space, to measure small changes in the strength and direction of the magnetic field, as well as how these changes coincide with events in the corona. One important product is vector magnetograms that illustrate variations in the strength of the Sun's magnetic field.

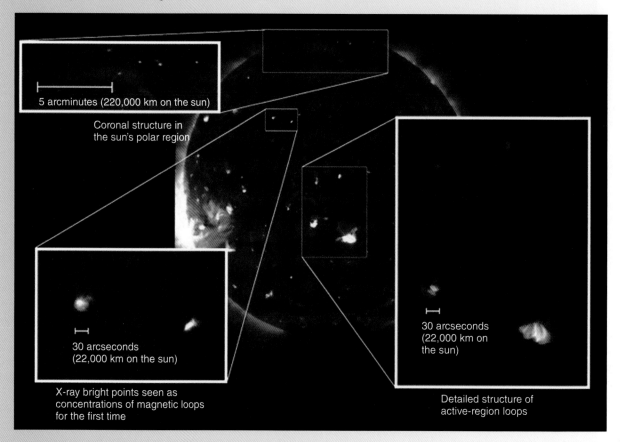

Figure 2.37   Hinode is equipped with the highest resolution solar X-ray telescope ever flown. This full disk image, taken early in the mission, shows features of the X-ray Sun with a spatial resolution of nearly 1 arcsecond. These include coronal activity within dark holes near the poles and coronal loops associated with active regions. (JAXA-NASA)

Also on board is an advanced version of the Soft X-ray Telescope flown on Yohkoh. The highest resolution solar X-ray telescope ever flown, it shows the structure and dynamics of the corona over a wide range of temperatures and a broad field of view. By combining optical and X-ray observations, it is possible to study how changes in the magnetic field trigger explosive solar events. An Extreme Ultraviolet Imaging Spectrometer (EIS) provides a key link in the data by observing the chromosphere and transition region that separate the photosphere from the corona. The EIS measures the velocity of solar particles, and the temperature and density of solar plasma.

Its high resolution images revealed gigantic arcing magnetic structures that dwarf the underlying sunspots. It found evidence for several mechanisms that may be contributing to the extraordinary heating of the corona, including twisted and tangled magnetic fields that snap and reconnect, extreme turbulence and acoustic waves in the lower atmosphere, magnetic Alfvén waves that propagate upwards at high speed, and X-ray jets and nanoflares that are continually exploding. Hinode also discovered extended structures at the edges of active regions where material is flowing rapidly outward, possibly contributing as much as a quarter of the total solar wind.

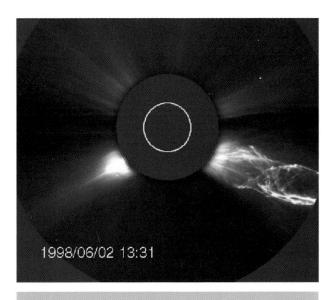

Figure 2.38 A LASCO C2 coronagraph image showing a spiral-shaped CME (lower right) erupting from the Sun on June 2, 1998. This CME was rather unusual since the width of the blast was fairly narrow and the strands of plasma were twisting. The LASCO instrument on SOHO blocks the Sun in order to observe coronal structures in visible light. The white circle represents the Sun. (NASA-ESA)

Figure 2.39 Coronal mass ejections occur when solar magnetic field lines snake around each other, often forming the letter "S". Usually, they go past each other, but if they connect, the mid-section breaks free and creates a mass ejection. (NASA/MSFC)

footprints of a cluster of loops located above a sunspot group. In 2008, Hinode observed a CME in which one end of the tube spun clockwise, whilst the other rotated counterclockwise. This unfurling action caused the field lines to rupture, producing a huge explosion that heated a huge cloud of material and propelled it away from the Sun in the form of a coronal mass ejection.

Although CMEs are so large, they are very sparse and spread out, containing only a few particles per cubic centimeter. Much of their ability to disrupt the flow of the solar wind and disturb planetary environs comes from their magnetic fields, particularly if their polarity is aligned in the opposite direction to a planet's field, leading to favorable conditions for magnetic reconnection to take place.

Traveling at up to 2 000 km/s, it usually takes a CME two to four days to reach Earth, although some may arrive in just over a day. During their journey, they can create major disturbances in the interplanetary medium and in Earth's magnetic field. If they reach Earth, they result in beautiful polar auroras but may also cause large-scale power cuts, for example, Quebec in 1989, and problems with spacecraft systems.

Much of our knowledge of CMEs has come from the Large Angle and Spectrometric Coronagraph (LASCO) on SOHO. Advance warning of a CME that could be heading toward Earth was provided when LASCO imaged a "halo event," when the entire Sun appeared to be surrounded by the CME. However, it was still not possible to definitively say if a CME was coming Earthward. Another viewpoint was needed to provide the third dimension.

Two extra viewpoints have been provided by NASA's twin Solar Terrestrial Relations Observatory (STEREO) spacecraft, launched

Figure 2.40 An S-shaped structure (sigmoid), in a solar active region is often a precursor to a CME. In this Hinode X-ray image, a bright sigmoid is observed on the right-hand side of the disk at the beginning of its eruption on February 12, 2007. The fine structure reveals that the sigmoid is really two opposing "J" shapes wrapping around each other. Sigmoid structures, defined by twisting magnetic fields, can often be observed for several days before a CME occurs. (University of Montana)

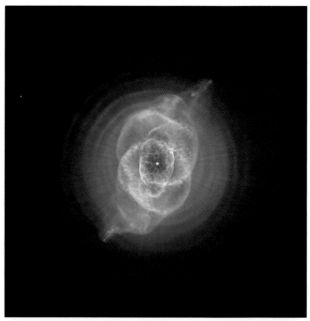

Figure 2.41 The Cat's Eye nebula in the constellation Draco may resemble the remains of our Sun some seven billion years from now. One of the first planetary nebulas discovered, the Cat's Eye is also one of the most complex examples of this kind of nebula, with eleven rings, or shells, of gas surrounding the dying star at the center. The star ejected a series of concentric shells in a series of pulses at 1 500-year intervals. (NASA, ESA, HEIC, and The Hubble Heritage Team STScI/AURA)

in October 2006. With one flying ahead of Earth and the other behind it, scientists can produce 3-D images of Sun-Earth space. These make it possible to pinpoint the location and speed of a CME, and study how it interacts with its surroundings. In this way, a CME can be tracked all the way to Earth and its arrival predicted at least a day in advance.

The STEREO data show that almost all CMEs have a common shape, similar to a croissant. This shape is explained by the twisted magnetic flux tubes being wider in the middle and narrower at one end.

The most recent spacecraft to study the Sun is NASA's Solar Dynamics Observatory, which was launched in February 2010. It was placed in a geosynchronous orbit around Earth, which enables it to observe the Sun without interruption. The observatory carries three instruments to determine how the Sun's magnetic field is generated, structured and converted into the solar wind, flares and coronal mass ejections.

The Sun's Future

The Sun will not continue to shine forever (see also Chapter 1). Like most stars, it is undergoing gradual evolution on the so-called main sequence. Calculations based on the amount of

hydrogen in the Sun and the rate at which it is being consumed, allied with theories of stellar evolution based on observations of other Sun-like stars, indicate that it is currently almost half way through its active life. Eventually, the Sun will leave the main sequence and swell into a red giant as it begins the final, dramatic phase of its existence.

Some seven billion years from now, the Sun will have grown to 2.3 times its present diameter and shine 2.7 times brighter than it does today. Within a few 100 million years, it will balloon outward, expanding to more than a 100 times its current size, so that it engulfs Mercury and extends almost to the present orbit of Earth. Meanwhile, the surface temperature will drop to around 4 000°C, making it appear red in color, and the bloated star will lose up to 25% of its mass in the form of a powerful solar wind.

The first red giant phase ends abruptly when there is no longer enough hydrogen available to continue the fusion process that leads to the creation of helium. As the core temperature soars to 100 million degrees Celsius, the intense heat and pressure initiate nuclear reactions that use helium as fuel. In this process, three helium atoms are fused to create one carbon atom.

The Sun will remain a fully fledged red giant for only a few hundred million years. During this time it will steadily burn its supply of helium, until this, too, begins to run out, leaving the star with a core made of nuclear by-products – carbon and

Figure 2.42 The evolution of the Sun will culminate in a red giant phase, beginning about seven billion years in the future. The Sun will expand enormously in the first phase as hydrogen burning accelerates, possibly reaching the orbit of Venus. Once helium fusion takes over, it will suddenly shrink. A few hundred million years later, when most of the helium is converted to carbon and oxygen, it will expand once more, marked by periodic outbursts known as helium flashes. When the star runs out of helium it collapses to form a white dwarf. (I-J Sackman, Caltech)

oxygen. As the energy output begins to drop, the star balloons outward again, once more extending to at least the present orbit of Venus. During this "asymptotic branch" stage on the sequence of stellar evolution, it will experience enormous pulsations known as helium flashes. However, these 10,000-year-long bursts of energy are almost the last throw of the dice.

The second red giant phase, which lasts for perhaps 100 million years, is marked by another huge mass loss as the remaining atmosphere is ejected into space, creating a beautiful, but short-lived, planetary nebula. Eventually, all nuclear fusion ceases. Without the outward push of radiation pressure, the stellar remnant collapses under its own gravity to form a white dwarf.

All that is left is an extremely hot (100,000°C) core no larger than Earth. This is composed of highly compressed matter which is so dense and hot that the atoms are stripped of almost all their electrons. One teaspoon of the degenerate matter would weigh around five tons.

Over hundreds of millions of years, the white dwarf will gradually cool and fade until it emits only a tiny amount of heat or light. The Sun will end its life as a black dwarf, a cold, dark object hidden amongst the surrounding stars and surrounded by what remains of its retinue of planets and satellites.

Questions

- In what ways does the Sun influence its planetary system?
- The Sun has been in existence for 4.5 billion years. How is such longevity possible?
- Name the internal layers of the Sun. Describe the main characteristics of each layer.
- Why was the extremely high temperature of the corona so unexpected? How might this temperature be accounted for? Briefly describe and explain: (a) solar flares; (b) coronal mass ejections; (c) sunspots; (d) spicules; (e) prominences; (f) solar eclipses.
- What is the solar cycle? Suggest reasons for the changes that occur during this cycle.
- Explain why most of the recent advances in solar science have been made by observatories in space, rather than ground-based observatories.
- Why will the Sun undergo major changes some 6–7 billion years from now? How will these changes affect Earth and the other planets?

THREE
Earth

Earth is our home, the most familiar planet in the Solar System. Of the four terrestrial planets, it is the largest and has the highest bulk density. It also has a strong magnetic field and thick atmosphere, which protect the planet from solar and cosmic radiation, and an ozone layer which acts as a barrier to ultraviolet light. Its unique nitrogen-oxygen atmosphere also contains greenhouse gases that maintain an equable temperature. Earth is also the only world where conditions are right for large bodies of liquid water to exist on the surface. As far as we know, it is the only planet that supports life.

Orbit and Rotation

Earth is the third planet from the Sun. It follows an almost circular orbit (eccentricity 0.017) at an average solar distance of 149,600,000 km. At perihelion, which occurs around January 3, the Sun-Earth distance is 147.1 million km. Aphelion, when Earth is farthest from the Sun, is reached around July 4. The planet is then 152.1 million km from the Sun. Moving at a mean orbital speed of about 30 km/s, Earth completes each orbit in one year or 365.25 days.

The planet rotates from west to east once every 24 hours.[1] This is called the solar day. As a result, Earth's equatorial region is moving eastward at a rate of 1 670 km/h, though this rate decreases to almost nothing near the poles.

The planet's rotation explains why the Sun, Moon and stars appear to rise in the east and set in the west. However, the time of sunrise and sunset changes by up to 4 minutes each day because the planet is also moving around its orbit. This means that Earth has to rotate for approximately another 4 minutes before the Sun returns to the same place in the sky where it was the previous day. On the other hand, Earth's sidereal day, measured relative to the position of the 'fixed' background stars, is 23 hours 56 minutes 4 seconds.

Earth's rotational axis is currently tilted about 23.5° to the ecliptic plane. The axis is aligned in the same direction relative to the stars throughout the year, with the North Pole pointing towards the star Polaris.

However, the orbit and rotation rate are not stable and unchanging. The planet's motion is influenced by its gravitational interaction with the Sun, Moon and other planets. There are four major variations over time:

- The eccentricity of Earth's orbit slowly changes from nearly zero to 0.07, i.e. it becomes slightly more elliptical and then less elliptical every 100,000 years. When the orbit is highly elliptical, the amount of insolation (incoming solar radiation) received at perihelion is about 20–30% greater than at aphelion.
- The direction of Earth's axis slowly changes because the planet wobbles on its axis as it spins – an effect known as precession. This wobbling motion does not affect the obliquity (tilt angle) of Earth, but the direction in which the rotational axis is inclined. The planet completes a full wobble every 25,700 years. At present, the rotation axis is pointing almost exactly at Polaris. However, 13,000 years from now the axis will be aligned toward Vega, the brightest star in the constellation Lyra.
- Another type of precession is associated with the sideways rotation of Earth's orbit. Known as the precession of the ellipse, this motion causes the positions of perihelion and aphelion to shift with regard to the stars.

[1] The braking action of the tides causes Earth's period of rotation to slow by between 1.5 and 2 milliseconds per century. In order to ensure that clocks remain linked to Earth's rotation, an extra second may be added to the year. This is known as a leap second.

Exploring the Solar System, First Edition. Peter Bond.
© 2012 by Peter Bond. Published 2012 by Blackwell Publishing Ltd.

- The inclination of Earth's axis, that is, its angle of tilt relative to the plane of its orbit, varies between 22.1° and 24.5° every 41,000 years. (It is currently decreasing). Such changes modify Earth's seasons. A higher inclination means more extreme seasons – warmer summers and colder winters. A lower inclination results in less severe seasons – cooler summers and milder winters.

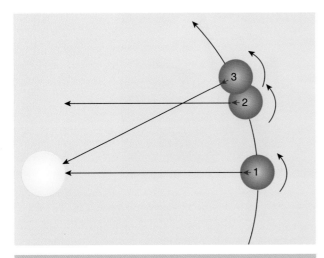

Figure 3.1 On a prograde planet like Earth, which rotates west to east, the sidereal day is shorter than the solar day. At **1**, the Sun and a certain distant star are both overhead. At **2**, the planet has rotated 360° and the distant star is overhead again. The time period between **1** and **2** is one sidereal day. Only when Earth has rotated a little further, is the Sun overhead again **3**. The interval between **1** and **3** is one solar day. (Wikipedia)

All of these changes affect the timing and duration of the seasons. Today, northern summer is the longest season and northern winter the shortest. 10,000 years from now, the length of the seasons will be reversed. The gradual modifications of the orbit are also thought to explain Earth's periodic ice ages (see below).

Seasons

Planets have seasons because their rotational axes are tilted, rather than perpendicular (upright) in relation to the plane of their orbits. Venus and Jupiter have negligible axial inclinations, so there is no difference between the amount of solar radiation arriving at their equator or poles throughout their year. This is not the case with planets such as Earth, Mars and Saturn, which have noticeable tilts. Even more unusual conditions can be expected on planets which are more or less spinning on their sides, such as Uranus and Pluto.

The northern hemisphere experiences summer when the North Pole is tilted toward the Sun. Six months later, when Earth has traveled half way around its orbit, the northern hemisphere is tilted away from the Sun and experiences winter. (The seasons are reversed in the southern hemisphere.)

The dates on which Earth's axis is most directly tilted toward or away from the Sun are known as the solstices. They occur on or around June 21, when the Sun is overhead at midday at the Tropic of Cancer (23.5°N), and December 22, when the midday Sun is overhead at the Tropic of Capricorn (23.5°S).

Not only are days longer in summer, but the Sun moves noticeably higher above the horizon at that time of year, providing more heat per square meter of surface area. The opposite is true in the winter. These two factors combined account for much of the difference in seasonal temperature.

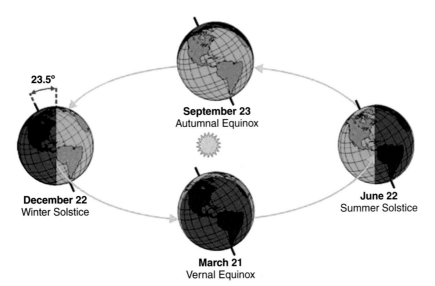

Figure 3.2 Since Earth's axis is inclined to its orbital plane, the amount of radiation any one place receives varies throughout the year. In June, the North Pole is tilted towards the Sun, resulting in long days and warmer temperatures (i.e. summer). In December, the North Pole is tilted away from the Sun, causing long nights and colder temperatures (winter). The seasons are reversed in the southern hemisphere. (NOAA)

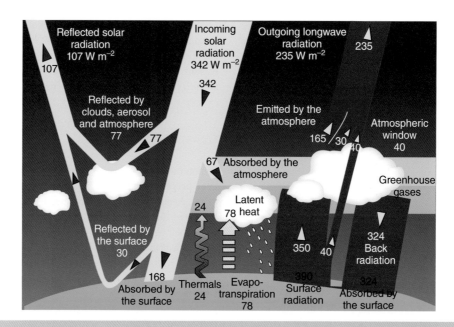

Figure 3.3 Earth's temperature is determined by its radiation budget – whether it receives more radiation from the Sun than it loses to space. This diagram shows what happens to incoming solar radiation and longer wavelength radiation emitted by the surface. Here, the overall amounts of incoming and outgoing radiation balance, so the temperature is in equilibrium. In reality, the global temperature has generally been increasing gradually since the 1970s. (NASA)

When one of the poles is tilted towards the Sun, the surrounding regions at high latitude are bathed in permanent sunlight (hence the term, "the land of the midnight Sun"). The opposite polar region endures 24-hour darkness and extreme cold.

However, even in December or June, the noonday Sun is never far from the zenith at the equator, so the amount of insolation received shows little variation throughout the year. Hence, the equatorial regions are always hot.

Midway between the solstices, on or around March 21 and September 23, the Sun is directly overhead at noon at the equator. On those dates, Earth's axis is inclined away from the Sun. Day and night are equal in length across the globe – so they are known as the spring (or vernal) and autumnal equinoxes.

Although Earth's orbit is slightly elliptical, the difference between the aphelion and perihelion distances – about 5 million km or 0.3% – is too small to cause any significant seasonal effects. The difference in distance amounts to about a 6% increase in insolation from July to January.

Earth reaches perihelion in early January, only about two weeks after the December solstice.[2] This means that the northern hemisphere winter and the southern hemisphere summer begin about the time that Earth is nearest the Sun. (Similarly, the southern hemisphere's winter and northern hemisphere's summer coincide with aphelion.)

The dates of perihelion and the December solstice will not always be so close, since the date of perihelion is not fixed, but slowly regresses (becomes later in the year). The rate of precession (see above) is about one full day every 58 years. There is some

evidence that this long-term change in the date of perihelion influences Earth's climate.

Surface Temperature

Earth is located in the habitable zone of the Solar System, where temperatures are neither too hot not too cold for life (as we know it) to exist. Although extremes do occur, notably in hot deserts and at the poles, the planet's mean temperature is a hospitable 7°C. (However, without the greenhouse gases in the atmosphere, notably water vapor, carbon dioxide and methane, Earth's temperature would be well below the freezing point of water.)

The global radiation budget is determined by the amount of incoming solar radiation and the amount of radiation absorbed or lost by the planet. When solar radiation passes through the atmosphere, some wavelengths are absorbed by gases, dust and cloud droplets. Some visible light is scattered by the gases, giving the sky its blue appearance. About half of the total is absorbed by the surface, compared with one third which is reflected back into space from the clouds and the surface (see Fig. 3.3).

Of course, the surface does not retain all of its heat; some is re-radiated at infrared wavelengths. A fraction of this is absorbed by the greenhouse gases and clouds, then re-radiated in all directions.

The role of clouds and aerosols is crucial in this planetary balancing act. (Up to 70% of the world is covered with cloud at any

[2] There are small year-to-year variations in the dates and times of solstice and perihelion due to the leap-year cycle in the calendar and the effect of the Moon on Earth's motion.

Figure 3.4   These images show the amount of long wavelength thermal radiation (heat) emitted to space from Earth's surface and atmosphere (left – **a**) and shorter wavelength solar radiation reflected by the ocean, land, aerosols and clouds (right – **b**). The largest flux of long wave radiation is from cloud-free tropical regions. Clouds over the equator and at high latitudes reflect a lot of shorter wavelength solar radiation back into space. Such images help to determine the factors that influence Earth's radiation balance. (NASA)

one time.) They help to cool the planet by reflecting a substantial amount of solar radiation back into space. However, they can also raise its temperature by absorbing or re-radiating downwards much of the infrared radiation that would otherwise escape. Scientists are keen to know how the climate would react to a long-term increase or decrease in this cloud and aerosol blanket.

The pattern of surface temperatures is also influenced by a number of other factors. In winter, the amount of incoming solar radiation (insolation) is insufficient to compensate for the heat that escapes into space at night. The same effect, of course, on a much smaller scale, applies to day and night temperatures.

The daily and seasonal variation in overall insolation greatly influences surface temperatures, particularly over the continents. This is because land surfaces respond more rapidly than water to daytime heating and night-time cooling. Oceans take a long time to warm, and an equally long time to cool. In effect, they act as storage heaters. (The lower "specific heat" of rock means that its temperature rises more quickly than that of water for the same input of energy.)

In Earth's southern hemisphere, which is largely covered in water, the slow response of the oceans to variations in insolation means that diurnal and seasonal temperature ranges are less

Box 3.1   Earth Observation Satellites

Since the early 1960s, numerous satellites have been launched to observe Earth. Some of them are fairly specialized, such as meteorology (weather) or radar satellites (see separate box). Others, such as the various spacecraft in NASA's Earth Observing System or ESA's Envisat, carry a wide range of sensors for multiple uses, such as environmental monitoring, land use mapping, altimetry etc.

Two main types of orbit are used. For large-scale or hemispheric obervations, geostationary orbits are favored. From an altitude of 35,780 km above the equator, such satellites circle the planet once every 24 hours, so they appear to hover over the same spot, providing continuous monitoring of the same part of Earth's surface. These orbits are ideal for weather satellites such as the U.S. Geostationary Operational Environmental Satellite (GOES) series and Europe's Meteosats.

More detailed observations require much lower orbits. These are usually provided by satellites in near-polar orbits, inclined almost 90° to the equator. However, polar orbiting spacecraft can only provide brief snapshots as they pass overhead. Such overpasses typically take place twice per day at any given spot on the surface. Global coverage is achieved by combining many swaths (strips) of data that are acquired as the planet rotates beneath the satellite.

One of the most popular types of polar orbit is one in which illumination conditions on the surface remain constant. Such Sun-synchronous orbits are typically inclined about 98° to the equator, enabling the satellite to cross the equator at the same local time on each orbit. Perhaps the most famous example is the US Landsat series, which has been providing continuous, medium resolution imagery of Earth's land masses in different spectral wavelengths since 1972.

Multiple observations of the same area by different instruments on different satellites are also invaluable in providing information about changes over short and long periods of time (i.e. multi-temporal data). The most enterprising example is the Afternoon Constellation, often known as the "A-Train," which includes up to seven satellites from the US and France that follow the same orbital path around Earth, one behind the other. The A-Train enables simultaneous coordinated measurements of the same regions, giving a more complete overview than would be possible from a single satellite.

Each satellite in the A-Train crosses the equator at around 1:30 pm local time, separated by intervals of seconds or minutes. Satellites such as Aqua and CloudSat observe many different surface and atmospheric phenomena, including hurricanes, clouds and aerosols.

extreme. (Places on the shores of large lakes experience similar moderating effects.) In the northern hemisphere, where land masses predominate, summer temperatures tend to be much higher, while thermometers plunge during the winter. As a result, places located in continental heartlands have a much greater variation in temperature than islands and coastal regions.

The nature of the surface also plays a part. Dark surfaces, such as those covered with vegetation and soil, have a low albedo (reflectivity). They heat up more quickly than lighter, more reflective surfaces, such as ice sheets. Ice and snow reflect some 80% of the solar energy they receive, compared with 20% for an area of grassland and 10% for a dry, black soil. This explains why scientists are concerned that the melting of polar ice sheets will contribute to global warming.

Atmosphere

Earth's atmosphere is unique in the Solar System, since nitrogen and oxygen together make up 99% by volume. Other gases which are naturally present in small, variable, amounts include carbon dioxide, methane, water vapor and ozone. Minor constituents include inert gases, particularly argon, along with human-made pollutants such as sulfur dioxide and nitrogen dioxide. Near-surface ozone and carbon dioxide are also generated by human activity.

Earth's original atmosphere was probably made up of hydrogen and helium from the solar nebula, but these light gases soon escaped. A secondary atmosphere was then created as the result of outgassing through volcanic activity. Its actual composition is uncertain, although its main constituents most likely included carbon dioxide, water vapor, ammonia (NH_3) and methane (CH_4). Large amounts of water vapor, organic (carbon-rich) compounds and other volatiles would also have arrived through the initial bombardment by comets and asteroids.

As the temperature dropped, condensation of water vapor caused the planet to be blanketed in cloud. The resultant global downpour led to the formation of the first oceans. Some carbon dioxide gas dissolved in the rain and oceans. At the same time, chemical reactions involving acidic rain and the first, primitive crust led to the formation of carbonate rocks and a further reduction in atmospheric carbon dioxide.

According to this scenario, nitrogen – a gas which is not very chemically active – continued to accumulate in the atmosphere as a result of outgassing and numerous impacts. Free oxygen was scarce, since it was soon removed through chemical reactions with rocks and other gases. The first single-celled organisms were probably also important in regulating the climate by generating substantial amounts of methane – a greenhouse gas.

The change to a more modern atmosphere began when bacteria and algae developed the ability to split water molecules by harnessing the energy of sunlight – a key part of photosynthesis. As a result, the amount of oxygen released into the atmosphere began to increase, although, for a long time, most of the gas was removed by chemical reactions with rock minerals. To this day, the majority of Earth's oxygen produced over time is locked up in ancient, oxidized rock formations.

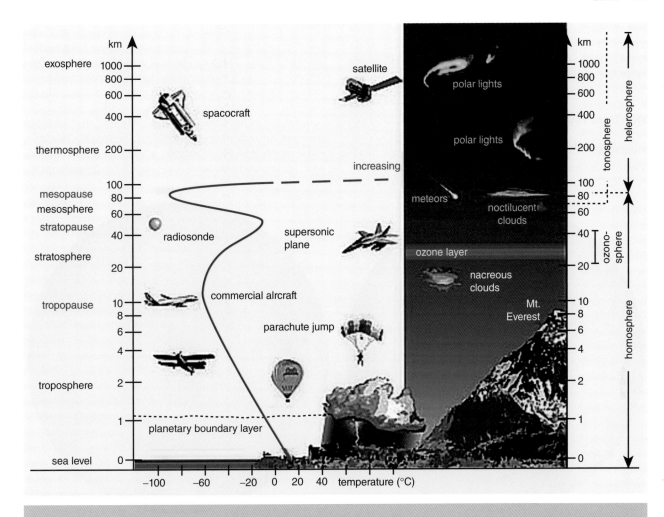

Figure 3.5 The vertical structure of Earth's atmosphere. Most of the atmosphere and weather is found in the lower region – the troposphere. Temperature decreases at a steady rate (the "lapse rate") until it reaches the top of the troposphere. It then increases in the stratosphere, where ozone gas absorbs solar radiation. In the ionosphere, about 100 km above the surface, positively charged particles (ions) reflect radio waves. (After Kowoma)

It was not until about one billion years ago that the rate of oxidation slowed sufficiently to enable free oxygen to stay in the air and the levels of oxygen in the oceans to rise. Once oxygen gas was available, ultraviolet light began to split the molecules, producing a layer of ozone (O_3) high above the planet that acted as a shield against ultraviolet light. Only at this point did life move out of the oceans and respiration evolve.

Troposphere

Earth's atmosphere extends from the surface to an altitude of about 500 km, when it effectively fades into the vacuum of space. The lowest part of the atmosphere is called the troposphere. This region is where most weather phenomena take place, for example, cloud formation. The troposphere contains about 80% of the atmosphere's mass, including almost all of its water.

Air temperature generally decreases at a steady rate (the "lapse rate") with altitude, until the tropopause, the upper boundary of the troposphere, is reached.[3] The height of this boundary extends up to 17 km near the equator, reducing to about 8 km above the polar regions.

Since the troposphere is characterized by a continuous decrease of temperature with increasing height, any parcel of air that is warmer than its surroundings is likely to become buoyant and unstable, sometimes rising to great heights. The result of this convection is towering clouds and heavy precipitation (rain, snow, hail etc.).

In the upper troposphere, the temperature drops to about −50°C. At this height, jet streams occur at the boundaries between air masses that have significantly different temperatures. The jet streams are relatively narrow bands of very high wind speeds (up to 360 km/h) which meander around the globe at about 60° latitude, above the polar fronts, and at 30° latitude. During the

---

[3] **The average decrease in temperature – the normal lapse rate or environmental lapse rate – is 6.5°C per 1 000 meters of altitude.**

winter, Arctic and tropical air masses create a greater temperature contrast, resulting in a stronger jet stream. However, during the summer, when surface temperature variations are less dramatic, the jet streams are weaker.

The Ozone Hole

The dust-free, almost cloudless stratosphere lies above the troposphere. It is characterized by an increase in temperature with altitude, due to the presence of ozone gas which absorbs solar radiation (and protects life on the planet's surface from harmful ultraviolet light). The ozone layer is particularly marked above the poles, where the stratosphere becomes much warmer in summer. However, the lack of insolation in winter means that the upper air then becomes very cold.

In the late 1970s, British scientists discovered that the ozone layer over Antarctica was getting thinner every spring. By 1985 it was clear that an "ozone hole" was opening up above the continent. A similar, but smaller, feature was later discovered over the

September 17, 1979

October 7, 1989

October 9, 2006

October 1, 2010

Figure 3.6 Data from satellites and stratospheric balloons shows that the spring ozone hole (blue-purple) over Antarctica grew rapidly from the mid-1980s. It is expected to take many years for the hole to disappear. (CSIRO / NASA-GSFC)

Arctic. These holes last for several months before disappearing. The largest ozone hole, detected over Antarctica in September 2000, covered more than 28 million km², an area approximately three times the size of the United States.

One major factor in the creation of these ozone holes was the injection of human-made pollutants, such as chlorofluorocarbons (CFCs), into the atmosphere. Through a complex series of chemical reactions, these compounds release chlorine into the air. When the Antarctic spring arrives, this chlorine destroys the ozone gas.

After the introduction of the 1987 Montreal Protocol, which aims to phase out production of ozone-destroying chemicals, levels of CFCs have leveled off. However, the chemicals can stay in the atmosphere for decades, so it will be a long time before the stratospheric ozone recovers.

Upper Atmosphere

At the top of the stratosphere, the rise in temperature ceases. In the mesosphere (50–80 km) there is a continuous decrease of temperature, reaching a minimum of almost −100°C. The temperature then rises rapidly as the atmosphere becomes ever thinner in the thermosphere.

The upper part of the mesosphere and much of the thermosphere (80–500 km) is also known as the ionosphere. In this region, high energy UV and X-ray solar radiation splits atoms and molecules – a process known as ionization. Radio stations take advantage of the fact that these free electrons and ions reflect radio waves.

Figure 3.7 Hydrogen is continuously escaping into space, creating a cloud of neutral atoms around the Earth. This halo of hydrogen is illuminated by reflected solar ultraviolet light, when it is known as the "geocorona." In this false color ultraviolet image, the geocorona is brighter on the sunlit side (top). The UV camera was operated by astronaut John Young on the Apollo 16 lunar mission. (NASA-JSC)

The outermost region, above an altitude of about 500 km, is known as the exosphere. Its main constituents are the lightest gases, particularly hydrogen and helium, with some atomic oxygen. The exosphere is characterized by the escape of these gases into space. Air pressure is so low that it is effectively a vacuum. However, the orbits of low-flying satellites are gradually lowered by slight atmospheric friction, particularly during periods of maximum solar activity when the upper atmosphere is heated more strongly and balloons outwards.

Atmospheric Circulation

On balance, regions above latitudes 35° N or S lose more energy than they receive. In contrast, the tropical regions receive more radiant energy than they lose. On a planet with no atmosphere or oceans, this would result in the poles becoming cooler and the tropics becoming hotter. Fortunately, Earth is very efficient at transferring heat towards the poles. One of the main mechanisms for doing this is through large-scale atmospheric circulation.

Surface temperature differences are largely responsible for Earth's winds. Warm air expands and rises. Colder air is dense and heavy, so it sinks. As a result, air is always in motion. This effect can be clearly seen in coastal areas where sea breezes blow. Air over the warm land rises, to be replaced at ground level by cooler air from offshore. This circulation is often reversed at night, when the land becomes cooler than the sea.

Similar atmospheric motions can also be seen on a global scale, in the form of six large cells of circulating air – three in each hemisphere. Nearest to the equator are the Hadley cells, named after the scientist who first proposed their existence. In temperate latitudes are the Ferrel cells, also named after their discoverer. Finally, there are the polar cells (see Fig. 3.8).

Each Hadley cell is driven by hot air rising through convection in equatorial regions, creating low pressure at the surface. This unstable air soars upward until it reaches the top of the troposphere, about 17 km above the surface. This results in towering thunderclouds and heavy downpours, particularly in the warmth of the afternoon.

The air at high level then spreads poleward before sinking at about 30° latitude. This descent warms and dries the air, resulting in cloud-free skies – perfect conditions for hot deserts, such as the Sahara. The large areas of stable, sinking air have light winds – so they were often dreaded by crews of sailing ships.

At the surface the air then diverges. Some moves back toward the equator, completing the Hadley circulation. Some moves toward higher latitudes until it meets up with colder air spreading from the poles. The boundary between the warm, subtropical air and the cold, polar air is known as the polar front.

The polar front is like a battleground, where a cool air mass is always trying to undercut and displace the lighter, subtropical air. Air forced upward at the polar front results in the formation of clouds and heavy rain. Low pressure areas known as depressions are driven eastward by the polar jet stream (see above).[4]

---

[4] It is this jet stream which causes flights from Europe to North America to take much longer than flights in the opposite direction.

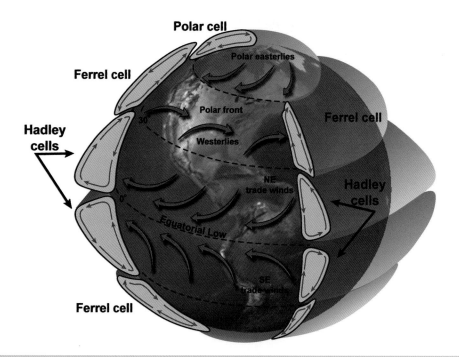

Figure 3.8 Earth's main zonal winds are associated with six large atmospheric circulation systems – three in each hemisphere. In the Hadley cells, warm air rises near the equator, moves poleward and then sinks in the subtropics. The rising section of the Hadley cell causes thick, deep convective clouds and thunderstorms. Over land, the sinking section is associated with deserts. Over oceans, sinking air results in clear, dry air and very light winds. Similar circulation systems, called the Ferrel cells and polar cells, occur nearer the poles. Where these cells meet, active uplift of air results in numerous low pressure systems (depressions) and rain belts. (After Barbara Summey, NASA Goddard VisAnalysis Lab)

Zonal Winds

The simple north-south atmospheric circulation is modified by the Coriolis effect, caused by Earth's west to east rotation. Surface air near the equator moves eastward, in the same direction as Earth's rotation, with a speed of nearly 1 700 km/h. At higher latitudes, the surface and overlying air masses move eastward at slower speeds. Consequently, a parcel of air carried away from the equator by Hadley circulation moves eastward faster than the air or ground below it. An observer on the ground would see the air veer on a curved path to the east, as if moved by some force.

On the other hand, air near the poles starts off moving slowly to the east, since Earth's rotation speed is slower at high latitudes. As it moves towards the equator, a parcel of polar air travels over a faster-moving surface. As a result, it appears to move along a curved path to the west.

In effect, winds flowing north-south are turned to the right in the northern hemisphere and to the left in the southern hemisphere. This explains, for example, why the tropical trade winds in the northern hemisphere blow towards the south west, whereas the trade winds in the southern hemisphere flow toward the north west (see Fig. 3.8).

The Coriolis effect also explains why winds move around pressure systems in opposite directions in different hemispheres. There is a clockwise flow around a high pressure area (anticyclone) in the northern hemisphere and a counterclockwise flow around an anticyclone in the southern hemisphere. Conversely, winds blow counterclockwise around low pressure areas (cyclones or depressions) in the northern hemisphere and clockwise in the southern hemisphere.

Tropical Cyclones

The region where the trade winds of the two hemispheres meet is known as the *Inter-Tropical Convergence Zone* (ITCZ). There, warm, moist air is forced upward. When it cools, large amounts of water vapor condense, leading to massive cloud formation and high rainfall almost every day. Tropical cyclones, Earth's most powerful and destructive storms are also associated with the ITCZ.

Huge amounts of energy are transferred by these short-lived storms as they form over the tropical oceans, then move westward before heading towards higher latitudes and decaying. Most of them are born between June and October, when the surface temperatures of the oceans exceed 27°C. (They are rare close to the equator, since the Coriolis effect is weak and rapid rotation is less likely.) The number varies, but on average there are 40–50 tropical cyclones each year, with an equal number of less intense tropical storms.

Figure 3.9 A map of tracks followed by tropical storms and cyclones that develop over tropical oceans. Red and orange show the most intense category 5 and 4 storms. As they move westward, driven by upper level trade winds, they increase in intensity, then die out when they move over land or cooler water. (National Hurricane Center / Typhoon Warning Center)

These storms are given different names by the local inhabitants, but the causes are the same.[5] They begin as the result of rapid upward movement of air caused by divergence and an absence of strong winds at high altitude. This causes surface pressure to fall rapidly. Warm, moist air flows inwards towards the developing center of low pressure. As the surface air pressure plummets, the wind speed increases dramatically. The strong updraught causes towering convectional clouds to form, causing heavy rain. The storms are also fed by an additional source of heat, which is released during the rapid condensation process (conversion of water vapor to water droplets). This is known as "latent heat."

A fully fledged storm can develop in less than a week. At its height, wind speeds around the central "eye" may exceed 120 km/h, occasionally producing sustained winds of 250 km/h in a Category 5 storm. Indeed, everything about tropical cyclones is extreme. The circulating system of strong wind and dense cloud may measure up to 1 000 km across. Spiral bands of convectional cloud tower 12 km or more above the ocean surface. A well developed tropical cyclone may pick up as much as two billion tonnes of water a day through evaporation and sea spray. 1 250 mm of rain may be dropped over any location along its track. As a result, flooding and mudslides are often more of a problem than damage caused by strong winds.

A central pressure as low as 880 millibars also causes the ocean surface to bulge upwards. When combined with huge waves whipped up by the wind and the torrential rainfall, the outcome is a storm surge, which can inundate low-lying coastal areas.

An average tropical cyclone transfers a huge amount of energy. The daily energy release from a cyclone is equivalent to the output of the entire United States' electricity grid over six months.

The tracks followed by tropical cyclones are very hard to predict. However, the average storm moves east to west along a curved path, traveling at 15–50 km/h. Fortunately, they rapidly lose their potency when deprived of their supply of warm, humid air, so they usually die out when they move inland. Others lose intensity when they head away from the tropics and move over cooler water. However, downgraded hurricanes may still bring strong winds and heavy rainfall to temperate regions, particularly if they merge with other low pressure areas or depressions.

Ocean Currents

The other major conveyor belts for heat from the tropics are the warm ocean currents. These fairly narrow ribbons of surface water display marked differences in temperature and salinity from the surrounding ocean.[6] They are also associated with differences in sea height, which can be measured by satellites. Warm currents are seen as rises and cold currents as valleys in the ocean surface.

Many of the currents are driven by surface winds blowing around the sub-tropical high pressure zones. This means that they flow clockwise in the northern oceans and the opposite direction in the southern oceans, traveling at speeds of between 0.4 and 1.2 m/s (35–105 km per day).

[5] They are called hurricanes in the Atlantic, typhoons in the Asian Pacific, and cyclones in the Indian Ocean and Australia.
[6] Colder or saltier water is denser and tends to sink.

Figure 3.10   Katrina was the first Category 5 storm of the 2005 Atlantic hurricane season. One of the most powerful hurricanes ever recorded, it formed over the Bahamas on August 23, 2005, then crossed southern Florida before strengthening rapidly over the warm waters of the Gulf of Mexico (orange-red). A storm surge from Katrina caused catastrophic damage and flooding along the coastlines of Louisiana, Mississippi and Alabama. (NASA)

Box 3.2   Radar

Most remote sensing of planets and moons is by passive sensors, which simply detect and record natural radiation emitted by the atmosphere or surface (see Chapters 6, 9 and 13). However, in recent years, a number of spacecraft have been equipped to carry out active remote sensing with lidars (laser imaging detection and ranging systems) or radar.

One of the main uses of radar is altimetry. The distance between a spacecraft and the planet below can be calculated by precise measurement of the time it takes for a microwave signal to return to the spacecraft from the surface, irrespective of whether it is an ocean, ice or land. Over time, such data can provide topographic maps of an entire planet.

The characteristics of the echoes also provide further information about the roughness of the surface, wave heights or wind speeds over the ocean. Such measurements can be made 24/7, regardless of cloud cover or night conditions.

One of the major uses of radar altimetry has been the investigation of variability in sea surface height and its impact on the general circulation of the oceans. Such measurements highlight the importance of eddies in shaping and controlling major ocean currents. They also reveal the growth and development of major climatic events, such as El Niño.

The accuracy of altimetry measurements depends on knowing the spacecraft's precise orbital position. Many Earth observation satellites, such as Envisat and Jason, carry a radio receiver for precise orbit determination. This radio positioning system makes it possible to calculate the orbit to an accuracy of a few centimeters by measuring the Doppler shift on signals broadcast from a network of more than 50 beacons spread around the world.

By combining two or more radar images of the same area, spaceborne systems also make it possible to produce maps showing surface change associated with both earthquakes and subsurface volcanic activity which are accurate to within a few millimeters. Another area of interest is the determination of changes in the surface height and area of global ice cover, particularly in the Arctic Ocean, Greenland and Antarctica. Measurements made over a number of years can reveal whether the ice sheets are losing or gaining mass and thickness – an important clue with regard to climate change.

Figure 3.11 Spacecraft such as ESA's CryoSat use radar to map surface topography. The first radar echo comes from the nearest point to the satellite. CryoSat can measure the angle from which this echo originates, so that the source point can be located on the ground. This, in turn, allows the height of that point to be determined. (ESA)

Figure 3.12 A multi-temporal radar image of the Bay of Naples in Italy as seen on three separate occasions by ESA's ERS-2 satellite. With its numerous buildings, the city of Naples (top center) is very radar reflective and bright, as are mountain ridges facing the radar. To the east of Naples is Vesuvius, one of the most explosive volcanoes in Europe. To the west is a much older volcanic region, the Phlegrean Fields. The colored patches north of Naples are fields of crops. The colors offshore indicate sea surface roughness on each viewing date. (ESA)

Ground-based radar has been used for many years to reveal rain and snowfall from storm systems. Radar instruments installed in spacecraft are now being used to give a wider perspective on major storms. For example, the US-Japanese *Tropical Rainfall Measuring Mission* (TRMM) satellite carried the first radar flown in space to measure precipitation. The instrument worked by measuring the echoes backscattered from rain. Since the strength of the echo is roughly proportional to the square of the volume of falling water, the instrument produced very accurate estimates of rainfall.

Another NASA spacecraft known as CloudSat carries a millimeter-wavelength cloud profiling radar which is over 1 000 times more sensitive than typical weather radar. It can not only take a vertical slice through clouds and storm systems, even in the polar winter, but also distinguish between cloud particles and precipitation.

Not only do the warm currents carry heat absorbed by sea water, but they also warm the air above them. (The oceans redistribute about half as much heat as the atmosphere.) The effects can be very noticeable at mid-latitudes. The most famous example is the North Atlantic Drift (commonly known as the Gulf Stream) which carries warm water from the Gulf of Mexico to the shores of Norway. As a result, winters in north western Europe are much milder than would be expected from their latitude (50–70°N).

Other places are cooled by the oceanic circulation. Cold currents moving from high latitudes towards the equator tend to cool

nearby coastal areas. Since cool air is relatively dense and stays near the surface, fog is quite common but clouds and rain are rare. As a result, places such as southern California, northern Chile and south western Africa experience desert conditions.

The foggiest places on Earth occur where warm and cold currents meet, for example, off the coast of Newfoundland where the interaction between warm, moist air and cooler air (also fairly moist) results in 150–200 days of sea fog per year.

Vertical motions in the oceans are also important in places such as Peru. Earth's rotation and strong winds push surface water

Figure 3.13 Earth's ocean currents are driven by the prevailing winds. Many of the currents circulate around the subtropical high pressure zones – so they flow clockwise in the northern hemisphere and counterclockwise in the southern hemisphere. Currents moving away from the equator are regarded as 'warm'; those moving away from polar regions are "cold." Huge amounts of heat energy are transferred from the tropics to higher latitudes by warm currents. This helps to keep the north western coasts of Europe and North America warmer than would otherwise be the case. (NASA)

away from some western coasts, so that cold water rises from the depths to replace it. This upwelling of nutrient-rich waters is a bonanza for marine life.

The cooling and sinking of cold water in the polar regions drives a much deeper, global circulation (see Fig. 3.14). This oceanic "conveyor belt" is set in motion when cold, dense water in the North Atlantic sinks and moves south. It circulates around Antarctica, and then moves into the Indian and Pacific basins, where it returns to the surface. Once at the surface it is carried back to the North Atlantic, and the cycle begins again. This circulation is extremely slow – water from the North Atlantic may take 1 000 years to find its way into the North Pacific.

El Niño and La Niña

Changes in oceanic circulation can lead to variations in heat transport and very different weather patterns. One of the most important of these periodic variations in the ocean current and atmospheric circulation is the El Niño – Southern Oscillation (ENSO) which occurs over the Pacific Ocean (see Fig. 3.15).

Under normal conditions, the trade winds blow towards the west across the tropical Pacific. These winds pile up warm water in the western Pacific, so that the sea surface is about 0.5 meters higher in Indonesia than off the coast of Peru. As a result, the sea surface temperature is about 8°C higher in the western Pacific. Moist air rising over the warm water results in convectional clouds and heavy rainfall. In contrast, the upwelling of cold water from depth off the western coast of South America causes the overlying air to be stable, so rainfall is low.

Every two to five years this pattern changes. During an El Niño[7], the trade winds weaken and the equatorial countercurrent strengthens, sending warm water eastward towards Ecuador and Peru. The storm pattern also shifts toward the east and the upwelling of cold water in the eastern Pacific is suppressed. Instead of drought conditions, the west coast of the Americas experiences heavy rainfall, but there is decreased rainfall in the western Pacific (e.g. Australia and Indonesia).

El Niño is often replaced by La Niña, which is characterized by unusually cold ocean temperatures in the equatorial Pacific and a tongue of cool water which extends farther westward than usual. The reasons for these major changes are not well understood.

Changes in the ENSO system of the Pacific Ocean also influence wind patterns, rainfall and temperature in many other parts of the world.

[7] In Spanish, El Niño means "the boy," a reference to the baby Jesus, since historically the phenomenon has been observed near Christmas. The reverse situation is called La Niña, "the little girl."

Figure 3.14   The oceanic conveyor belt. Warm water is carried by ocean currents to higher latitudes, where the stored heat is released into the atmosphere. As the water cools it becomes denser. In regions where the water is also very salty, such as the North Atlantic, the water becomes dense enough to sink to the bottom. This travels at depth around Antarctica and into the Indian and Pacific basins before returning to the surface. Carbon dioxide is also transported during this circulation. Cold water absorbs carbon dioxide from the atmosphere, and some is carried to great depth. When deep water returns to the surface in the tropics and is warmed, the carbon dioxide is released back into the atmosphere. (NASA)

Figure 3.15   (a) Normal conditions in the Pacific Ocean. The trade winds blow steadily toward the west, piling up warm ocean water. Moist, warm air rises, causing heavy convectional rain. In contrast, conditions are dry in the eastern Pacific, where cold water rises from depth and cools the air above. (b) During El Niño, the trade winds weaken and warm water moves eastward, causing wet conditions in the Americas and dry conditions in the western Pacific. The red and orange areas correspond to higher sea surface. (NOAA)

Monsoons

One of the most predictable atmospheric cycles is associated with the monsoons, particularly those of south east Asia, where the surface winds change with the seasons over a vast area. Less pronounced monsoons are found over eastern Africa, North America, northern Australia, China and Japan.

The primary cause of the Asian summer monsoon is convection, when warm, rising air over the extremely hot Indian subcontinent creates an area of low atmospheric pressure. The rising air is replaced by extremely moist air flowing from the Indian Ocean, where pressure is relatively high. The result is a season of torrential downpours and floods, often made worse by tropical cyclones. Rainfall is particularly high over the southern slopes of the Himalayan mountains – the wettest place on Earth – where moist air is cooled as it is forced upward. The arrival of the summer monsoon usually occurs around the same date each year.

In winter, the air flow is reversed. The interior of the Asian land mass becomes very cold, creating an area of intense high pressure. The dense, sinking air flows away from the land and out to sea, resulting in dry conditions over most of continental Asia.

Ice Ages

As overdeepened, U-shaped valleys, gouged rock outcrops and plains strewn with unsorted boulders or pulverized rocks show, large areas of the Earth were once covered with sheets of moving ice. It is clear that, from time to time, extended periods marked by long, cold winters allowed a huge build-up of snow and ice. Under such ice age conditions, Earth's average temperature may drop by 5°C and polar ice sheets can advance over much of northern Eurasia and North America, covering up to 30% of the planet's land surface.

Major glaciations are known to have taken place during the late Proterozoic (between about 800 and 600 million years ago), the Pennsylvanian and Permian (between about 350 and 250 million years ago), and the Quaternary (the last 4 million years).[8] There is evidence that at least two dozen warm-cold cycles have occurred during the past 1.6 million years, with the most recent glacial advance peaking about 20,000 years ago and ending about 10,000 years ago. Today, most of Earth's fresh water is locked up in the ice sheets that persist over Antarctica and Greenland.

Various theories to explain the ice ages have been proposed, including the emergence of supercontinents due to continental drift, reductions in solar activity, asteroid impacts, enormous volcanic eruptions and cosmic dust clouds obscuring the Sun. However, the most widely accepted explanation for the recent ice advances is the astronomical theory popularized by the Serbian scientist, Milutin Milankovitch, between 1920 and 1941.

According to Milankovitch, Earth's current glacial-interglacial cycles are mainly the result of slow, but significant, changes in its orbit (see above). Three orbital parameters are especially important in causing the waxing and waning of ice sheets:

- Changes in the eccentricity of Earth's orbit over a period of 100,000 years;
- Changes in the tilt of Earth's axis over a period of 41,000 years;
- Precession of the orbit over a period of 22,000 years.

Milankovitch developed a mathematical model that was able to calculate how these variables would influence the amount of insolation reaching the planet's surface at different latitudes. He showed that Earth's changing orbital geometry can reduce the amount of solar radiation reaching high latitudes (around 65°N) by 10–15%. These periods of lower summer heating (and melting) coincide with advances of the polar ice sheets and mountain glaciers. Nevertheless, the Milankovitch theory does not explain all of the fluctuations in ice cover during the Quaternary or in the distant past, and other factors almost certainly played a part.

What we do know is that, at the height of the last major advance, ice sheets about 2 km thick covered much of Europe and North America. Enough water was locked up in these ice sheets to cause sea levels to drop more than 100 meters below present levels. Such dramatic changes in ice cover inevitably led to major shifts in surface drainage and vegetation cover. The subsequent melting of the massive ice sheets has caused the crust to rebound – a process called isostatic uplift which is still going on today.

Climate Change

By comparing historical records and modern measurements from weather stations and satellites, it has become clear that Earth's average temperature has generally been increasing since the 1970s. Studies suggest that the period from January 2000 to December 2009 was the warmest decade in the past thousand years. Furthermore, the planet's overall temperature is forecast to rise by at least 1.4°C during the 21st century.

Such global warming is generally attributed to an increase in greenhouse gases, particularly carbon dioxide and methane. Most of these gases have been added to the atmosphere by human activity, notably the clearing of forests by slash and burn farming, changes in agricultural land use, increased use of fossil fuels and rampant urbanization / industrialization. Since about 1860, levels of atmospheric carbon dioxide have increased by more than 30% and methane levels have more than doubled.

The role of greenhouse gases in influencing Earth's temperature seems to be confirmed by measurements of air bubbles trapped in ancient ice cores. These indicate that on the ice age Earth of 20,000 years ago there was 50% less carbon dioxide and an even greater reduction in methane compared with current levels. As the orbital geometry initiated warmer conditions, it seems that the warming was accelerated by the spread

---

[8] There is some evidence that, during at least two periods between 550 and 800 million years ago, ice sheets reached all the way from the poles to the equator, a situation known as "snowball Earth."

of plant cover and the release of carbon dioxide and methane. The minor warming by these gases also triggered a larger increase in water vapor, the most common greenhouse gas.[9] In recent times, the process has been further accelerated by human activity.

Nevertheless, the warming trend since the end of the last glacial advance has been punctuated by some major fluctuations. Reconstructions of past climate show relatively warm conditions around the year 1000, which enabled the Vikings to settle in Greenland and even reach North America, followed by a relatively cold period, or "Little Ice Age," from roughly 1500–1850.

It is also worth noting that there are remarkably strong correlations between measures of past solar activity and global temperature. For example, the Little Ice Age coincided with a period when no sunspots were observed. Moreover, the rise in temperature over the past 100 years has occurred when the Sun increased its output to its highest levels in the last millennium.

Although radiant heat from the more active Sun is not enough to explain the rise in 20th century temperatures, it has been suggested that a change in solar activity can have important side-effects. For example, an active Sun generates a more powerful magnetosphere. This provides a stronger shield from high-energy cosmic radiation. Scientists theorize that these cosmic rays affect our climate by ionizing particles and gases in our atmosphere. The ionized molecules act as nucleation points for water droplets and lead to the formation of clouds. These clouds reflect sunlight back into space but also trap heat radiated from the ground.

Scientific views on climate change will continue to be revised as new data become available, but the present consensus is that man-made CO_2 in the atmosphere is modifying the climate and is the main cause of global warming.

Size and Density

With an equatorial diameter of 12,756 km, Earth is the largest of the Solar System's four rocky planets. Although it is generally portrayed as a sphere, the planet is very slightly oblate, that is, it has a slight outward bulge at the equator and flattening at the poles. This means that the polar diameter is 43 km less than its equatorial diameter. As is the case with other non-spherical planets, the reason for this difference is Earth's fairly rapid rotation, which causes it to bulge outward at the equator.

It is interesting to note that Earth's shape, defined by the planet's gravity field, or geoid, also changes over time. Satellite data have shown that the shape is influenced by major earthquakes and climatic events that cause changes in the mass of water stored in oceans, continents and atmosphere.

Earth's bulk density is 5.52 g/cm^3, higher than any of the other planets in the Solar System. Since the density of the rocks in Earth's crust is much lower than this, it follows that the material in the interior must compensate by having a much higher density.

Interior

Seismology – the study of earthquake waves – provides a window into the planet's interior. Careful measurement of the way these

Figure 3.16 Satellite data show a significant decrease in the area of summer Arctic sea ice since the late 1970s. These maps show maximum sea ice extent in 1979 (left) and 2005 (right). On September 21, 2005 – sea ice cover decreased to 5.3 million km², an estimated decline of 8.5% per decade over the 27 year satellite record. Since 2002, springtime melting has begun earlier in much of the Arctic. The 2004–2005 winter also showed a smaller growth of sea ice than any previous winter in the satellite record and the earliest onset of melting throughout the Arctic. These changes, which affect how much solar radiation is reflected back into space, have been attributed to global warming. (NASA)

---

[9] It has been estimated that a 1% rise in water vapor could raise Earth's global average temperature by more than 4°C.

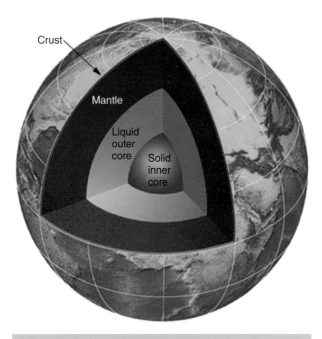

Crust

Mantle

Liquid
outer
core

Solid
inner
core

Figure 3.17 Earth's interior is composed of four main layers. At the center is a solid, iron-rich core, surrounded by a liquid outer core, which is the source of Earth's magnetic field. Above this is the solid mantle, where giant convection cells are thought to exist. The low density, solid crust "floats" on top of the "plastic" upper mantle. (Lawrence Livermore National Laboratory)

waves are bent inside the Earth and the time they take to travel through the planet (and, hence, their speed) has enabled geophysicists to determine the density and physical properties of the interior. The data show that the interior is rather like an onion, with a number of concentric layers.

At the surface is a thin, rocky crust which is 20–60 km deep beneath the continents and 8–10 km deep on the ocean floor. The continental crust is less dense (average 2.7 g/cm³) than the oceanic crust (average 3.0 g/cm³). Whereas the continents have a very complex structure and variable composition, the oceanic crust has a simple, layered structure and a uniform composition of basaltic lava. The continental crust is generally much older, with parts of Canada and Australia dating back more than 3500 million years. In contrast, the oceanic crust is geologically young, with a maximum age of about 200 million years.

At the base of the crust is the so-called Mohorovičić discontinuity (usually shortened to "Moho"). This boundary marks a sudden change to the mantle, which is composed of dense rock rich in the mineral olivine – a silicate containing magnesium and iron.

Geophysicists refer to the crust and the solid portion of the upper mantle as the lithosphere. The lithosphere is divided into many separate slabs, known as plates, that move independently. These lie on top of a mainly solid, "plastic" layer known as the asthenosphere, which extends to a depth of about 250 km.

The temperature of the rocky materials that make up the asthenosphere tends to be just below their melting point. This gives them a plastic-like quality comparable to glass. An increase in the temperature or pressure causes the material to deform and flow.

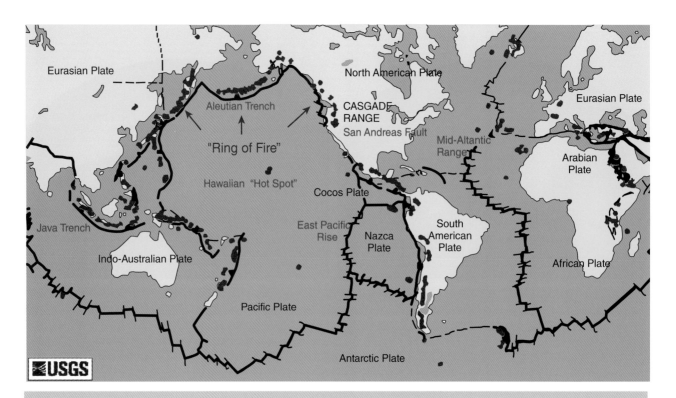

Figure 3.18 Earth's lithosphere is divided into seven large plates and a number of smaller ones which are in continual motion. Volcanoes and fold mountains form along plate boundaries. Most of the world's earthquakes also occur where crustal plates meet. (USGS)

If the pressure on the material is greatly reduced, so is its melting point, and the material may begin to melt quickly, providing a supply of magma for volcanoes.

The solid silicate mantle extends to a depth of 2900 km, where it meets the even denser, metallic core. Large convective cells in the mantle circulate heat very slowly and are probably partly responsible for the motions of the crustal plates (see below).

The outer core, which is about 2300 km thick, seems to be composed mainly of an iron-nickel alloy with trace amounts of lighter elements, such as sulfur. Swirls and eddies in the liquid outer core are responsible for generating Earth's magnetic field.

The solid inner core, which is composed of iron with some nickel, makes up the final 1200 km to the planet's center.

Surface Features

Earth's extremely varied surface ranges from high mountains to deep valleys. This varied topography is largely an expression of crustal processes associated with plate tectonics.

About 70% of the planet is covered by water, that is, below sea level. Although the continental shelves bordering the major land

Box 3.3 Flat Earth or Round Earth?

Before the Space Age, there were three main ways to prove that the Earth is round:

1. By watching ships sailing towards the horizon. A ship's hull follows the curve of the Earth and drops out of view first. Only later do the sails or masts disappear from sight. If Earth was flat, a ship should simply get smaller and smaller until it is no longer visible.
2. People traveling south see different constellations rise higher above the horizon. At the same time, the familiar northern constellations disappear below the horizon. If Earth was flat, travelers would always see the same star patterns.
3. During a lunar eclipse, the Moon passes through Earth's shadow. The edge of the shadow seen on the Moon is always curved, no matter how high the Moon is over the horizon. Only a sphere casts a circular shadow in every direction.

If Earth is spherical, how big is it? Once again, the ancient Greeks found an answer. It was clear from the size of its shadow during a lunar eclipse that our planet is bigger than the Moon. The actual size was first calculated by Eratosthenes (276–195 BC).

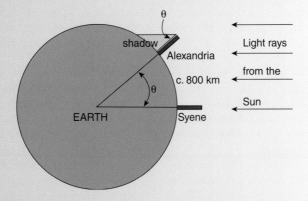

Figure 3.19 Simple geometry was used by Eratosthenes to calculate the Earth's size. The first step was to calculate the distance between Syene and Alexandria. He then measured the length of the shadows at these two places at midday on June 21, and calculated the angle of the Sun (θ) above the horizon at Alexandria (a little over 7°). Since there are 360° in a circle, he was then able to calculate Earth's circumference. (Peter Bond)

The first step was to place sticks in the ground at two places, 800 km apart. On the day of the summer solstice (June 21), he obtained measurements of the length of their shadows at noon. At one location, Syene (modern Aswan), the Sun was overhead and cast no shadow. At Alexandria, the Sun was about 7° away from the zenith, so there was a noticeable shadow. A simple calculation using the length of the shadows and the distance between the two sticks enabled him to work out Earth's circumference and diameter. (If a difference of a little over 7° is 800 km, then a full circle of 360° must be about 40 000 km. The diameter is equal to the circumference divided by 3.14, the value of π ["pi"]. This gives a diameter of about 12 700 km – very close to the actual figure.)

Later astronomers attempted to repeat his observations and obtained smaller figures. The belief in this "shrunken" Earth prevailed for centuries, and resulted in Christopher Columbus believing that the westward journey from Europe to Asia would be much shorter than it really is. His expedition was only saved by the unexpected discovery of the Americas.

Figure 3.20 Earth's surface exhibits a difference in height of about 20 km from the highest mountain to the deepest ocean trench. The highest land (apart from the ice sheets of Greenland and Antarctica) is found in ranges of fold mountains – shown in red-brown. Most of the continental lowlands (green) are drained by large river systems and covered by sedimentary deposits. The ocean floors exhibit major ranges of volcanic mountains, isolated volcanic islands and deep, narrow trenches. (NGDC)

masses are very shallow, they rapidly drop away to a deep abyssal plain. Much of the ocean floor lies at depths of about 5 km, but it is much deeper in narrow ocean trenches. The record holder is the Marianas Trench in the western Pacific, with a maximum depth of 10,911 m.

Although it is often blanketed by thick sediment, the ocean floor is not completely flat and low-lying. Running along the spine of the Atlantic Ocean is a ridge of submarine, volcanic mountains, 2–4 km in height and hundreds of kilometers wide. In places, the ridge rises above sea level, most notably in Iceland. Similar underwater mountain ranges are found in the other oceans, where crustal plates are moving apart.

High mountain ranges are also found on the continents, often near the coasts. These fold mountains, largely made up of highly compressed sediments, are usually associated with collision zones where two plates are inexorably pushing into each other. The highest of Earth's mountains are the Himalayas, between India and China, with Mt. Everest towering 8 850 meters above sea level.

Large areas of the continents are fairly flat and close to sea level. The higher plains have been lowered and smoothed by weathering and erosion over millions or billions of years. Lowland plains are typically covered in thick layers of sediment. These may have been deposited by glaciers (either as unsorted debris or meltwater deposits), rivers, wind or ocean waves.

Plate Tectonics

If you look at a map of the world, it is easy to see how the shape of Africa's west coast appears to mirror the eastern coast of South America. Studies of the geological structure and rock types on either side of the Atlantic also indicate that the two continents were once joined, for example the Appalachians and the Scottish Highlands.

Evidence such as this was used by a German meteorologist, Alfred Wegener (1880–1930), to develop a theory of continental drift. Although his vision of moving land masses was too revolutionary to be accepted at the time, it is now accepted that Earth's lithosphere is divided into a number of huge, jostling plates that are in perpetual motion. This is known as the theory of plate tectonics.

According to this theory, the solid lithosphere is "floating" on the denser mantle. The plates are slowly, but continuously jostling against each other. Some are moving apart (diverging), others are colliding (converging) or sliding laterally past each other. This movement, typically a few centimeters per year (comparable to the growth rate of finger nails), is responsible for earthquakes and volcanoes.

It is also responsible for the gradual growth or shrinkage of the oceans. For example, Europe and North America are creeping further apart as new crust is created on the floor of the Atlantic Ocean. In contrast, the Pacific Ocean is gradually shrinking as oceanic crust is destroyed at the margins.

The lithospheric plates grow at the mid-ocean ridges, where two slabs of crust move apart. Hot material slowly rising beneath the ridge experiences a reduction in pressure which lowers its melting temperature. This enables rock from the asthenosphere or upper mantle to melt beneath the rift area, forming basaltic lava flows which solidify on or beneath the ocean floor. Over millions of years, as the new oceanic crust piles up to form a huge mountain range, the ocean gradually becomes wider.

An example of a new ocean experiencing its birth pangs can be seen in East Africa, where the crust is pulling apart to form a huge rift valley. The northern end of the Great Rift Valley has already been flooded to form the Red Sea.

Since we know that Earth is not increasing in size, it follows that sea-floor spreading must be balanced by the destruction of crustal material. This is now known to occur at ocean trenches, where a thick, relatively dense oceanic plate is melted as it dives beneath a neighboring plate which is less dense.

The descending slab contains a significant amount of water. As the plate descends deeper and deeper, encountering greater temperatures and pressures, this water is released into the overlying wedge of mantle. The addition of water has the effect of lowering the mantle's melting point. Magma produced in this way varies from basalt to andesite in composition.

As it rises to the surface, the magma feeds arcs of volcanic islands, such as Japan, Indonesia and the Philippines, that lie parallel to the oceanic trench. If the plate sinks beneath a continent, the magma creates a similar belt of volcanoes near the coast, for example, the Andes of South America, or the Cascades in North America.

These destructive plate margins, often known as subduction zones, are also associated with the deepest (and some of the most severe) earthquakes on the planet, such as the one which caused the Indian Ocean tsunami disaster on December 26, 2004.

Not all plate boundaries involve the formation or destruction of crust. In some places, known as transform faults, two plates

Figure 3.21 Oceanic plates grow at mid-ocean ridges, where magma rises from the mantle and creates new crust. Oceanic crust is destroyed at subduction zones, where a dense plate dives beneath a less dense plate and is slowly destroyed. Water from the melting plate helps create magma that rises to the surface and feeds nearby volcanic islands. The erratic descent of the subducted slab also sets off deep, sometimes catastrophic, earthquakes. Some volcanoes also form above hot spots in the mantle. (USGS)

Figure 3.22 A map of ocean floor age shows the youngest areas (red) along the mid-ocean ridges, where new rock is forming from rising magma. The oldest oceanic crust is furthest from the ridge, often along the continental margins (blue). Oceanic crust is destroyed at ocean trenches. Nowhere is it more than about 180 million years old. This contrasts with the much older continents, parts of which date back at least 3 500 million years. (NOAA)

slide horizontally past each other. The most famous example is the San Andreas fault system in western California.

The San Andreas fault zone, which is about 1300 km long and sometimes tens of kilometers wide, separates the North American Plate from the Pacific Plate. These have been grinding horizontally past each other for 10 million years, at an average rate of about 5 cm per year. Land on the western side of the fault zone (on the Pacific Plate) is moving in a north westerly direction relative to the land on the eastern side (on the North American Plate) – so Los Angeles and San Francisco are slowly moving away from each other. Sudden releases of built-up stress cause powerful earthquakes in this region.

By reversing current movements and studying the magnetization of minerals in the rock, geologists can trace back the positions of the continents through time. On several occasions in the past, the present land masses have come together to form a supercontinent.

Approximately 550 million years ago, Africa, South America, Australia, Antarctica, and India were joined to form one gigantic land mass known as Gondwana. Over the next 300 million years, Gondwana was further enlarged by the addition of Europe and North America, so that all of the world's major land masses were eventually combined in a supercontinent known as Pangaea.

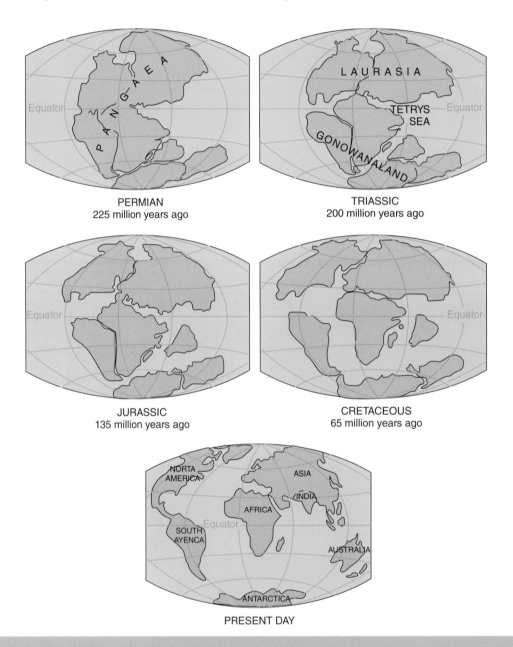

Figure 3.23   A series of maps showing the gradual break-up of the Pangaea supercontinent over the last 225 million years. Note how South America and Africa drifted apart with the opening up of the Atlantic Ocean, followed by Europe and North America. India moved north to collide with Eurasia, but Australia remained attached to Antarctica until quite recently. (USGS)

Pangaea began to break into two large masses, Gondwana and Laurasia, around 170 million years ago. Eventually, crustal rifting enabled the South Atlantic Ocean to open up as South America began to drift slowly westward away from Africa. Then India began to move northward, eventually colliding with Eurasia to form the Himalayan mountains and the Tibetan plateau. Other young fold mountains are growing around the Mediterranean Sea, where Africa is plowing into Europe, and along the west coast of the Americas, where oceanic plates are diving beneath the continents.

The driving forces behind the plate movements are not completely understood. It was formerly believed that the plates are like rafts, passively drifting on top of the moving asthenosphere. However, more recent studies suggest that they contribute to their own movement.

One hypothesis suggests that the weight of a cold slab of crust descending into the asthenosphere pulls the entire plate along. This is supported by normal faulting in the ocean crust, evidence that these plates experience tensional stresses. There is also a suggestion that gravity causes plates to slowly slide downhill, away from the high mid-ocean ridges. However, during initial rifting, plates begin to move without the help of such an elevated ridge.

It seems that convection cells also exist in the solid mantle. In this scenario, warm, rising rock from the mantle reaches the lithosphere and then spreads out laterally. The drag applied to the lithosphere causes it to move. The size of these convection cells is uncertain. It may be that they are confined to the asthenosphere and upper mantle, down to a depth of 700 km.

Alternatively, the convection cells could extend much deeper, reaching all the way down to the core boundary. There is considerable evidence that thermal plumes, vertical columns of upwelling mantle 100–250 km in diameter, lift the overlying lithosphere and spread laterally at divergent plate margins (e.g. Iceland).

Plates probably move as a result of a combination of these mechanisms. Plate separation may be initiated by mantle convection or a plume, but the subsequent formation of a topographically high spreading ridge may then drive the plates apart.

Earth is the only planet in the Solar System where such complex plate tectonics occur at the present time. On the Moon, Mercury, Venus and Mars the rigid lithosphere forms a single layer, rather than separate slabs whose motions are driven by convection in the mantle.

Volcanoes

There are estimated to be more than 1500 active or dormant volcanoes in the world, about 550 of which have erupted in historical times. This makes Earth one of the most volcanically active places in the Solar System.

The vast majority of these volcanoes are located on islands or near to coastal regions. They are also generally located close to active plate boundaries, either along mid-ocean spreading ridges or alongside subduction zones at ocean trenches (see above).

The main exceptions are chains of volcanic islands which occur above hot spots in the mantle, far from any plate boundary. The most famous example of hot spot volcanism is the island chain of Hawaii. Each of the islands has been created, one by one, as the Pacific Plate passed over a stationary plume of hot magma that has been in the same place for about 80 million years. The individual volcanoes erupt for a few million years before the movement of the plate carries them away from the rising plume. This scenario is confirmed by the ages of the islands, which become progressively older with distance from today's center of volcanic activity on the main island of Hawaii. A bend in the seamount chain was probably caused by a shift in the direction of movement of the Pacific Plate, from a northward to a more northwesterly direction, about 47 million years ago.

Mauna Loa, which makes up half the area of the island of Hawaii, is the largest volcano on Earth. However, only a small part

The Hawaiian Islands and Emperor Seamount Chain formed as the Pacific Plate drifted over a hot spot in the mantle. Today, only the Big Island of Hawaii – the youngest island – has active volcanoes. Mauna Loa, which makes up half of the Big Island, is the largest shield volcano on Earth. The age of each volcanic sea mount increases with distance from Hawaii. The bend in the chain may have been caused by a change in the Pacific Plate's direction of movement. (NGDC/USGS)

of the huge shield volcano is visible above sea level. With an estimated volume of about 75,000 cubic km, and a base that is more than 145 km in diameter, Mauna Loa is so massive that it deforms the oceanic plate on which it sits. Although its overall height from base to summit is about 17 km, the average gradient of its slopes is quite gentle, a mere 4–5°. Since it began to form nearly one million years ago, the enormous structure has grown from accumulated layers of fluid, basaltic lava. At its summit is a large caldera, a crater-shaped basin created when subterranean magma drained away and caused subsidence.

The Hawaiian volcanoes are characterized by frequent eruptions of very hot, non-viscous, basaltic lava which sometimes produces fountains of liquid rock and small spatter cones. Since any gas the lava contains can easily escape, explosions are rare.

In other parts of the world, eruptions of shield volcanoes are associated with basaltic lava pouring quietly from long fissures, instead of central vents. The fluid lava repeatedly floods the surrounding land, creating multiple layers which build into broad plateaus. Lava plateaus of this type can be seen in Iceland, north western USA and the Deccan of India.

Eruptions are more violent where the lava is less fluid and gases can escape less easily. Pressure may build inside the volcano until the gases escape in a massive explosion, destroying part or all of the summit. In the case of volcanoes such as Vesuvius in Italy and Mount St. Helens in the USA, explosive activity is quite common, resulting in huge clouds of steam, gas, solid fragments and ash which can rise several kilometers into the air and blanket the surrounding terrain. Glowing clouds of red hot rock fragments, known as nuées ardentes, may flow downhill, incinerating everything in their path.

The combination of alternating layers of ash and lava results in composite volcanoes or stratovolcanoes. The classical shape of such a volcano is a cone, like Mt. Fuji in Japan, but the symmetry is often spoiled by smaller cones and lava flows on the flanks.

Many high volcanoes have summits covered with ice and snow, and heavy rainfall often accompanies large ash clouds. The plentiful supply of water may turn the ash deposits into mud flows (also known as lahars) which bury the surrounding landscape.

In extreme cases, the products of an explosive eruption may travel around the planet and even modify the climate. The most notable example of modern times was the 1991 eruption of Mount Pinatubo in the Philippines, which ejected vast amounts of ash and gas as high as the stratosphere (see Fig. 3.25). Pinatubo also injected about fifteen million tonnes of sulfur dioxide into the stratosphere, where it reacted with water to form a hazy layer of aerosol particles composed primarily of sulfuric acid droplets. Over the course of the next two years, strong winds in the stratosphere spread these particles around the globe. The result was a drop of about 0.6°C in Earth's average temperature for a period of almost two years.

Mountains

Although volcanism accounts for the extensive mid-ocean ridges and some large islands, such as those of the Philippines, Japan and Indonesia, Earth also exhibits extensive mountain ranges which have been created by tectonic forces.

At the present time, the western side of the Americas, southern Europe and much of southern Asia are dominated by relatively young, high fold mountains. Many of these have grown within the last 60 million years, and they are still increasing in height today.

The tallest mountains on the planet are the Himalayas, which are continuing to rise at an average rate of about 40 mm per year as the Indian tectonic plate slides north beneath the Eurasian Plate.

Like the Himalayas, all of the continental fold mountains have been created by compressional tectonic activity involving a collision between two lithospheric plates. There are three types of tectonic convergence: ocean-continent, arc-continent or continent-continent.

When oceanic and continental plates collide, marine sediments are scraped off the descending oceanic plate, squeezed and accreted to the edge of the continent, forming a range of fold mountains parallel to the coast. Magma formed by melting of the descending plate feeds numerous volcanoes. This is the case with the Andes mountains of South America.

When a volcanic island arc collides with the edge of a continental plate, sediments are also scraped off the descending oceanic plate, but the volcanic islands collide with the continent and are also accreted to the margin of the continent. This type of collision may have been responsible for the creation of the Sierra Nevada range in California.

The most dramatic type of convergence occurs when an ocean basin closes and two continental plates collide. In this case, sediments that once lay on the old ocean floor and rocks in the continental crust are folded, fractured and changed by heat and pressure.

Over time, the mountains are weathered and eroded. As material is removed by glaciers and rivers, the weight of the crust becomes progressively less. The continental crust makes an isostatic adjustment, causing it to rise slowly – rather like removing a weight from a floating raft. Such adjustment places considerable strain on the crust, causing it to fracture along fault lines or break up into separate blocks. A block of crust may rise to form a horst, or sink between two faults to form a graben or rift valley.

Over hundreds of millions of years, weathering and erosion lower the once-towering mountains and make them more rounded in shape. The modern-day Appalachians and Scottish Highlands are examples of this process, remnants of a period of mountain building that ended about 250 million years ago (see plate tectonics above).

The Blue Planet

The presence of vast quantities of liquid water at the surface makes Earth unique among the bodies of the Solar System. 71% of the planet is covered by the saline oceans, equal to 97% of the world's free liquid water. In comparison, the volume of fresh water is very small. Most of this (2% of the total) is locked up in the ice sheets of Greenland and Antarctica, while the remaining 1% occurs in rivers and lakes.

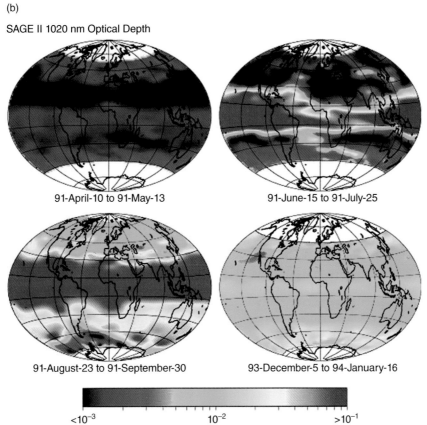

Figure 3.25 (a) A false color radar image, taken from the Space Shuttle on April 13, 1994 – shows the after-effects of the 1991 eruption of Mount Pinatubo in the Philippines. The red color shows deposits of ash on the higher slopes. The dark areas are mudflows which filled nearby river valleys after heavy rain. The area shown is approximately 45 × 68 km. (b) The eruption sent a cloud of fine particles and gases into the stratosphere and around the world, causing a noticeable cooling of the planet. A large amount of sulfur dioxide (green to red) was injected into the upper atmosphere. (NASA-JPL / NASA-GSFC)

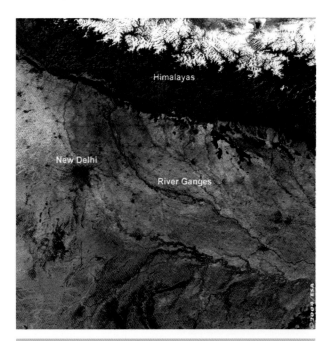

Figure 3.26 The Himalayas are the highest mountains on Earth. They are still increasing in height as the Indian plate moves north beneath the Eurasian plate. Weathering and erosion by glaciers and rivers carves deep valleys and removes large amounts of material which is eventually carried to the sea as sediment. This false color image shows the Ganges river, which rises in the snow-covered Himalayas and flows 2 510 km to the Bay of Bengal. (ESA)

Water plays a key role in the planet's evolution, since the temperature range on Earth enables it to exist as a gas (water vapor), a liquid or a solid (ice). Even in the driest deserts there is likely to be some morning dew, night frost or an occasional rain storm.

Water plays a subtle role in breaking down Earth's rocks into small particles which can then be removed by wind or rain. This weathering process takes many forms, ranging from the expansion of ice in cracked rocks to chemical solution of rock minerals.

However, the impact of water is most noticeable through its ability to erode the land, i.e. wear away and remove surface material. In high mountains, frozen water in the form of moving glaciers slowly carves very wide and deep valleys. More commonly, innumerable streams and rivers flow from upland areas towards the ocean (or occasionally, large, landlocked lakes, such as the Caspian Sea).

The force of the flowing water removes loose material – anything from sand and silt to large boulders – from the stream's bed and banks. This in turn acts like sandpaper, further wearing away the surface. After millions of years, this process is capable of wearing down mountains and forming broad valleys.

When the water slows down, it drops much of its load of sediment. This may happen during floods, or when the river enters the sea. Over thousands of years, the build-up of sediment at river mouths may lead to the growth of large deltas, such as those of the Mississippi and the Nile.

Large-scale erosion and deposition also occur in coastal regions through the action of waves and currents. This may result in the growth of new land through the accumulation of mud, sand and gravel, or the collapse of coastal cliffs which are undercut by wave action, especially during storms.

Deserts

Most places on Earth receive more than 25 cm of annual precipitation, with coastal mountains exposed to moist winds receiving at least 10 times that amount. However, there is considerable variation between different latitudes and between coastal or continental locations.

In temperate regions, where depressions are the major source, rainfall is usually spread throughout the year, with a modest increase in winter (when the low pressure systems are deeper and more frequent).

At the equator, rainfall is also found throughout the year, though the maximum coincides with the time of strongest solar heating and convective uplift. This typically results in thunderstorm activity during the afternoon.

Between these two wet zones, the rainfall is more seasonal and unreliable. Many of the driest places occur at 20–25° latitude, where dry, sinking air results in a belt of hot deserts (see zonal winds section above). Other deserts occur in the center of Asia, in rain shadow regions which are cut off from moist, maritime air by high mountains.[10]

Deserts are characterized by cloudless skies, high daytime temperatures, low nighttime temperatures and an annual precipitation of less than 25 cm. Some places may experience several years with no rain at all. This inevitably results in an absence of vegetation. When rain does fall, it usually comes in the form of short-lived, but torrential storms, which often lead to flash floods. However, any surface water soon sinks into the sand or is evaporated.

Although major rivers, such as the Nile and Colorado, carry sufficient water from their mountain sources to reach the sea, most desert river beds remain dry throughout the year.

Under extreme temperatures, rock minerals expand and contract. This process may also be aided by occasional wetting from rainfall or ground water. As time goes by, the rock disintegrates into particles of sand or gravel. (90% of the Sahara is made up of rocky or stony areas, with only 10% covered by the more familiar sand – derived from the breakdown of sandstone rock.)

With no surface moisture and no vegetation to slow the wind or bind the loose material together, wind becomes the main agent of erosion and transportation. The smallest dust particles (less than 0.2 mm) are carried in suspension in the air, while larger sand grains more commonly bounce across the surface, displacing other particles – a process known as saltation.

---

[10] The Arctic and Antarctic are also extremely dry. Only the slow rate of evaporation prevents these regions from being classed as deserts.

Figure 3.27 Earth's water cycle. Liquid water on the surface is warmed and evaporates, turning to water vapor. When this gas rises, it cools and condenses, turning back into cloud droplets of liquid water. These droplets collide, growing in size until they are heavy enough to fall towards the surface in the form of precipitation (rain, sleet, snow etc.). Rain seeps into the ground or flows over the surface, feeding streams and rivers which flow downhill into lakes or the ocean. Melting snow and ice also feed the surface drainage systems. (NASA-GSFC)

These grains, often traveling at high speed, act like sandpaper when they collide with bare rock surfaces. Wind erosion shapes a wide variety of strange rock formations, such as rock pedestals which have been eroded at their base to leave narrow necks, and yardangs – elongated ridges up to 8 meters high, separated by troughs where less resistant material has been removed.

The loose material is always shifting location, though the finest dust travels the furthest. Sometimes the grains are removed by the wind, leaving deflation hollows. If these fill with temporary ponds or lakes, which then evaporate, layers of salt may be left behind. The most famous example is the Bonneville salt flats in Utah, USA, the remains of an ancient lake that formed during the last ice age and subsequently evaporated. The salts on the surface contain potassium, magnesium, lithium and sodium chloride (common table salt).

When it meets an obstacle, moving sand piles up, forming dunes. These may take different shapes, depending on the variability of the wind. Barchans are crescent-shaped dunes with one steep side and one more gently sloping side. They form when the wind blows from one dominant direction. The sand moves up the gentler slope until it reaches the crest, then falls down the lee side. This process also drives the entire dune forward. Barchans can be up to 30 meters high and 400 meters long.

Longitudinal dunes lie parallel to the prevailing wind. They form symmetrical ridges up to 100 meters high, 600 meters wide and 80 km long. They generally form on the lee side of an obstacle where sand is abundant and the wind is constant and strong. At least some of these are derived from barchans which have been modified by cross winds. Transverse dunes lie

Figure 3.28 This false color Landsat image shows the Grand Canyon of northern Arizona which has been created by the Colorado river after millions of years of erosion. In this westward-looking image, the river cuts through the salmon-colored rock of the Colorado Plateau. The Little Colorado river enters from the east (bottom) across the Painted Desert. Nearby uplands are covered in green vegetation. The fairly rapid downward erosion by the river, combined with the lack of rainfall in this desert region, has preserved the canyon's remarkably steep sides. (NASA-GSFC)

Figure 3.30 A satellite image of a dust storm sweeping from the Algerian desert over the Canary Islands. Such storms may carry dust and aerosols all the way across the Atlantic Ocean to the Caribbean and Florida, blocking sunlight and affecting surface temperatures. (SeaWiFS Project, NASA-GSFC and ORBIMAGE)

star dune

barchan

Figure 3.29 The Issaouane Erg (sand sea) is located in eastern Algeria. This part of the Sahara desert, which is made up of vast areas of moving sand with little or no vegetation, covers an area of about 38,000 km². The most common landforms in the image are star dunes and barchan (crescent) dunes. Star dunes are formed when sand is transported from various wind directions, whereas barchans form when the wind is predominantly from one direction. The superimposition of both types suggests that wind regimes have changed over time. Heavy rain occasionally fills basins between the dunes. As the water evaporates, salt deposits are left behind (bluish-white areas). (NASA-GSFC)

at right angles to the prevailing wind direction. They have a gently sloping windward side and a steeply sloping leeward side.

Huge amounts of loose material can be removed from deserts or dry regions where there is little vegetation. Extensive deposits of fertile, windblown sediment are found in a broad zone from western Europe to China. This material, known as loess, was moved during the cold, dry periods of the recent ice age. On a smaller scale, dust from the Mongolian desert often blankets the Chinese city of Beijing, and Saharan dust frequently moves across Europe, with occasional excursions as far as North America.

Biosphere

Earth is unique in possessing a biosphere – an environment dominated by living organisms. These organisms, which range from microscopic bacteria to plants, animals and humans, have adapted so that at least some of them can survive in every type of environment on or near Earth's surface. Some microorganisms even exist within the crust, several kilometers underground.

The presence of these organisms greatly influences the planet's surface and atmosphere. Studies of newly formed, barren, volcanic landscapes show how plants and animals are able to colonize even the harshest environments. First, mechanical and chemical weathering processes break down the rock into fragments of inorganic material. Later, plant roots also

Figure 3.31 The global biosphere. Different plant ecosystems on the continents are shown in shades of green. Chlorophyll measurements (blue – purple) indicate the amount of phytoplankton in the ocean. Both vegetation and phytoplankton consume atmospheric carbon dioxide and release oxygen during photosynthesis. Note the large area of barren land (tan color), where there is little, if any, vegetation. (NASA-GSFC)

intrude into every crack and pore in a rock, helping to break it apart.

Once a small amount of inorganic debris develops, plants, then animals, begin to invade the area. Over time, their waste and dead remains decompose, providing water, carbon dioxide, organic acids and salts such as nitrates, which are then incorporated into the soil by fungi and bacteria. Earthworms and insects mix the soil as they consume and spread the dead organic material.

Certain bacteria, such as those which occur on the roots of leguminous plants like peas and beans, can take gaseous nitrogen from the air and convert it into protein. Other bacteria then break this down into ammonia and nitrates, which can be absorbed by plants. This nitrogen cycle is only one method of circulating soluble nutrients around the biosphere. Micro-organisms also play key roles in the cycling of elements such as sulfur and phosphorus.

Vegetation also plays an important part in the planet's water cycle, since plants remove large quantities of moisture from the soil and release large amounts through their leaves into the atmosphere during the growing season. An increase in vegetation cover is generally associated with an increase in regional rainfall, and vice versa. Plants also modify the local climate, for example by slowing the wind, thereby reducing erosion, and reducing the likelihood of frosts. High latitude forests reflect less solar radiation than grasslands or tundra blanketed by ice and snow.

Perhaps the most important process of all is the ability of many organisms – plants, algae and some bacteria – to take up carbon

Figure 3.32 The lower reaches of the Betsiboka, the largest river in Madagascar, transport a huge amount of red sediment to the sea. The amount of sediment has increased in recent years as a result of large-scale deforestation inland. The removal of forest cover accelerates soil erosion. Once the soil has been removed it cannot be replaced. Very little of the original forest now remains. (NASA)

dioxide from the atmosphere and produce oxygen through photosynthesis. Using sunlight as an energy source, they are able to fix (convert) carbon dioxide into carbohydrates, producing oxygen as a byproduct. It is this process which largely explains why Earth is the only object in the Solar System to exhibit an abundance of atmospheric oxygen and an ozone layer in the stratosphere.

Vegetation cover protects and stabilizes the soil, reducing erosion by heavy rain and wind. When this ecological balance is upset, perhaps by logging or clearance for agriculture, the exposed soil may suffer large-scale erosion, causing rivers to fill with sediment. This results in landslides, increased flooding and transport of soil to the sea.

Impacts

Like all of the terrestrial planets (and the icy satellites in the outer Solar System), Earth has been subjected to innumerable impacts – large and small – by asteroids and comets. As the lunar craters show, Earth and the Moon underwent a very heavy bombardment prior to about 3.8 billion years ago. Indeed, the formation of the Moon is generally thought to be associated with a collision between Earth and a Mars-sized intruder (see

Chapter 4). Such a violent episode would have been characterized by global melting of Earth's crust, perhaps delaying the establishment of life.

A glance at the cratered Moon shows that there must have been many millions of impacts during Earth's 4.6 billion year history. Yet, at the present time, there are only 178 confirmed impact structures on Earth's surface. They range in size from the 2000 km wide Vredefort feature in South Africa to a 15 meter diameter hollow in Kansas.

Since about 70% of Earth is covered in water, it is logical to assume that most impact craters are hidden beneath the sea. However, this alone can not account for the extremely modest number of visible impact structures compared with the Moon.

The main factor is Earth's much younger crust. Plate tectonics, combined with active weathering, erosion and deposition, mean that the ancient crust, which recorded impacts prior to 200 million years ago, has largely been destroyed or covered by sediments (see above). Only a few of the biggest impact structures, some dating back more than two billion years, can still be found on the continents.

Box 3.4   The Story of Life

No one knows how and when life began on Earth. Some believe it originated in a random manner, possibly in a chemical "soup" enriched by early atmospheric gases such as methane and carbon dioxide, with energy supplied by lightning or solar ultraviolet light. Others suggest the raw materials, or life itself, were delivered by comets and meteorites – a theory known as "panspermia." Yet another possibility is that the first life forms lived deep beneath Earth's surface, where the temperature was far above 100°C, with sulfur in the rocks as their source of nutrition. Warm environments around deep sea volcanic vents, known as black smokers, are another alternative.

Once the surface began to cool and life was able to establish a foothold on Earth, it was probably extremely precarious and short-lived. The massive bombardment by incoming asteroids and comets may well have wiped it out on a number of occasions until the rain of planetary debris tailed off about 3.8 billion years ago.

Nevertheless, the fossil record indicates that life was flourishing 3.5 billion years ago. Microfossils thought to be the remains of blue-green algae have been found in rocks of this age in parts of Australia. If this is the case, it shows that photosynthesis began very early in Earth's history.

Although micro-organisms seem to have spread quite rapidly across the oceans and the land surface, the oxygen produced did not build up in the atmosphere, since it was involved in chemical reactions with rocks, particularly oxidation of iron. Even today, most of the oxygen produced over time is locked up in ancient rock formations. Levels of oxygen in the atmosphere seem to have remained quite low until about one billion years ago. Only when the crust became saturated and the rate of oxidization dropped dramatically did the oxygen content of the atmosphere increase.

At the beginning of the Devonian period, about 400 million years ago, there was a wide variety of plants and animals in the oceans, including the first fish and sharks. Despite the steady evolution of life in the oceans, there seems to have been little change on the continents until the invasion of the land by algae, lichens and mosses. Exactly when this occurred is uncertain, although fossils suggest land plants may have existed by 475 million years ago.

At that time, Earth's surface underwent a major change as North America collided with Europe, creating the Appalachian mountains of the USA and the highlands now found in Norway, Scotland and Greenland. Meanwhile the huge land mass of Gondwanaland rested over the South Pole. This coincided with an explosion in the number and variety of land creatures.

Once large plants such as ferns moved onto the land, animals soon followed. The first were the arthropods, creatures such as crabs, spiders and scorpions. Then came amphibians, fish-like animals that were able to breathe air and spend short times out of water. Insects began to appear about 325 million years ago, evolving from land-dwelling arthropods.

During the next 50 million years, there were major evolutionary changes in both plants and animals. By 280 million years ago, large areas were colonized by swampy forests with trees up to 30 meters high and 1.8 meters in diameter. The final break of life's tie with water came with the arrival of the first reptiles, which laid hard-shelled eggs on the ground. These eventually developed into the most fearsome (and successful) creatures that ever roamed the Earth – the dinosaurs. Their reign came to an end 65 million years ago, opening the way for the dominance of warm-blooded mammals.

Since the end of the heavy bombardment, the number and size of the impacts has decreased dramatically. Even so, Earth is still exposed to occasional collisions, with an estimated population of 1,000 near-Earth objects (NEOs) that are more than one kilometer in diameter. However, a more likely threat comes from smaller objects, which are much more numerous. By February 2011, about 1200 NEOs had been classified as potentially hazardous (see Chapter 13).

An example of the potential threat was provided in 1908 when a 60 meter wide object – probably a meteorite – exploded near Tunguska, Siberia. Although it disintegrated 6–10 km above the ground, the blast leveled 80 million trees over an area of 2000 km².

Extinctions

Life on Earth has suffered periods of mass extinction, when many of the living organisms were wiped out within a short period of time – perhaps a few thousand years. Many of these undoubtedly occurred too long ago to be recognized in the fossil record, but 5–20 such events are thought to have taken place in the last 540 million years. (The precise number depends on the definition of "mass extinction.")

Probably the most catastrophic of these events occurred about 250 million years ago. Sometimes known as the "Great Dying," the extinction marks the boundary between the Permian and Triassic periods in geological history. It is estimated that about 60% of all land species and 90% of all ocean species abruptly died out at that time.

The most famous extinction took place 65 million years ago, and was responsible for ending the age of the dinosaurs. The most recent major extinction took place at the end of the Eocene period, about 35 million years ago.[11]

The causes of these mass extinctions are not well understood. One possibility is that a major impact, or periodic increases in the rate of asteroid or comet impacts, were to blame. The most likely example of this is thought to be the event which coincided with the demise of the dinosaurs, as well as many other species of flora and fauna.

In 1980, a team of geologists, led by Luis and Walter Alvarez, discovered a relatively high concentration of iridium in a strip of clay that runs through rocks around the world. Iridium is an element that is very rare on Earth but common in meteorites and asteroids. This clay-iridium layer marks the end of the Cretaceous period in geological history.

Searches for likely impact candidates revealed a crater up to 320 km across, buried beneath the Yucatan peninsula of Mexico. Geophysical surveys indicate that the multi-ringed Chicxulub crater was formed on the ocean floor. The explosive impact would have caused huge clouds of debris to blanket the planet and rain down from the sky, disrupting the climate and the food

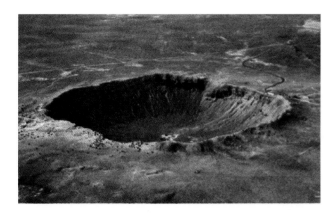

Figure 3.33 Meteor crater (also known as Barringer crater) in Arizona is one of the most recent impact craters on Earth. It was created about 50,000 years ago, when an iron meteorite excavated a hollow some 1.2 km in diameter and 180 meters deep. The crater has a simple basin shape with no central peak or rim terraces. Surrounding the basin is a wall of material 30–45 meters high where the rocks have been uplifted and, in some cases, overturned. The 30 meter wide meteorite probably weighed about 100,000 tons and struck the surface at a speed of around 12 km/s. The energy released was equivalent to about 2.5 megatons of TNT. (D. Roddy / USGS, Lunar and Planetary Institute)

Figure 3.34 A computer-generated gravity map reveals the Chicxulub impact crater (center) buried beneath Mexico's Yucatan peninsula. The impact structure is widely thought to be the cause of a mass extinction 65 million years ago. Gravity measurements show a multi-ring basin with a fourth, outer ring about 300 km in diameter. It is one of the largest impact structures on Earth. (Virgil L. Sharpton, University of Alaska, Fairbanks)

[11] The disappearance of species from the Earth at the present time, resulting from human activity, is often regarded as another mass extinction.

Box 3.5 Tides

Most of Earth's seas experience two high and two low tides each day. There is also a less noticeable tide in its solid crust, which causes a variation in height of about half a meter.

The tides are caused by the combined effects of the gravitational pull of the Moon and the Sun. Clearly, the pull of lunar gravity is greatest where Earth's surface is nearest the Moon. On the opposite side of Earth, the pull of lunar gravity is weakest. The overall result is that there are equal-sized oceanic bulges on opposite sides of the planet. The tidal bulges are fixed with respect to the direction of the Moon. However, the solid surface rotates beneath the oceans much more rapidly than the Moon orbits Earth. This gives rise to two high tides each day.

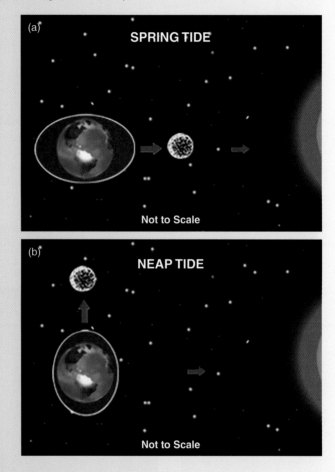

Figure 3.35 (a) The highest tides, known as spring tides, occur when the Moon and Sun are aligned so that their combined gravitational pull increases the size of the tidal bulges in the oceans. (b) When the Sun and Moon are 90° apart in the sky, their gravity pulls in different directions and the tides are least noticeable. These are the neap tides. (Office of Naval Research)

Precisely the same effects are created by the interaction of the Sun and Earth, although the tidal bulges are slightly less than half the size of those created by the Moon. About twice a month, the Sun and Moon line up and then their gravitational effects reinforce each other to produce the largest tides, the spring tides. When the Sun and Moon are 90° apart in the sky, the tidal bulges are at their minimum. These are known as the neap tides.

Although this simple pattern predominates in the deep ocean, the times of the tides can be delayed by several hours, even for places that are quite close together, in the complex waters of the continental shelf and river estuaries.

In reality, Earth's rotation carries the tidal bulges slightly ahead of the point directly beneath the Moon. This means that the force between the Earth and the Moon is not acting exactly along the line between their centers. This causes a net transfer of rotational energy from Earth to the Moon, slowing the planet's period of rotation by about 1.5 milliseconds per century. It also causes the Moon to raise its orbit by about 3.8 cm per year. Over a period of 100 million years, the Moon will recede from Earth by 3800 km.

chain, while surrounding shorelines were devastated by tsunamis. This interpretation is supported by finds of shocked rocks, tektites (small pieces of natural glass) and widespread deposits of iridium.

Whether similar impacts are to blame for other extinctions remains uncertain. It has been suggested that most mass extinctions over the past 300 million years coincided with large-scale volcanic activity. At the time of the Permian extinction, major eruptions were taking place in Siberia, where 1.5 million cubic kilometers of lava flowed from huge fissures in the crust. Other possible contenders include high energy radiation from a nearby supernova and environmental changes brought about continental drift.

Even the impact theory for the Cretaceous event is disputed by some scientists, who point out that the extinction also coincides with the formation of the Deccan Traps in India, one of Earth's largest eruptions of fluid lava. Geologists have identified a 600 meter thick lava sheet that may have piled up in as little as 30,000 years – potentially causing a deadly shift in atmospheric gases and global climate (see volcanoes section above).

Magnetic Field

For centuries, explorers and navigators have relied on the presence of Earth's magnetic field – hence the value of the magnetic compass. In fact, this magnetic field is by far the strongest of all the terrestrial planets. Theory suggests that the field is generated by a dynamo effect created by currents circulating in the planet's liquid outer core.

Figure 3.36 Earth's magnetic field creates a huge "bubble" – the magnetosphere – around the planet. This acts like a shield, diverting electrons and protons in the solar wind around the planet. The motion of the solar wind shapes the magnetosphere like a tadpole, with a blunt "head" and a very long "tail." Where the solar wind slows, upstream of the magnetosphere, a bow shock is formed. Some solar particles pierce the shield at the polar cusps, above Earth's magnetic poles. These particles collide with the upper atmosphere and cause auroras. Other particles are trapped in the van Allen radiation belts. Also shown are ESA's four Cluster spacecraft, which were launched to explore the magnetosphere and its interaction with the solar wind. (ESA)

The magnetic field resembles a dipole, with one North Pole and one South Pole. At these magnetic poles, which are fairly close to Earth's geographic poles, a compass needle will point straight down, or up, respectively.

However, the magnetic field is very dynamic. Studies of ancient magnetized rocks, both on the continents and the ocean floor, show that the magnetic poles have wandered across the surface throughout Earth's history. This motion is continuing today. Since the 1800s, Earth's magnetic north pole has moved over 1100 km across the Canadian Arctic. By the year 2030, it will be almost on top of the geographic North Pole, after which it will probably head towards Siberia.

Every so often, the magnetic field also reverses, so that the magnetic poles switch polarity, causing a compass needle to point towards the south instead of north.[12] This sequence of reversals is recorded in the once-molten rocks of the ocean floor, forming "stripes" that run parallel on either side of a mid-ocean ridge. As such, they provide key evidence supporting sea floor spreading, continental drift and plate tectonics.

Although the last magnetic reversal occurred almost 800,000 years ago, the rate of reversals during the last 10 million years has averaged four or five every million years. However, the interval has often been much longer. For example during the Cretaceous era, 70 million years ago, the time between magnetic reversals was about one million years.

The magnetic field may also experience an "excursion" – a large, but temporary, decrease in its overall strength – rather than a reversal. Today, the field strength is steadily declining at a rate of about 5% per century, leading to suggestions that a reversal may be on the way. Even so, the current strength of the magnetic field is comparable to what it has been in the last 50,000 years.

Although the area of space dominated by the magnetic field – the magnetosphere – acts as a protective shield against solar radiation and high-energy particles that flow from the Sun and other sources in the Galaxy, there is no evidence to suggest that a reversal has any major impact on the planet's atmosphere or biosphere. Even in the absence of a magnetic field, the atmosphere is still able to filter out most cosmic radiation.

The magnetosphere itself takes the shape of a tadpole or windsock. Between the Sun and Earth is a blunt "head," shaped by the incoming solar wind. Forced around the planet by the magnetic field, the solar wind shapes the magnetosphere into a long tail which stretches far beyond the Moon's orbit.

However, the magnetic shield is not impenetrable. Its weakest points are the cusps above the magnetic poles, where protons and electrons from the solar wind cross into the magnetosphere. These build up in the magnetotail on Earth's night side, until gusts in the solar wind or coronal mass ejections from the Sun make the plasma unstable. This triggers a magnetic substorm, which causes plasma to flow away from the disturbance. At such times, accelerated electrons spiral down field lines towards the magnetic poles and bombard the thin upper atmosphere. Such magnetic storms often result in spectacular auroras, more commonly known as the northern and southern lights.

Close to Earth are two donut-shaped regions, one inside the other, where high energy electrons and protons have been trapped

Figure 3.37 NASA's Dynamics Explorer spacecraft imaged both the aurora borealis ("northern lights") and the aurora australis ("southern lights"). These glowing ovals, centered above the magnetic poles, are caused by high energy particles striking the upper atmosphere. Each oval is about 500 km wide and 4500 km in diameter. Green lines show outlines of land areas. Australia is at lower left. (NASA)

by the magnetic field. They are known as the Van Allen radiation belts, after their discoverer.

The inner belt was discovered by a Geiger counter carried on the first US satellite, Explorer 1. It is located between 1000 and 5000 km above the equator, and contains particles which have been captured from the solar wind or which originate from collisions between cosmic rays and atoms in the upper atmosphere. Since Earth's magnetic field is offset from the axis of rotation, the inner belt dips down towards the surface over the South Atlantic Ocean, off the coast of Brazil. This South Atlantic Anomaly poses a threat to astronauts and satellites in low-Earth orbit.

The outer belt lies between 15,000 and 25,000 km above the equator, though it curves closer to the surface towards the magnetic poles. This region mainly contains particles from the solar wind.

The magnetosphere is constantly changing in response to levels of solar activity. Coronal mass ejections from the Sun and solar flares can prompt a sudden intensification of the magnetic field at ground level together with a rapid shift in orientation. Ground-level electrical currents induced by geomagnetic storms may affect power grids and electronics, causing blackouts and interference with radios and telephones.

---

[12] At present, Mercury is the only planet with the same polarity as Earth.

Questions

- Summarize at least six major differences between Earth and the other terrestrial planets of the Solar System.
- Describe the processes that take place in Earth's water cycle.
- Outline the main characteristics of plate tectonics.
- What differences, if any, does the biosphere make to Earth's physical characteristics?
- Earth is sometimes described as a "Goldilocks" world, where conditions have been just right for life to evolve. With reference to examples, explain whether this is a reasonable description.
- Describe the main factors that influence:
  1. Earth's surface temperature
  2. Earth's zonal winds
  3. Earth's atmospheric precipitation.
- How is human activity changing the physical characteristics of the planet?
- Describe some of the ways in which satellite technology being used to enhance our knowledge of planet Earth.

FOUR

The Moon

Earth is the only terrestrial planet to be orbited by a large satellite, the Moon. As we saw in chapter 3, the Moon has had a considerable influence on Earth's history and evolution. Since the Earth-Moon system has been in existence for more than four billion years, both bodies have experienced similar rates of impact and solar activity. Whereas most of this history has been erased on Earth, it has been preserved on the Moon. Much of what we know about the early Solar System has been derived from studies of lunar samples brought back by the Apollo astronauts. Today, a second era of lunar robotic exploration is under way, providing new opportunities to increase our understanding of its origin and the dramatic changes that have taken place in the Earth-Moon system over the eons.

The most obvious characteristic of the Moon, apart from its close proximity, is that the same hemisphere is always visible. Until 1959, when the Luna 3 spacecraft sent back pictures from behind the Moon, the "far side" had never been seen.

We never see anything other than the familiar features of "the man in the Moon" because the orbital period of Earth's only natural satellite is 27 days 7 hours and 43 minutes – identical to its period of axial rotation.[1] This is a particular form of synchronous rotation, known as spin-orbit coupling (see also Chapter 2).

Rotating west to east, the same direction as its orbital motion, the Moon always keeps the same hemisphere towards Earth. However, a number of minor libration effects mean that some regions that are normally just out of view may occasionally become visible. Hence, 59% of the Moon's surface can be observed from Earth at some time or other.

The most significant libration effects result from the slight tilt of the Moon's axis and the elliptical shape of its orbit. Although the plane of the Moon's orbit is inclined about 5.1° to the ecliptic, its equator is inclined about 6.6°, resulting in a 1.5° inclination of the Moon's spin axis to its orbital plane around the Sun.

Libration of longitude – an apparent sideways "wobble" – occurs because the Moon moves faster around its orbit near perigee and more slowly near apogee, whereas its rotation rate remains the same. As a result, regions beyond the eastern or western limbs sometimes come into view. Libration of latitude occurs because the Moon's equator is slightly inclined, allowing the poles to tilt alternately towards and away from Earth.

A third, smaller effect, known as diurnal libration, enables observers to see a little more of the surface at the Moon's western limb when it is rising and more at the eastern limb when it is setting. There are also some slight irregularities in the Moon's motion caused by its non-spherical shape.

Spin-orbit coupling is common in all major planetary satellites. It comes about because tidal interactions with the nearby planet modify a satellite's rotation (see also Chapter 5). In the distant past, the Moon rotated more quickly than it does today. However, the gravitational pull of the Earth on the slightly ellipsoidal Moon has slowed down the satellite's rotation over billions of years.

At present the Moon's distance from Earth (center to center) varies from 356,410 km at perigee to 406,700 km at apogee. However, this separation has also varied over time. Once again, tidal effects are to blame. The Sun and Moon create tidal bulges in Earth's oceans (see Chapter 3). However, frictional forces cause the bulges to be carried around the planet slightly forward of the line joining the centers of Earth and the Moon. In this position, the bulges exert a turning effect on the Moon that accelerates it

[1] On average, the Moon rises (and sets) about 50 minutes later each night because of its motion around the Earth. It moves eastward against the background stars by about 13° per day.

Exploring the Solar System, First Edition. Peter Bond.
© 2012 by Peter Bond. Published 2012 by Blackwell Publishing Ltd.

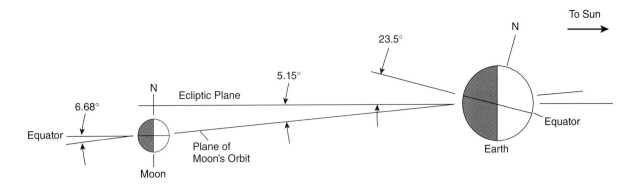

Figure 4.1 On average, the Moon's orbit is inclined 5.15° to the ecliptic plane, while its spin axis is tilted 1.5° to its orbit. This means that the Moon's axis of rotation is inclined about 6.6° relative to Earth's north-south axis. As a result, the lunar north and south poles are alternately tilted slightly towards Earth. (Paul Spudis)

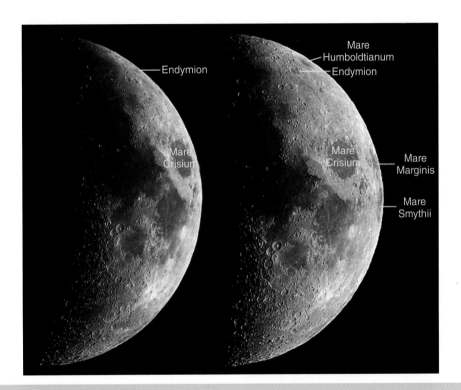

Figure 4.2 Libration effects sometimes allow parts of the far side to become visible. (David Haworth)

slightly. This raises the Moon's orbit by about 4 cm per year. (It also slows the rotation of the Earth by about 1.5 milliseconds per century.) In a few hundred million years the Moon will be too far away to cause a total eclipse of the Sun. Only annular eclipses will be possible.

Phases

The Moon shines by reflecting sunlight. Only half of its surface can be illuminated at any one time, but the Moon's orbital motion results in a continuously changing position in relation to the Sun and Earth. As the days go by and the Moon moves around Earth, different proportions of its sunlit side are visible to us – it goes through phases (see Fig. 4.4).

When the Moon lies in the direction of the Sun, its unlit side is facing us. Since this night side is very dark and the Moon is very close to the Sun in the sky, it is impossible to see the Moon at all. This is called the "New Moon."

A couple of days later, when the Moon has moved around its orbit a little, a narrow, sunlit crescent becomes visible on the its eastern side. Sometimes the remaining dark part of the face can be seen bathed in faint Earthshine – light that is reflected from the nearby, fully illuminated Earth.

Figure 4.3 The apparent size of the full Moon varies considerably between apogee and perigee. (Anthony Ayiomamitis / NASA)

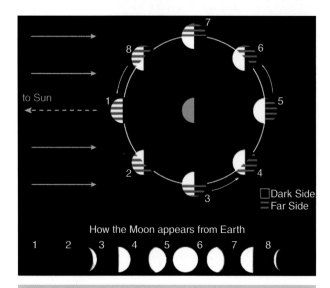

Figure 4.4 Lunar phases are caused by different amounts of the Moon's dayside being visible from Earth as the satellite moves around its orbit. (UCAR)

The Moon continues to "wax" or show more of its illuminated side. A week after New Moon, it has gone a quarter of the way round its orbit. At "First Quarter" we see half of the sunlit side, that is, a quarter of the entire surface. Over the next few days the sunlit area grows larger than a semicircle and the phase is called "Gibbous." Two weeks after New Moon, when the entire visible hemisphere is illuminated, a "Full Moon" occurs.[2] The Moon has now moved half way around its orbit and is on the opposite side of the Earth to the Sun. In the northern hemisphere, this means that the Full Moon is due south in the sky at midnight.

Over the following two weeks the pattern of phases is reversed as the Moon appears to shrink or "wane." It becomes Gibbous again for a few days, then passes through "Last Quarter" to become a shrinking crescent, visible in the eastern sky just before sunrise. Finally it vanishes in the morning twilight to become a New Moon again.

The complete cycle of phases, from New Moon to New Moon (known as the synodic month) lasts 29.5 days – more than two days longer than the Moon's orbital period (the sidereal month) (see Fig. 4.5). The extra two days occur because the Earth is orbiting the Sun. As the Moon goes through its phases, the Earth moves approximately one-twelfth of the way around its orbit. This means that the Moon has to make slightly more than one full orbit of Earth before it is back in line with the Sun.[3]

Lunar Eclipses

Earth casts a cone-shaped shadow across space that measures about 1.4 million km in length. From time to time, the Moon passes through this shadow and becomes darker. This can only

---

[2] A Full Moon is bright enough to cast shadows, but this is deceptive. The Moon's surface is actually quite dark. Its albedo, the fraction of the total light that is reflected from it, is on average only 0.067, lower than all the planets except Mercury.

[3] The term "month" comes, of course, from the word Moon.

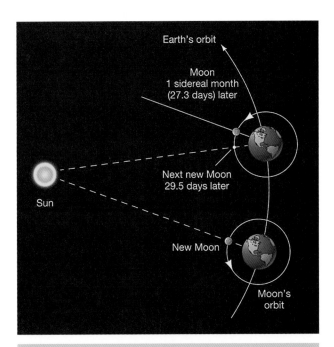

Figure 4.5 The Moon completes one orbit around Earth every 27.3 days. This is called the sidereal month. However, the motion of Earth (and Moon) around the Sun means that the period between each New Moon or Full Moon is 29.5 days. This is the synodic month – sometimes called the "lunation." (Prentice Hall)

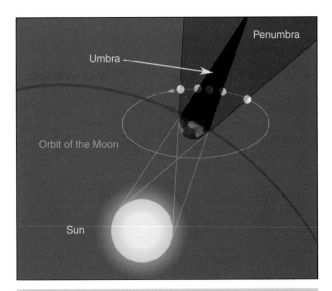

Figure 4.6 Geometry of the Sun, Earth and Moon during a lunar eclipse. The inclined orbit of the Moon means that it does not enter Earth's shadow at each Full Moon. In some eclipses, the Moon enters only the penumbra. In others, it is partly within the main shadow (umbra). There are typically two lunar eclipses each year. Sizes and distances are not to scale. (Wikipedia)

happen at the time of Full Moon, when it is directly opposite the Sun. However, Earth's shadow is not very wide – about two-and-a-half times the Moon's diameter at the lunar distance – and the satellite's orbit is inclined to the ecliptic. Hence, a lunar eclipse does not take place at every Full Moon. There are typically two or three each year.

For a lunar eclipse to occur, the Moon must be at or near one of its nodes – the point in space where it crosses the ecliptic. During a total eclipse, the Moon passes first into the area of partial shadow, the "penumbra," then into the darker, full shadow, the "umbra." The duration of totality (when it is fully immersed in the umbra) may last up to 1 hour 44 minutes.

Lunar eclipses are visible on Earth wherever the Moon is above the horizon. Even during totality, the Moon does not disappear. Instead, it remains partially illuminated by sunlight scattered by Earth's atmosphere, sometimes taking on a blood red hue. Some eclipses, when the Moon only enters the penumbra, are barely noticeable with the naked eye.

Day and Night on the Moon

The slow rotation of the Moon means that most latitudes have two-week nights followed by two-week days. In the absence of any oceans or atmosphere, there is no mechanism to distribute heat from the equator to the poles, or from the day side to the night side. There are also no clouds or aerosols to reflect or scatter incoming solar radiation. As a result, the bare, arid terrain is strongly heated by the Sun, but heat radiates rapidly into space during the long night. Under these conditions, surface temperatures at the equator may soar to 123°C in the day, but plummet to −170°C at night.

However, in the polar regions, the Sun always appears close to the horizon. Solar energy spreads out over a larger surface area and daylight temperatures are a little less extreme. Some deep craters at the poles harbor areas that lie in permanent shadow. In these dark, cold traps, a thermometer would consistently register −233°C, allowing any water ice to remain deep frozen for millions of years (see below).

Physical Characteristics

The Moon is the fifth largest satellite in the Solar System. With a diameter of 3476 km (over a quarter of Earth's diameter), it would just about cover Australia. Its average density, 3.34 times that of water, is also considerably lower than Earth's, so its overall mass is one eightieth that of our planet.

These figures may not sound too impressive, but, with the exceptions of Pluto's satellite Charon and some Kuiper Belt objects, the Moon is the largest satellite in relation to its primary in the Solar System. This has led to the two neighbors being considered as a "double planet" and to suggestions that the Earth-Moon system originated in an unusual way.

As a consequence of its lower density and smaller size, surface gravity on the Moon is approximately one sixth of that on Earth. (This proved to be both a bonus and a problem for the Apollo

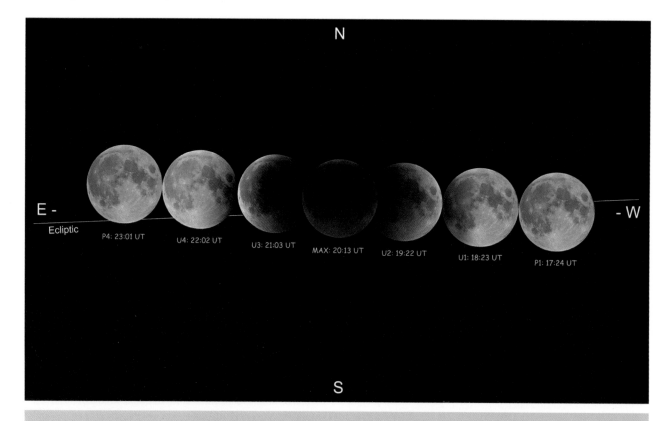

Figure 4.7 A sequence of images showing the lunar eclipse of June 15, 2011. Note the orange coloration of the fully eclipsed Moon, caused by sunlight passing through Earth's atmosphere. (Kosmas Gazeas)

astronauts. Lifting large loads was relatively easy, but they had to learn new techniques of locomotion, such as "kangaroo hops.")

Escape velocity is also low (2.37 km/s), which means that it is unable to hold onto any appreciable atmosphere. However, it is much easier to launch a rocket from the Moon to Earth than the other way around. A number of lunar meteorites have also been discovered on the Earth, having been blasted into space by large impacts (see Chapter 13).

Near Side, Far Side

Until the arrival of the Space Age, no one had ever seen the lunar far side, whereas the near side had been carefully sketched and photographed for some 350 years. The Earth-facing hemisphere was characterized by relatively flat, dark regions (called "maria" – meaning "seas") and lighter, heavily cratered uplands. As the Moon went through its phases, it was possible to observe changing shadows along the terminator and determine that the rugged lunar surface has many mountains and valleys, as well as impact craters.

Mankind's first sight of the hidden hemisphere came in 1959, when the Luna 3 spacecraft sent back a series of fuzzy pictures. The far side proved to be very different from the familiar near side, with few mare regions among the expansive, cratered highlands.

The brighter highlands cover nearly 85% of the lunar surface as a whole, including almost the entire far side and 65% of the near side. The rest of the surface is covered by smoother, less cratered mare material, divided into 22 separate "seas."

The reason for this dichotomy is related to the thickness of the lunar crust. Spacecraft data indicate that the crust is 60 km thick on the Earth-facing side and up to 110 km thick on the far side. Early in the Moon's history, large asteroid impacts on the near side excavated giant basins, fracturing the crust and enabling magma to flood the basins from below. On the far side, the thicker crust acted as a shield, largely preventing magma from reaching the surface.

Asymmetry of the Moon also shows up in topographic maps, which show that the overall range of topography is about 19.8 km. The highest regions occur on the north-central far side, perhaps due to impact ejecta. Although the near-side maria are below the general surface level, the deepest depression occurs on the far side, near the lunar south pole. The highest point (+10.75 km) is on the southern rim of the Dirichlet-Jackson Basin, located on the lunar far side, while the lowest point (−9.06 km) is inside Antoniadi crater in the South Pole–Aitken Basin.

Although it is 2500 km in diameter, making it the largest impact feature known anywhere in the Solar System, the South Pole-Aitken Basin is not deep enough to contain mare material.

The Moon has a much "bumpier" gravitational field than Earth, with small anomalies due to mass concentrations

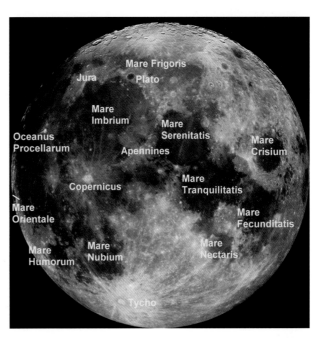

Figure 4.8 Earth and Moon shown to scale. Earth's diameter is almost four times greater than the Moon's, so 50 Moons would fit inside Earth. Note the much lower albedo (surface reflectivity) of the Moon. (NASA)

Figure 4.9 The near side of the Moon, showing the contrast between the dark maria and the lighter highlands of the near side. Several bright ray craters are also obvious, particularly Tycho in the far south. The image was taken by the Galileo spacecraft. (NASA)

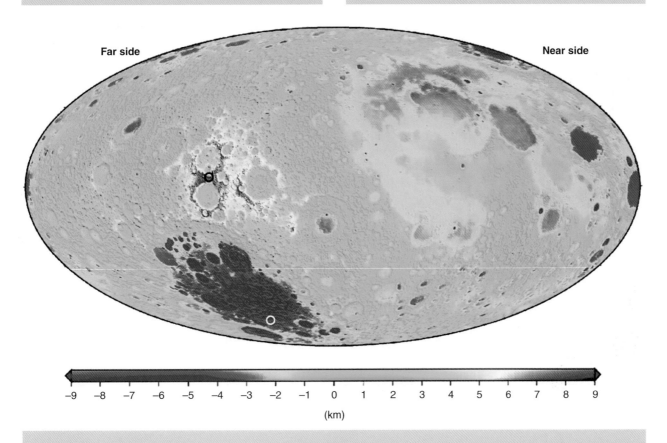

Figure 4.10 Lunar topography, based on data from Japan's Kaguya orbiter. The mare areas of the near side are revealed as lowlands, contrasting with the much higher, ancient terrain of the far side. However, the deepest and largest basin of all, the South Pole–Aitken Basin, dominates much of the far side. The lowest and highest points are marked by circles. (JAXA)

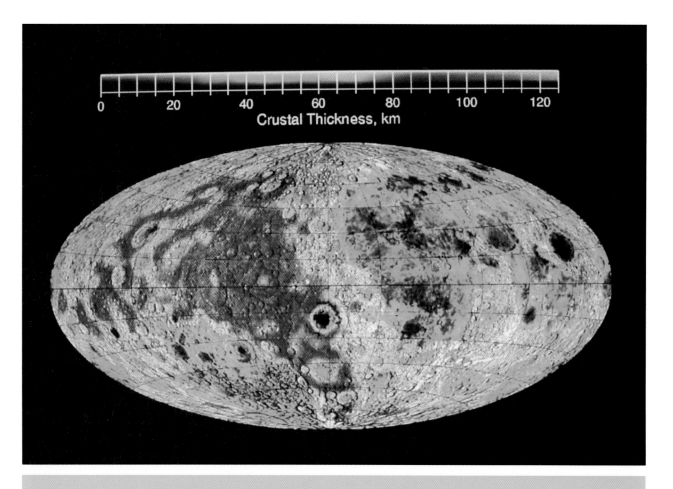

Figure 4.11 The thickness of the lunar crust. The crust is thickest beneath the highlands of the far side and thinnest beneath some of the large multi-ring basins. The mantle may be only a few tens of kilometers beneath the surface in some areas, for example, the South Pole-Aitken Basin. (NASA)

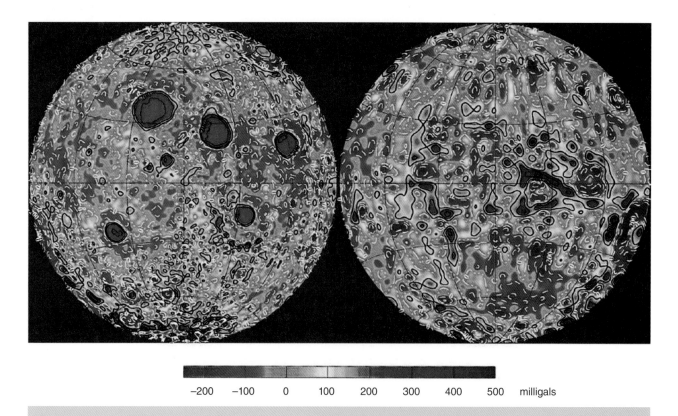

Figure 4.12 A map of lunar gravity on the near side (left) and far side (right), based on Lunar Prospector data. Note the higher gravity associated with the mascons beneath the basalt-filled maria, particularly on the near side. (NASA)

("mascons") on the surface, and a large overall asymmetry due to the fact that the crust is thicker on the far side of the Moon. Surprisingly, the mascons – areas of increased gravity – are particularly associated with the large, lava-filled impact basins in both hemispheres, rather than tall mountains or highlands, where the crust is thicker.

Almost all of the mascons are thought to be associated with regions of denser volcanic rock underlying the maria, although one region, beneath Mare Orientale, consists of a core of high gravity surrounded by lower than normal gravity. Since it cannot be readily explained by volcanic activity, this mascon might be caused by part of the denser mantle being thrust up into the crust by a meteorite impact. Older, irregular basins (such as Mare Nubium and Mare Tranquillitatis) have thinner lava deposits and do not contain extensive mascons.

Major Impact Basins

Impact features more than 300 km across are termed impact basins. More than 40 such basins have been identified on the Moon, many of them filled with the dark, basaltic lavas that form the maria. In addition to these basalts, which are mainly confined to the central topographic depressions, the impact basins are associated with widespread faulting and other forms of deformation.

The maria are broad plains, pockmarked with a few large craters and many smaller impact hollows. They are concentrated on the near side of the Moon, where they cover 35% of the surface. In contrast, they cover only 5% of the hidden side, where even the largest, Mare Muscoviense, pales into insignificance compared with its near-side cousins. Material ejected from these basins (ejecta) is distributed over wide regions, serving as a useful marker in trying to determine the broad chronology of events that shaped the Moon's surface. For example, if a crater lies on top of the ejecta, then it must be younger than the impact basin from which the ejecta originated. On the other hand, if a feature is partially buried by an ejecta blanket, it must be older than the corresponding impact basin.

Although the original floors of many of the basins are now hidden by lava infill, various features are visible in relatively young impact structures such as Mare Imbrium and Mare Orientale. Both of these display several concentric rings of mountains that formed when the basin floors bounced back upward and shock waves rippled outward immediately after the monumental impact events.[4]

The outermost ring in the Imbrium Basin has a diameter of 1300 km and rises roughly 7 km above the mare surface. It is divided into several different ranges – the Carpathians to the south, Apennines to the southwest, and the Caucasus to the east – though it is less marked to the north and west. The middle ring is marked by the Alps in the north, and some isolated uplands near the large craters Archimedes and Plato. Least conspicuous is the 600 km wide inner ring, which is now largely buried by lava and marked only by low hills that protrude through the plains and a series of curved mare ridges.

Extending some 800 km from the Imbrium Basin is a huge blanket of ejecta, known as the Fra Mauro formation, which was visited by the crew of Apollo 14. Also encircling the basin is a pattern of radial grooves – deep, linear furrows that were gouged by large projectiles that were blasted from the basin and skimmed across the surface.

The force of the colossal impact literally shattered the crust, creating a global pattern of faults, both radial and concentric to the basin. Directly opposite the basin, on the far side of the Moon, is a region of chaotic terrain associated with the Van de Graaff crater. This is thought to have been formed by focusing of seismic waves from the impact that traveled through the Moon's interior.

The slightly younger Orientale Basin suffered less flooding by mare basalts, so its impact features are easier to recognize. Once again, the collision caused huge ripples in the crust, creating three concentric upland rings – the Cordillera on the outside and two ranges of the Rook Mountains. Ejecta from the basin extend up to 500 km beyond the outer ring. Linear patterns in the rough, hummocky material radiate away from the basin.

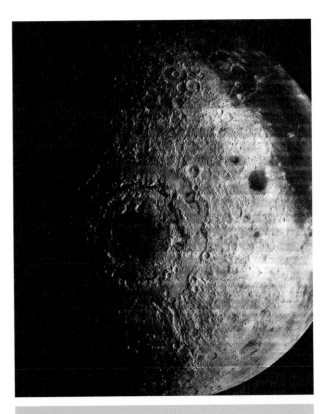

Figure 4.13 A Lunar Orbiter mosaic showing the three concentric mountain rings of the Orientale Basin. The outermost ring is 930 km in diameter. The fairly small area of dark mare material in the center rose to the surface long after the original impact. (NASA)

---

[4] Almost all of the mountains on the Moon are associated with impacts and ejecta blankets. With heights up to 7800 m, they are comparable to the highest mountains on Earth, though generally not as steep.

However, the most surprising impact feature is the South Pole-Aitken Basin on the Moon's far side. Little was known about it until 1994, when Clementine topographic maps revealed an ancient basin about 2500 across and 12 km deep, making it the largest impact structure in the Solar System. Since its creation, the basin has been much modified by smaller impacts, hence the difficulty in recognizing it. Although it contains only modest pools of dark mare material, its floor is markedly darker than the surrounding highlands and shows an enrichment of iron and titanium.

The reason for this unusual composition is uncertain. It may be due to a mixture of material from the lower crust and rock dug up from the lunar mantle. However, the fact that mantle material seems to be unexpectedly scarce suggests that it was not a typical, high-velocity impact. One hypothesis suggests that it was formed by a low-velocity projectile that hit at a low angle (about 30° or less) and so did not plow very deeply into the Moon. If so, much of the debris would have been blasted into space, possibly providing a source of projectiles that gouged out other lunar basins between 3.85 and 3.95 billion years ago.

Maria

The floors of the maria, which vary from flat to gently undulating, are covered by a thin layer of powdered rock (regolith) that darkens them and accounts for the Moon's low albedo. The maria consist largely of basalt and were created between 3.9 and 3 billion years ago, when fluid lavas originating in the mantle periodically filled the major, pre-existing impact basins. The oldest maria are irregular in shape and found in broad lowlands (e.g. Mare Tranquillitatis or Mare Fecundidatis).

The largest of all is Oceanus Procellarum, on the west side of the Earth-facing hemisphere, which stretches 2500 km from north to south and covers roughly four million km². Unlike the other lunar maria, Procellarum is not contained within a single, well-defined impact basin. Around its edges are many smaller "bays" and "seas", including Mare Nubium and Mare Humorum to the south.

The more recent maria are roughly circular (such as Crisium, Serenitatis, Nectaris, Imbrium and Orientale). They lie at successively lower levels to the east, with Mare Smythii – the easternmost of the mare basins on the near side – lying almost 5 km below the average lunar radius.

Various distinctive features favor a volcanic origin for the mare material, e.g. many broad, low domes with summit craters that closely resemble terrestrial shield volcanoes. In some places, irregular, steep-sided volcanic mountains can be seen. Also visible are broad, lobate flow fronts that mark the furthest limits of individual lava flows. These flow fronts may be several hundred kilometers long and up to 100 meters high.

The maria are also crossed by numerous sinuous rilles and wrinkle ridges. Many of these rilles originate in craters near the higher margins of the mare basins and flow into the lowlands.

Figure 4.14 A view westward, across southwestern Mare Tranquillitatis, from the Apollo 11 orbiter. Tranquility Base (TB), where Armstrong and Aldrin made the first manned landing, lies on the terminator. As in most maria, the smooth surface is interrupted by craters, numerous ridges (R), straight rilles or grabens (G), sinuous rilles (S), and islands of unburied terrae (T). Craters such as Maskelyne (M) are surrounded by extensive aprons of ejecta. Maskelyne is 210 km from the landing site. (NASA / LPI)

Samples collected from the margins of Hadley Rille by the crew of Apollo 15 confirmed that these landforms were carved by basaltic lava flows.

Although the mare material is only a few hundred meters thick near the basin rims, the central regions may contain 2–4 km depth of ancient lava flows, for example, Mare Humorum. This mass of dense rock has deformed the underlying crust, creating fractures and fault-like depressions (graben) around their outer edges and compressed wrinkle ridges, where the mare surface has buckled. The most spectacular fault is the 110 km long Straight Wall (Rupes Recti), which lies on the edge of Mare Nubium. Land on one side of the fracture is a few hundred meters higher than on the other side.

Mare Rocks

Mare and highland rocks differ in both appearance and chemical content: for example, most of the mare basalts are richer in iron and titanium, but poorer in aluminum than highland rocks. However, analysis of the Apollo samples shows that the chemical composition of mare materials varies from place to place. By correlating these differences with the subtle color changes in spectra obtained from orbit, it is possible to map surface composition.

The shapes of the mineral grains and the way they are intergrown in mare basalts confirm that they formed in lava flows. As on Earth, some of these were only one meter thick, others much thicker (perhaps up to 30 meters). Many of the basalts also contain holes, called vesicles, that were formed by gas bubbles trapped when the lava solidified. The composition of these gases was probably a mixture of carbon dioxide and carbon monoxide, with some sulfur-rich gases. The absence of water or water-bearing minerals shows that little water vapor was present.

When the gases escaped in explosive fashion, the liquid rock formed tall lava fountains. Fragmented rock cascaded to the ground, forming deposits of pyroclastics. One such deposit was sampled by the crew of Apollo 17 and found to consist of numerous, small, orange glass beads. The molten rock cooled so rapidly that there was not enough time for crystals to form and grow, so it simply turned to glass.

Experiments conducted on mare basalts and pyroclastic glasses show that they formed when the interior of the Moon partially melted, typically at a temperature of 1000–1200°C. The melting, caused by heat from decaying radioactive isotopes, took place at depths of 100–500 km. The basalts were largely composed of the silicate minerals plagioclase (a type of feldspar) and pyroxene (containing iron, calcium and magnesium), with varying amounts of olivine (another magnesium-iron silicate) and ilmenite (an iron-titanium oxide).

Highlands

The lunar highlands (terrae) are brighter and more rugged areas than the maria. They occupy two thirds of the Earth-facing hemisphere and about 85% of the entire lunar surface. The greater age of the highlands is indicated by a much higher density of large craters and confirmed by radiometric dating of rock samples.

Figure 4.15 A mare basalt, showing numerous hollows (vesicles) formed by bubbles of dissolved gas escaping from the molten lava. Rock 15016 was collected at the Apollo 15 landing site near Hadley Rille. Radiometric dating shows it is 3.3 billion years old, younger than many other lunar basalts. (NASA)

Figure 4.16 Beads of orange, volcanic glass found at Shorty Crater in the Taurus-Littrow valley. The glass particles, intermixed with black crystals of ilmenite, are 20–45 microns across – about the same size as silt particles on Earth. The orange color is caused by a titanium oxide (TiO_2) content of 8%. Other constituents included oxides of silicon (39%), iron (22%) magnesium (14%) and calcium (8%). The glass formed 3.64 billion years ago from material that melted about 400 km below the surface and then erupted onto the surface in a lava "fire fountain." (NASA)

Figure 4.17 When the Moon first formed its surface was probably composed mostly of feldspar-rich igneous rocks called anorthosite. Today, this rock type makes up the lunar highlands. This 4.4 billion-year-old sample was collected by the Apollo 16 crew. (NASA)

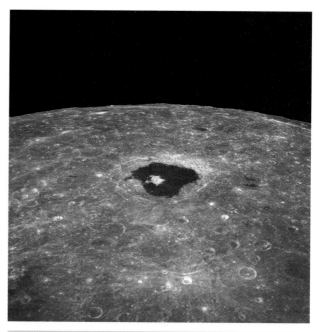

Figure 4.18 An Apollo 8 image showing the contrast between a dark mare region and surrounding highlands. Tsiolkovsky crater, on the lunar far side, is 198 km in diameter. Its crater is partially filled by lava and the internal peaks are offset from the center. Note the extensive landslips around the rim. (NASA)

Most of the light-colored highlands are largely composed of anorthosite – a type of igneous rock that is rich in aluminum and calcium, and relatively poor in iron, magnesium and titanium, compared with the mare rocks. They appear to have formed when feldspar crystallized and floated to the top of a global magma ocean that surrounded the Moon soon after it formed (see Geological History below).

One peculiar type of highland rock has been dubbed "KREEP," because it is rich in potassium (K), rare earth elements (REE) and phosphorus (P). It is also rich in the radioactive elements thorium and uranium, making it easy to map from orbit. These incompatible elements cannot fit into the crystal structures of the common lunar rock-forming minerals of plagioclase feldspar, pyroxene, and olivine. As a result, when lunar magma slowly crystallizes, the KREEP elements get left behind in the dwindling reservoir of melt.

The unique elemental signature suggests that the rock was formed during the final stages of the formation of the primitive lunar crust, when the cooling of a global magma ocean was nearing completion. The last stage in the crystallization of the magma produced rocks that are strongly enriched in KREEP material.

This ancient highland material was later spread over the mare regions by major impacts, hence the ability of the Apollo 12 astronauts to pick up samples from the Oceanus Procellarum. The distribution mapped by Lunar Prospector supports the theory that the impact which formed the Imbrium Basin excavated KREEP-rich rocks and spread them over the Moon. The huge impact that created the South Pole-Aitken Basin also exposed KREEP-rich material.

Since their formation, some 4.4 billion years ago, the ancient highland rocks have been drastically modified by countless impact events. The cumulative effect of these collisions has transformed the nature of the surface, with the result that few samples recognizable as original crustal material were recovered by the Apollo astronauts.

Most of the highland rocks are classified as breccias – conglomerates formed from anorthosite that was shattered by an incoming meteorite and then welded together by the heat and pressure generated by the impact. These are often studded with small, glassy spheres that probably formed when the spray of molten rock from the impact solidified in midflight. In addition to metamorphism of the rock minerals by sudden impact shocks, huge amounts of ejecta were spread over huge distances.

Other processes that have modified the highlands are tectonism and mass wasting. Tectonism is visible in numerous linear faults or fractures, sometimes in the form of grabens (rift valleys). Some of the largest linear structures, radiating from the edge of the Imbrium Basin, are obviously related to its formation (see above).

Smaller linear or curvilinear fractures are widespread. These lobate scarps have steep slopes and are often segmented. They are similar to, but much smaller than, the lobate scarps found on Mercury and Mars, with a maximum relief of less than 100 meters and lengths of only a few tens of kilometers. Their global distribution indicates that they may have resulted from the cooling and shrinking of the Moon, perhaps within the last one billion years.

Mass wasting has subdued the rugged highland landscape by moving loose regolith downslope. In the absence of running water the rate of movement is often very slow, but it may be hastened by moonquakes or shock waves from impacts. One photo taken by the Apollo astronauts clearly showed a track left by a boulder that had rolled down a steep slope.

Volcanism in the highlands is hard to identify, though a few areas, notably the Gruithuisen domes west of Mare Imbrium, have been suggested as evidence of volcanic activity.

Lunar Craters

On Earth, craters are associated with both volcanic cones and meteorite impacts. For many years, scientists could not agree on the origin of the lunar craters, until close-up images and other data from spacecraft closed the debate.

Today, we know that volcanic craters on the Moon are rare and comparatively small, whereas the landscape is covered by impact features of all sizes, ranging from pits less than a meter across to huge hollows more than 200 km across. It is estimated that there are at least 300,000 craters with a diameter larger than 1 km on the near side of the Moon alone. Most of these were created more than 3.8 billion years ago, during a heavy bombardment by large planetesimals left over from the formation of the Solar System.

The shape of lunar craters varies with their size and age. The smallest hollows, typically less than about 15 km in diameter, have a simple, bowl-like form. Larger examples display more complex forms, including shallow, flat floors made of solidified lava, one or more central peaks, and terraces on the inner-rim walls (see Box 4.1).

The impact structures become more complex as they increase in size. When they reach between 180–300 km in diameter, they change from craters with central peaks or groups of peaks to circular basins surrounded by two or more rings.

Large scale crustal faulting has produced small, linear rilles that slice through many craters. Volcanic activity has also filled many

of the largest craters with lava, completely covering the original floors and, sometimes, their central peaks. Craters within or on the fringes of the maria have often been breached and flooded with lava, producing bays and "ghost rings."

Since larger impacts excavate larger basins and eject more debris, the major craters are surrounded by extensive debris aprons and numerous secondary craters created by falling ejecta. Relatively young craters, such as Copernicus and Tycho, also display bright rays of fragmented material that radiate outward for hundreds or thousands of kilometers. One explanation for the brightness of the rays is that they are composed of light-colored highland material that is conspicuous against the darker mare material. The powdery nature of some rays also makes them more reflective. However, as time goes by, the constant bombardment by the solar wind and innumerable micro-impacts causes them to fade.

Copernicus, 93 km in diameter, is one of the most prominent features on the Moon's near side and displays most of the features associated with a large lunar crater. The well defined crater is believed to have formed less than one billion years ago, by which time the crust had thickened too much to allow its interior to be flooded with lava. In the center are several peaks made of material that rebounded from deep in the crust as a result of the impact. Much of the crater floor is hummocky or uneven, while the inner part of the crater walls is broken up into a series of terraces marked by debris flows. Just beyond the crater rim is a continuous ejecta blanket, breaking up into discontinuous clumps as the distance increases. Chains of secondary impact craters are also prominent on the surrounding plains.

Like Copernicus, most of the craters on the Moon are circular, but there are a few exceptions, notably the neighboring Messier and Messier A in the Mare Fecunditatis. The former feature is elongated, measuring about 11 km long and 8 km wide. About 20 km to the west is the double crater, Messier A, where one crater is superimposed on another. Two bright rays stretch 120 km towards the west. These craters and the rays were probably created by a very shallow (1–5°) grazing impact.

Two crater chains on the Moon are thought to have been formed by the impact of asteroids or comets that broke apart before they arrived. The first is the Davy chain, a 47 km long sequence of 23 craters, 1–3 km in size. More prominent is the 250 km long Abulfeda chain, which comprises 24 craters with diameters of 5–13 km. Neither of these features seems to be associated with debris ejected from large, nearby impact craters (see also Chapters 8 and 13).

Volcanic Domes and Pyroclastics

The lunar maria were flooded by huge volumes of magma, but not all volcanic landforms are on such a grand scale. Many of the maria exhibit small, dome-shaped, shield volcanoes, that were formed by quiet, repetitive outpourings of non-viscous (runny) lava. This makes their flanking slopes quite gentle, no more than 1–8°. The domes are often less than 10 km across and a few hundred meters high. At their summits are pits around 1.5 km wide. Their modest dimensions make them quite hard to observe unless they are near the day-night terminator.

Figure 4.19 Copernicus, one of the youngest major impact craters on the Moon, displays several central peaks, terraced walls and a raised rim. The crater is 93 km wide and more than 3 km deep. The image was taken by the crew of Apollo 17. (NASA)

Box 4.1   Impact Craters

Impact craters are found on the surfaces of all the terrestrial planets and satellites – with the exception of Io, which is being resurfaced continuously by volcanic activity. The craters are created by collisions involving an incoming projectile – an asteroid, comet or meteorite. In the case of the Moon, these objects have widely ranging speeds, but average about 20 km/s.

On Earth, impact craters are not easily recognized because of weathering and erosion, but on worlds with no atmosphere, such as the Moon, they provide an impact record going back more than four billion years.

The shape and size of craters depends on many factors – the size, density, speed and angle of arrival of the impactor, together with the physical and chemical properties of the target surface. One key factor is the amount of kinetic energy possessed by the impacting object.

Kinetic energy (energy associated with motion) is described mathematically as: $KE = \frac{1}{2}(mv^2)$ where m = mass and v = velocity. During impact, the kinetic energy of an asteroid is transferred to the target's surface, shattering rock and moving debris around. Generally speaking, the greater the kinetic energy, the larger the crater.

When a solid object strikes a planetary surface at high speed, an extremely powerful shock wave – millions of times Earth's normal atmospheric pressure – spreads out from the impact site. This compresses the rock and makes it deform so that it flows like a fluid.

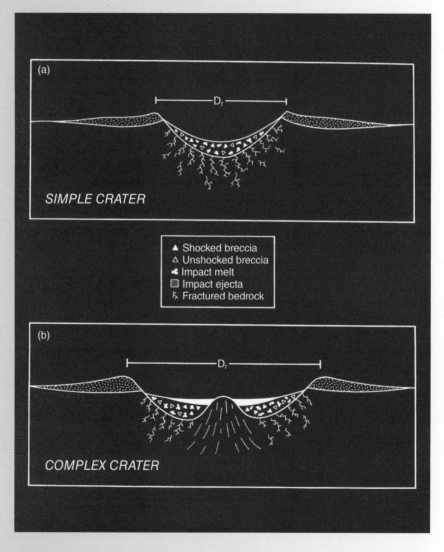

Figure 4.20   Schematic cross sections of **(a)** a simple impact crater and **(b)** a complex crater. Simple craters are typically bowl-shaped and small in size (less than 20 km across on the Moon or 4 km on Earth). Larger craters are more complex, with central peaks, relatively shallow floors and terraced or slumped rims. In both cases there is a raised rim, a surrounding blanket of ejecta and a scattering of smaller, secondary impact craters. (LPI)

As the shock wave radiates outward, decompressed material behind it is thrown upwards and outwards, excavating a volume of material much larger than that of the impactor. Rock strata that were once horizontal may even be bent back or turned upside down, so that the older rocks are now on top of younger rocks. Beneath the crater floor is a lens-shaped body of breccia – rock that has been broken and pulverized by the shock wave. The incoming object itself is melted or partly vaporized.

As a general rule, small (100 meters) craters tend to be about 20 times the diameter of their impactor, whereas medium-sized craters (1–10 km) are about 10 times wider, and large (100 km) craters are about five times wider.

The critical diameters at which one type of crater changes to another depend inversely on the surface gravity. For example, the transition between simple and complex craters occurs at about 20 km diameter on the Moon but at only 2–4 km on Earth. The transition from a central-peak structure to a ringed structure occurs at about 150–200 km on the Moon, but at only 20–25 km on Earth.

Small craters are bowl shaped, with smooth walls, a raised rim and a surrounding apron of hummocky ejecta. Their depth is between one fifth and one tenth of the diameter. Immediately after its excavation, a simple crater is filled, to perhaps half its original depth, by a mixture of redeposited (fallback) ejecta and debris that slumps inward from the walls and rim.

Lunar craters with diameters of more than 20 km are shallower and more saucer-shaped. They develop one or more central peaks, caused by rebound of the surface and reflection of shock waves from depth. Steep internal slopes are marked by terraces where slumping has occurred. The rim may tower more than one kilometer above the surroundings, and an ejecta blanket spreads far from the crater in all directions.

Impacts by large pieces of debris excavate numerous secondary craters. Finer, more powdery ejecta travel hundreds or thousands of kilometers from the impact site, forming bright, radial ray deposits.

Figure 4.21 Lying amid a scattering of small, bowl-shaped craters is 18 km wide Dawes crater, imaged from Apollo 17. Located between Mare Serenitatis and Mare Tranquilitatis, Dawes is an example of a simple, slightly oval-shaped, lunar crater. Note the deposits of slumped or fallback material on its floor, the steep walls and sharp rim. (NASA / USGS)

The largest craters, with diameters over 200 km are known as basins. The central peak is replaced by a cluster of peaks (a peak ring). The largest craters contain a series of concentric mountain rings, which may be caused by a combination of slumping and shock waves traveling across the surface.

Most of the domes are found in clusters, often close to the edges of the maria, where the lava flows are relatively thin. This implies that their magma sources were near the surface, where pressures were low.

At the opposite extreme are pyroclastics, deposits of ash and cinders created during explosive volcanic eruptions. When gases trapped within viscous (sticky) magma suddenly escape, they expand rapidly, shredding the magma and producing clouds of fine material that are deposited around the source vent in a circular patch.

Lunar pyroclastics are not easy to detect, but they can be identified by their relatively dark color compared with typical mare lavas, for example, on the floor of Alphonsus crater. The dark coloration is caused by tiny black beads of glass and crystals that formed from cooling of the basalt fragments as they were blasted from the volcanic vent. The glass resulted from rapid cooling of liquid droplets, while the crystals condensed more slowly.

Some 75 pyroclastic deposits have been recorded on the Moon. Most of them are small, covering an area of less than 100 km², but a few cover enormous areas. The largest of all, which occurs on and around the Aristarchus plateau, blankets an area of 50,000 km², while the Imbrium ejecta south of Sinus Aestuum rille has a 10,000 km² coating of pyroclastic material. In this and several other cases, no obvious vents are visible, but sinuous rilles link the deposits to volcanic activity (see below).

Orbiter images of probable pyroclastic deposits prompted scientists to support the landing of Apollo 17 in the Taurus-Littrow valley, and their forecasts were confirmed when the astronauts brought back samples of black and orange glassy beads (Fig. 4.16).

Sinuous Rilles

Snakelike valleys on the Moon (called sinuous rimae or rilles) have been the focus of much debate. One possibility is that they were cut by flowing water, but the absence of water in lunar samples makes this very unlikely. They were probably formed either by faulting and subsidence of the crust, or represent lava channels or collapsed lava tubes. The wide variety in shapes suggests that all of these processes have been involved. The valleys may have originated as fault troughs that were later modified and obscured by lava flows, impact ejecta, or landslides on the walls.

Those rilles that are broken into separate sections are almost certainly collapsed lava tubes. Molten lava flows often develop a solid crust, forming tunnels beneath the surface. When the lava later drains from the tunnels, a long, hollow, cylindrical tube remains. All or part of the roof may subsequently collapse, leaving steep-sided troughs or chains of near-circular pits.

Some sinuous rilles contain a small channel within a larger valley, indicating at least two episodes of valley erosion. Other meandering channels are not associated with any obvious craters

– they occur on lava plains that are so flat it is difficult to tell the direction of flow. Still other channels have formed on mountainsides covered by hummocky ejecta thrown outward from major impact basins, suggesting that they may have been formed by some other process. Perhaps the best known example is Hadley Rille, a V-shaped gorge that starts in a volcanic crater on the flank of the Apennine Mountains and flows along the eastern edge of Mare Imbrium, eventually merging with a second rille about 100 km to the north. Its average width is about 1.5 km and depth is about 400 m.

The Apollo 15 astronauts explored Hadley Rille and found that layered basaltic lava flows are exposed in the valley walls. The walls themselves are quite steep, with an average slope of about 25°. The floor of the channel is very irregular, suggesting that eruptions of lava at different places along the channel may have added to the material flowing from the source crater. Its meandering shape indicates a great deal of erosion and modification by flowing lava of what may originally have been a fault trough.

Linear Rilles

There are a few dozen large trenches or linear rilles on the Moon. Unlike their sinuous cousins, these valleys have straight, steep walls and flat floors – features characteristic of grabens, valleys created when crust drops down between two faults. These grabens form in regions of horizontal extension, where the crust is forced apart.

How could such extension take place on a world where there are no crustal plate movements or fold mountains? One important clue is that at least one third of the linear rilles are associated with nearby volcanic features, such as domes, cones and

Figure 4.22 Apollo 15 astronaut James Irwin with the lunar rover at the edge of Hadley Rille. The view is looking northwest from the flank of St. George Crater. Loose blocks are visible on the sides and the floor, evidence that the rille has been widened and made shallower over time by wasting along its walls. (NASA)

Figure 4.23 Ariadaeus Rille cuts across the bright plains to the west of Mare Tranquillitatis. The crater south of the rille (left of center) is 13 km wide Silberschlag. (NASA)

Box 4.2   Dating the Moon

Crater densities are used for relative dating of planetary surfaces. Assuming that impactors strike all regions of a planet at approximately the same rate, the crater density – the number of craters in a given area – should always be the same. However, in many cases, another, more recent, process has covered or removed some of the oldest craters. Hence, regions with a higher crater density are generally older than regions with a lower crater density.

A common way for large numbers of craters to be removed is through volcanism. Early in the Moon's history, for example, lava flows flooded large areas, completely burying many craters. These areas were essentially wiped clean of craters about 3.5 billion years ago and thus their crater density only tells what happened to them after the lava flows ceased. The lunar highlands, in contrast, were rarely flooded by lava. Their crater density dates back to when they were formed, more than 4 billion years ago, and thus is much higher.

Crater size distributions can also be used to estimate the absolute age of a surface. The cratering rate was very high after the planets finished forming about 4.5 billion years ago and there was much leftover interplanetary debris to cause impacts. However, there is considerable evidence that nearly all of the major lunar impact basins were excavated by 3.8 billion years ago. Since then, the frequency (and size) of impacts has been steadily decreasing as the amount of debris has declined due to continuous collisions and impacts (see Geological History below).

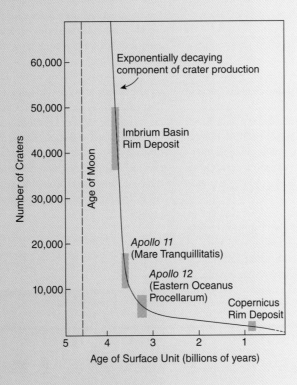

Figure 4.24   The rate at which craters were formed on the Moon has varied tremendously. During the first 800 million years there were frequent impacts. Near the end of this intense bombardment, several large impacts created the mare basins, but, for the last 3.5 billion years, the impact rate has been only one ten-thousandth that during the most intense bombardment. (Eugene Shoemaker)

By measuring the crater density on different areas of the Moon, and using radiometric dating techniques to determine the ages of rocks returned from different regions by Apollo astronauts, scientists can calibrate the cratering rate to actual surface ages. For example, rocks from the Fra Mauro region, ejected by the impact which created the Imbrium Basin, have been dated at 3.75–3.85 billion years. (The most common radiometric method compares the amounts of argon isotopes in lunar rocks. This ratio changes at a fixed rate over time.

This relationship of cratering rate v time can then be extrapolated to other places in the Solar System, although other modifying factors also have to be taken into account, for example, distance from the asteroid belt (Mars, located near the asteroid belt, may have a rate of crater formation roughly twice that of the Moon) and proximity to the Sun or large planets, such as Jupiter, whose gravitational field attracts impactors.

pyroclastic deposits. One form of extension related to volcanic activity is the intrusion of magma, where sheets of liquid rock flow into near-vertical cracks in the crust and solidify. Magma rising towards the surface in such dikes pushes the crust apart.

If the magma reaches the surface, an eruption occurs and the dike is buried beneath its own lava flows. Where the eruption of magma onto the surface is limited, small cones and pyroclastics form at a few locations along the dike. If the dike does not reach the surface, no surface volcanism is visible.

Some linear rilles are very large and easily visible in ground-based telescopes. Ariadaeus Rille, for example, measures 220 km long and 4 km wide, with near-vertical, parallel walls that drop a few hundred meters to a flat floor (see Fig. 4.23). Nearby are magma collapse pits and a volcanic dome.

Other linear valleys, such as the Sirsalis Rille, are associated with the formation of major basins and maria. They are roughly radial to the curved rim of the Oceanus Procellarum, and probably helped to feed the vast volumes of magma that filled the huge basin.

Present-day Volcanism?

The Moon is generally regarded as a geologically dead world. However, hundreds of observations of minor changes, usually flashes or glows, have been reported by amateur astronomers over the years. The most famous of these "transient lunar phenomena" took place on November 3, 1958, when Soviet scientist Nikolai Kozyrev obtained a spectrum of Alphonsus crater which he interpreted as proof of a gaseous emission from its central peak. Other red glows were reported near the crater Aristarchus by astronomers at Lowell Observatory.

Although no confirmation of any of these apparent glows or gaseous outbursts has been obtained, it has been shown that the phenomena are most frequently reported around the time of lunar perigee, when tidal forces are at their greatest and moonquakes are common. Orbital observations also revealed peaks in radon gas over Aristarchus, Grimaldi and the fringes of the maria – presumably associated with gas seeping through surface fractures in these regions.

Brief flashes caused by small objects striking the surface have also been observed during meteor storms (e.g. in 1999 and 2005).

Regolith and Moon Dust

Without protection from an atmosphere or a magnetic field, the lunar surface is bombarded continuously by micrometeorites, cosmic rays – mostly very high speed protons (hydrogen nuclei), helium nuclei (alpha particles) and carbon nuclei – and the charged particles of the solar wind. Over the Moon's lifetime, these tiny impacts have broken up the rocks and minerals on the surface, producing a layer of fine-grained regolith. Mixing of the regolith by the constant, small scale, bombardment is sometimes called "lunar gardening".

Over the ancient highlands the regolith may be more than 15 meters thick, compared with 2–8 meters on the maria. However, the material is very fine-grained, mostly in the form of dust particles less than 0.1 mm across. During the Apollo landings, rocket exhaust from the lunar module kicked up dust clouds that often became visible at altitudes of 30–50 meters, largely obscuring the surface during the final 10–20 meters of descent.

The famous photo of Armstrong's bootprint in the lunar surface showed that the regolith easily supported the weight of the astronauts and their equipment. Even the lunar module footpads sank only 2–20 cm into the soil. However, the astronauts found the compacted regolith extremely difficult to penetrate, both when erecting the Stars and Stripes or drilling to obtain core samples. For sampling at greater depths, they used a battery-powered drill which could penetrate up to 3 meters.

The all-pervasive, fine, gritty dust seeped into pressure suits and equipment. Some of the particles became electrostatically charged so that they clung to any objects with which they came into contact. The dark grains also absorbed sunlight, with the result that dust-coated equipment sometimes became extremely hot.

When inhaled, the astronauts likened its smell to that of gunpowder. Laboratory studies suggest that lunar dust could pose a health threat. Unlike particles on Earth, lunar dust is coarse and jagged. If inhaled, the grains could embed themselves in the lungs, eventually causing a long-term respiratory ailment similar to silicosis.

Future visitors hoping to set up an astronomical observatory on the Moon may have to cope with "streamers" of suspended dust reported by the crew of Apollo 17. Scientists surmise that the carpet of dust must be electrostatically charged by periods of high ultraviolet solar radiation and by the solar wind. As a result, the dust may rise 100 meters or so in a phenomenon dubbed "fountaining." The particles then discharge, sink back to the surface and the cycle starts again. At night, the particles in the solar wind that curve around behind the Moon and hit the surface drive a similar process.

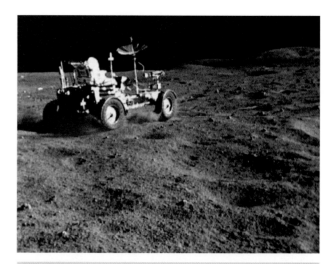

Figure 4.25   Plumes of dust spurt from the wheels of the Lunar Roving Vehicle driven by Apollo 16 commander John Young. Note the numerous small craters and the rounded shape of the nearby uplands. (NASA)

Internal Structure

There are a number of factors that must be accounted for when modeling the lunar interior. The most fundamental of these is the bulk density (3.34 g/cm^3), which is very similar to the density of the surface rocks (3.3 g/cm^3). This implies that – in contrast with the Earth – the Moon does not possess a large, iron-rich core.

Nevertheless, studies of lunar rocks, surface features and seismicity (moonquakes) make it clear that the Moon has a layered structure, and that the interior has a different composition from the surface. Furthermore, heat flow experiments set up by the crews of Apollo 15 and 17 show that the interior has a high temperature, presumably due to the presence of radioactive elements such as potassium, uranium and thorium.

Some of the most compelling evidence comes from seismometers that were left on the Moon. The instruments measured less than 3 000 events per year on average, none of them exceeding a value of 3 on the Richter scale – very weak by terrestrial standards. They recorded three different moonquake events: meteorite impacts, internal movements and impacts by abandoned rocket stages or lunar modules.

The most surprising seismic records came from the artificial impacts, which set the Moon "ringing like a bell" for more than an hour – implying that most of the lunar mantle is solid. Other information came from an impact on July 17, 1972, which was thought to involve a meteorite weighing about 1000 kg. The seismic waves indicated that the lunar rocks are hot enough to be molten below a depth of 1000–1200 km.

Some natural moonquakes originated deep inside the satellite and are thought to have been generated by tidal stresses created by the Sun and Earth. Shallow moonquakes are more likely to be caused by expansion and contraction of the surface rocks as a result of the large diurnal temperature range.

The velocities of the seismic waves provided useful information about the lunar interior. The upper few hundred meters are rubble generated by eons of meteorite bombardment. There is a change from regolith to firmer, more compacted material at a depth of 1 km.

The speed of seismic waves continues to increase as far down as 20 km, indicating fractured basaltic material whose density increases with depth. Below this, to a depth of 60 km, there may be a rock layer rich in anorthosite, the most common material in the lunar highlands. Deeper still, the sparse data suggest that there is a transition from the crust to the mantle.

The mantle rocks, which are relatively dense compared with the crust, are believed to be rich in olivine, pyroxene, iron and magnesium. The rigid mantle – which extends to a depth of about 1000 km – and the crust make up the lunar lithosphere. This thick layer makes it almost impossible at the present time for molten rock to reach the surface and for plate tectonics to take place.

The question of whether the Moon has a dense core, and whether this is still liquid, remains unanswered. Studies of seismic waves suggest that they slow down when they pass through the deep interior, offering the possibility of a partially molten layer below a depth of 1000 km. Although this layer may extend all the

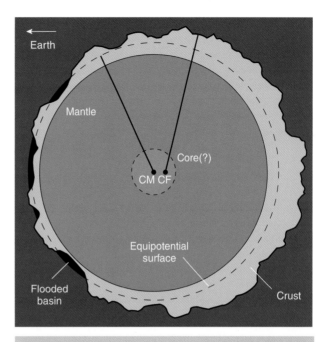

Figure 4.26 The internal structure of the Moon. The thickness of the crust varies from 60 km on the near side to about 110 km on the far side. Beneath the crust is the lunar mantle, which reaches to a depth of about 1000 km and makes up the bulk of the Moon. At the center there is probably a small, solid iron core, no more than 680 km in diameter. The center of mass (CM) is offset from the center of figure (CF) by 2 km. (Susan Pullan, Geological Survey Canada)

way to the center, most geophysicists believe that it forms a thin shell surrounding a solid metallic core. Remanant magnetism in surface rocks indicates that this may be composed of iron or iron sulfide at a temperature of about 830°C. The radius of the core has been estimated as 340 km ± 90 km. For an iron-rich composition, this represents a mere 1–3% of the Moon's total mass.

Magnetic Fields

Although the Moon probably has a small, iron-rich core, it does not possess a global magnetic field. However, there is evidence that fluid motions in the core once created a dynamo that generated a weak, global field.

Isolated magnetic fields exist, both on a local and regional scale.[5] The magnetic field strength measured at four Apollo landing sites varied from a low of 6 nanoteslas for Apollo 15 to a high of 313 nanoteslas for Apollo 16. (Earth's field is about 30,000 nanoteslas). In places, the field is strong enough to deflect solar wind particles. Older rocks are generally more magnetized, suggesting that the field was stronger in the distant past, gradually fading as the core cooled and solidified.

---

[5] **The unfortunate result is that it is not possible to navigate around the Moon using a compass.**

The presence of strong, isolated fields – producing mini-magnetospheres only 100 km in diameter – supports the theory that shock waves melted the crust, causing iron particles in the rocks to be magnetized by the global field that still existed about 3.6–3.8 billion years ago. In addition, high speed impacts would have vaporized rock, producing hot gas that was partly ionized into electrons and positive ions. The ionized gases may also have amplified the magnetic field.

Surface Resources

Future astronauts visiting the Moon for long periods will be encouraged to "live off the land" as much as possible. The most obvious local resource is the regolith, a fine-grained, cohesive mixture made up of glass spheres, dust and coarse breccia. This could be used to protect any future habitation against ionizing radiation, extreme temperatures and meteorites simply by piling it onto the roof. A straightforward sintering (heating and compressing) process could produce bricks for construction.

The composition of the regolith is generally similar to the Earth's crust, but with very different element ratios. In particular, the lunar soil is rich in refractory elements (i.e. ones with high boiling points) and low in volatile elements (i.e. ones with low boiling points).

As on Earth, oxygen (in the form of oxides) is the most abundant element on the Moon's surface. However, unlike most soils on Earth, the lunar regolith has high concentrations of sulfur, iron, magnesium, manganese, calcium, and nickel (again, often in the form of oxides).

Elements such as magnesium, iron and titanium are enriched in the surface of the maria, often in the form of the mineral ilmenite ($FeTiO_3$). The highland rocks consist mainly of the mineral feldspar, which is enriched with calcium and aluminum.

There are also traces of sulfur, phosphorus, carbon, hydrogen, nitrogen, helium and neon in the soil. The carbon, hydrogen, helium and nitrogen found in the regolith are almost entirely due to implantation resulting from constant exposure to the solar wind.[6]

An enhanced level of hydrogen found at the poles has been interpreted as evidence of trapped deposits of water ice (see below). The water ice could be used either for production of oxygen, water supply and growing crops, or for manufacturing hydrogen rocket fuel.

Solar energy will be a valuable resource, particularly on crater rims in the polar regions, which are almost permanently

Box 4.3   Lunar X-rays

It has been known for many years that the Moon's sunlit side emits X-rays. These X-rays are generated by energetic solar electrons that strike the surface and cause the atoms in the rocks to fluoresce.

The emissions can be used to determine the composition of the lunar surface. By measuring the precise energy of each X-ray, the element that is emitting it can be determined. For example, the Chandra X-Ray Observatory has been used to detect oxygen, magnesium, aluminum and silicon. ESA's SMART-1 spacecraft provided more detailed X-ray maps of the surface and made the first detection from orbit of the element calcium. By studying the abundance and distribution of such elements, scientists can gain a better understanding of how the Moon was formed.

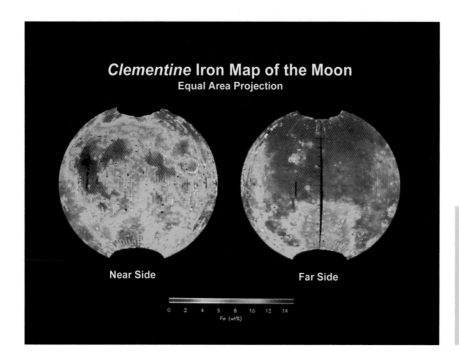

Figure 4.27 These Clementine maps show that iron-rich rocks are concentrated in the mare basalts of the near side, with very low amounts in the ancient, anorthositic rocks of the far side (with the exception of the South Pole–Aitken Basin). (NASA / LPI)

[6] The helium-3 isotope, rare on Earth, has been proposed as a future fuel for nuclear fusion reactors.

illuminated by the Sun. One "peak of eternal light" has been identified close to the lunar north pole, between three large impact craters called Peary, Hermite and Rozhdestvensky. Another is on the rim of Shackleton crater near the south pole.

Polar Ice

The lunar surface is extremely dry. This was confirmed when rocks brought back from the Moon contained no moisture and revealed no evidence of alteration by aqueous processes. On the other hand, there is no doubt that comets and asteroids have delivered vast quantities of water to the Moon over billions of years. Since Earth's oceans may have been largely derived from water delivered by large planetesimals some four billion years ago, it would be illogical to assume that Earth's neighbor escaped this hydration process.

So what happened to the Moon's reservoir of water? Most water molecules are split by sunlight into their constituent atoms of hydrogen and oxygen, then lost to space. However, there is another possibility.

Some water released from impacting comets and water-rich asteroids may migrate by "hopping" to places where it could be stable and survive over long periods. This hopping process would involve sublimation and migration in the daytime, followed by condensation on the surface at night. Eventually, a small fraction of the incoming water could be deposited in craters filled by permanent, deep shadows – the only places to escape direct sunlight and soaring daytime temperatures.

The first direct evidence that such icy deposits may exist came in 1995, when the Clementine spacecraft confirmed that substantial areas of permanent shade exist at the bottom of deep craters near both lunar poles. Their temperatures never rise above −170°C and sometimes plummet to −249°C, the coldest temperatures yet measured in the Solar System. Under such conditions, water ice could exist for billions of years.

Data from Clementine's radar experiment also suggested that the radio signals were being reflected by water ice. Similar results have been returned recently by other spacecraft, for example, India's Chandrayaaan-1 (Fig. 4.28). Radar echoes from 40 craters near the north pole displayed the signatures of water ice, with an estimated content of at least 600 million tons.

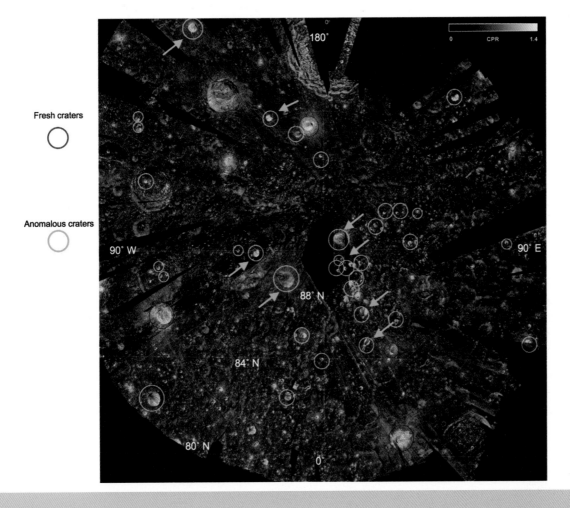

Figure 4.28 A radar map of the lunar north pole obtained with NASA's Mini-SAR radar instrument on India's Chandrayaaan-1 lunar orbiter. The 40 craters circled in green may each contain water ice at least two meters deep. (NASA)

Other supporting data have come from spacecraft such as Lunar Prospector, which found concentrations of hydrogen close to the poles. The favored explanation was that the hydrogen was derived from water ice, which appeared to be mixed in with the lunar regolith at low concentrations, perhaps 0.3–1%.

Analysis indicated the possible presence of discrete, near-pure water ice deposits buried beneath 40 cm to 1 meter of dry regolith. The hydrogen signature was stronger at the Moon's north pole, where the ice was thought to be spread over 10,000 to 50,000 square km, compared with 5 000–20,000 square km around the south pole.

Confirmation of the presence of small amounts of water near the south pole was provided in October 2009 by the LCROSS space-craft and its Centaur rocket stage, which both impacted a crater called Cabeus. Spectral measurements led to estimates of about 100 kg of water in the ejecta cloud from the newly formed crater.

Atmosphere

The Moon does have a sparse atmosphere – more correctly referred to as an exosphere. The entire atmosphere, altered to match Earth's surface temperature and pressure, would fit into a volume of 65 cubic meters. As mentioned above, the escape velocity is so low that gases are easily lost to space.

Atoms of helium, argon, neon, sodium, and potassium have been identified as present in the atmosphere. Of these, the hydro-gen and neon come from the solar wind, as does 90% of the helium, while most of the argon originates from radioactive decay of rocks in the lunar interior. The discovery of sodium and potas-sium shows certain similarities with the atmospheres of Mercury and Io.

Geological History

Geologists recognize a number of major geological periods in lunar history (see Table 4.1). This chronology is based on radio-metric dating of returned samples.

- The oldest rocks on the Moon, found in the heavily cratered highlands, are 4.4 billion years old, so they belong to the Pre-Nectarian period. Since the Solar System formed 4.56 billion years ago, this implies that the accretion of the Moon took no more than 200 million years
- The rocks found in the ejecta near ancient craters are up to 4.2 billion years old
- The rocks close to the lunar maria are 3.9 billion years old
- The basaltic rock that comprises the lunar maria is typically 3.7 billion years old
- No rocks younger than 2.9 billion years have been found, evidence that the Moon has been geologically dead for most of its history, although small-scale volcanism may have con-tinued until about 1 billion years ago.

The sequence of events began with the birth of the Moon when a large planetary embryo collided with the Earth about 4.5 billion years ago. Material from both bodies recondensed to form the Moon. The Moon was probably molten to a depth of at least 500 km during this accretion phase. As the accretion slowed, the magma ocean began to cool, allowing minerals to crystallize.

Since it was less dense than the magma, the cooled anorthosite rock began to float to the top, eventually accumulating to produce the first lunar crust. By about 4.4 billion years ago, the highland crust was in place.

After the surface of the Moon solidified, the interior heated up and melted because of energy released by the radioactive decay of elements, such as uranium and thorium. Differentiation of the interior into a mantle and a core took place, and molten lava welled up to fill the low-lying regions.

At first, during the intense bombardment by incoming aster-oids and comets, many impacts pierced and fractured the thin crust. As the crust thickened, it became heavily cratered and marked by numerous multi-ringed basins. The bombardment shattered, melted and mixed the highland crust, creating wide-spread breccias. Volcanism during this period produced lava flows of KREEP basalt, representing the final residue of the magma ocean. However, the extensive lava flows were destroyed and reprocessed or blanketed by ejecta from the incessant impacts.

Table 4.1
Lunar Geological Systems

System	Age (billion years ago)	Main characteristics
Pre-Nectarian	3.92–4.5	Basins and craters formed before the Nectaris Basin. 30 multiring basins identified.
Nectarian	3.85–3.92	Period between formation of Nectaris basin and Imbrium Basin. Includes 12 large, multiring basins and some buried maria.
Imbrian	3.1–3.85	Period between formation of Imbrium Basin and youngest known mare lavas. Includes the Orientale Basin, most visible maria and many large impact craters (e.g. Plato, Archimedes).
Eratosthenian	1.1–3.1	Includes youngest mare lavas and craters without visible rays (e.g. Eratosthenes).
Copernican	Present–1.1	Period since the formation of Copernicus crater. Includes other fresh ray craters (e.g. Tycho).

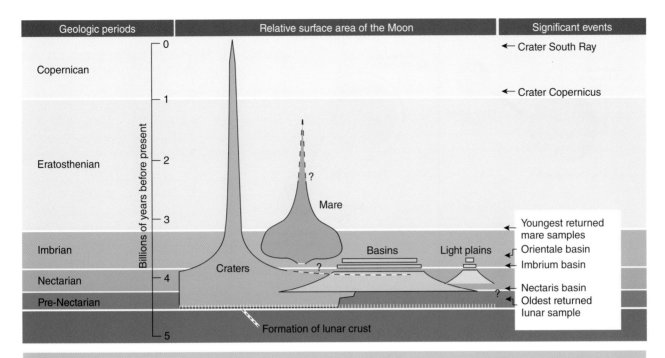

Figure 4.29 The geological history of the Moon, showing the principal events that have shaped its surface. (D Wilhelms)

About 3.85 billion years ago, a huge impact formed the Imbrium Basin, blanketing much of the near side with ejecta as the Fra Mauro formation.[7] Shortly after, the bombardment eased sufficiently for the extensive lava flows that flooded the major impact basins to retain their dark color and their record of cratering. The eruptions of basaltic lava that formed the maria continued on a large scale until about three billion years ago, after which they died down. Volcanism ceased around one billion years ago. Since then, the surface has changed very little, apart from occasional impacts that have created fresh ray craters, such as Copernicus, Kepler and Tycho.

The Birth of the Moon

Any explanation of the Moon's origin must take into account a number of unusual characteristics: the high angular momentum of the Earth-Moon system, the strange orbit inclined at 5.1° to the ecliptic, its large size and mass relative to the Earth, and a bulk density which is considerably lower than that of the four terrestrial planets.

One early theory favored the co-accretion of the Moon and Earth, in which both objects formed at the same time and in the same vicinity of the solar nebula, similar to the way the planets formed around the Sun. However, this double planet theory does not take into account the very different density and composition of both bodies. It was also difficult to account for the high angular momentum of the system and the Moon's orbital inclination.

Another possibility was that the Moon formed elsewhere and was captured by Earth during a close approach. Unfortunately,

dynamic studies have shown that the capture of such a large passing body is extremely unlikely.

The most widely accepted explanation is the single impact hypothesis. This envisages a collision about 4.5 billion years ago between a Mars-sized planetary embryo and the young Earth. There is overwhelming evidence that such large embryos existed in the early Solar System, during the final stages of planet formation.

Both the intruder and Earth are assumed to have already differentiated into a silicate mantle and an iron-rich core. The grazing blow blasted into orbit a mixture of largely vaporized material from both the impactor and Earth. Temperatures exceeded 10,000°C in some parts of the cloud.

The impactor's core was accreted by Earth within a matter of hours, as was the majority of the material from Earth's mantle. The remainder, a metal-poor mass of silicate largely derived from the impactor, rapidly collapsed to form an orbiting disk, then coalesced into a molten Moon within only a few years.

This hypothesis accounts for the geochemical evidence that at least half of the Moon was molten shortly after accretion. Since most of the material in the disk would have been white-hot vapor, the proto-Moon would have been depleted in volatiles (including water), and enriched in refractory elements (those that remain solid at high temperatures). The absence of metals in the accretion disk would explain the Moon's relatively low density. The hypothesis also accounts for Earth's relatively high spin rate and the Moon's inclined orbit.

The hypothesis assumes that the impactor came from the same broad region of the solar nebula as Earth. It also requires the Moon to condense from a debris disk orbiting quite close to

[7] Some studies of argon isotopes in Apollo 14 rock samples suggest an age of 3.75 billion years.

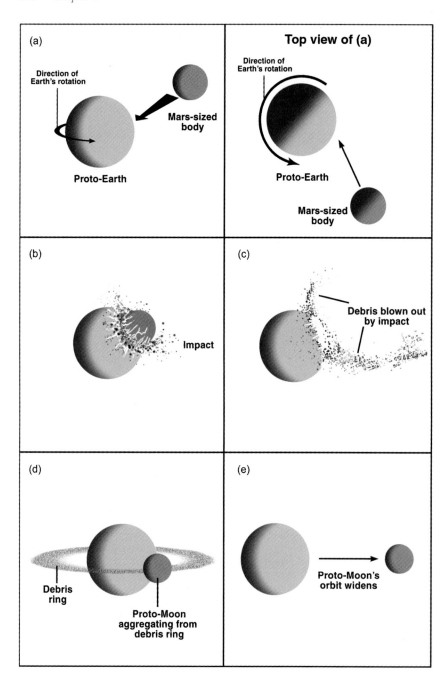

(a)

Direction of
Earth's rotation

Mars-sized
body

Proto-Earth

Top view of (a)

Direction of
Earth's rotation

Proto-Earth

Mars-sized
body

(b)

Impact

(c)

Debris blown out
by impact

(d)

Debris
ring

Proto-Moon
aggregating from
debris ring

(e)

Proto-Moon's
orbit widens

Figure 4.30 A simplified version of how the Moon might have been created by a grazing collision between a Mars-sized planetesimal and Earth. The vaporized material derived from the impactor and the Earth's mantle was ejected into orbit, but most of the iron fell back to Earth. The ejected rocky matter formed a disk that coalesces to form the Moon. (LPI)

Earth. Support for this assumption comes from laser measurements showing that the Moon is currently drifting away from the Earth at a rate of 3 cm per year.

Lunar Exploration

Some 70 lunar missions have been launched by the USA, Russia (formerly the USSR) and other countries since 1959. Activity was particularly frenetic during the 1960s, when the race to send men to the Moon resulted in both superpowers sending armadas of robotic craft to map the Moon and learn more about its surface.

The first detailed images came from three American Ranger probes which impacted the Moon 1964–1965. They were followed by various orbiters and soft-landers. Five US Surveyor craft (1966–1968) sent back panoramic and close-up images, dug trenches and conducted studies of the magnetic environment and elemental chemistry. NASA also launched five Lunar Orbiters (1966–1967) to image most of the lunar surface at high resolution.

The effort culminated with the American Apollo program, in which nine missions and 27 astronauts were sent to the Moon between 1968 and 1972. After two dress rehearsals in lunar orbit, there were six landings involving twelve crew. For safety reasons they were directed to sites fairly close to the lunar equator. The

Figure 4.31 The locations of all the successful landings on the Moon. Apollo missions are green, Lunas are red and Surveyors are yellow. Note the concentration near the lunar equator. (NSSDC)

Box 4.4 **Lunar Meteorites**

Samples of lunar rocks are available without actually visiting the Moon. Fragments ejected from the Moon by impacts are sometimes found on Earth as meteorites. It is possible to tell that they have come from the Moon by comparing their chemical compositions, isotope ratios, mineral content and textures with the samples brought back by the Apollo astronauts and automated Lunas. They are an important source of information because they come from all over the surface, including the far side. However, it is very difficult to pin down their precise place of origin. (One meteorite found in Oman has been traced back to a small crater on the Mare Imbrium.)

Any rock on the lunar surface that is accelerated by the impact of a meteoroid to lunar escape velocity or greater will leave the Moon's gravitational influence. Calculations suggest that impacts excavating craters as small as 450 meters in diameter can launch lunar meteorites. Some of this ejected material is captured by Earth's gravitational field and lands on our planet within a few hundred thousand years (sometimes much less). Other ejected material enters an orbit around the Sun, returning to strike the Earth much later. Lunar meteorites Yamato-82192/82193/86032 and Dhofar 025 remained in space for perhaps 10 and 20 million years respectively before finally landing on Earth.

The first lunar meteorite to be identified was found in the Allan Hills of Antarctica in 1981. Today, more than 130 examples are known, although many of these are fragments of larger rocks, so the number of actual meteoroids is probably about 65. The largest single example is Kalahari 009, with a weight of 13.5 kg, but others weigh only a few grams.

last three missions carried battery-powered rovers which enabled the astronauts to travel much further from their lunar module and to transport more rock samples. These missions, primarily focused on science, were sent to Hadley Rille, Cayley Plain at Descartes and the Taurus-Littrow valley. A total of 381.7 kg of rock samples were brought back, three quarters of which are still preserved for future researchers.

Although the Soviets had lost the race, they continued to show interest in the Moon. During the early 1970s, three unmanned spacecraft (Luna 16, 20 and 24) recovered 320 g of regolith from

Box 4.5   Apollo Experiments

Thirty-eight scientific experiments were conducted during the Apollo missions, 16 from the orbiting Command-Service Module and the remainder on the surface. The Apollo 11 crew deployed a solar wind collector and the Early Apollo Scientific Experiments Package (EASEP), which included a passive seismic experiment and laser retroreflector. Subsequent missions, which stayed longer on the surface, deployed the more advanced Apollo Lunar Surface Experiments Package (ALSEP). A central station equipped with a radioisotope thermoelectric generator (RTG) supplied power for the ALSEP instruments and communications so that data could be relayed to Earth.

One of the most significant data sets came from active and passive seismic experiments. The Apollo 11 seismometer returned data for only three weeks but more advanced seismometers deployed at the Apollo 12, 14, 15, and 16 landing sites transmitted data to Earth until September 1977. These were used to detect vibrations from natural moonquakes and impacts involving rocket stages or lunar modules, or small explosions set off in the vicinity of the sensors. The data provided key information about near-surface layering and the structure of the deep interior.

Figure 4.32   The Apollo Lunar Surface Experiments Package deployed by the crew of Apollo 17 at Taurus-Littrow valley. In the foreground is the Lunar Ejecta and Meteorites experiment. In the background are (L − R) the Lunar Surface Gravimeter, the Central Station, and the RTG. Note the rubbish pile at left. North Massif mountain rises in the distance (right). (NASA)

More information on the interior was provided by heat flow experiments on Apollo 15 and 17. The experiment involved drilling two holes about 2 meters into the regolith and measuring the temperature at several depths. The results indicated a heat flow 18–24% of Earth's average heat flux of 87 milliwatts per m^2.

Various instruments were used to measure the composition of the tenuous lunar atmosphere and the solar wind. A lunar dust detector measured the effects of dust accumulation, while a portable magnetometer measured the local magnetic field. (The build-up of dust proved to be much lower than expected.)

Among the more unusual experiments conducted on Apollo 17 were a surface electrical properties experiment, in which electrical signals were transmitted through the regolith and recorded on the lunar rover, and a portable gravimeter experiment.

One passive experiment is still in use today – the laser ranging retroreflectors deployed on Apollo 11, 14, and 15.[8] These consisted of a series of corner-cube reflectors that return an incoming light beam in the direction it came from. (Similar reflectors were included on the Soviet Lunokhod rovers.) Decades later, scientists still calculate the distance to the Moon by sending laser pulses and measuring the precise time taken for them to return. Results show that the Moon is moving away from the Earth by 38 mm per year and that it also wobbles a little.

The Command-Service Modules of Apollo 15, 16 and 17 also carried some remote sensing experiments. These included panoramic and mapping cameras, laser altimeters, X-ray, alpha-ray and gamma-ray spectrometers and mass spectrometers. Apollo 15 and 16 also carried a small subsatellite which was released in lunar orbit to measure gravitational anomalies, magnetic fields and the interaction of solar plasmas (charged particles) with the Moon.

[7]However, US funding for the the McDonald Laser Ranging Station in Texas was axed in 2009.

the near side, while two automated Lunokhod rovers spent many months exploring the surface.

The next major advances came when two small American spacecraft, Clementine and Lunar Prospector, went into orbit around the Moon in 1994 and 1998. Clementine produced multispectral maps that showed the composition of the surface rocks, and carried a laser altimeter to map global topography. Lunar Prospector also produced detailed maps of surface composition and gravity. Both spacecraft found circumstantial evidence for the existence of water ice in permanently shaded craters at both poles.

Since then, there has been a renewal of interest in lunar exploration. The European Space Agency's SMART-1 spacecraft used an ion drive to spiral out to lunar orbit in November 2004. On board were seven miniaturized instruments, which studied the lunar environment until the spacecraft was deliberately crashed into the Moon in September 2006.

The United States, China, Japan and India have also initiated in-depth lunar surveys. NASA's Lunar Reconnaissance Orbiter has been sampling the radiation environment, searching for water ice, making topographic maps and imaging the entire surface. The LCROSS mission sent a small spacecraft and its upper stage booster into the Moon in a successful attempt to detect water in the debris cloud.

India's Chandrayaan-1 was launched in 2008 on a year-long mission to map the Moon and deliver a small impactor to the surface. On board was a US radar instrument, which found evidence of water ice in polar craters, and a number of European instruments similar to those on SMART-1. A follow-on mission will deliver another orbiter and a lander/rover.

China also has plans to follow the launches of its two Chang'e lunar orbiters with a lander and an automated sample return mission. The ultimate objective is the delivery of Chinese astronauts to the Moon.

Having already sent two small spacecraft to the vicinity of the Moon in 1980, Japan launched its Kaguya spacecraft in 2007. Kaguya was the largest spacecraft sent to the Moon since the Apollo era. The main orbiter carried a high resolution camera, various spectrometers, a laser altimeter and a radar sounder. Three-dimensional tracking of two small subsatellites enabled the first direct gravity measurements of the lunar far side. No missions have yet been sent to the far side or the polar regions, which have quite different geological histories.

Questions

- The Apollo program is said to have revolutionized scientific knowledge of the Earth-Moon system and the Solar System as a whole. Explain why you agree or disagree with this statement.
- The Moon is generally regarded as an arid world, with a minuscule amount of water. Explain why the Moon is so dry, and why water ice is thought to exist on the Moon.
- Discuss the main theories of lunar formation and the arguments for and against each of them.
- Many countries currently have programs to explore the Moon with automated orbiters, landers, rovers and sample return missions. Explain why there is this renewal of scientific interest, despite almost five decades of intensive research.
- Describe and explain the major differences between the lunar highlands and the maria.
- Describe and explain the various impact structures that are found on the Moon.
- How important has volcanism been during the lifetime of the Moon?
- How have scientists been able to learn about the lunar interior? What evidence is there for an iron-rich core?

FIVE
Mercury

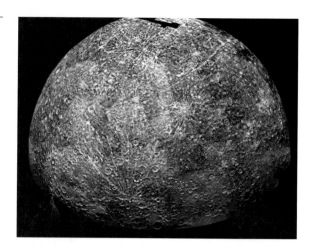

At first glance, the innermost planet in the Solar System, Mercury, looks like a close cousin of the Moon. Both are small, rocky worlds that are covered in impact craters. With no atmosphere worth mentioning, surface temperatures soar in the daytime, but plummet at night. However, first glances can be deceptive, particularly where planetary studies are concerned.

Of the five planets visible to the naked eye, Mercury is the most elusive since it is always low above the horizon and close to the Sun in the sky. Its existence has been known since at least the 3rd millennium BC, when the Sumerians and Babylonians kept careful records of its periods of visibility and motion across the sky. Yet, although this little world has probably been recognized since prehistoric times, it has remained one of the most mysterious inhabitants of our Solar System until recent years.

Apparitions of Mercury

Mercury is one of the four terrestrial planets, regularly approaching to within 80 million km of Earth – closer than any planet other than Venus and Mars. At such times it reaches magnitude –1.3 and rivals Sirius, the brightest star in the sky.

To an observer on Earth, the planet becomes visible for about two months as a "morning star," appearing just before dawn, to the west of the rising Sun. It then disappears from view, only to reappear a few weeks later as an "evening star," which lingers for another two months. The cycle then begins again. During 2011, for example, Mercury could be observed during four different apparitions in the morning, separated by three evening apparitions.

Even at its greatest elongation (its angular distance from the Sun in the sky) Mercury never moves more than 28° from the Sun –a little more than the width of fully spread hand held at arm's length. At such times, it appears above the horizon for about two hours after sunset or before sunrise although for much of this short window of opportunity it is invisible in the solar glare. However, Mercury's highly elliptical orbit means that some elongations are less favorable, with the planet straying only 18° from the Sun.

Orbit

Mercury is termed an "inferior planet," because it orbits nearer to the Sun than Earth. Indeed, its twilight apparitions and rapid motion across the sky show that it is closer to the Sun than any of the other planets. Its mean distance from the Sun is only 57.9 million km, or about 0.39 AU. However, its orbit is much more eccentric than those of the other planets.

As a result, it approaches to within 46 million km of the Sun at perihelion, while its furthest distance (aphelion) is about 69.8 million km – half as far again. This unusual orbit means that an observer on Mercury would see the Sun grow from twice to three times its apparent size, as seen from Earth, over a period of just six weeks.

Not only is the orbit far from circular, but it is inclined by 7° to the ecliptic, so that it dips far above and below the paths of all the other planets. The reason for Mercury's unusual orbital inclination and eccentricity is linked to periodic, long-term oscillations in its orbital characteristics, caused by perturbations from the other planets.

In order to avoid being pulled into the Sun, Mercury must also travel around its orbit faster than the other planets – in accordance with Kepler's third law of planetary motion (see Chapter 1).

Exploring the Solar System, First Edition. Peter Bond.
© 2012 by Peter Bond. Published 2012 by Blackwell Publishing Ltd.

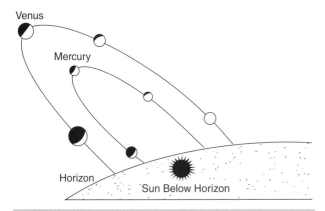

Figure 5.1 The changing positions and phases of Mercury and Venus in the evening sky. (NASA)

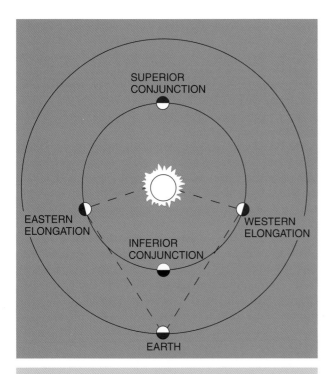

Figure 5.2 Key orbital positions of Mercury. When Mercury (or Venus) is between the Sun and Earth, it is at inferior conjunction. Unless it is transiting the Sun's disk, it is then invisible. On the opposite side of the Sun from the Earth, Mercury (or Venus) is at superior conjunction. The planet is most easily observed at greatest elongation, when it is furthest from the Sun in the sky. (NASA)

By traveling at an average speed of 48 km/s, it is able to overtake the more distant planets on the inside track, completing one journey around the Sun in only 88 Earth days. Mercury finishes four orbits before Earth completes one. However, the elliptical orbit means that its orbital velocity varies greatly, ranging from a

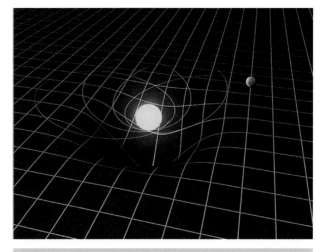

Figure 5.3 Mercury's elliptical path around the Sun shifts slightly with each orbit, so that its closest point to the Sun (perihelion) moves forward (precesses). Newton's theory of gravitation predicted an advance only half as large as the one actually observed. According to Einstein's theory of relativity, the precession of Mercury's orbit is caused by the massive Sun warping the space around it. (Hong Kong Space Museum)

relatively modest 39 km/s at aphelion to 56 km/s around perihelion.

Another orbital peculiarity is the substantial forward drift of the position of its perihelion. All of the planets undergo some shift of perihelion as a result of the gravitational influence of their neighbors. However, the advance of Mercury's perihelion, known as precession, is by far the largest of any planet.

In fact, the motion is still very small – less than 2° per century. At this rate it would take a quarter of a million years for Mercury's perihelion position to complete a full circle around the Sun. However, even this rate of precession is too large to be explained by Newtonian physics. When the effects of Earth's precession (the changing orientation of its axis in space) and the gravitational pull of the other planets are taken into account, 43 arc seconds of Mercury's perihelion movement remain unexplained.[1]

One possibility, put forward in 1859 by Urbain Le Verrier, was that an unknown planet was pulling Mercury off its path. (Neptune had been discovered only 13 years earlier as the result of a similar calculation.) However, we now know that such a planet does not exist.

The explanation had to wait until Albert Einstein published his General Theory of Relativity in 1915. One of its predictions was that massive objects such as the Sun significantly warp the space around them. In the case of Mercury, the planet is traveling in space that is much more curved at perihelion than at aphelion. The predicted advance of Mercury's perihelion by moving through this warped space is 0.1 seconds of arc per orbit (43 arcsec per century), exactly in agreement with the observed value.

---

[1] All of the planetary orbits precess. Newton's theory of gravity predicts these effects, which result from the planets pulling on one another. With the exception of Mercury, this precession can be understood using Newton's equations.

(One second of arc = 1/3600 degrees). The success of this prediction has been cited as one proof of the validity of Einstein's groundbreaking theory.

Phases

When seen through a telescope, Mercury, like Venus and our Moon, appears to go through phases (see Fig. 5.2). At superior conjunction, on the far side of the Sun from Earth, it appears as a full disk, though it can never be properly viewed because of the intervening glare of the Sun.

As it steadily approaches our world, Mercury displays a smaller illuminated area but grows in apparent size. At greatest elongation, the planet resembles a quarter Moon, then, as it comes ever nearer to the Earth, it shrinks to a slim crescent. By the time it reaches inferior conjunction no part of the visible surface is bathed in sunlight and the planet cannot be seen – unless it happens to transit the Sun.

Transits

Although it is the closest planet to the Sun, transits of Mercury occur only a dozen times or so per century. This is because of the 7° inclination of the planet's orbit to the ecliptic, which means that Earth, Mercury and the Sun are rarely aligned precisely when the innermost planet is at inferior conjunction.

2003-May-07
05:27:18
dt = 45.0

Figure 5.4 Two views of the transit of Mercury on May 7, 2003. The time sequence (top) was obtained by the SOHO spacecraft, while the image of Mercury against the Sun's photosphere (bottom) was obtained by TRACE. The entire transit lasted about 5½ hours. (NASA/ESA)

Transits of Mercury can only occur close to 8–9 May or 10–11 November, when its orbit crosses the ecliptic – either at the ascending or descending node (see Chapter 1). The May events take place when Mercury is near aphelion, while the autumn transits occur near perihelion. Although the latter events are twice as frequent as those in May, they last for a shorter time. The interval between successive crossings varies from 3 to 13 years, most recently with transits in May 2003 and November 2006, followed by another pair in 2016 and 2019.

The sharp outline of Mercury's disk during transits – in contrast to the blurry outline of Venus – provided important evidence that the planet does not possess much of an atmosphere. They also made it possible to determine its orbit and diameter.

Spin-Orbit Resonance

Until the early 1960s, it was thought that Mercury was locked into a synchronous rotation which caused the planet to keep its same face toward the Sun – just as the Moon does when it orbits Earth (see Chapter 4). Under these circumstances, its rotation period would last 88 days, the same length as its year. One side of Mercury would always be exposed to the heat of the Sun, with the other hemisphere in perpetual darkness. Drawings and maps made during favorable viewing opportunities suggested that the same surface was always visible.

The first doubts arose in 1962, when studies of radio waves from Mercury suggested that the night hemisphere was warmer than expected. Then, in 1965, Gordon Pettengill and Rolf Buchanan Dyce used the 330 meter Arecibo dish in Puerto Rico to bounce radio waves off the planet's surface and detect the echoes. These Doppler radar observations showed that Mercury's rotational velocity was consistent with a rotation period of about 59 days, a figure subsequently refined to 58.65 days.

The error of early observers was soon explained by Giuseppe "Bepi" Colombo, a specialist in celestial mechanics, who noted that the 59 day rotation period was two thirds of the Mercurian year. In other words, Mercury rotates three times during two orbits of the Sun – a type of spin-orbit resonance found nowhere else in our Solar System.

This 3:2 ratio explains how the telescopic observers came to wrong conclusion. After three complete orbits, the same face of the planet is visible during the same phase, so observations made at the most favorable viewing times frequently revealed the same surface features. Presented with familiar, dusky markings at many apparitions, astronomers tended to ignore any occasional variations and jumped to the wrong conclusion.

In order to explain how Mercury entered this unusual, but stable, resonance, most studies assume that it originally rotated much faster than it does today. A rapid rotation rate – perhaps as short as eight hours – would cause the planet to bulge outwards at the equator. A combination of tidal friction – when the Sun raises tides in the solid body of the planet – and differential motions in the liquid core relative to the mantle would enable the Sun's gravity to slow the rotation.

Recent studies have also shown that the orbital eccentricities and inclinations of the inner planets are very unpredictable on timescales of millions of years. This makes it likely that Mercury's orbit was even more eccentric in the past, varying from near zero to 0.45 or more.

Research by Correia and Laskar shows that, when the eccentricity exceeds a value of 0.325, solar tidal effects lead to corresponding changes in the planet's spin rate. The result is that some resonant states – including the 3:2 spin–orbit resonance – are passed through many times, so that the probability of eventual capture into that resonance is greatly increased. Indeed, models suggest that, over a four billion year period, the 3:2 spin-orbit resonance becomes the most likely outcome.

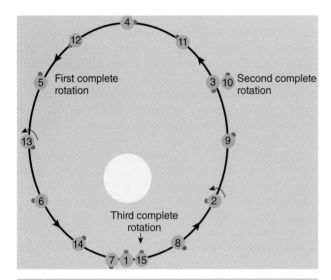

Figure 5.5 The spin-orbit coupling of Mercury means that Mercury rotates three times during two orbits of the Sun. (Calvin Hamilton)

Two-Year Days and Double Sunsets

The spin-orbit resonance means that Mercury's solar day (from sunrise to sunrise) is equal to two Mercurian years, or 176 Earth days. As a result, it is the only planet to have its day longer than its year. The unusual spin-orbit coupling, combined with the high orbital eccentricity, would produce some very strange effects for anyone standing on Mercury's surface.

At the 0° and 180° meridians, an observer would see the Sun rise above the eastern horizon and then gradually increase in apparent size as it slowly moved toward the zenith over the next 1½ months. During perihelion, at local noon, the Sun would slow to a complete halt in its westward motion across the sky and then move a little in a retrograde direction (eastward) for eight days.

The reason for this reversal is that the orbital angular velocity exceeds the spin angular velocity near perihelion. In other words, Mercury's orbital motion is so rapid around perihelion that its already slow rotational motion is unable, in effect, to keep up.

The Sun's disk – three times larger than on Earth – would dominate the sky, remaining almost overhead for several weeks before resuming its path toward the horizon and decreasing in apparent size. The period of continuous daylight would last 88 days.

Observers elsewhere on Mercury would see different, but equally bizarre motions during a single perihelion passage. Someone standing at the 90° longitude would see a double sunrise, while another at 270° (on the opposite side of the planet) would witness a double sunset. At the next perihelion, the situations would be reversed.

The first sunrise occurs, as usual, when the planet's prograde spin brings the Sun above the horizon. However, near perihelion, the planet's orbital speed exceeds its spin rate, bringing the

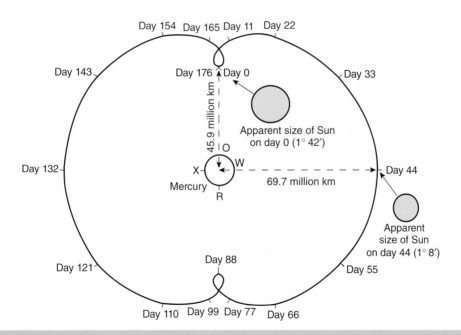

Figure 5.6 The Sun as seen by an observer on Mercury. The Sun's disk appears 66% larger at perihelion, so the surface is much hotter at such times. The Sun's motion across the sky appears to reverse direction for a time around perihelion. (Peter Francis)

observer back into darkness as the Sun sets. Once the planet is past perihelion, the spin rate becomes the dominant factor once more, causing a second sunrise.

Hot Spots

Mercury's spin axis is almost perpendicular to the plane of its orbit (unlike most of the other planets), so there are no seasonal variations with latitude. Every part of the surface, apart from some deep, permanently shaded, craters at the poles, suffers from lengthy exposure to daytime heat and extreme cold at night.

Its proximity to the Sun means that Mercury receives up to 11 times more solar radiation than Earth, so its day side is much hotter than a domestic oven set at maximum. The slow rotation period and the lack of an atmosphere mean that Mercury has the harshest thermal environment of any planet in the Solar System, with a day-night variation of up to 610°C. The noon temperature on the sunlit side at perihelion soars to 430°C, hot enough to melt zinc, tin and lead. With no atmosphere to transfer warmth from the day side, heat radiates into space during the long night and the temperature plummets to −183°C just before dawn.

However, the spin-orbit coupling means that some parts of the planet, centered on 0° and 180° longitude, suffer more extreme thermal conditions than others. Since the planet rotates 1½ times during each orbit, these longitudes face the Sun at alternate perihelion passages. Hence, the hemisphere centered on 0° longitude is subjected to prolonged, maximum heating by the noon Sun at one perihelion. This experience is then repeated at 180° longitude during the following perihelion.

The places where the Sun's zenith coincides with its closest approach are sometimes known as "hot poles" or "hot spots". These regions receive 2½ times more radiation overall than places at 90° and 270° longitude, which suffer the noonday Sun when the planet is near aphelion. As a result, maximum temperatures are noticeably lower, so the 90° and 270° sub-solar points are often called the "warm poles."

Images returned by spacecraft show that the 180° hot spot coincides with the location of a gigantic impact basin, appropriately known as Caloris ("Hot"), while its antipodal equivalent is marked by a "peculiar terrain" of hills and linear features.

Surface Observations

Until the visits of Mariner 10 in 1974–1975, very little was known about Mercury's surface. Apart from the difficulties caused by its phases and twilight apparitions (see above), observations were also complicated by its modest size. Mercury is the smallest planet in the Solar System. With a diameter of 4878 km, it is dwarfed by Jupiter's moon Ganymede and Saturn's moon, Titan. Even at its closest to Earth, Mercury's apparent diameter is only about 11 arc seconds, less than all but Uranus and Neptune.

As a result, features less than about 300 km across cannot be resolved in even the largest telescopes. Not surprisingly, the best maps of Mercury until 1974 merely showed vague variations in shading, though irregular markings along the terminator – the boundary between night and day – suggested a mountainous or cratered surface.

Evidence that Mercury's surface might resemble that of the Moon was supported by similarities in the way it reflected sunlight and radar pulses, and in its emission of infrared and radio waves. These suggested a similar blanket of fragmented material generated by eons of meteorite bombardment.

Some observers noted whitish features that they described as mists or thin clouds. Others reported spectroscopic observations

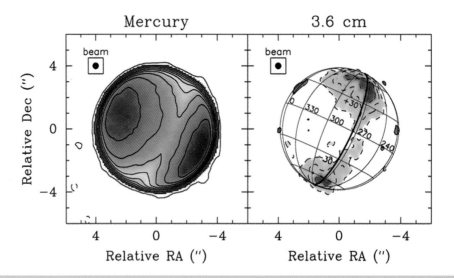

Figure 5.7 Radio emissions at a wavelength of 3.6 cm show the two hottest regions on Mercury (dark red in left image). The hot spots appear in both the day (left) and night (right) hemispheres, and coincide with regions which receive maximum solar radiation around perihelion. The Caloris Basin is located near the spot on the right – hence its name. The spot at left lies in a region of "peculiar" hilly terrain. The right hand image shows cooler (blue) areas at the poles and along the morning terminator (bold line, Sun is to the right). They are caused by shadows cast by hills and craters. (David Mitchell)

of a thin atmosphere. However, any gaseous blanket was clearly very sparse, since no blurring of the planet's limb is observed during transits. Furthermore, with a mass only 5% that of the Earth and an escape velocity of 4.25 km/s it seemed clear that Mercury could not hold on to a substantial atmosphere, even if it originally possessed one. Under these circumstances, the sky would appear black and full of stars even in the middle of the day, and there would be no natural protection from the Sun's ultraviolet radiation.

Size, Mass and Density

Prior to the Space Age, the absence of a satellite made it difficult for ground observers to determine Mercury's mass or calculate its density. One of the most precise estimates was obtained by studying the way the planet perturbed the large asteroid Eros. The problem was finally solved by Mariner 10, which confirmed that its diameter is only 4878 km, about one-third the diameter of Earth. Approximately 18 Mercurys would fit inside our world.

Measurements based upon the way it modified Mariner 10's trajectory also showed that Mercury is very massive for its size. Although it is 30% smaller than Mars, this is compensated by its large mass, which results in the two planets having similar surface gravity and escape velocities. Similarly, although its diameter is only 40% greater than that of the Moon, its surface gravity is almost twice that of Earth's satellite.

Calculations of Mercury's density – obtained by dividing its mass by the volume – were even more surprising. The planet's average density (5.4 g/cm^3) is second only to Earth in the Solar System. However, materials are much more compressed at the center of the larger Earth, thus increasing their density. Mercury would have a density of 5.3 g/cm^3 in an uncompressed state, compared with only 4.4 g/cm^3 for Earth (see Chapter 1).

The Iron Planet

How can Mercury have the greatest uncompressed density of any planet? The only possible explanation is that it contains a lot of dense, metallic elements. Since iron is the most abundant heavy element in the Solar System, and seismic data indicate that it is the dominant constituent of Earth's core, it seems reasonable to suppose that Mercury is also largely composed of iron. This theory is supported by the presence of a magnetic field and by studies of the probable composition of the early solar nebula.

Iron probably accounts for over 60% of Mercury's mass, with rocky, silicate-rich material accounting for the remainder. This is twice as much iron per unit volume as any other planet or satellite. Assuming that most of this metal is concentrated in the planet's center, the iron core fills about 75% of Mercury's diameter and accounts for about 42% of its volume.

In absolute terms, the iron core is probably about 3600 km across, and overlain by a silicate mantle and crust. This rigid outer shell, or lithosphere, is approximately 600 km thick. Prior to the arrival of NASA's MESSENGER spacecraft, many scientists expected the lithosphere to be mainly composed of anorthosite, a type of igneous rock that is made of calcium-rich feldspar, one

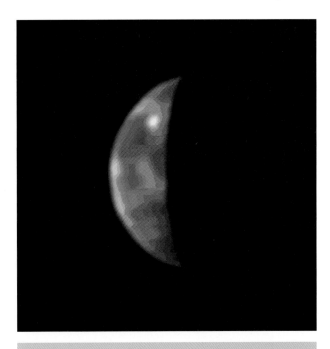

Figure 5.8 One of the best ground-based images of Mercury, obtained at the Mount Wilson Observatory in August 1998 by Boston University astronomers. This view shows a region not imaged by Mariner 10. Hundreds of thousands of pictures taken with short time exposures (1/60 second) were examined to find the 30 images with the clearest surface markings, taken during instances of "perfect seeing." There was little obvious correlation between bright features in the visible image and radar-bright features. (Baumgardner et al., Boston University)

of many common silicate minerals found on Earth and the Moon (see Chapter 4). However, MESSENGER's instruments have shown that its surface is not dominated by feldspar-rich rocks.

The existence of such a massive, metallic core is a major consideration for any theories attempting to explain the planet's origin. It also has a significant bearing on the most surprising discovery by Mariner 10 – the presence of a significant magnetic field.

Origin

How did an iron planet come about? It is generally accepted that the planets accreted from a solar nebula of gas and dust, and that the composition of this nebula varied with temperature. However, the fact that non-volatile, rocky elements would be expected to condense in the hotter, innermost part of the nebula is not sufficient to account for Mercury's huge iron core.

There are three main theories to explain why Mercury is so different from Venus, Earth and Mars. Each theory predicts a different composition for the rocks on Mercury's surface.

In the selective accretion model, drag by nebular gas favored accretion of dense particles as the planet formed from the cloud. These particles (such as metallic iron) condensed and were preferentially retained in the innermost regions near the

Box 5.1 Mariner 10

Until recently, the source of most of our knowledge about Mercury was the Mariner 10 spacecraft, which flew past the planet on three separate occasions 1974–1975.

Figure 5.9 Mariner 10 science payload. Visible at the bottom are the steerable high gain antenna and the large sunshade. The "eyes" at the top are the TV cameras. The magnetometers that discovered Mercury's magnetic field are on the long boom (right). (NASA-JPL)

Launched on November 3, 1973 – Mariner 10 swept past Venus on February 5, 1974. The extremely accurate flyby, at a distance of 5768 km, modified the spacecraft's trajectory and speed, putting it on course for an encounter with Mercury.

Its first images of half-lit Mercury were acquired on March 24, 1974, from a distance of 5.4 million km. As the range closed, it became clear that the planet did, indeed, have an intensely cratered surface that closely resembled the ancient lunar highlands. To everyone's surprise, the plasma and particle data indicated the presence of substantial magnetic field. Closest approach, at a distance of 703 km, took place on the planet's night side. Further mosaics of a different hemisphere were returned on the outward leg.

The spacecraft's orbit was such that, with only a few minor adjustments, it could make repeated flybys of Mercury. The only drawback was that these took place at the same point in Mercury's orbit, when the same hemisphere was illuminated. (Mariner's orbital period was almost exactly twice Mercury's period – see spin-orbit coupling above.)

The second encounter was dedicated to imaging the southern hemisphere, while the third flyby primarily concentrated on magnetic field and particle measurements. In order to image the entire southern hemisphere, terrain that would link the two regions seen during the first encounter, the flyby on September 21, 1974, took place on the sunlit side at an altitude of 48,000 km. Although their spatial resolution was a modest 1–3 km, the 360 images extended the coverage of the illuminated hemisphere from about 50% to 75%.

The third and final encounter occurred on March 16, 1975. Closing to only 327 km, the spacecraft was expected to send back high resolution images of some intriguing geological features. Unfortunately, a problem with the ground station in Canberra, Australia, meant that only one quarter of each picture could be received. Nevertheless, the measurements of the magnetic field confirmed that Mercury generates an internal field that is weaker, but similar to, that of Earth.

Altogether, more than 2300 black and white images of Mercury were taken, many at moderate resolution (3–20 km/pixel) but a limited number with a resolution of only 140 m. Altogether, 45% of the planet was imaged.

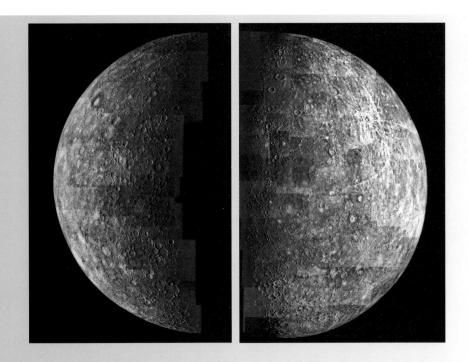

Figure 5.10 Mariner 10's first close encounter with Mercury took place on March 29, 1974. The left hand mosaic was assembled from 18 photographs taken at 14-second intervals at a distance of about 200,000 km, six hours before closest approach. The outgoing view (right) is composed of 18 images taken six hours after closest approach at a distance of 210,000 km. The most notable feature, on the sunrise terminator, is the giant Caloris impact basin. (NASA-JPL)

Figure 5.11 Relative sizes of the terrestrial (rocky) planets – left to right, Mercury, Venus, Earth and Mars. Mercury is much smaller than its neighbors, but has a surprisingly high bulk density. (LPI)

Sun, whereas the lighter particles were swept away. As a result, Mercury became enriched in iron and other metals, but there was no comparable enrichment in silicate minerals. In this case, surface rock composition would be similar to that of the other terrestrial planets.

Another theory suggests that intense radiation from the young Sun (during its T Tauri stage) vaporized much of the silicates in proto-Mercury, while allowing the planet to hold on to its metal-rich core. This idea predicts a surface enriched in aluminum, calcium, and magnesium, but poor in easily vaporized elements

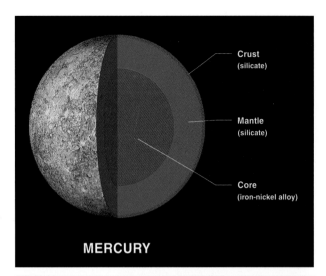

Figure 5.12   The interior structure of Mercury is dominated by a metallic core which has a diameter about 75% that of the entire planet. This accounts for its surprisingly high bulk density, 5.44 times denser than water, which is second only to the much larger Earth. The reason for this remarkably large core remains uncertain, but it is clearly related to the surprising strength of Mercury's magnetic field. (LPI)

like sodium and potassium. It also predicts rocks low in ferrous iron (FeO).

The third theory suggests that a giant impact took place between the young Mercury and a large planetesimal. This proto-Mercury may have consisted of a small iron core and larger rocky (silicate) mantle, with at least twice the mass of Mercury today. A high speed collision would strip away most of the primordial crust and upper mantle, leaving a much smaller planet that is dominated by its iron core. Under these conditions, we would expect the present-day surface to be made of rocks highly depleted in those elements (silicon, aluminum, and oxygen) that were concentrated in the primordial crust and mantle.

With our current knowledge of Mercury's crustal composition, choosing between these options is not possible, although there is some spectroscopic evidence that FeO content is low (approximately 3%) MESSENGER data also indicate less aluminum and more magnesium than in lunar rocks and terrestrial basalts. Mercury does not seem to be depleted in volatile elements, such as potassium. However, support for the giant impact hypothesis has been provided by computer simulations, which suggest that the final stages of terrestrial planet evolution involved huge collisions with marauding planetesimals.

Magnetic Mercury

Mariner 10 was expected to find a geophysically dead world. A planet without a liquid core should not be able to support the fluid, internal dynamo that generates a magnetic field, and Mercury's small size and slow rotation indicated a Moon-like world with a metallic core that had cooled and solidified long ago. (Small planets have a high surface area compared with their

volume. Hence, other factors being equal, smaller bodies radiate their internal heat to space more rapidly. This is why a pebble cools down faster than a large rock.)

However, Mariner 10 detected a weak magnetic field (about 300nT) with a strength at the surface about 1% of Earth's magnetic field. With the exception of periods of high solar activity, this is sufficient to deflect the solar wind and prevent it from reaching the planet's surface. However, major, rapid fluctuations are caused by gusts in the solar wind, solar flares and coronal mass ejections.

Data from NASA's MESSENGER mission confirm that Mercury's field is dipolar and tilted about 10° to its spin axis. This means that on a global scale the planet behaves as if there is a giant bar magnet at its core. However, unlike on Earth, the magnetic equator on Mercury is offset about 480 km (0.2 Mercury radii) north of the geographic equator. The polarity is the same as Earth's, with the south magnetic pole in the northern hemisphere. A marked bow shock occurs above the sunward side of the planet, where the solar wind is slowed as it comes into contact with the planet's magnetosphere.

Observations of low energy solar wind electrons reveal that the magnetosphere resembles a scaled down version of Earth's magnetic bubble. Although the velocity of the solar wind is not much faster at Mercury than at Earth, the Sun is much closer, so its density is eight times greater. When combined with a weaker planetary magnetic field, this compresses the dimensions of the magnetosphere to about 5% that of Earth.

The solar wind compresses the field so much on Mercury's sunward side that there is no possibility of radiation belts like those of Earth or Jupiter. Solar wind particles are guided around the planet and sent down the magnetic tail rather than trapped. Like Earth, the magnetotail is divided by a magnetically neutral sheet.

The magnetic fields of terrestrial planets are thought to be produced by an electrically conducting, fluid outer core surrounding a solid inner core. Hence, if Mercury's field is generated by an internal dynamo, it is likely that part of the iron-rich core – probably an outer shell – is molten.

Recent evidence suggests that Mercury may, indeed, possess a lingering remnant of a molten core. Jean-Luc Margot of Cornell University and colleagues used radar to precisely measure variations in Mercury's rotation. A radar pulse was repeatedly beamed at Mercury from the 70 meters antenna at Goldstone, California, and the reflected signal was picked up at both Goldstone and the 100 meters antenna at Greenbank, West Virginia, 3200 km to the east. Matching up the distinctive pattern in the signal received at each station, they could gauge the time lag of reception between stations and thus calculate the rotation rate to an accuracy of one part in 100,000. The results showed that the planet is unlikely to be solid throughout.

Since there is no evidence for local magnetic anomalies – like those found on both the Moon and Mars – that would signify patches of magnetized crust, it is almost certain that the internal dynamo is still active. However, the northerly location of the magnetic equator means that the dynamo mechanism in the outer core has a strong north-south asymmetry.

This raises the question of how the core has managed to remain molten for more than four billion years. There are several possibilities:

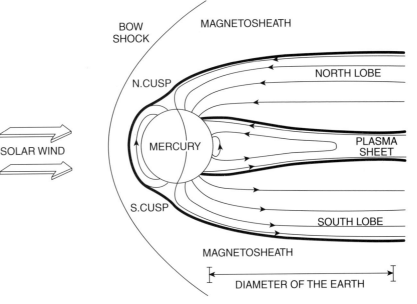

Figure 5.13 Despite its slow rotation, Mercury has a magnetic field which is stronger than those of Venus, Mars or the Moon. Its strength at the equator is about 300 nanoteslas, or about 1% of Earth's surface magnetic field. Although the density of the solar wind is 10 times higher at Mercury than Earth, the modest magnetic field is usually strong enough to deflect most of the particles and create a teardrop-shaped magnetosphere which is 5% the size of the magnetic bubble surrounding the Earth. The dipolar field is inclined about 10° to the rotational axis and centered about 0.2 Mercury radii, or 480 km, northward of the planet's center. (NASA-JPL)

- The core is enriched in the radioactive elements thorium and uranium, which provide additional heat.
- The mantle prevents heat easily escaping to the surface.
- The presence of an element such as silicon or sulfur in the core lowers the melting point of iron.

The third of these is considered to be the most likely explanation. Calculations indicate that, with a sulfur abundance of 0.2% or less, the entire core should now be solid. At the other extreme, an abundance of 7% sulfur would enable the entire core to remain fluid at the present time. The most likely situation is somewhere between the two extremes, with a pure iron inner core and an outer core of iron and sulfur, with a temperature of around 1000°C.

A recent laboratory study suggests that the convection currents in the outer core may be the result of differentiation of the molten, iron-sulfur mixture. As the liquid slowly cools, iron atoms condense into "flakes" that fall toward the planet's center. When the iron sinks and the lighter, sulfur-rich fluid rises, convection currents are created that power the dynamo and produce the magnetic field.

A Varied Surface

Until the MESSENGER mission, little was known about the surface properties of Mercury. Most of our knowledge was inferred from ground-based optical, thermal infrared, and radio observations. Such techniques relied on comparisons with the known properties of lunar and terrestrial rocks and soils.

Earth-based radar provided the first indications that Mercury is covered in craters and closely resembles the Moon. Subsequent radar observations showed small, highly reflective regions at the poles that may be related to surface deposits – possibly water ice that survives in deep, permanently shaded craters.

On a much smaller scale, the reflection of sunlight as viewed from different angles suggests a surface layer composed of fine,

dark granular rock. Thermal infrared observations from Mariner 10 also showed that the surface is a good insulator, probably covered by a layer of fine silicate dust and sand at least tens of centimeters thick.

As is the case with the Moon, this mantle of loose material, or regolith, is thought to have been formed by the repeated pulverizing of surface rocks by the high-speed impacts of micro-meteorites. However, projectiles, including microscopic dust, strike Mercury much faster than the Moon, presumably resulting in a surface that contains more glass and material condensed from impact-produced vapor.

Not everywhere is the same. Infrared observations also show areas of compacted soil, boulders or bedrock, indicating places where the regolith is much thinner or where fresh, unweathered surfaces are exposed.

The brightest areas are found on the floors of a few craters, where some 50% of the incident sunlight is reflected. The darkest features are not as dark as the basaltic maria on the Moon (see Chapter 4), but the overall albedo of many regions is comparable to that of the heavily cratered lunar highlands, which reflect about 15% of the incoming sunlight.

Remote sensing observations indicate that Mercury has a silicate crust that contains minerals such as feldspars, pyroxenes and olivine. However, many questions remain. The amount of ferrous (iron) oxide (FeO) tied up in the surface materials is of particular interest, since it has important implications for theories attempting to explain the formation of a planet with a huge iron core (see above).

For many years it was believed that the surface rocks contained no more than about 6% of iron by weight, which is comparable to the lunar highlands and Earth's continental crust, with an average content of only 2–3%. It seemed that the average Fe abundance in the surface rocks, and by inference the crust and mantle, was the lowest of all the terrestrial planets. This was very difficult to explain for an iron planet.

However, this picture has now been overturned by data from the neutron spectrometer on MESSENGER, which has shown

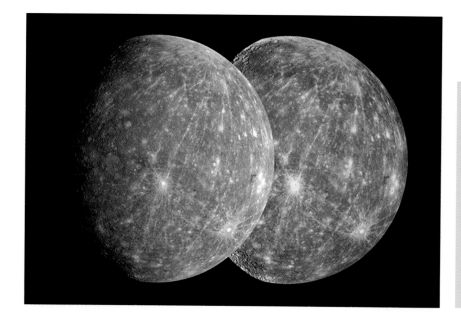

Figure 5.14 Two views taken during MESSENGER's second Mercury flyby. The image at left is approximately true color, the image at right accentuates subtle color differences. The bright Kuiper crater lies near the center of the image. Bright rays extend over 4500 km from the relatively young Hokusai crater in the north. Debussy, another prominent rayed crater, lies south east of Kuiper. (NASA/JHU-APL/Carnegie Institution of Washington)

that the outer silicate shell of Mercury has more iron than previously believed. The low amount of neutrons escaping from the surface indicates a high abundance of iron and titanium – elements that efficiently absorb these elementary particles. The iron and titanium content is similar to that found in the basaltic rock of the lunar maria.

Recalibration of Mariner 10 images by Mark Robinson and Paul Lucey generated the first three-color images that gave tantalizing clues to the elements, minerals and other properties of the surface materials. Some images suggested the presence of lava flows, pyroclastic (fine ash) volcanic deposits and distinctive subsurface layers. Clearly, Mercury's surface is compositionally heterogeneous and complex.

Much of the planet is now known to have been resurfaced by volcanic activity, implying that large sources of magma once existed in Mercury's upper mantle. The most glaring example is the Caloris Basin, which is completely filled with smooth plains material that appears to be very similar to the mare basalt flows on the Moon (see below).

Multispectral imaging by MESSENGER has confirmed that the surface displays three major color units: relatively high-reflectance smooth plains, average-reflectance cratered terrain, and low-reflectance material.

The widespread low-reflectance material is particularly enigmatic. It is found in the giant Caloris and Tolstoj basins, as well as in southern, heavily cratered regions. In some places, impacts have excavated the material from depth and then thrown it beyond the newly formed crater as an ejecta blanket. On the Moon, dark volcanic rocks in the maria have a high abundance of iron-bearing silicate minerals, but neutron spectrometer results show that the iron on the surface of Mercury rarely occurs in silicates. One possible solution is that the iron is hidden within the chemical structure of minerals such as ilmenite (FeTiO$_3$). However, this is hard to explain, since such minerals are not common on the other planets.

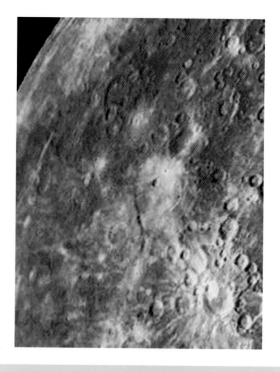

Figure 5.15 This false color image, centered at 12°S 31°E, was created from recalibrated Mariner 10 data. Distinct color units provide clues about mineral abundance and the nature of surface deposits. Here, the small, bright Kuiper crater (lower right) shows fresh material excavated from rock that may have an unusual composition. The fairly dark, blue areas (top) suggest enhanced titanium content. The bright red areas may represent primitive crustal material, while the brown plains (lower left) are probably lava flows. (Mark Robinson, Northwestern University/NASA-JPL)

Surface History

The history of Mercury seems to be very similar to that of the Moon (see Chapter 4). After the early period of planetesimal accretion, a thin crust formed over the molten planet. This initial crust may have been removed by a giant impact or vaporization in the hot solar nebula, but a thicker crust soon formed as the planet gradually cooled.

During the subsequent heavy bombardment, the planet experienced internal differentiation. Elements of comparatively low density, such as aluminum, rose into the upper crust. At the same time, iron and other heavy elements sank toward the center and formed a massive core.

The impact craters and basins which can still be seen today reflect the final phase of the late heavy bombardment, which may have ended about 3.8 billion years ago. These features were then partly covered by eruptions of magmas of varying composition over an extended period of time.

The record of gigantic impact events is clearly visible today. One of the earliest examples is the 500 km diameter Tolstoj Basin in the southern hemisphere. This was followed by the most dramatic impact of all, which formed the Caloris Basin. Shortly after the Caloris impact, massive extrusions of flood lavas formed broad, smooth plains.

Thereafter, the number of impacts decreased rapidly. During the last three billion years, Mercury's surface has continued to be modified by small volcanic eruptions and occasional impacts. The current low rate of cratering continues to produce a regolith of dust grains of different sizes that covers all surface features.

Meanwhile, a slight shrinkage of the planet caused by its gradual cooling has resulted in the formation of lobate scarps and wrinkle ridges.

Surface Features

Many different types of physical feature are seen on the surface, each with its own system of nomenclature:

- Impact craters are usually named for men and women who have made contributions to the arts and humanities.
- Scarps (rupes) are named for ships used in exploration on Earth.
- Ridges (dorsa) and valleys (valles) are named for astronomers and radio observatories, respectively.
- Plains (planitia) are often named for the planet or god Mercury in various languages.
- Mountains, usually the rims of large impact craters, are named for the nearest plains.

Together these features form the different physiographic provinces that reflect the geologic evolution of Mercury. Three principal types of terrain are found globally: heavily cratered terrain, intercrater plains, and smooth plains. Some distinct, regional landscapes are also found, notably those associated with the enormous impact feature called the Caloris basin and an area of highly disrupted hills and depressions on the opposite side of the planet to Caloris.

Figure 5.16 MESSENGER's multi-spectral images show compositional variations on Mercury. The image on the left enhances the largest color differences, most notably the Caloris basin (C). Dark craters within Caloris are indicated by white arrows. Mozart crater (M), the Tolstoj Basin (Tsp) and material ejected from it (T) also stand out. In the color processed image (right), white arrows identify possible "young" volcanic plains. Black arrows show possible volcanoes. Most of the color differences probably indicate variations in the mineral composition and physical state of the rocks. (NASA-JHU-APL)

Impact Craters

As is the case with most solid bodies in the Solar System, impact craters are seen everywhere on Mercury – evidence of a bombardment by meteoroids and comets over billions of years.

The preservation of this impact record is not surprising. Mercury has no plate tectonics or atmosphere to erase the ancient landscapes. It has also been subjected to a more severe pounding than Earth or Mars. This is because the Sun's gravity accelerates incoming projectiles, causing them to impact the surface at higher velocities than on more remote planets. On average, asteroids strike Mercury at a velocity of about 34 km/s, compared with 22 km/s on the Moon and 19 km/s on Mars. Comets on parabolic paths arrive at even faster velocities – typically 87 km/s, compared with 52 km/s for the Earth-Moon and 42 km/s on Mars.

Collisions with comets arriving from the outer Solar System should also be more frequent, partly because they often reach perihelion close to Mercury's orbit, and partly because Mercury's distance from the asteroid belt reduces the likelihood of an asteroid strike. Hence, about 41% of the craters on Mercury are likely to be created by comet impacts, compared with about 10% for the Earth-Moon and less than 3% for Mars.

Higher velocity impacts release more energy, so incoming objects of similar size will excavate larger craters on Mercury than on the Moon. The morphology of fresh craters and their ejecta blankets is also influenced by gravity. Mercury's surface gravity is 2.3 times greater than that of the Moon, and this permits ejecta to travel only 65% of the distance it would reach on the Moon.

Studies of similarly-sized craters on the two worlds show that the amount of ejected material is roughly the same, but Mercury's stronger gravity causes a higher concentration of secondary craters and a thicker ejecta blanket close to the main craters. These factors also reduce the amount of modification of the surrounding terrain by ejecta that cover or degrade adjacent craters.

Crater counts on different terrains show remarkable similarities to lunar cratering. Although it is not certain that the same number of craters was formed at the same rate on both Mercury and the Moon, their histories have probably not been dramatically different.

Studies of lunar cratering suggest that the heavily cratered terrain on Mercury retains the imprint of the late heavy bombardment which ended about 3.8 billion years ago. This terrain is so crowded with closely grouped and overlapping craters more than 30 km in diameter that ejecta blankets and individual fields of secondary craters are difficult to detect.

Many of the craters appear degraded, with smoother walls, more rounded rims and small craters on their floors. Their appearance probably reflects their greater age, the effectiveness of erosion and burial by ejecta from adjacent craters. In some cases, lava infilling also seems to have occurred, with breached walls and smooth, interior plains. The youngest craters, such as Kuiper, display a sharp rim and prominent ejecta deposits, sometimes with bright rays that spread large distances from the impact site (see Chapter 4).

Some craters (e.g. Berkel and Basho) contain dark material in the center and in a ring immediately surrounding them, as well as rays of ejecta. Other craters, such as Debussy, have the rays but lack dark halos.

Figure 5.17 Impact structures of different ages and sizes are visible in this Mariner 10 view of Mercury's ancient cratered terrain. A relatively young crater lies in the center of an older, larger crater. The newer crater is about 12 km across. The photo covers an area 130 by 170 km. (NASA-JPL)

Figure 5.18 85 km diameter Debussy is a relatively young impact crater with a central peak and rays that extend for hundreds of kilometers. It was first identified by ground-based radar. (NASA/Johns Hopkins University Applied Physics Laboratory/Carnegie Institution of Washington)

Mercury, the Moon and Mars all have similar crater distributions, the main exception being the number of craters less than 50 km in diameter. The marked deficit of such craters on Mercury is due to burial by intercrater plains material during the period of heavy bombardment or flooding by lava.

Below a diameter of about 20 km, the abundance of craters increases sharply. The crater population superimposed on the smooth plains within and surrounding the Caloris Basin shows the same size/frequency distribution as the lunar highlands and the ridged plains of Mars over the same diameter range, but the crater density is much lower. On the other hand, the crater density on Mercury's smooth plains is much greater than on the lunar maria.

Impact Basins

Mercury's ancient surface shows evidence of enormous collisions with asteroids and comets in the distant past. The largest impact feature is the Caloris Basin, about 1500 km in diameter (see below), followed by Rembrandt (715 km), Beethoven (625 km) and Tolstoj (510 km). Many other heavily degraded and buried basins have been proposed from inferred remnants of basin rings.

One of the most unusual examples is Rembrandt (33°S, 88°E), which may have formed about 4 billion years ago, making it one of the youngest large impact basins on Mercury – comparable in age to Caloris. Smooth plains in the center and near the northern rim are evidence of partial flooding by lava. It also displays a wheel-and-spoke–like pattern of radial and concentric wrinkle ridges and graben. The wrinkle ridges are crosscut by the graben, suggesting that deformation due to contraction occurred before the smooth plains were stretched and fractured due to crustal extensional deformation. The youngest deformation of the basin involved the formation of lobate scarps during the global cooling and contraction of Mercury.

Some parts of the crater floor seem to contain unusually high amounts of iron and titanium. These elements indicate that some exposed materials have not been covered by more recent lava flows, and so might date from a very early epoch in Mercury's formation. The data from MESSENGER indicate that volcanic activity and surface tectonics have also modified many other impact basins on the planet.

Not all large, ancient basins have flat floors. Beethoven (20°S, 124(W) shows a 1 km deep trough near the subdued rim, then a broad topographic rise towards the interior and a depression in the centre. One interpretation is that the basin was flooded by lava, then underwent differential subsidence in the centre and near the margins.

Dozens of multiple ring impact basins larger than 200 km across (e.g. Bach) have also been recognized – far more than on

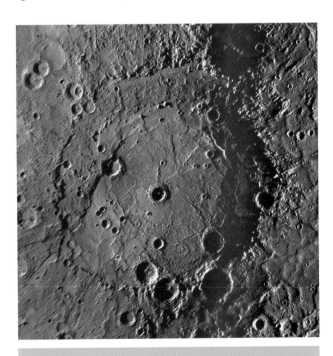

Figure 5.19 The multi-ringed Rembrandt basin has a diameter of about 715 km and may be the youngest large impact basin on Mercury. High cliffs and small craters are visible inside the basin. A 1000 km long lobate scarp cuts through one of the large internal craters (left). (NASA/ Johns Hopkins University Applied Physics Laboratory/ Carnegie Institution of Washington)

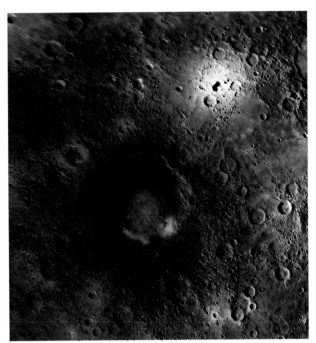

Figure 5.20 An enhanced color MESSENGER view of the 290 km wide Rachmaninoff double ring impact basin. The bright yellow area (top right) is centered on a rimless depression that may have been an explosive volcanic vent. The basin's smooth interior may be the result of flooding by lava. Smooth plains, thought to be a result of earlier episodes of volcanic activity, also cover much of the surrounding area. (NASA/Johns Hopkins University Applied Physics Laboratory/Carnegie Institution of Washington)

the Moon. Once again, they are often associated with evidence of various types of volcanic activity.

One of the youngest examples is Rachmaninoff, a remarkably well preserved double ring basin. The low numbers of superposed impact craters and marked differences in color across the basin suggest that the smooth area within the innermost ring may be the site of some of the most recent extrusive volcanism on Mercury. There is also evidence of an explosive volcanic vent beyond the outer ring.

Caloris Basin

The best preserved impact basin on Mercury, and one of the youngest, is the Caloris Basin (30.5°N, 170°E). With a diameter of about 1500 km, it is also one of the largest impact basins in the Solar System.

The rim is marked by a ring of smoothly rounded mountain blocks, up to 2 km high and 50 km square, bounded on the interior by a relatively steep escarpment. The ring of mountains is almost continuous, with the exception of a 300 km wide gap. Smooth plains appear to intrude into this gap from outside the basin. Some valleys – possibly fault-bounded depressions (graben) – within the mountain chain also seem to be filled with plains material. A second, much smaller and less continuous escarpment occurs about 100 to 150 km farther out from the centre of the

Figure 5.21 Although the Caloris Basin (left) was partly hidden by shadow during the Mariner 10 flybys, the spacecraft's cameras revealed an outer ring of smooth, rounded block mountains rising 1–2 km above the surrounding terrain, and an interior crossed by ridges and fractures. Smaller craters were also visible in the basin. (NASA-JPL)

basin. This may be evidence of block faulting associated with basin deformation.

The interior of Caloris is occupied by smooth plains that show prominent wrinkle ridges and fractures, evidence of forces that deformed Mercury's crust long ago. The wrinkle ridges are arranged in a roughly polygonal pattern and probably formed as the result of horizontal compression when the basin floor subsided under the weight of inflowing lava. The largest examples are a few hundred kilometers long, 1.5–12 km wide, and 500–700 meters high. Although they bear some similarity to ridges found on the lunar maria, their radial alignment and large number makes them quite distinct.

The fractures in the center of the basin cross the wrinkle ridges at various angles, clear evidence that they formed at a later date, presumably when uplift created tensional forces in the crust. This uplift may have been the result of magma migrating towards the surface. However, they must be ancient features, since they do not deform the impact craters which they cross.

The series of over 200 radial fractures in the center of the basin is known as Pantheon Fossae. Near the center of these fractures is the 41 km diameter crater Apollodorus. Their size tends to increase towards the centre of the basin. The largest of them are flat-floored grabens, about 700 meters deep.

Other fractures that occur near the outer edge of the basin are roughly concentric. This rare combination of fractures is only seen in large impact basins on Mercury. They probably formed when the floor of the basin was uplifted, causing horizontal stretching and breaking apart of the material that filled the basin. Also visible are many smaller craters that have been excavated more recently.

Besides lava flows, Caloris contains spots with diffuse boundaries which appear to be volcanic in origin. Sometimes the spots are centered on rimless depressions, possibly indicating that they were caused by explosive, pyroclastic eruptions.

Beyond the basin rim is so-called lineated terrain, characterized by long, hilly ridges separated by radial grooves that cut through pre-Caloris craters. This type of terrain can be traced for about 1000 km from the rim. It bears a close resemblance to the sculptured terrain around the lunar Imbrium basin, and may represent fault troughs or chains of coalescing secondary craters formed by ejecta produced during the huge impact that created the basin.

Also visible beyond the main escarpment are several areas of smooth and hummocky plains. The hummocky terrain comprises low hills, some close together, others scattered, and less than 1 km across. They may represent basin ejecta composed of solid fragments and impact melt. The smooth plains are probably ancient lava flows (see below).

The enormous impact that created Caloris seems to have modified the surface on the antipodal (opposite) side of the planet, with the result that the rims of all the old craters in this region have been broken up into a jumbled landscape of hills and depressions.

Termed hilly and lineated terrain, it is similar to formations antipodal to the Imbrium and Orientale basins on the Moon. These have been attributed to the transport and deposition of material ejected by the impacts on the opposite side of the satellite, but this process does not seem likely on Mercury, which is considerably larger and has a much higher surface gravity.

Figure 5.23 This Mariner 10 image shows part of the hilly and lineated terrain that was crated in the aftermath of the Caloris impact. Hills, depressions and valleys disrupt pre-existing landforms. The floor of the crater (left) was flooded by lava after the impact. (NASA-JPL)

Figure 5.22 (Top) A map of the numerous linear features that formed when the rocks within the Caloris basin were subjected to large horizontal forces. Hundreds of extensional troughs (black lines) show where the surface has been pulled apart and faulted. The Pantheon Fossae region around Apollodorus crater (located in the white box and shown in more detail at bottom) has over 200 such troughs that radiate outward, but near the outer edges of the basin they are broadly concentric. The basin interior also has many wrinkle ridges (red lines), which formed when the surface was compressed horizontally. (NASA-JHU-APL)

Instead, studies suggest that the seismic waves from the huge impact were focused as they passed through the large core, resulting in a sudden deformation of the rocks exactly opposite the Caloris Basin. The result was a 500 km wide landscape marked by roughly polygonal hills 5–10 km across and up to 1.5 km high. The localized nature of these features and the lack of similar damage to younger craters indicate a one-off, catastrophic event, in which the ground underwent vertical motions of more than 1 km in a matter of minutes.

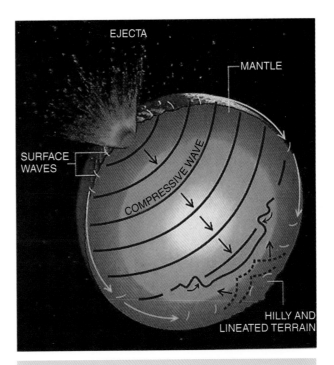

Figure 5.24 The probable cause of the chaotic terrain on the opposite side of Mercury from the Caloris Basin. Seismic waves generated by the Caloris impact passed around and through the planet, and were focused at the antipodal point. This resulted in large vertical movements which created hilly and lineated terrain. (Peter Schultz, Brown University)

Plains

Relatively flat or smoothly undulating plains cover large parts of the surfaces of the terrestrial planets. In most cases, they are the result of resurfacing, the covering or destruction of an earlier, rough topography.

This resurfacing may take place in at least three different ways:

- An increase in temperature may reduce the strength of the crust and its ability to retain high relief. Over millions of years, the mountains sink and the craters rise.
- The movement of material down slopes and along valleys leads to infilling of low areas. This may be through lava flows or erosion and deposition of sediments.
- Fragmented material is deposited on the surface from above, eventually covering the rough topography, for example, blanketing by impact crater ejecta or volcanic ash.

All three of these processes may have modified Mercury at different times, particularly during its early history, although volcanic flooding seems to have been most significant over a wide area.

Two general types of plains units are usually recognized on Mercury: intercrater plains and smooth plains. The relative abundance of craters on the intercrater plains indicates that they are older.

The intercrater plains are widespread and, as their name suggests, they are found between the areas of heavily cratered terrain. Although they have relatively smooth or low, undulating relief, they are generally covered with many small secondary craters (mostly less than 10 km in diameter), which were created by ejecta from large impacts. However, the plains material partially infills some of the large craters and covers ejecta blankets. Many craters less than 50 km in diameter have been buried and destroyed by the intercrater plains formation.

Dating based on crater counts suggests that the intercrater plains formed between 4.0 and 4.2 billion years ago, during the period of heavy bombardment. As time went by, the amount of new intercrater plains material being created seems to have diminished.

The smooth plains cover about 40% of the surface. They have a lower density of impact craters and typically fill low-lying areas such as impact craters and basins. Individual deposits range from hundreds of square kilometers to millions of square kilometers (e.g. in the interior of the Caloris Basin and on the plains around the north pole), rivaling the sizes of the largest flood basalt units on Earth or the Moon. The majority of smooth plains are probably volcanic in origin, although impact melt and basin ejecta may account for some of them. The relative absence of craters indicates that they are among the youngest features on Mercury.

In most respects, the smooth plains resemble the lunar maria, with abundant ridges and scarps. The main differences are that the Mercurian terrain has a similar albedo and color to its surroundings, and it displays more impact craters.

Lobate Scarps and Global Contraction

All of the major terrains on Mercury exhibit lobate scarps, long, sinuous cliffs with rounded fronts that are unique to the planet. They range in length from about 20 km to more than 500 km, often reaching heights of several kilometers.

Images of craters and other landforms sliced through by these scarps show surface lateral displacement of up to 10 km, clear

Figure 5.25 This high resolution view of Mercury's intercrater plains is centered at 3°N, 20°W. Most of the numerous small craters are secondary impact features created by ejecta from young, primary impact craters. The prominent Santa Maria Rupes scarp cuts both intercrater plains and older craters. The large crater (top right) has been flooded by intercrater plains material. (NASA-JPL)

Discovery Rupes
schematic

Figure 5.26 With a length of about 550 km and a height of 1.6 km, Discovery Rupes is one of the largest lobate scarps on Mercury. This steep, sinuous feature was created by a thrust fault which disrupted an older impact crater. (Mark Robinson, Northwestern University)

evidence that they are faults or major fractures in the crust. The appearance of landforms such as Discovery Rupes and Hero Rupes suggests that they are created by thrust faults, where one block of crust is pushed up over another (Fig. 5.26). The overthrust block, unable to support its weight, then collapses. The surface outline seems to meander because the fault plane slopes at a low angle and reaches the surface at different places according to elevation.

Since Mercury has not experienced any plate tectonic activity (see Chapter 3), how did these huge fractures form? Their global distribution and more-or-less random orientation indicates that they are the result of a major episode of contraction and compression that encompassed the entire planet.

Studies of surface features indicate that the crust was fractured after the formation of the large, degraded craters of the highlands and the intercrater plains. Only the younger craters are seen to disrupt the scarps. It seems clear that the cliffs date from the period of late heavy bombardment, possibly around 3.8 billion years ago.

Calculations based on the overall amount of crustal shortening associated with these faults indicate that the surface area of Mercury shrank by about 0.1%, as its diameter decreased by 1–2 km. The cause of such a unique contraction was probably cooling of the planet's interior, particularly its huge iron core.

There is also some evidence that the underlying crust may have been fractured by despinning of the planet. If Mercury once rotated much faster, then was subjected to tidal forces which slowed it until it was trapped in a 3:2 spin-orbit resonance (see above), the planet's shape would have changed dramatically from an oblate spheroid (like a slightly squashed ball) to a sphere. As the equatorial diameter decreased and the polar regions became less flattened, stresses in the crust would have created a series of compressional and tensional stress fractures.

Curiously, such a fracture pattern is not seen, with the possible exception of some linear ridges that trend northeast and northwest. It seems that any fractures caused by despinning occurred very early in Mercury's history, before the formation of the intercrater plains.

Polar Ice?

As on the Moon, the low axial inclination of Mercury, combined with rugged polar relief, result in some large craters having permanently shaded floors (see Chapter 4). These near-polar craters have been associated with unusual radar images obtained by ground-based radio telescopes.

In each case, a very high fraction of the incident radar was reflected back to Earth. Analysis showed that the echoes were very similar to those obtained from some icy satellites of the outer planets and from the residual water ice cap of Mars. If the strong reflection is due to water ice, it must be fairly uncontaminated and perhaps 2–20 meters thick. However, the data indicate that it may be covered by a thin layer of dust or soil, or else does not completely cover the crater floors.

High resolution (1.5 km) radar images of the north polar region have been able to pin down the highly reflective surfaces to localized areas that (with a few exceptions) coincide with fresh,

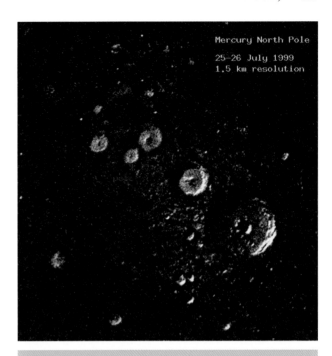

Figure 5.27   A radar image of Mercury's north polar region. While most of the craters linked with highly reflective material are located at latitudes above 85°N, some are located down to 72°N. This has led to suggestions that a thin layer of regolith may cover and protect water ice deposits. Alternatively, the reflections may be caused by enhanced concentrations of sulfur. (John Harmon, National Astronomy & Ionosphere Center, Arecibo Observatory)

deep craters, some less than 10 km across. The donut shape of the deposits is explained by the presence of central peaks in the craters, while bright arcs coincide with their shaded rims.

How can the presence of ice be explained on a planet that is roasted by the nearby Sun? In fact, temperature is not a problem, since the interiors of large, permanently shaded craters remain perpetually cold, below −212°C. Mercury's surface is extremely dry, but water could be supplied by an occasional comet or meteorite impact, or through outgassing from the interior. Some of this water vapor would be lost through a variety of processes, but a significant proportion would condense on the surface at night. Eventually, some of the water vapor might reach a dark haven in a deep polar crater. There, it could remain for billions of years.

Although the nature of the polarization of the radar waves strongly suggests the presence of water ice, other possibilities cannot be completely ruled out. Some of the apparent ice deposits are in craters too small and too far from the pole to be in permanent shadow. Thermal models suggest that 10 km diameter craters should not be capable of supporting even insulated ice. Another challenge involves the discovery of several "ice" features at latitudes between 71–75°N.

An alternative suggestion is that elemental sulfur could be the cause. Outgassing of sulfur and volatilization of sulfide minerals by meteorite impacts could release sulfur into the atmosphere. Over time, this could interact with the surface and migrate to

Figure 5.28   Mercury's south pole is located inside Chao Meng-Fu, the 180 km wide crater on the planet's limb (lower center). Although the crater's floor is in shadow, sunlight is reflected from its far rim. The image also shows a double-ringed basin just above and to the right of the south pole, and bright rays ejected from a 50 km crater (upper right). The horizontal line near the top was caused by a drop out in data. (NASA-JPL)

Figure 5.29   An image of Mercury's gaseous tail obtained by a camera aboard the STEREO-A spacecraft. The planet is over-exposed, so it appears much larger than its actual size. The tail extends anti-sunward from the planet. It is composed of sodium and other atoms sputtered off Mercury's surface. (Boston University Center for Space Physics)

permanently shaded polar regions. Estimates suggest that such a process could deposit some 35 meters of sulfur over a period of one billion years. MESSENGER observations have revealed substantial amounts of sulfur on Mercury's surface, indicating that sulfide minerals are present and that sulfur-containing gases may have contributed to past explosive volcanic activity on Mercury. On the other hand, although sulfur is stable at a higher temperature than water ice, there are no high backscatter radar signatures from other regions where temperatures are within the stability range of sulfur.

Atmosphere

Ground-based searches for a Mercurian atmosphere were unsuccessful during the 1960s and 1970s. This was not too surprising, since Mercury's weak gravity could not be expected to hold onto any atmosphere generated early in its history.

However, ultraviolet observations by Mariner 10 revealed that there is a very thin atmosphere. (The proper term is exosphere, since the few gaseous atoms on Mercury rarely collide, but bounce from place to place on the surface. In an atmosphere, the gaseous atoms are in constant collision.) The surface pressure is a mere 10^{-12} bar or ~1/1,000,000,000,000 the sea level pressure on Earth. However, the planet's highly elliptical orbit and proximity to the Sun lead to marked seasonal variations in exospheric density.

Gases detected include atomic hydrogen, helium, oxygen, sodium, magnesium, potassium and calcium. All of the species were identified by searching for faint spectral features as sunlight reflected from the surface is scattered by the atoms. There may also be trace amounts of argon, carbon dioxide, water vapor, nitrogen, xenon, krypton and neon.

High daytime surface temperatures cause atoms to travel so quickly that they exceed the escape velocity of 4.2 km/s and escape into space. Solar ultraviolet radiation also ionizes some of the

atoms, enabling them to be carried away by the magnetic field in the solar wind.

In addition, Mercury's intrinsic magnetic field interacts with the interplanetary medium to modify the spatial distribution and density of solar wind plasma and high-energy charged particles that impact and sputter materials from the surface (see Fig. 5.30). This interaction leads to changes in exospheric densities that vary on time scales as short as a few hours. Impacts also supply both meteoritic and volatilized surface materials to the exosphere.

This depletion of Mercury's atmosphere is counterbalanced by constant replenishment. The solar wind acts as a source of ions and atoms which are captured by the planet. Atomic hydrogen and helium come, at least partly, from the stream of hot, ionized gas emitted by the Sun. Some of the hydrogen and oxygen may also come from collisions with comets and meteorites.

Several different processes also liberate material from the surface and are responsible for supplying the exosphere: evaporation of elements due to strong ultraviolet light, meteoroid impacts, excitation of atoms on rock surfaces due to solar radiation, sputtering by impacts with solar wind ions, or emission of gases from the planet's interior.

MESSENGER

Many of the mysteries involving Mercury should be answered in the next decade, after three orbital spacecraft arrive at the planet. The first of these is NASA's MESSENGER (MErcury Surface, Space ENvironment, GEochemistry and Ranging) spacecraft, which was launched on August 3, 2004 and entered a near-polar orbit on March 17, 2011. Two more orbiters are planned in the

Mercury's Surface-Bounded Exosphere

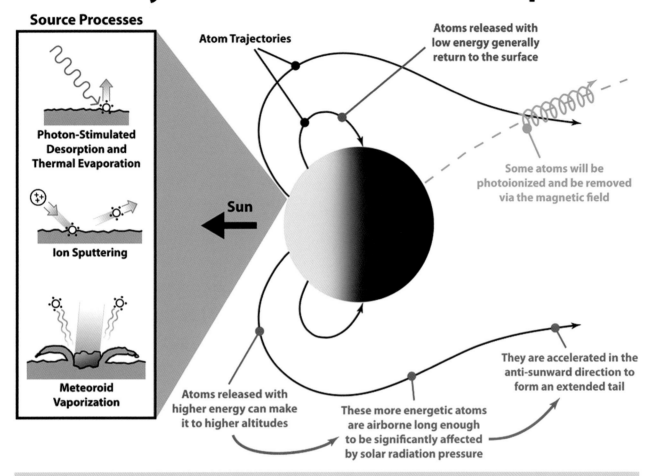

Figure 5.30 The major processes that generate and maintain Mercury's exosphere. The panels (left) summarize the three primary sources. Photon-stimulated desorption occurs when sunlight excites surface-bound atoms or molecules, releasing them to the exosphere. Solar heating causes atoms and molecules to evaporate. During both of these low energy processes, most of the released material reaches only low altitudes and usually returns to the surface. Ion sputtering occurs when charged particles from the solar wind or Mercury's magnetosphere impact the surface, "knocking off" atoms and molecules. Vaporization occurs when incoming meteoroids, generally small dust particles, impact the surface at high speeds. Both ion sputtering and meteoroid vaporization are high-energy processes, and the released material can reach high altitudes. All material in the exosphere is accelerated away from the Sun by radiation pressure; atoms and molecules at high altitudes escape the planet's gravitational influence and enter Mercury's neutral tail. Some of the ions are returned to the surface by Mercury's magnetosphere. (NASA-JHU-APL)

European-Japanese BepiColombo mission, due for launch in 2014, with arrival scheduled for November 2020.

During a 7.9 billion km journey that included 15 trips around the Sun, MESSENGER received gravity assists from Earth, Venus and Mercury itself before entering orbit. Three Mercury flybys in January 2008, October 2008 and September 2009 enabled the spacecraft to match the planet's speed and location prior to the orbit insertion maneuver. During these encounters, the spacecraft mapped almost the entire planet and gathered data critical to planning the year-long orbit phase (equivalent to four Mercury years, or two Mercury solar days).

During the primary science mission, MESSENGER operates in a highly elliptical orbit around Mercury, 200 km above the surface at the closest point (periapse) and 15 193 km at the farthest (apoapse). The plane of the orbit is inclined 82.5° to Mercury's equator, and the low point in the orbit occurs at 60°N. MESSENGER orbits Mercury twice every 24 hours; for one third of this period it is oriented to send data to Earth. The instruments are focused on global mapping during the first Mercurian solar day, with an emphasis on targeted science investigations during the second.

MESSENGER carries seven instruments to examine the planet's surface, atmosphere and near-space environment. The multispectral imaging system has wide- and narrow-angle CCD imagers to map the surface. Nearly all of the planet is being imaged in stereo to determine topographic variations and landforms.

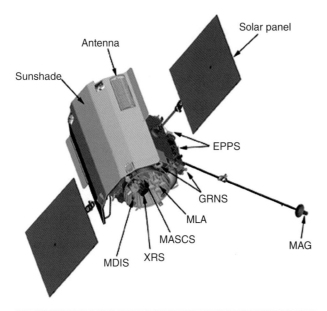

Figure 5.31 MESSENGER carries seven scientific instruments, including cameras that will map the entire surface, a laser altimeter to gather topographic information, a magnetometer and several spectrometers. In order to survive where solar radiation is up to 11 times more intense than on Earth, the spacecraft carries a sunshield made of heat-resistant Nextel fabric. (JHU-APL)

Its X-ray, gamma-ray, and visible-infrared spectrometers determine the elemental and mineral composition of the rocks on the surface. Just as importantly, they also show which elements and minerals are conspicuously absent. Another key objective is a search for evidence of water ice or other volatiles in the polar craters.

The neutron spectrometer maps variations in the neutrons that Mercury's surface emits when struck by cosmic rays. This can be used to estimate the amount of hydrogen – possibly locked up in water molecules – and other elements. The X-ray spectrometer is able to detect emissions from the elements magnesium, aluminum, silicon, sulfur, calcium, titanium and iron.

A laser altimeter maps Mercury's landforms and other surface characteristics using an infrared laser transmitter and a receiver that measures the round-trip time of individual laser pulses. The altimeter can view the planet from up to 1000 km away with an accuracy of 30 cm. The data are also used to track the planet's slight forced libration – a slight wobble about its spin axis – that provides information about the size of its core and whether the core is partially liquid. Altimeter data, when compared with gravity field measurements gathered by tracking MESSENGER's orbital motion, help to determine the thickness and structure of the crust and map the planet's gravitational field.

The magnetic field is studied by a magnetometer that is mounted on a 3.6 meter boom to keep it away from the spacecraft's own magnetic field. The sensor determines the field's exact strength and how it varies with position, altitude and time.

The composition of Mercury's tenuous atmosphere is measured by ultraviolet and energetic particle spectrometers. By comparing these data with X-ray and gamma-ray measurements of the surface rocks, scientists learn more about the origin of each element in the exosphere.

Questions

- Mercury is sometimes called "the iron planet." Explain why this is so. How might this unusual composition have come about?
- Mercury and the Moon are often said to be near twins. To what extent is this true? (Make a list of the similarities and differences between the two objects.)
- What evidence is there to suggest that plate tectonics do not take place on Mercury?
- Valleys are relatively rare on Mercury. Explain how those that have been imaged may have formed.
- Mercury's atmosphere is extremely sparse. Why?
- Why were scientists surprised to find that Mercury has a significant magnetic field? How can such a magnetic field be accounted for?

SIX

Venus

Familiar to humans since prehistoric times, Venus is the brightest object in the night sky after the Moon. Unsurprisingly, the "morning star" or "evening star" played an important role in the astrological and religious beliefs of many ancient civilizations. It is named after the Roman goddess of beauty, love, laughter and marriage. For centuries after the invention of the telescope, the cloud-shrouded planet was regarded as Earth's twin, a misconception which only ended after visiting spacecraft revealed its true, extremely inhospitable, nature.

Orbit and Size

Venus follows the most circular orbit of any planet around the Sun, at an average distance of only 108 million km – closer than all of the planets except Mercury. Since it is located inside Earth's orbit, it is never too far from the Sun in the sky. Whether it appears to the west of the rising Sun in the morning, or to the east of the setting Sun in the evening, Venus is a wonderful sight to behold. At its brightest, the planet reaches a dazzling magnitude −4.5, bright enough to cast a faint shadow on a clear moonless night. Venus is also observable during daylight under certain conditions.

Since there is little variation in its distance from the Sun, Venus' orbital velocity is also fairly constant. Traveling at an average speed of 35 km/s, it completes one circuit of the Sun every 225 Earth days. It comes closest to Earth every 584 days during inferior conjunction. (Since five intervals of 584 days are almost exactly equal to eight years of 365/366 days, apparitions of Venus in the sky repeat every eight years.)

By comparing measurements of its angular size with its distance, astronomers were able to calculate that Venus has a diameter of 12,104 km, only slightly smaller than Earth.[1] Determination of its mass and density was complicated by the absence of any moons, but it was generally assumed that Venus is a rocky planet, with an overall composition and internal structure very similar to that of our world. Subsequent measurements of its gravitational influence on visiting spacecraft have confirmed this view.

Phases

When seen through a telescope or binoculars, Venus resembles Mercury and the Moon by displaying phases (see Chapters 4 and 5). This means that the planet's appearance changes from a small full phase around superior conjunction on the far side of the Sun, to a large, slim, crescent close to inferior conjunction (see Fig. 6.1

Of all the planets, Venus is by far the brightest and easiest to observe with the naked eye. It is best placed for observation when at its greatest elongation (angular distance) of 47° from the Sun in the sky. Venus then appears above the horizon for about 5½ hours after sunset or before sunrise. However, it is actually at its brightest when seen as a broad crescent at an elongation of 39°, about 36 days before and after inferior conjunction. At its closest, the planet's apparent diameter is more than 65 arc seconds – providing easily the largest telescopic disk of any planet.

Transits

From time to time, both Mercury and Venus appear to cross the face of the Sun, as seen from Earth, at the time of inferior conjunction (see Chapter 5). If the conjunction occurs when the orbit of Venus crosses the ecliptic (the ascending or descending node), then it appears as a small dark spot moving from east to west across the Sun's disk. However, the slight inclination of the planet's orbit to the ecliptic means that such transits are extremely

[1] The oblateness of Venus is zero. This means that, unlike other planets, there is no appreciable difference between the equatorial and polar diameters of Venus. This is due to its extremely slow rotation.

Exploring the Solar System, First Edition. Peter Bond.
© 2012 by Peter Bond. Published 2012 by Blackwell Publishing Ltd.

1910 SEPT 27   1910 JUNE 10   1927 OCT 24

1919 SEPT 25   1964 JUNE 19

Figure 6.1 The phases of Venus. In this sequence, it appears fully illuminated, but very small, when it is at superior conjunction, on the opposite side of the Sun from Earth. As the distance between the two planets shrinks, Venus seems to grow in size and moves through its phases. Around inferior conjunction, when the crescent is narrowest and Venus is nearest Earth, its apparent diameter is about seven times larger. (Lowell Observatory)

rare. The most recent transit took place on June 8, 2004 – the next will occur on 5–6 June, 2012.

As Table 6.1 shows, transits of Venus only take place in June or December. However, the duration of the transit – typically over six hours – varies according to the distance to be traversed across the Sun's disk.

The Venus transits have played an important role in the history of astronomy. In the 18th and 19th centuries many international scientific expeditions were dispatched to remote corners of the world in order to note the time the transit began and the apparent position of Venus on the solar disk from widely separated locations. By careful analysis of observations and the use of

trigonometry, astronomers attempted to determine the distances of Venus and the Sun. Perhaps the most famous observation was made from Tahiti in 1769, during Captain James Cook's expedition to the South Pacific and circumnavigation of the globe.

Unfortunately, the accuracy of the measurements was adversely affected by a phenomenon known as the "black drop," which occurred at the beginning and end of each transit (see Fig. 6.2). Observers found that the silhouette created by Venus appeared to lengthen and link to the solar limb. By the time the effect disappeared, the transit was already well under way. The black drop is now generally regarded as an optical effect, linked to the quality of a telescope's optics and its resolving power.

Table 6.1
Transits of Venus 1631–2125

Date	Start Time (UT)	Duration
1631, December 7	03.49	3 h 00 m
1639, December 4	14.56	6 h 58 m
1761, June 6	02.01	6 h 35 m
1769, June 3–4	19.15	6 h 20 m
1874, December 9	01.50	4 h 36 m
1882, December 6	13.57	6 h 13 m
2004, June 8	05.15	6 h 13 m
2012, June 5–6	22.22	6 h 31 m
2117, December 11	00.03	5 h 41 m
2125, December 8	13.19	5 h 35 m

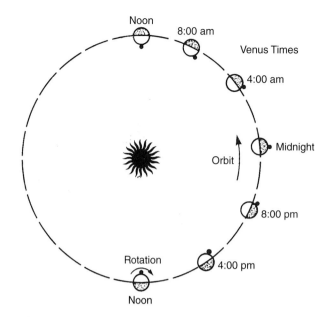

Axial Rotation: 243.01 Earth days
Orbital period: 225 Earth days (= Venus year)
Noon to Noon: 116.8 Earth days (= Venus year)
Venus year: 1.93 Venus year

Figure 6.3 Venus spins in a retrograde (east to west) direction, completing one full turn every 243.01 days. Since each orbit of the Sun takes 224.7 Earth days, a solar day on Venus – from one sunrise to the next – lasts for 116.75 Earth days. (NASA)

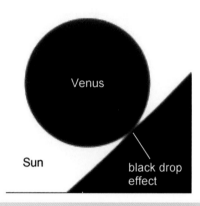

Figure 6.2 Venus at the western limb of the Sun, imaged by NASA's TRACE spacecraft during the transit of June 8, 2004. Unlike airless Mercury, Venus is outlined by a faint, glowing ring during transits, caused by scattering of sunlight in its thick atmosphere. The "black drop" effect was observed for about four minutes. It is thought to be caused by the "point-spread function" of the telescope and the light-scattering properties of Venus' atmosphere. (NASA/ Stanford-Lockheed Institute for Space Research)

Rotation Period

Since Venus comes closer to Earth than any other planet, it should be extremely easy to observe. The first telescopic observers recorded pronounced markings that suggested a cloud-free view of Venus' surface, but it was soon realized that it is covered by an impenetrable blanket of cloud. Even with the most modern optical telescopes, it is only possible to record gray, nebulous markings that change from day to day.

In the absence of any visible surface features, it was impossible to determine its rotation period. The first clue came in 1956, when Doppler measurements of light reflected from Venus indicated a very slow retrograde rotation. If true, this meant that it must be

spinning from east to west – the opposite direction to Earth and most other planets.

Further Doppler studies in the early 1960s supported the "backward" rotation, but suggested that Venus was spinning remarkably quickly – once every four days. This was later shown to be due to hurricane force winds that swept high level clouds around the planet every four days (see below).

The first information about the hidden surface came in 1961, when ground-based radar was used to pierce the clouds. By measuring the Doppler shift of the radio waves reflected from Venus, the period of spin was finally revealed. In contrast to the high-level winds, the solid planet rotates very slowly, completing one retrograde revolution in 243 Earth days. Venus has by far the longest rotation period of any planet in our Solar System.

When combined with a 224.7 day orbital period, this sluggish spin has a remarkable effect. The Venusian "day" from sunrise to sunset lasts for 116.75 Earth days. If the Sun could be seen from the surface, it would rise in the west and set in the east.

A number of theories have been put forward to explain Venus' strange behavior. Perhaps Venus was struck a glancing blow by a large impactor early in its history, causing it to topple over, so that the direction of its rotation appeared to reverse.

Alternatively, the planet may have simply slowed to a halt and then begun to rotate in the opposite direction. This could result from the chaotic effects created by several factors: gravitational tidal forces generated by the Sun combined with atmospheric thermal tides; friction between the core, mantle and crust; and the angle of obliquity between the planet's equator and the plane of its orbit around the Sun. In this scenario, the Sun's gravitational pull on the solid body of the planet (possibly working in tandem with internal core-mantle friction), caused the spin to slow down. As the planet slowed and the influence of the atmospheric tides grew, its direction of rotation reversed.

In 2001, Alexandre Correia and Jacques Laskar (CNRS Institute of Celestial Mechanics) simulated the evolution of Venus' rotation and concluded that this type of reversal mechanism was most likely if Venus had a slow initial spin and a modest axial inclination.

Another possibility is that precession of the axis of rotation and gravitational perturbations from the other planets might cause chaotic shifts in its axial inclination, reaching a variety of positions throughout the planet's evolution (see also Chapters 3 and 6). If Venus reached a high obliquity, for example, an equatorial inclination of more than 70° in relation to the plane of its orbit, and it was also spinning rapidly, the axis could have eventually flipped completely over. (Without the stabilizing influence of the Moon, Earth might also have experienced a similar trauma.)

It is interesting to note the relationship between the orbits of Venus and Earth and the rotation of Venus. At every inferior conjunction, the planet presents the same hemisphere to Earth. (This was obviously an obstacle to those who wanted to study its entire surface with ground-based radar.)

Clouds

The entire planet is blanketed by yellowish clouds, with no glimpse of the surface. It is the ability of these clouds to reflect sunlight which accounts for its brilliance in the night sky.

For centuries, astronomical sketches of Venus merely displayed the planet's phases, upon which were superimposed vague, transient markings. The most noticeable variations were the horns or cusps of the crescent Venus, which appeared brighter than the rest of the planet.

Photography made little difference – pictures taken in visible light resembled views of a fog bank on Earth. However, images taken at near-ultraviolet wavelengths did reveal clearly defined cloud features, presumably at high level, that were only hinted at in ground-based views. These features survived for some time, enabling astronomers to calculate that they traveled around Venus at high speed, completing one circuit in a mere four days.

These observations were confirmed by NASA's Mariner 10, which passed by Venus in February 1974 (see Chapter 2). Ultraviolet images taken seven hours apart showed that the upper clouds were speeding around Venus at 360 km/h. This motion from east to west – the same direction as the planet's spin – was 60 times faster than the rotation of Venus. Scientists refer to it as an example of super-rotation.

Since the zonal wind speed at the equator is faster than near the poles, this produces a symmetrical, V-shaped cloud pattern, centered on the equator, that is visible in many UV images. Broad bands of cloud diverge from the sub-solar point, where the Sun is overhead, towards thick "polar collars."

Mariner's camera also spotted a mottling of cellular cloud structures in the region of maximum solar heating, indicating

Table 6.2
Venus and Earth Comparison

	Venus	Earth
Mass	4.87×10^{24} kg	5.98×10^{24} kg
Equatorial radius (km)	6052	6378
Mean density	5250 kg/m^3	5520 kg/m^3
Average distance from Sun (million km)	108.21 (0.7 AU)	149.60 (1 AU)
Rotation period	243 Earth days (retrograde)	23 hours 56 minutes
Length of year (orbital period)	224.7 Earth days	365.2 days
Axial inclination (degrees)	177.3	23.5
Mean surface temperature	465°C	15°C
Highest point on surface	Maxwell Montes (17 km)	Mount Everest (8.8 km)
Reflectivity (albedo)	0.76	0.37
Surface gravity (at equator)	8.9 m/s^2	9.8 m/s^2
Surface atmospheric pressure	90 bars	1 bar (sea level)
Major atmospheric components	96% carbon dioxide, 3% nitrogen	78% nitrogen, 21% oxygen, 1% argon

that large convectional plumes were thrusting upwards from the main cloud deck on columns of rising gas.

Until the early 1970s, the composition of the clouds remained the subject of intense speculation. One of the favorite explanations was dust, stirred up by global, hurricane-force winds. However, Mariner 10's measurements, together with infrared spectroscopic studies, showed that the clouds resemble the output from a hazardous chemical plant.

The dominant constituent is highly concentrated sulfuric acid (H_2SO_4), together with traces of hydrochloric acid (HCl) and hydrofluoric acid (HF). The tiny droplets (about 1 micron across) of sulfuric acid are in a concentrated (75%) acid solution. Such corrosive clouds can exist over a much wider range of temperatures than the water clouds of Earth.

Where did the sulfur come from? One of the major gaseous products of volcanic eruptions on Earth is sulfur dioxide (SO_2). Since Venus is now known to possess many hundreds of volcanic structures (see below), it seems likely that these are the original sources of the sulfur. Volcanic eruptions would also replace any sulfur lost during chemical reactions with surface rocks.

When combined with water vapor (H_2O) in the upper atmosphere, the sulfur dioxide eventually produces sulfuric acid.

However, any sulfuric acid rain falling beneath the main cloud deck soon evaporates in the high temperatures, producing acid vapors and water vapor.

When these constituents return to high levels, they are affected by photochemistry – chemical reactions involving ultraviolet sunlight. The UV photons break up SO_2, starting a series of chemical changes which results in the creation of H_2SO_4 once more. Some sulfur is also produced, and this may account for the yellowish color of the clouds. The substance that causes the ultraviolet markings remains unknown.

Changes in Mariner 10's radio signals as they passed through the atmosphere, showed the presence of at least two separate zones of cloud at altitudes between 35 and 60 km. Above the main cloud deck were layers of tenuous haze.

Subsequent data from the Soviet Venera landers and the American Pioneer Venus probes have refined this view of the vertical cloud structure (see Fig. 6.6). Above the main cloud deck, extending from about 90 to 70 km, is a layer of sulfuric acid haze. There are three main cloud layers between 70 and 48 km. The upper cloud deck comprises one-micron-sized sulfuric acid droplets but the particle size is slightly larger in the middle and lower levels. At the level of the lower clouds, the temperature has already reached 95°C and atmospheric pressure is comparable to sea level on Earth.

Beneath the clouds the atmosphere is once again fairly hazy, but there is good visibility below about 30 km, except for the bottom 10 km, which become increasingly murky due to scattering of light by the dense atmosphere, bathing the landscape with a reddish hue. An observer on the surface would not be able to see the Sun through the dreary blanket overhead.

Atmospheric Circulation

At the cloud deck, the circulation is dominated by the four-day, retrograde super-rotation, a phenomenon which has yet to be

Figure 6.6 The horizontal structure of Venus' atmosphere, based on data from the Venera and Pioneer Venus missions. 360 km/h winds sweep the higher clouds around the planet in four days. Three layers of sulfuric acid cloud occur 48–70 km above the surface. In the gloom below, the atmosphere is stable, becoming increasingly dense and hot towards the surface, where the temperature is about 460°C. Surface pressure is equivalent to that found almost 1 km down in Earth's oceans, so low-level winds are slow. (NASA)

explained satisfactorily. The only comparable winds on Earth are the ribbon-like jet streams, but they are limited to narrow zones and much closer to the surface. In contrast, the Venusian zonal winds cover almost the entire planet, although their velocity quickly decreases towards higher latitudes.

Up to three east-west jets have also been identified: one above the clouds, between 50 and 55° latitude, with a maximum speed of 140 m/s; a second with a speed of 95 m/s, centered at 70° latitude and 60 km altitude; and a possible third jet at 15° latitude and 65 km altitude, with a speed of 100 m/s.

The only viable source of energy for these high velocity jets appears to be the absorption of solar energy by the clouds, hazes and relatively thin upper atmosphere, perhaps combined with turbulence.

A much more leisurely, north-south circulation – only 5 to 10 m/s – also occurs within the troposphere. This Hadley cell circulation (see Chapter 3) takes the form of heated gas rising at the equator and spreading out at an altitude of about 50–65 km. At latitudes of about 60° N and S, the gas then sinks before returning toward the equator.

On Earth, the Hadley cells only extend to mid-latitudes before being disrupted by Earth's rapid rotation (see Chapter 3). On Venus, the Hadley circulation is much simpler because of its small temperature variations, the absence of oceans and the slow planetary rotation. However, several additional north-south circulation cells may also occur at lower levels.

The polar edge of the Hadley cell is marked by a "collar," where a wide (~1000 km), shallow (~10 km) jet stream of fast-moving cold air circles the poles at about 70° latitude. The clouds that mark each collar are colder and higher than their surroundings.

Inside each collar is a vast vortex, several thousand kilometers across, which often takes the form of a complex, double "eye" that rotates every 2.5–2.8 Earth days. The "eyes" are connected by S-shaped cloud structures. These vortices probably resemble the winter polar anticyclones that occur on Earth, especially over Antarctica. The circulation involves heavy, sinking air in the center and rising air in the cold polar collars.

The troposphere, where most convective motion takes place, extends to an altitude of about 60 km, about four times higher than on Earth. As on Earth, convectional motion has been

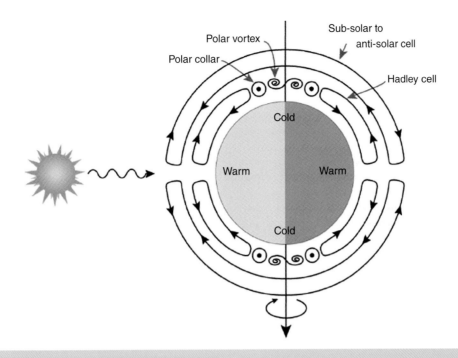

Figure 6.7 The general meridional (north-south) circulation of the atmosphere. The main feature is a convectively driven Hadley cell, in which the atmosphere moves away from the equator at an altitude of about 60 km, then sinks at about 60° latitude before returning towards the equator at a lower level. Above about 100 km, the circulation changes completely, with a sub-solar (midday) to anti-solar (midnight) pattern. Smaller vertical motions are associated with convection close to the subsolar point, where atmospheric heating is greatest. The overriding motion at the level of the cloud deck is the super-rotation from east to west (not shown). (ESA)

observed near the top of the Venusian clouds when huge bubbles of heated gas thrust upwards. These are particularly noticeable where the Sun is overhead.

Closer to the surface, where the atmosphere becomes extremely dense and there are no major variations in temperature, the atmospheric motion is fairly sluggish, generating only light breezes of about 1–2 m/s, in contrast to the super-hurricanes raging up above.

Although these modest surface winds should have enough momentum to move dust and fine sand, the thick air will certainly inhibit saltation – the bouncing motion of sand grains across the surface – and the erosion of rocks by abrasion or sandblasting.

Atmospheric Composition

Prior to the Space Age, there was some uncertainty over the composition of Venus' atmosphere. Spectroscopic data indicated that carbon dioxide was a major constituent, but there were conflicting views over the surface temperature and the amount of water that was present.

In 1918, the Swedish Nobel prize winner Svante Arrhenius declared that Venus was a wet planet covered in swamps and tropical vegetation. Others argued that the presence of plentiful carbon dioxide and the uncertainty in the amount of water vapor favored a desert planet swept by dust storms.

By the 1960s, observations of radio emissions from Venus clearly showed that its surface temperature is extremely high, removing any possibility of a watery environment. Spacecraft measurements confirmed that carbon dioxide makes up about 96% of Venus's atmosphere, compared with 0.03% on Earth (see Table 3). Apart from carbon dioxide and water vapor, the

Table 6.3
Composition of the Terrestrial Planet Atmospheres (% volume)[1]

	Venus	Earth	Mars
Carbon dioxide	96	0.03	95
Nitrogen	3.5	77	2.7
Argon	0.007	0.93	1.6
Water vapor	~0.01 (variable)	~1	~0.03
Oxygen	0.13	21	~0

[1] Some minor constituents are not shown.

atmosphere consists primarily of inert gases, particularly nitrogen and argon. Other minor constituents include sulfur dioxide, carbon monoxide, hydrogen chloride and hydrogen fluoride.

As is the case with the other rocky planets, the present-day atmosphere of Venus is almost certainly very different from the primordial atmosphere which existed more than four billion years ago (see Chapter 3). This gaseous blanket was probably lost long ago when the young Sun increased in surface activity and luminosity.

Today, the atmosphere is remarkably massive and dense. It contains 100 times as much gas as Earth's atmosphere, with a surface pressure of 92 bars. (One bar is roughly equivalent to sea level air pressure on Earth). This crushing pressure is equivalent to what a diver would experience at a depth of more than 900 meters in Earth's oceans.

The present atmosphere is the product of outgassing from the interior, a process which almost certainly continues today through active volcanism, and the influx of cometary and meteoritic material – rich in carbon, nitrogen and water. With similar sources of volatiles, Earth and Venus would be expected to have similar atmospheric compositions. However, at first sight the two "sister" worlds appear to possess very different amounts of carbon and water.

Runaway Greenhouse

One of the most intriguing mysteries about Venus is the reason for its completely hostile environment, with a sulfurous atmosphere, sweltering temperature and crushing atmospheric pressure.

Why is Venus so arid? The amount of atmospheric water vapor and water bound up with sulfuric acid and other compounds in the clouds is 100,000 times less than exists in Earth's oceans and atmosphere. On the other hand, the amount of deuterium (chemical formula ^2H or D, a "heavy" form of hydrogen) is about 100 times greater on Venus than on Earth. Although both deuterium and normal hydrogen escape from the atmosphere while there is free water on the surface, the heavier isotope escapes less efficiently – hence its relative abundance today. This suggests that Venus initially had much more water than is observed today, but most of it has been lost.

What about the present abundance of carbon dioxide on Venus, compared with our planet? Both worlds probably started out with similar amounts of CO_2, but on Earth most of it was dissolved in the oceans and eventually ended up in carbonate-rich rocks, such as limestone (see Chapter 3). In contrast, any hot ocean that existed on the young Venus would soon have evaporated.

Estimates of how long the proposed ocean may have survived vary considerably. One factor is the output of solar energy, which was perhaps 30% lower in the early history of the Solar System than it is today. Eventually, as the Sun became more active and surface temperatures inexorably rose, the ocean began to disappear. In the absence of liquid water, the carbon dioxide remained in the atmosphere, trapping heat like the windows of a greenhouse. The ocean would have totally evaporated, possibly within 600 million years of Venus' birth. This is known as the "runaway greenhouse" effect.

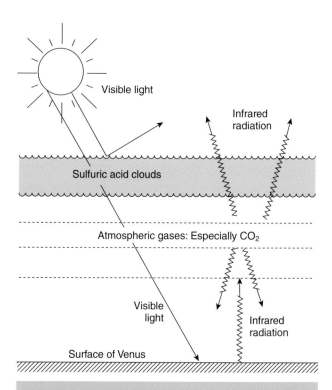

Figure 6.8 About 75% of the sunlight arriving at Venus is reflected by the high level clouds. However, some light filters through to the surface, where it is absorbed. The warmed ground emits infrared radiation, most of which is absorbed by carbon dioxide and other greenhouse gases in the atmosphere. (Tobias Owen)

An alternative scenario suggests that that the surface water could have persisted for up to two billion years. According to this theory, the global cloud blanket would have reflected a lot of incoming sunlight back into space, thereby limiting the rate of surface heating. At the same time, as the oceans slowly evaporated, perhaps 50% of the atmosphere was composed of water vapor.

Under those conditions, the vapor pressure was so high (i.e. the atmosphere was saturated with water vapor) that further evaporation was inhibited – rather like water in a pressure cooker. The continuing loss of hydrogen at the top of the atmosphere was steadily compensated by further evaporation from the ocean. Only after a very long period, perhaps several billion years, would the ocean have disappeared altogether. This is known as the "wet greenhouse theory."

Both scenarios are based on the fact that, although carbon dioxide is transparent to incoming solar radiation, enabling sunlight to warm the surface of the planet, it is also an efficient absorber of infrared radiation which is emitted from the hot ground (see also Chapter 7). Together with other greenhouse gases, such as water vapor and sulfur dioxide, the carbon dioxide built up in the atmosphere and formed a heat barrier which transformed the troposphere into a sweltering greenhouse.

Water vapor (H_2O) was carried to high levels, where it combined with sulfur compounds or was dissociated (split) into hydrogen and oxygen atoms by solar ultraviolet radiation. The light hydrogen gas then escaped into space, or was carried away by interaction with the solar wind, while the oxygen combined with sulfur dioxide or oxidized surface rocks, a process similar to rusting on Earth.

Over billions of years the atmosphere was affected by widespread, massive volcanic eruptions and changes in solar output, but today it has reached a state of equilibrium, so that Venus is losing as much heat as it gains from the Sun and the surface temperature is no longer increasing.

Box 6.1 The Vega Balloons

Both of the Soviet Vega spacecraft dispatched to explore Halley's comet also carried payloads that were dropped at Venus. During flybys in June 1985, each Vega released a spherical capsule into the planet's atmosphere, eventually deploying a surface lander and a balloon – the only time meteorological balloons have so far been used in Solar System exploration.

Constructed of Teflon fabric, they weighed 12 kg and, when inflated, measured 3.4 meters in diameter. Each balloon carried a payload of 6.9 kg on a gondola suspended beneath a 13 meter tether. The gondola carried pressure, light, and temperature sensors, a nephelometer cloud sensor, and an anemometer (wind speed gauge).

The balloons alternated between 25 minutes of data gathering (often taking measurements every 90 seconds in order to conserve power) and five minutes of data relay to Earth. They were released in the night hemisphere, and, after filling with helium, floated to their equilibrium altitude of 53.6 km. This altitude was chosen in part for its benign temperature – it was feared that they would explode in higher daytime temperatures.

Tracking by an international array of 20 radio telescopes showed that their flights began at 180°E and 7°N and S of the equator, respectively. Both balloons traveled more than 11 000 km and operated for 46.5 hours, passing into the daylight hemisphere, 33 hours after deployment, before their lithium batteries failed.

The Vega 1 balloon was blown around the planet at speeds up to 250 km/h. Strong vertical winds bobbed the craft up and down 200–300 meters through most of the journey. The atmospheric temperature averaged 40°C and pressure was 0.5 Earth atmospheres. The nephelometer found no clear regions in the surrounding clouds.

The second balloon began in a rather placid environment, but turbulence increased later. When it passed over a 5 km high mountain on the "continent" of Aphrodite Terra, it was pulled 2.5 km towards the surface by a powerful downdraft. The temperature measured by the Vega 2 balloon was consistently 6.5°C cooler than found by its counterpart. Unfortunately, its nephelometer failed to function, so no cloud data were returned. No evidence for lightning was sent back by either balloon.

Physical Characteristics

Venus is a rocky planet with a bulk density of 5.24 g/cm³, and a mass 81% that of Earth. This indicates that Venus formed in the inner region of the solar nebula, where it was too hot for volatiles such as ice to condense. As a result, the region was dominated by refractory elements, including iron, nickel, silicon and aluminum, which condense and solidify at high temperatures.

These physical characteristics suggest that the internal structures of Venus and Earth are very similar, though the lower bulk density indicates that Venus has a slightly smaller iron-rich core. This core may be divided into a solid iron inner region, surrounded by a molten outer core of iron mixed with nickel or sulfur.

One significant difference seems to be that Venus has a thicker crust than Earth. Comparisons of gravity data with surface topography show that the crust seems to be typically 25–40 km deep, increasing to perhaps 50 or 60 km under the older "tesserae" regions and the larger volcanoes. This thick, dry, rigid crust prevents Earth-like plate tectonics from taking place.

The Hidden Surface

The most promising technique for observing the surface is to transmit radar signals which penetrate the dense clouds and are then reflected by the rocky terrain. The echoes give important information about topography, surface texture and composition, for example, the more the return radar signal is scattered, the rougher the surface (see Box 6.2).

By the late 1970s, pioneering work with ground-based radar made it possible to detect features 10–20 km across on the hemisphere centered on 320°E. Particularly notable were the isolated,

Figure 6.9 Venus probably has a very similar internal structure to Earth, with a large, iron-rich core, a mantle and a rocky crust. The core may be rather smaller than that of Earth, but similarly divided into a solid inner region and a liquid outer core. (Calvin Hamilton)

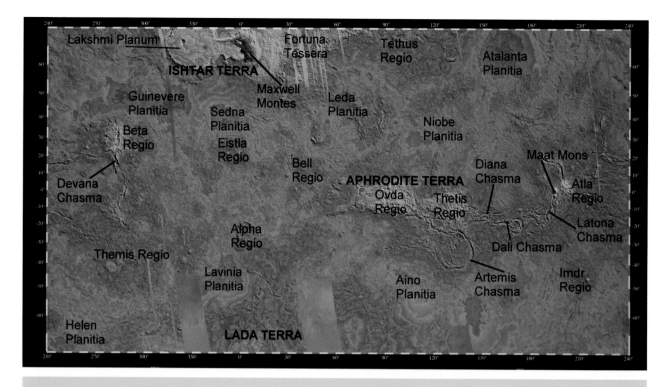

Figure 6.10 A topographic map of Venus obtained with the Magellan radar instrument. Highest land is red and yellow, lowlands are blue. Apart from the two upland regions of Ishtar (top) and Aphrodite, Venus is remarkably flat and low-lying. Most of the surface is within 500 meters of the planet's mean radius. Zero degrees longitude on Venus is measured from a small feature known as Eve Corona, on the south west slopes of Alpha Regio. Blank areas show where no data were obtained. (NASA-Ames)

radar-bright regions named Alpha, Beta and Maxwell. These were considered to be rough terrain, probably volcanic highlands or mountains. Also visible was a large canyon and a few craters, but most of the surface seemed fairly smooth and featureless.

The major breakthrough came with the arrival of NASA's Pioneer Venus orbiter in 1978. More detailed radar maps have since been provided by the US Magellan spacecraft (see Box 6.2).

The satellite data make it clear that Venus is very smooth, with few uplands or low-lying basins compared with Earth (or Mars). 80% of the surface mapped by Pioneer Venus lay within 2 km of the mean radius – the equivalent to sea level – and 60% within 500 meters of this level. Although there are several upland regions which rise 2 km or more above the general level, these only cover 5% of the total area. Most of Venus is made up of flat, rolling plains, such as Guinevere Planitia and Kawelu Planitia.

This is not to say that there are no extremes of relief. The highest topography, in a northern region known as Maxwell Montes, has peaks that rise 10.8 km above the average level, dwarfing Mount Everest on Earth. At the other extreme is a rift valley in the southern hemisphere known as Diana Chasma, which is

2.9 km below the average level – about one quarter the depth of Earth's Marianas ocean trench.

Ishtar Terra

The most prominent features on the radar maps are two large upland regions: Ishtar Terra in the north and Aphrodite Terra along the equator.[2] Ishtar, the size of Australia, is made up of three major units. At its core is Lakshmi Planum, a plateau which lies 4–5 km above the mean level – about the same height as the Tibetan plateau on Earth, but twice as large. On its northern and western flanks are the mountain ranges of Akna Montes and Freyja Montes, while to the south is Danu Montes.

However, the most spectacular mountains of Ishtar are Maxwell Montes, which lie to the east. Within these towering, radar-bright mountains is a 100 km diameter, double-ring impact basin known as Cleopatra. Other notable features on Ishtar are a large shield volcano known as Sacajawea, and Colette Patera, another volcanic center in the west of the plateau. The southern edge of Ishtar drops very steeply, suggesting possible faulting and tectonic

[2] **Almost all of the surface features on Venus are named after famous women and mythological goddesses (e.g. Ishtar was the Babylonian goddess of love).**

Figure 6.11 Sacajawea Patera is a 233 km wide caldera located in western Ishtar Terra. The central caldera is 1–2 km deep, 120 × 215 km in diameter, and bounded by a zone of graben and fault scarps. Extending up to 140 km toward the south east is a system of linear structures thought to be a rift zone along which the injection and eruption of magma may have occurred. (NASA-JPL)

activity on a large scale. The rough surface seems to be covered in loose debris.

Aphrodite Terra

Although Aphrodite is generally lower than Ishtar, it covers twice the area and is equal in size to Africa. It contains several highland regions to the east and west separated by a broad, lower area. The most notable of these mountains are Ovda Regio and Thetis Regio. However, the most intriguing features lie further east in the "scorpion's tail," where the surface is marked by a series of enormous, northeast-trending curved ridges and rift valleys, known as Diana and Dali Chasma. Diana Chasma, for example, measures 2500 km long and 280 km wide.

These valleys extend to a 3 km high upland region called Atla Regio. Also associated with these "chasmata" are impressive volcanic structures, such as 8 km high Maat Mons, and circular structures, called "coronae," which are marked by radial or concentric fractures. The origin of these features is a matter of debate (see below).

Weathering and Erosion

The absence of water and seasonal temperature variations limits the amount of weathering that takes place on the surface of Venus. Since wind speeds are low near the surface (1–2 m/s), the rapid movement of loose material and subsequent erosion through sand-blasting is also minimal. Erosion and resurfacing through

meteorite impacts is also reduced by the protective blanket of the atmosphere.

On the other hand, the dense atmosphere does possess sufficient momentum to transport loose material, as evidenced by radar images of dune fields and wind-blown streaks. Images from the Venera landers also showed pebbles and fine-grained debris, as well as large individual slabs, suggesting that solid rock can be broken down by mechanical or chemical processes. (An alternative theory is that this debris originated from a large volcanic eruption.)

The most prominent wind-related features in the Magellan images are streaks that form in the lee of large impact craters and other topographic obstacles as a result of deposition or removal of sand and dust. They can be used to calculate the direction of the most powerful or prevailing winds. However, some may be caused by the violent blasts generated during an impact event or a slower redistribution of fine impact debris by the wind.

Landslides are also quite common, particularly in the "scorpion's tail" region between Atla Regio and Aphrodite. Significant rock falls have occurred on the steep slopes of large volcanoes. Other landslides occur on the steep slopes that flank the major rift valleys or "chasmata." They are typically larger than terrestrial landslides but smaller than their Martian counterparts. Possible causes of these events may be oversteepening of slopes, perhaps by intrusion of magma, or Venusian quakes.

Impact Craters

About 1000 impact craters have been found on Venus. Although this far exceeds the known impact features on Earth, it is well below the number found on most of the other solid bodies in the Solar System. This is not too unexpected, since most meteorites would be expected to burn up in the dense atmosphere, while any ancient craters should be masked by billions of years of weathering, erosion and resurfacing. However, their modest number and fairly regular distribution also suggests that the entire visible surface may be geologically young.

Although they are widely scattered over the surface, fewer impacts are found on the uplands. The least common, but largest, impact structures are multi-ringed craters. They have smooth floors and are typically more than 100 km across. Largest of all is 275 km diameter Mead crater, located on the plains north of Aphrodite (Fig. 6.15).

More than one third of the craters are intermediate in size, with radar-bright central peaks or mounds and smooth floors. No impact craters smaller than 3 km across have been observed, presumably because any small incoming meteorites burn up or break apart in the dense atmosphere.

Volcanic Activity

About 1100 volcanic centers have been detected on Venus. They are much more widely distributed than on Earth, where the majority of volcanoes are restricted to linear zones near plate boundaries. However, very few of the large volcanoes are found in the lowland plains or the highland regions, with relative concentrations to the east and north east of Aphrodite. The reason

Box 6.2   Revealing Radar

Ground-based and spacecraft radar instruments have been powerful tools for studying the hidden Venusian surface. These observations take advantage of the fact that radio waves can penetrate the dense clouds and bounce back from the solid surface.

The echoes received on Earth give important information about the roughness and composition of the terrain, for example, the more the radar signal is scattered in all directions, the rougher the surface. Slopes that reflect the signal back in the direction from which it came appear bright, while slopes that face a different direction appear dark. Topographic information is obtained by accurate measurement of each signal's return time – an early echo shows an area of highland.

An early success for this radar technique was the discovery in 1962 that Venus rotates very slowly, so that one complete revolution takes 243 Earth days, in a retrograde (east to west) direction. It was also possible to build up a picture of one hemisphere from the radar echoes picked up by the Arecibo radio telescope in Puerto Rico.

Figure 6.12   This Magellan mosaic of Venus, centered at 180°E, was composed from radar images taken 1990–1994. Magellan imaged 98% of the surface at a resolution of about one hundred meters. Gaps were filled with images from the Earth-based Arecibo radar. White and brown areas are high, generally rough terrain; dark blue areas are low and usually smooth. To the west (left) is the upland of Aphrodite, with huge shield volcanoes such as Maat Mons in the "scorpion's tail" just right of center. At the top (north) is Ishtar Terra. (NASA-JPL)

It eventually became possible to detect features 10–20 km across on the hemisphere centered on 320°E. Particularly notable were the isolated, radar-bright regions named Alpha, Beta and Maxwell (in honor of British physicist James Maxwell). However, most of the surface seemed fairly smooth and featureless.

One limitation was the inability to accurately measure altitude. The breakthrough came through the radar altimeter on board the Pioneer Venus orbiter, which was capable of height measurements accurate to within 0.2 km. Over a period of two years, the instrument surveyed most of the planet between 73°N and 63°S, with a spatial resolution of 75 km (see Box 6.4).

Even before Pioneer Venus concluded its survey, both NASA and the Soviet Union launched spacecraft to map the surface in even greater detail. First to arrive were the Soviet Veneras 15 and 16, each equipped with a side-looking radar, which operated at a wavelength of 8 cm and was able to reveal structures less than 2 km across. Unfortunately, their elliptical orbits meant that only the northern hemisphere above 30°N could be mapped.

The images clearly showed linear ridges and folds skirting the edge of the Lakshmi Planum plateau, with Maxwell Montes dominating eastern Ishtar. Most controversy focused on strange circular features. Were they craters of volcanic or impact origin? Some, dubbed "arachnoids" because of their spider-like shape, were a total mystery.

Clearly, images with higher spatial resolution were required to clear up the mysteries. These were provided by NASA's Magellan spacecraft, which entered orbit around Venus in August 1990 (see Box 6.5).

Figure 6.13 A Magellan radar image, centered at 45.2°S, 201.4°E, showing landslide deposits 70 km across which apparently result from the collapse of a volcanic structure. The bright deposit probably consists of huge blocks of fractured volcanic rock up to several hundred meters across. The remnant of the volcano, about 20 km across, is seen in the center. Numerous small volcanic domes can be seen in the northern half of the image. Bright linear features trending to the north-northwest are ridges caused by tectonic deformation of the upper crust. (NASA-JPL)

Figure 6.14 A Magellan radar image showing a bright wind streak next to a 5 km diameter volcano located at the western end of Parga Chasma (9.4°S, 247.5°E). The north-east trending streak was probably formed by deposition of material on the lee of the volcano. The streak is 35 km long and 10 km wide. (NASA-JPL)

for this grouping is not fully understood, but it may reflect a cluster of closely spaced hotspots in a region of thinner, weaker crust.

The most notable volcanic region, and the first to be identified, is the radar-bright Beta Regio, which is centered at 27°N, 282°E. This is made up of two, 5 km high shield volcanoes known as Theia Mons and Rhea Mons.

Theia Mons is the largest volcano on Venus, reaching over 4 km high, with lava flows covering an area more than 800 km wide. At its center is an oval caldera roughly 75 km long and 50 km wide. The volcano also lies at the junction of three major rifts. The most notable of these, Devana Chasma, is over 200 km wide in places, and up to 3 km deep on the flanks of the volcano. Another rift cuts through Rhea Mons.

Like most volcanic centers on Venus (and the Hawaiian Islands on Earth), Beta Regio is thought to be associated with a hot spot in the mantle. According to this scenario, the region was uplifted and pulled apart by a plume of rising magma that could have formed at a depth of around 3000 km, which approximately corresponds to the core-mantle boundary.

The vast majority of Venus' volcanoes are small shields, less than 20 km across. Numbering more than 100,000, these dome-shaped hills are often found in swarms on the lowland plains and lower uplands. Although they may have contributed significantly to the resurfacing of the planet in the relatively recent past, a large proportion have been partly buried by later lava flows. Many of the shield fields cluster around larger shield volcanoes, reflecting a concentration of volcanic activity in these regions.

There are more than 150 large shield volcanoes. With diameters of between 100 and 600 km and rising up to 5.5 km above the surrounding terrain, they are generally broader and flatter than their counterparts on Earth, such as Mauna Loa (see Chapter 3). On the other hand, although some of the Venusian shields cover nearly the same area as Olympus Mons, which has a basal diameter of approximately 800 km, the Martian volcano towers to a much greater altitude and contains far more lava and ash.

Like their terrestrial counterparts, the Venusian shields have broad, gentle slopes and summit calderas, indicating that they are composed of non-viscous, basaltic lavas which flowed long distances from the central vent before they solidified.

The highest shield volcano on Venus is Maat Mons, in eastern Aphrodite. At its summit is a large caldera about 30 km across. Within the caldera there are at least five smaller collapse craters, up to 10 km in diameter. A chain of small craters 3–5 km in diameter extends some 40 km along the south west flank of the volcano. They may also have been formed by surface collapse, due to withdrawal of subsurface magma.

The variety of volcanic landforms is evidence that the viscosity of lava on Venus is quite variable, although the high temperature

Figure 6.16 Maat Mons is the highest shield volcano on Venus, rising 5 km above the surrounding terrain. It may still be active. Lava flows spread out from the volcano for hundreds of kilometers. Vertical scale has been exaggerated ten times. Simulated hues are based on color images obtained by Venera 13 and 14. (NASA-JPL)

Figure 6.15 Multi-ringed Mead crater, the largest impact feature on Venus, is located north of Aphrodite Terra (12.50°N, 57.20°E). Its flat, radar-bright floor has several fractures visible as brighter lines. The patchy outer ejecta are rougher than the crater floor, so they appear brighter. The surrounding plain is covered by fine debris that appears darker in the image. (NASA-JPL)

Figure 6.17 Most of the large volcanoes and calderas on Venus, outlined in yellow, are located near the equator, particularly to the east of Aphrodite, where there are major fault structures and rift valleys. The major uplands have very few large shield volcanoes. (LPI)

of the atmosphere may extend the cooling time of surface flows, enabling them to travel further than on Earth or Mars.

At one extreme are numerous channel-like features which appear to have been formed by fluid lava, rather than liquid water. 1–3 km wide and hundreds or thousands of kilometers in length, they resemble mature rivers with meanders, cutoffs and abandoned channel segments. However, Venus' channels are not as tightly sinuous as terrestrial rivers, and most are partly buried by younger lava plains, making their sources difficult to identify.

A few are associated with vast, radar-dark plains, suggesting large flow volumes. These appear to be older than other channel types, as they are crossed by fractures and wrinkle ridges, and are often buried by other volcanic materials. In addition, they appear to run both up and down slope, suggesting that the plains were warped by regional tectonism after channel formation.

The largest of all is Baltis Vallis, which is approximately 1.8 km wide and extends for about 6800 km across the northern lowlands. Several hundred kilometers longer than the Nile, Earth's longest river, Baltis is the longest known channel in the Solar System.

Its remarkable length and almost constant width imply that the channel was carved by huge volumes of fast-flowing, fluid lava. Even in the hot Venus environment, the lava must have had an unusual composition, perhaps a high temperature form of basalt, or lava containing a lot of sulfur or calcium carbonate.

On the other hand, a few volcanic features seem to have formed from silica-rich, viscous lavas, which solidified close to their magma source. They fall into three types.

Pancake domes are circular in shape and typically 750 meters high. Complex fractures on top of them suggest that a solid outer layer formed and then further lava flowing in the interior stretched the surface. Each dome also shows a small central pit. These pits look like the central vents found on shield volcanoes, but they seem to have formed after the domes themselves. After the domes grew during one large, slow eruption of viscous lava, the central pits seem to have formed as gases escaped from the magma or as the dome cooled and shrank. Many domes lie near corona structures in the lowland plains.

Peculiar "tick" shaped structures are often much larger, but have fairly flat, concave summits from which ridges and valleys radiate. The origin of these lateral ridges and troughs is unknown. They may mark the scars of landslides, or possibly volcanic dykes running away from the central summit. Like the pancake domes, most ticks occur in the lowland plains far from the major shields. Many are found near the lowland coronae, and some are quite close to one or more pancake domes.

A few volcanoes have thick, fan-shaped or banded flows. Since most basalt lava flows are very fluid and fairly thin, these volcanoes seem to be made from other material, perhaps granitic or rhyolitic lavas rich in quartz. Most examples are located in one area of the southern plains.

Whether any of the volcanoes on Venus are active remains a mystery. Marked variations in the amount of sulfur dioxide in the atmosphere from 1978 to 1983 have been attributed to large eruptions, but this has not been confirmed. Thermal measurements by Venus Express have also shown evidence of possible volcanic hot spots (see Fig. 6.30).

Figure 6.18 This Magellan radar image shows a 600 km long segment of Baltis Vallis, the longest sinuous lava channel on Venus (marked with arrows). Both ends of the channel are obscured, so its original length is unknown. (NASA-JPL)

Figure 6.19 Volcanic features such as this are dubbed "ticks," since they resemble tiny insects. This example lies near the edge of Eistla Regio (18°S, 5.5°E) and is 66 km across. It has a broad, flat, slightly concave summit with a central vent. The sides are characterized by radiating ridges and valleys of uncertain origin. To the west, the rim has been breached by dark lava flows. (NASA-JPL)

Figure 6.20 A chain of pancake domes located near the eastern edge of Alpha Regio (30°S, 11.8°E). The seven circular hills average 25 km in diameter, with maximum heights of 750 m. They are thought to have been formed by very viscous lava flows. The bright margins possibly indicate rock debris on their slopes. Some linear fractures on the plains cut through the domes, while others appear to be covered by them. (NASA-JPL)

Figure 6.21 As the name suggests, "arachnoids" are spider-like, circular or oval structures, with a central dome or depression surrounded by a complex network of radial and concentric fractures. They range from 50 to 230 km across and may be linked with volcanic domes that collapsed, forming cracks in the surface. This example is part of an arachnoid cluster in Fortuna Regio. (NASA-JPL)

Coronae and Arachnoids

Some enigmatic landforms, known as coronae and arachnoids, are apparently unique to the lowlands of Venus, with no counterparts on any other world.

Coronae are circular or oval features that are surrounded by rings of ridges and troughs. Radial fractures extend outwards, sometimes from the central regions. Some coronae are up to 2000 km across, but they differ greatly in size and detailed morphology – they may be domes, plateaus, plateaus with interior depressions, or rimmed depressions. There may be a central depression which lies above or below the surrounding plains, and a possible raised rim or a moat outside the rim. They may also contain one or more volcanic domes.

About 400 coronae have been observed on Venus. The largest of these is Artemis Corona, to the south of Aphrodite, which is almost surrounded by a huge trough, Artemis Chasma. The coronae tend to occur in linear clusters along the major tectonic belts and may form over localized upwellings of magma. When the subsurface magma retreats, the crust subsides, forming a depression surrounded by radial cracks.

Arachnoids are similar, but smaller, features which resemble spiders' webs in radar images. They may be precursors to corona formation. Cooling and contraction of magma may have caused the centers of volcanic mounds to collapse, cracking the crust and producing a distinctive pattern of radial and concentric fractures around a central depression. In some cases, lava may be extruded onto the surface as radial dykes.

In some cases, radial fracture patterns can be seen without a central dome or depression. Called "novae" these are believed to represent an even earlier phase of corona formation, when magma rises toward the surface, bulging it upward. Fractures radiate out from the center of the bulge, but subsidence due to cooling and withdrawal of magma has not yet created a central depression.

Tectonics

The crust of Venus is mainly basaltic, although the major uplands may bear some resemblance to Earth's continents, which contain granitic rocks high in silicon and oxygen. The surface is dominated by volcanic processes, but radar images also reveal extensive ridged plains, faulting, large rift valleys and towering fold mountains.

The most widespread geological terrain consists of ridged plains, volcanic areas which have been deformed by wrinkle ridges. These are concentrated in the lowlands and cover 60–65% of the surface. Since they are overlain by younger material in some areas, the ridged plains must have once occupied an even larger area.

Few impact craters have been deformed by the ridges, implying that the formation of the plains and most of the wrinkle ridges took place soon after the apparent resurfacing of the planet about 600 million years ago (see below). They were created over a fairly short period – no more than 100 million years – after which they were pockmarked by new impact craters.

As is the case with similar structures on the other terrestrial planets, wrinkle ridges are thought to be the products of

Figure 6.22   This area, which covers 945 km² is centered at 29.8°S, 274.3°E. It contains three features along a rift zone that illustrate the possible evolutionary sequence of corona structures. With the rise of a hot magma body a few hundred kilometers across in the mantle, the surface domes and cracks in a radial fashion (profile A-A'). As the domed structure begins to collapse, concentric faults and a surrounding trough begin to form (profile B-B'). When collapse is complete, it creates a central plateau or depression with a raised rim, surrounded by a circular trough (profile C-C', Gertjon Corona). (LPI)

horizontal shortening of the lithosphere, the outer layer of the planet. This would account for the consistent alignment of most wrinkle ridges over areas covering thousands of kilometers and the tendency for the ridges to encircle the major highlands.

The plains also contain regions of more marked tectonic deformation, known as "ridge belts." These often take the form of broad, linear hills rising a few hundred meters above their surroundings.

About 8% of Venus's surface is made up of tesserae. These are areas of very rough, elevated terrain which show evidence of a complex deformational history – ridges, fractures and rift valley structures (graben). Several large elevated areas, such as Alpha, Ovda and Thetis, are composed largely of tesserae. In some cases they display parallel ridges and troughs that cut across one another at many different angles.[3]

The deformation of tesserae is sometimes so complex that it is difficult to determine whether horizontal compression of the crust followed or preceded extension. However, it is widely accepted that they precede the major volcanic resurfacing event and represent the oldest terrain on Venus.

Tesserae are sometimes bordered by mountainous fold or thrust belts, further evidence of crustal convergence. The most impressive of these is Maxwell Montes, which rises to an altitude

of 12 km above the mean level. Mountain ranges also occur elsewhere around the edge of the Lakshmi plateau. The slopes of these mountains are very steep, with gradients of up to 30° over distances of 10 km, despite the searing heat which would tend to weaken the rocks and reduce the slopes.

It seems that these towering mountains are only able to persist through continuing compression of the crust. If this was removed, the mountains would quickly subside. One possibility is that convection currents in the mantle are dragging two regions of crust towards each other. Where the currents converge and sink, the crust piles up into a thick, compressed mass, forming the mountain ranges.

Data based on perturbations of orbiting spacecraft also show that high gravity correlates with thick crust and high topography. This suggests that the mountains of Ishtar Terra are made of low density rock that rests on the denser mantle and not uplifted by magma plumes thrusting upwards from beneath.

Uncertainty also exists over the origin of the deep, narrow canyons called "chasmata." These troughs often have a raised rim and may be many thousands of kilometers long, up to 100 km wide and up to 10 km deep. Some chasmata occur in major volcanic highlands, such as Atla Regio and Beta Regio. Many are associated with volcanism and are located around the margins of major coronae, such as Artemis and Latona.

[3] The term "tessera" comes from the Latin term for "mosaic tiles."

Box 6.3   The Venera Landers

Most of the *in situ* observations of the Venusian atmosphere and surface have been provided by the Soviet Venera probes. After a number of failures, Venera 4 succeeded in transmitting data during its October 1967 descent, recording a maximum temperature of 277°C and a pressure of 22 Earth atmospheres. However, Soviet claims that it had reached the surface had to be withdrawn after subsequent analysis of Mariner 5 data. The first direct sampling of the atmosphere found 90–95% carbon dioxide, plus small amounts of oxygen and water vapor.

The first successful landing took place on December 15, 1970 – when Venera 7, equipped with a smaller parachute, was able to make a more rapid descent on the planet's night side, eventually touching down at 5°S, 351°E. Unfortunately, contact was lost almost immediately after arrival.

In 1972, Venera 8 landed on an upland plain not far from its predecessor. Fifty minutes of data from the surface confirmed the dense carbon dioxide atmosphere and 470°C temperature. Measurements of natural gamma radiation indicated a surface made of granite-like rock. This implied the presence of water long ago, since granite is formed on Earth when basaltic rocks are driven down into the mantle by plate movements, mix with water and then slowly re-emerge.

A simple light meter indicated that lighting was equivalent to pre-dawn illumination on Earth. It was later determined that the gloom was caused by the Sun lingering only 5° above the horizon.

Figure 6.23   TV cameras on Venera 10 produced a 180°, black-and-white panoramic scan of a rock-covered, rolling plain to the south east of the Beta Regio volcanic rise. The sketch shows the area scanned. (NASA)

The next generation of landers was better equipped to survive the hostile environment. Fitted with an internal cooling system and a partially collapsible "donut" base attached to shock absorbers, Veneras 9 and 10 completed their descent by aerobraking in the dense atmosphere, rather than using parachutes all the way down. Their instruments studied the cloud decks and atmosphere all the way down to the surface.

Both spacecraft touched down on the slopes of Beta Regio. Powerful floodlights on each craft were not needed, since illumination was comparable to a cloudy day on Earth. In contrast to the hurricanes raging high above, surface wind speeds were only 1.4 to 2.5 km/h. Suggestions that Beta is a giant volcanic complex were supported by gamma ray spectrometer readings indicating a basalt lava composition. However, the most notable breakthrough was the return of the first images from the surface.

The 180° black-and-white panoramic view from Venera 9 showed a possible rock slide, with 30–40 cm flat, angular boulders as far as the horizon, about 100 meters away. Venera 10, which landed some 2200 km away, revealed an older, flat plain covered with numerous rocky slabs, possibly evidence of weathered lava flows. Between the slabs were pebbles, sand and dust.

The next pair of landers did not return any images, and few details of the surface conditions were returned. The most intriguing observations involved low frequency radio bursts that were thought to be associated with extensive lightning activity. (Venera 9 had previously recorded a brief glow in the night sky, suggestive of lightning.)

In March 1982, Veneras 13 and 14 landed about 1000 km apart, to the east of an upland area known as Phoebe. Venera 13 set down on rolling plains, recording a temperature of 457°C and a pressure of 89 atmospheres, while Venera 14 landed in a low-lying basin at 465°C and 94 atmospheres. They sent back the first (and so far, only) color pictures from the surface, revealing rust-colored landscapes draped beneath a lurid orange sky.

Venera 13 landed with a tilt of about 8°. Its 360° panorama showed a flat area with scattered angular rocks and pebbles of all sizes. A new instrument, known as an X-ray fluorescence spectrometer, was used to analyze a small sample of rock drilled from a depth of 3 cm. The results indicated the presence of an unusual type of basalt that is usually associated with continental volcanism on Earth. Venera 14 came down on a tabular terrain marked by continuous layers of bedrock divided into slabs. Analysis indicated a composition similar to the fluid basalt lavas found on Earth's ocean floors.

The final Soviet landings took place in June 1985, when Vegas 1 and 2 dropped off capsules and balloons *en route* to comet Halley (see Box 6.1). No surface images were returned.

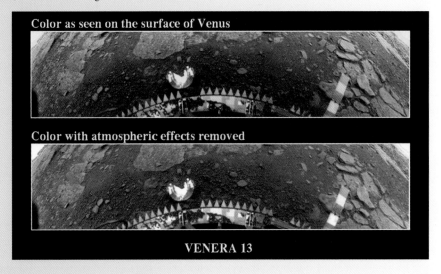

Figure 6.24   Venera 13 carried a twin camera system that imaged a 360°, panorama, half of which is shown here. The color images showed a surface with an orange hue as a result of the atmosphere absorbing light at blue and green wavelengths. When this effect is eliminated, the rocks appear gray. In the foreground is a discarded lens cap. The saw-toothed object at the bottom of each image is the base of the lander. The teeth are about 5 cm apart. Also visible is a color bar for image calibration. Note the flat slabs separated by small stones and other loose debris. (James Head, Brown University/USSR Academy of Sciences)

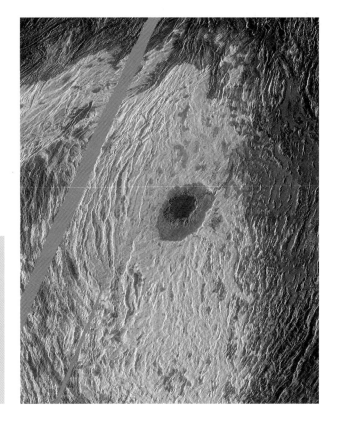

Figure 6.25   Magellan radar image of western Maxwell Montes, the highest mountains on Venus, which rise up to 12 km above the mean planetary surface. The mountains were formed by compression of the crust. Although they reflect radar strongly, this may be due to the presence of an unusual radar-reflective mineral, such as iron pyrite, rather than a rough surface. The 105 km diameter Cleopatra impact crater can be seen at upper right. The distance from Cleopatra to the steep south western slopes of Maxwell Montes is about 350 km. (NASA-JPL)

The shapes and sizes of these features suggest that they are rift valleys caused by the crust pulling apart – rather like the East African Rift Valley on Earth. However, some chasmata appear to be associated with crustal compression – more like oceanic trenches on Earth.

A One Plate Planet?

On Earth, fold mountains, volcanoes and rifts are generally located in linear zones along plate boundaries. No such pattern exists on Venus. It seems evident, therefore, that Earth-like plate tectonics, involving destruction of crust along subduction zones and creation of crust along ocean ridges, do not occur on Venus. Indeed, Venus has been described as a one-plate planet, in contrast to the dozen or so rigid lithospheric plates that exist on Earth.

Assuming that Venus has an Earth-like, mobile mantle, the absence of plate tectonics could be accounted for by a thicker crust which has difficulty in sliding over the mantle. It may also be linked to the fact that Venus is much drier, deprived of the water that lubricates Earth's plate tectonics. (Water lowers the melting point of lithospheric rocks and makes them more malleable.)

Scientists hoped that gravity and magnetic field measurements by orbiting spacecraft would provide definitive evidence about the thickness of the crust. Unfortunately, the data are open to different interpretations, so that estimates of the thickness of the Venusian lithosphere range from perhaps 60 km to about 300 km.

However, radar images have provided strong evidence that the average age of the surface is 300–750 million years, indicating that the entire planet may have been resurfaced by volcanic eruptions in the geologically recent past. Not only are the volcanic

Box 6.4 Pioneer Venus

In contrast to the sustained efforts of the Soviet Union to send landers to Venus, NASA concentrated its early efforts on the Pioneer Venus mission, which involved an orbiter and five atmospheric probes.

The drum-shaped orbiter, sometimes known as Pioneer 12, carried a dozen instruments to explore the planet's interaction with the solar wind, the composition of the clouds and the upper atmosphere, the shape of its gravitational field and the topography of its hidden surface. Its most innovative experiment was a radar mapper/altimeter to reveal the nature of the surface after centuries of speculation.

In contrast, the Multiprobe mission of Pioneer 13 was designed to deliver four atmospheric probes which were attached to the top of a 2.5 meter diameter "Bus." After separating from the Bus, the probes, protected by heat shields, were designed to send back data during their descent to the surface.

Figure 6.26 Pioneer Venus comprised two separate missions. The Multiprobe mission (left) included the main Bus, which was not designed to survive atmospheric entry, and four atmospheric probes, each protected by a heat shield. The large Sounder probe carried seven instruments, compared with three on the North, Day and Night probes. The Pioneer Venus Orbiter (right) entered a near-polar, elliptical orbit, observing the atmosphere and mapping most of the surface with radar. (NASA-Ames)

The orbiter set off from Cape Canaveral on May 20, 1978 – following a long, looping trajectory that minimized the amount of onboard propellant required. Pioneer 13 took a faster route, so although the probes left Earth 2½ months after the orbiter, they slammed into Venus' atmosphere only five days after Pioneer 12 arrived on December 4.

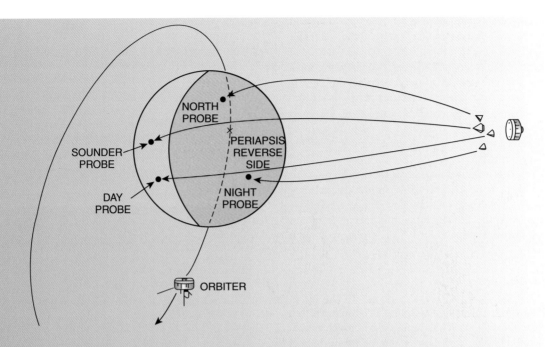

Figure 6.27 The Pioneer Venus Multiprobe mission began on November 15, 1978, with the release of the large Sounder probe. Its three small companions followed four days later, when the Bus was 9.3 million km from Venus. They were targeted to investigate different regions of the Earth-facing hemisphere – two on the night side and two on the day side. The Sounder probe slammed into the atmosphere on December 9, followed a few minutes later by the other probes. Each hit the surface at 32–35 km/h, but only the Day probe sent back any meaningful data after impact. (NASA-Ames)

The Sounderprobe entered the atmosphere at 42 000 km/h. Once its main parachute opened, the probe continued to descend for 16 minutes before the chute was cut loose. The final free fall lasted 39 minutes, ending with a loss of contact on impact.

The smaller probes followed diverging paths: one of them plunged into the atmosphere on Venus' day side, while the other two headed into the night. They survived their ballistic descent, and by the time they hit the ground, the dense atmosphere had slowed them to a speed of 35 km/h. The only useful data from the surface came from the Day probe, which continued operating for 67 minutes.

Meanwhile, Pioneer 12 went into a near-polar orbit, which enabled it to survey almost the entire planet. The spacecraft survived for 14 years before it ran out of fuel and was destroyed in the dense atmosphere. During its long life, it mapped most of the surface, studied the clouds and upper atmosphere, and confirmed that Venus has a negligible magnetic field.

landforms and impact craters uniformly distributed, but all signs of any older, heavily cratered, terrain have been destroyed, with the possible exception of the major upland regions.

This has led scientists to propose a number of different scenarios:

1. Venus may undergo episodic plate tectonics. For a long period of time, creation of new crust and destruction of existing crust ceases, so that heat builds up in the interior. Eventually, the surface becomes disrupted and catastrophically sinks into the mantle. The result is rapid overturning of the interior and virtually planet-wide volcanism, creating a vast magma ocean. This extracts so much heat from the interior that the mantle cools and the activity dies down, enabling a solid surface to be re-established. After a period of at least 500 million years, the process repeats itself. In this scenario, the visible surface is simply a record of the last resurfacing episode, some 600 million years ago.

2. Venus was continuously active, experiencing plate tectonics similar to those found on Earth. However, the interior then cooled sufficiently to shut down the crustal recycling process about 600 million years ago. In this case, convection in the mantle will continue to exert stresses on the lithosphere, but these are relieved in hot spots or relatively small compression zones instead of being concentrated at plate boundaries.

Magnetic Field

Spacecraft data show that the planet has a magnetic field no more than 0.09% the strength of Earth's. Since, like Earth, Venus almost certainly has a liquid, iron-rich core, the explanation must lie in its very slow rotation, which prevents the generation of an internal dynamo.

This means that there is no magnetic shield opposing the incoming solar wind. Instead, the solar protons and electrons strike the upper atmosphere, splitting neutral atoms to produce a region of positively charged particles known as the ionosphere (see Fig. 6.28).

The height of the ionosphere's upper boundary, the ionopause, varies with the strength of the solar wind, dropping to 250 km and rising to a height of 1 500 km. During the continuous battle with the solar wind, this region of the upper atmosphere is able to slow and divert the flow of particles around the planet, creating a magnetosphere shaped rather like a comet's tail.

Although there is a well-developed bow shock in Venus' outer atmosphere, there is no evidence of charged particles being trapped. As a result, there are no radiation belts above the planet. The negligible magnetic field also means that only feeble auroras occur on the planet's night side.

The Latest Missions

Venus has been visited by more spacecraft than any other planet. The most recent missions are ESA's Venus Express, which entered orbit on April 11, 2006, and Japan's Venus Climate Orbiter (Akatsuki).

Venus Express carries seven scientific instruments to investigate the structure, composition and dynamics of the Venusian atmosphere from a polar orbit that ranges between 250 and 66,000 km.

It confirmed the presence of a suspected double vortex above the south pole – a mirror image of the vortex in the northern hemisphere. Images making use of "infrared windows" in the atmosphere showed complex cloud structures, highlighted by thermal radiation rising from different atmospheric depths.

Figure 6.28 Venus has no internal magnetic field, so the solar wind interacts directly with the upper atmosphere. The solar particles are slowed when they reach the ionosphere, with a bow shock forming upstream. The solar wind then flows around the planet, forming a long magnetotail. This process slowly removes electrified (ionized) gases from the upper atmosphere. (After UCAR)

Box 6.5   Magellan

NASA's Magellan orbiter, which entered a near-polar orbit around Venus in August 1990, provided the most detailed radar images and surface maps of the planet. During each orbit, its synthetic aperture radar (SAR) pierced the cloud blanket, imaging the landscape below in narrow strips about 24 km wide and 16 000 km long. By studying radar echoes from the surface, scientists were able to determine the nature of the terrain and map the topography.

Figure 6.29 Magellan followed an elliptical, near-polar orbit around Venus between 1990 and 1994. When the orbit was close to Venus, the synthetic aperture radar imaged a swath between 16 km and 27 km wide, beginning near the north pole and continuing to the southern hemisphere. Subsequent swaths slightly overlapped, enabling the instrument to map 98% of the planet. Near apoapsis (its furthest distance from Venus), the high-gain antenna was turned toward Earth and data were transmitted. Magellan acquired radar data for about 37 minutes during each three-hour orbit. (NASA-JPL)

Over a period equal to three Venus rotations (roughly two Earth years), Magellan mapped 98% of the surface with a spatial resolution of 120–250 m. During repeat passes, images taken from different viewing angles made it possible to obtain stereo views and search for surface changes. In September 1992, a fourth observation cycle was devoted to mapping the planet's gravity and internal structure.

By the end of its mission in October 1994, Magellan had returned more data than all previous US planetary missions. Perhaps the most notable discovery was the importance of volcanic activity on Venus. There were volcanoes of all shapes and sizes, as well as unique features termed "coronae" and "arachnoids". The relative absence of impact craters suggested that Venus had been almost entirely resurfaced by lava flows about 600 million years ago.

Figure 6.30 Idunn Mons volcano in Imdr Regio (46°S, 214.5°E). The brown area is 3-D terrain data generated by the Magellan radar instrument, vertically exaggerated 30 times. Bright areas are rough or have steep slopes, dark areas are smooth. The colored overlay shows heat patterns derived from surface brightness data collected by the Visible and InfraRed Thermal Imaging Spectrometer on Venus Express. Temperature variations due to topography were removed. The brightness indicates the composition of the minerals. Red-orange is the warmest area and purple is the coolest. The warmest area – probably recent lava flows – is centered on the summit, about 2.5 km above the plains. Idunn Mons has a diameter of about 200 km. (ESA/NASA-JPL)

Other instruments detected hazes up to an altitude of 105 km, and oxygen ions being stripped away from the upper atmosphere by the solar wind.

Venus Express also made the first map of Venus's southern hemisphere at infrared wavelengths, providing some clues about the chemical composition of the rocks. The new data support suspicions that the highland plateaus of Venus are ancient continents, possibly composed of granite.

It has also provided the best evidence yet that the planet remains volcanically active. Although no unusual hot spots have been found on the surface, studies of the amount of infrared radiation emitted by different surfaces (the emissivity) show that Idunn Mons, and two other suspected volcanic areas in Themis and Dione Regio, are markedly different from the surrounding terrain (see Fig. 6.30). This may indicate that fresh lava flows spread across the surface only a few thousand or tens of thousands of years ago.

Akatsuki, Japan's first mission to Venus, was launched on May 21, 2010, but it failed to enter orbit around the planet on December 7, 2010. It may be possible to make a second attempt after flying around the Sun for another six years. The science payload includes advanced infrared cameras, an ultraviolet camera, a high-speed lightning sensor and a radio science experiment. It is designed to undertake comprehensive observations of the planet's atmosphere, surface and interaction with the solar wind.

Questions

- Venus and Earth are often described as "twin" or "sister" planets. Summarize the reasons for this description.
- Briefly describe the ways in which the physical characteristics of Earth and Venus are different, giving reasons for these differences.
- Why is Venus said to have experienced a "runaway greenhouse" effect? Give reasons why such an effect could or could not take place on Earth.
- What evidence supports the theory that Venus was almost completely resurfaced about 600 million years ago? How might this resurfacing have taken place?
- Briefly describe the appearance and possible formation process of these Venusian landforms: coronae; tesserae; arachnoids; wrinkle ridges; chasmata.
- Write a proposal for a future space mission to Venus, including a brief description of the method of exploration and the hardware you would use. Give reasons for your choices.
- Explain why Venus has no magnetic field. How does this influence its near-space environment?

SEVEN

Mars

Mars is the fourth planet from the Sun and one of Earth's closest neighbors. Seen with the naked eye as a bright, reddish-orange "star," Mars is named after the ancient Roman god of war. However, its color is not caused by human blood, but the presence of weathered, iron-rich minerals in its surface rocks and dust. In the late 19th and early 20th centuries, science fiction writers (and some serious observers) believed that the planet might harbor vegetation, possibly cultivated by intelligent beings, but the advent of space probes has made it clear that, if any life forms do exist on the red planet, they can only take the form of simple microorganisms.

As one of the "wandering stars" identified in prehistoric times, Mars appears in the night sky every 26 months or so, moving fairly rapidly through the constellations and increasing in brightness until it outshines everything apart from the Moon, Venus and Jupiter. Observations of its looping path in the sky around these times of opposition caused by Earth overtaking Mars on the inside track, helped to confirm the Copernican system of a heliocentric Universe.

Since it lies beyond Earth's orbit, Mars travels more slowly around the Sun, with an orbital period (its year) of 687 Earth days. The average distance from Mars to the Sun is about 227,920,000 km, roughly 1½ times the Earth-Sun distance. The orbit is tilted to the ecliptic by less than 2°.

The times of opposition and the brightness of the planet at such times are quite variable, because Mars follows a more eccentric orbit than most of the planets. The Sun-Mars distance can range from 206.6 million km to 249.2 million km. Consequently, the minimum distance between Earth and Mars at opposition also varies considerably, ranging from 55.7 million km to 101 million km. During the most favorable oppositions, Mars reaches visual magnitude −2.9, outshining Jupiter in the night sky.[1]

The apparent size of the planet's disk changes correspondingly. It is largest (more than 25 arcsec) during close oppositions, when considerable surface detail is visible with modest telescopes, and smallest (less than 14 arcsec) during distant oppositions, when surface observations are seriously limited.

Since it is further from the Sun and receives less than half as much solar radiation as Earth, Mars is much colder. Surface temperatures vary from −130°C near the poles during the winter to over 30°C at midday near the equator. The average temperature is −63°C, comparable to severe conditions in Antarctica.

Sols and Seasons

Like most of the planets, Mars rotates on its axis from west to east. Visitors would see a much smaller, dimmer Sun in the pinkish sky, but otherwise conditions of illumination would be quite familiar. The Martian solar day (sol) – the time in which the planet rotates once with respect to the Sun – lasts 24 hours 39 minutes 35 seconds, slightly longer than a day on Earth.

Since the axes of both planets are tilted by similar amounts (see Table 7.1), relative to the plane of their orbits, they also have seasons. The axial inclination causes the amount of sunlight falling on certain parts of the planet to vary widely: regions near the poles experience continuous sunlight in summer, but nonstop darkness in winter.

[1] The closest opposition of historic times took place on August 28, 2003, when Earth and Mars were only 55.758 million km apart. We will have to wait until August 28, 2287, for a closer encounter.

Exploring the Solar System, First Edition. Peter Bond.
© 2012 by Peter Bond. Published 2012 by Blackwell Publishing Ltd.

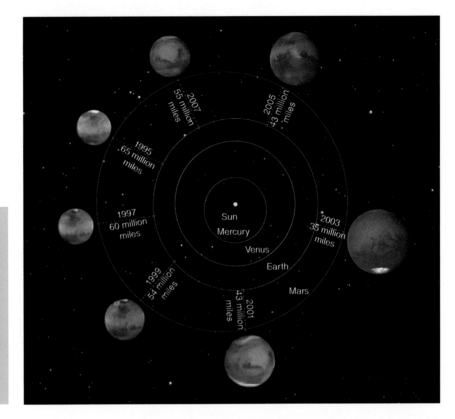

Figure 7.1 Mars follows a much more elliptical orbit than Earth. As a result, the Earth-Mars distance at each opposition varies considerably. In the period 1995–2007, the gulf between them ranged from 55.76 million km in 2003, when Mars was near perihelion, to 101.08 million km in 1995, near aphelion. The images compare the planet's apparent size at each opposition. (Z. Levay/STScI)

Table 7.1
Comparison of Mars and Earth

	Earth	Mars
Diameter (km)	12 756	6792
Rotation period	23 h 56 m	24 h 37 m
Density (water = 1)	5.52	3.91
Mass (Earth = 1)	1	0.11
Surface gravity (Earth = 1)	1	0.38
Inclination of axis to orbit	23.45°	25.19°
Orbital period (days)	365.24	686.93
Average distance from Sun (million km)	149.6	227.9
Average temperature (°C)	7	−63
Satellites	1	2

Table 7.2
Length of Seasons on Mars and Earth (Northern Hemisphere)

	Earth (days)	Mars (approx. Earth days/sols)
Spring	93	199/194
Summer	94	184/178
Autumn	89	146/142
Winter	89	158/154

The situation is complicated, however, by the eccentric orbit of Mars. On Earth, spring, summer, autumn, and winter are all similar in length, because the orbit is nearly circular and the planet moves at an almost constant speed around the Sun. However, the elongated orbit of Mars causes its seasons to vary significantly in length (see Table 7.2 and Box 7.1). Since it travels faster along its orbit near perihelion, the southern spring and summer are noticeably shorter than the southern autumn and winter (and *vice versa* in the northern hemisphere). Martian months are defined as spanning 30° in solar longitude, but, due to the eccentricity of Mars' orbit, their length varies from 46–67 sols.

It also means that Mars receives 44% more insolation at perihelion than at aphelion. Hence, the pole which is tilted toward the Sun at perihelion has warmer summers than the other pole. At present, the south has warmer summers, but this situation is not permanent. As a result of precession – a slow change in the direction of tilt of the rotational axis – the hot and cold poles switch places every 51,000 years.

The obliquity of Mars is subject to large, chaotic changes, whereas Earth experiences only minor changes. The axial tilt of Mars generally ranges between 15° and 35°, but it may occasionally reach 60°, when the planet is almost lying on its side, relative to its orbital plane. Summer temperatures at the poles change

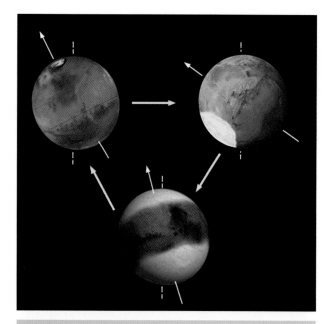

Figure 7.2 The evolution of Martian surface ice during an obliquity cycle. The angle between the white arrows and the dotted line denotes the obliquity (axial tilt). At high obliquity, the northern cap sublimates, losing a few centimeters of ice each year. The water vapor is then deposited as snow or frost in the cooler equatorial zones. When the obliquity decreases, ice sublimates from the lower latitudes and freezes out at high latitudes, resulting in growth of the polar caps. (ASD/IMCCE-CNRS, adapted from Jim Head/ Brown University and NASA/JPL)

dramatically during this obliquity cycle, becoming much warmer when the axial tilt is high than when it is fairly small. This means that Mars undergoes major climatic changes, shifting from ice ages to relatively warm interglacials.

At present, the fairly low obliquity means that it is coming out of an ice age which occurred between 500,000 and two million years ago. During this interglacial period, the ice cover has been degrading in the mid-latitudes as water ice sublimates and freezes out at the poles.

Physical Characteristics

With an equatorial diameter of 6792 km, Mars is considerably smaller than both Earth and Venus, but larger than Mercury. Its fairly rapid rotation causes it to bulge outward at the equator, so that its diameter is about 40 km less at the poles than at the equator.

It is also quite lightweight compared with Earth, which is about 10 times more massive. With an average density about 3.9 times that of water, Mars has roughly 70% of Earth's density and is clearly made of rocky material.

Because Mars is so much smaller and less dense, its surface gravity is only about 38% of that on Earth. Objects would fall to the ground more slowly than on Earth and lifting of heavy objects would be much easier.

Interior

Theories about the nature of the interior are based on modeling, Martian meteorites, comparisons with Earth, and the shape, mass, and moment of inertia of the planet.[2] It is almost certainly layered as the result of differentiation of lighter and denser materials very early in its history (see Fig. 7.5).

The average thickness of the crust is probably about 50 km – more than on Earth – since there is no evidence of craters that have punched through to the mantle, or of surface deformation due to the weight of overlying ice caps. On Earth, the weight of a similar stack of ice would cause the surface to sag, so the Martian lithosphere – its crust and upper mantle – must be markedly thicker and colder.

This is not surprising, since the outer layers of the smaller planet should cool faster and solidify to a greater depth. However, the crust is far from uniform in thickness, being thicker and older in the south than in the north (see below).

The mantle is probably similar in composition to Earth's mantle, and composed mainly of peridotite, a rock made up chiefly of silicon, oxygen, iron, and magnesium. The most abundant mineral in peridotite is olivine. The average temperature of the mantle may be about 1500°C, with the radioactive decay of elements such as uranium, potassium and thorium as the main source of internal heat.

The dense core is probably composed of iron, nickel, and sulfur. Models of how the planets condensed from the solar nebula suggest that Mars, being farther from the Sun, should have incorporated a higher proportion of moderately volatile elements, such as sulfur, sodium, and potassium, than Earth.

The relatively low density of the planet, compared with Earth, indicates that the core makes up a smaller proportion of the interior. If it is mainly composed of iron, then the minimum core radius would be about 1300 km. If it comprises less dense material, such as a mixture of sulfur and iron, the maximum radius may be up to 2000 km.

Until recently, it was widely assumed that – since Mars has a relatively small volume – its core has largely cooled and solidified, but recent modeling suggests that the outer core may still be liquid, with plumes of hot material rising from the core-mantle boundary.

Impact Features

Mars has many surface features that are common on Earth: plains, canyons, volcanoes, valleys, gullies and polar ice. The main exception is impact craters. When the Mariner spacecraft sent back the

---

[2] The moment of inertia is a measure of the way in which the planet rotates on its axis, which is influenced by the distribution of material within it. The measurements reveal that the central core is dense while the outer regions are less dense.

Box 7.1   The Martian Calendar

A Martian year lasts for 668.6 sols (Martian days). Unlike on Earth, there is no large satellite that can be used to create a calendar based on "months." Scientists mark the progress of the year on Mars by referring to solar longitude (Ls). Martian months are defined as spanning 30° in solar longitude. Ls is 0° at the vernal equinox (beginning of northern spring), 90° at summer solstice, 180° at autumnal equinox, and 270° at winter solstice.

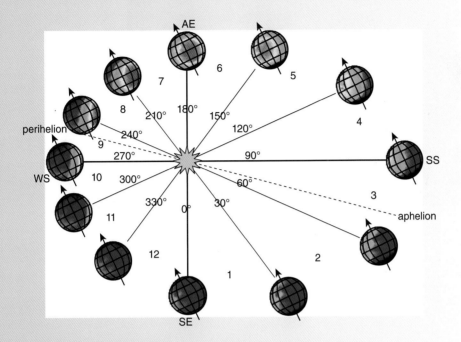

Figure 7.3   Martian seasons are based on the solar longitude, abbreviated Ls. Ls is 0° at the spring equinox (SE), the beginning of northern spring, 90° at summer solstice (SS), 180° at autumnal equinox (AE), and 270° at winter solstice (WS). The orbit is quite eccentric, so the planet's distance from the Sun is 206.6 million km at perihelion and 249.2 million km at aphelion. (Laboratoire de météorologie dynamique)

Mars currently reaches aphelion at Ls = 71°, near the northern summer solstice, and perihelion at Ls = 251°, near the southern summer solstice. The Martian dust storm season usually begins just after perihelion, at around Ls = 260°.

Figure 7.4   Mars and Earth shown to scale. With a diameter of 6792 km, Mars has a surface area comparable to all the land masses on our planet. Whereas Earth is largely covered by oceans, Mars is covered by ochre deserts. (NASA)

Crust
(silicate)

Mantle
(iron oxide-rich, silicate)

Core
(iron-nickel alloy,
iron sulfide)

MARS

Figure 7.5 Mars has a layered interior, with a solid crust, mantle and core. The outer core may still be at least partially liquid and conductive (though incapable of generating a global magnetic field), while the inner core may be solid. (LPI)

Box 7.2   Dating Mars

Mars appears to have experienced a similar cratering history to the Moon, although we do not know if the timing coincides precisely. As with all rocky planets, the age of the surface is estimated by counting craters: a higher number and density of craters indicates older terrain (see Chapter 4). Complications arise because Mars has undergone significant periods of volcanism, as well as erosion by glaciers and running water, and widespread deposition of sediments that may bury older craters.

Based on the presence of the largest impact structures, the highest crater densities and the impact history of the inner Solar System, the southern highlands of Mars coincide with the oldest crust. They probably formed prior to 3.8 billion years ago. The more sparsely cratered northern plains must be younger, since they have fewer and smaller craters, having formed after the end of the great bombardment.

Amazonian
Hesperian
Noachian

Figure 7.6 A map showing the ages of the major surface units. Most of the younger (Amazonian) surface is in the north, while the oldest terrain, the Noachian, is concentrated in the southern highlands. Major impact craters and their ejecta are marked in white. (Sean C. Solomon et al.)

The geological history of Mars has been divided into four main periods: pre-Noachian, Noachian, Hesperian, and Amazonian.

The geological record of the earliest events in Martian history, the so-called pre-Noachian era, has been largely erased by subsequent impacts and other resurfacing events.

The *Noachian period* probably lasted from about 4.1 billion years ago until about 3.8 billion years ago. This was a period of heavy bombardment, intense volcanic activity and warmer climate, with lakes, or possibly a shallow ocean. The Hellas and Argyre basins in the southern highlands date from this era. The Tharsis bulge and the Valles Marineris are also thought to have formed in this period. The atmosphere may have been thick enough for liquid water to survive for lengthy periods on the surface, and even for rain to fall from the sky. It is named after the ancient uplands of Noachis Terra in the southern hemisphere.

The *Hesperian period* (3.8 to perhaps 3 billion years ago) is considered to be a time of increasing cold and dryness as the greenhouse gases dispersed and thick permafrost formed. However, the planet may have remained warm enough for lakes – perhaps covered by ice – to have persisted until at least three billion years ago. There were also periodic and localized, but catastrophic, floods, caused by melting of ice. Most of the large outflow channels were probably carved and then abandoned at this time. The period was also marked by increased volcanic activity (including the birth of Olympus Mons). It is named after the plains of Hesperia Planum, not far from Noachis in the southern hemisphere.

Figure 7.7 Global topography, based on data from the Mars Orbiter Laser Altimeter. Note the dichotomy between the heavily cratered southern highlands (yellow/orange) and the smooth, sparsely cratered northern lowlands (blue/green). Vertical accuracy is less than 5 meters. Note the giant Hellas impact basin at right (purple, surrounded by mountains). Also the Tharsis rise (red and white) and Valles Marineris (left image). (David Smith/NASA/JPL)

The *Amazonian period* includes most of Martian history, extending from the Hesperian era to the present day. Landscapes of this age have few impact craters, all of them small. The climate was generally dry and cold. Some landscape changes occurred, e.g. eruptions on Olympus Mons and widespread lava flows elsewhere, landslides in Valles Marineris and formation of broad plains and sand dunes near the poles. The period is named after Amazonis Planitia, a broad area of plains in the northern hemisphere.

first images of its surface, they showed a heavily cratered landscape rather like the Moon. The density of cratering is much higher in the southern hemisphere, where most of the large impact features are found.

The largest of these structures is the Hellas basin, in the southern uplands (see Fig. 7.8). Its diameter from rim to rim is about 2300 km, while its depth is more than 8 km – making it the deepest place on the planet. Overall, Hellas is only slightly smaller than the South Pole-Aitken basin on the Moon – the largest recognized

impact structure in the Solar System. It is surrounded by a huge volume of excavated material, which, distributed evenly, would cover the continental US to a depth of 3 km.

Many smaller craters are found on its floor and in the surrounding uplands, indicating that it dates from the heavy bombardment of the Noachian era, just over 4 billion years ago. The mountains of ejecta along its rim are surprisingly low and dissected by valleys, indicating that they have been destroyed by erosion. Any surface drainage would be expected to travel

Figure 7.8 A topographic map of the Hellas impact basin, the largest on Mars, based on laser altimetry data. The upper panel shows a cross section through the basin. Hellas is about 2300 km wide and over 8 km deep. (NASA-JPL)

down slope, away from the rim, but many valleys flow into the basin.

The basin floor is covered with layers of sediments deposited by wind and water during the Hesperian and Amazonian periods. These weakly consolidated deposits of sand and rock have been carved into remarkable curvilinear shapes. Some of the strange landforms may be created by glaciers of water ice buried beneath layers of dirt and rock (see below).

Hellas appears bright even in Earth-based telescopes, not just because of its great size and sand-colored sediments, but also because of its proximity to the south pole. Its low temperature in winter results in the formation of surface frost and low level, morning haze.

The second largest impact basin, also located in the southern hemisphere, is Argyre. Once again, the rim of this ancient structure has been eroded and dissected. One of the most striking features is an apparent dry river channel that runs from the northern rim toward the equatorial lowlands. Water may once have filled the basin until it overflowed, allowing floodwater to travel northwards until it eventually disgorged onto the northern plains (see below).

There are hundreds of thousands of craters on Mars, many of them without names. The vast majority of these have similar morphologies to those on the Moon and other planets, the main exception being pedestal craters (see below). Small craters, no more than 8–10 km across, are simple, bowl-shaped depressions with constant depth-to-diameter ratios. Larger craters, up to about 140 km across, are more complex, with central peaks and a lower depth-to-diameter ratio. Largest of all are the multi-ringed basins.

Erosion rates at low latitudes for most of Mars' existence seem to have been very low. However, rates of erosion (and deposition) were much higher early in its history, when craters were modified by running water and glaciers, as well as wind. As a consequence, in the southern uplands, they range from sharply delineated circular structures to barely discernible, rimless, circular depressions. In contrast, almost all of the craters on the volcanic plains of the equatorial regions appear fresh, even though they may be billions of years old.

Frequent cratering is still taking place (see Fig. 7.11). Many of these small craters are excavated in the hidden, subsurface ice. The bright hollows soon darken as freshly exposed ice sublimates into the atmosphere.

Figure 7.9 A view of the southern part of the Argyre basin, taken by Viking Orbiter 1. The fragmented, eroded rim (right) may have been breached when water filled the basin and overflowed. Note the orange, dust-laden atmosphere above the horizon. (NASA-JPL)

Figure 7.10 Victoria crater, in Meridiani Planum, is about 750 meters in diameter and 75 meters deep, with a 5 meter high raised rim. This HiRISE image shows a degraded, scalloped rim and sand dunes on the crater floor. Layered sedimentary rocks are exposed along the inner wall, and boulders have fallen into the crater. It has been infilled by sediments and widened by erosion over billions of years. (NASA/JPL/University of Arizona)

Figure 7.11 HiRISE images of a recent, 6 meter wide impact crater, taken on October 18, 2008, (left) and January 14, 2009. Each picture is 35 meters across. The bright, freshly exposed ice darkened rapidly as it sublimated into the thin atmosphere. (NASA/JPL-Caltech/University of Arizona)

North-South Divide

One of the most surprising surface features is the dichotomy between the less cratered, lowland plains that cover much of the northern hemisphere and the ancient, heavily cratered, highlands which dominate the southern hemisphere. (see Fig. 7.7.) The much lower number of craters in the north indicates that the plains are probably hundreds of millions of years younger than the highlands.

Altimetry maps show a difference in altitude of a few kilometers between the two regions, with a clearly delineated, fairly steep, boundary between them. Gravity data obtained by studying minute changes in the paths of orbiting spacecraft indicate that the mean crustal thickness is about 60 km in the south and 35 km in the north.

The reason for the dichotomy remains uncertain. One suggestion is that a huge impact early in the planet's history removed much of the original crust in the north. The absence of an obvious impact basin is explained by the subsequent superimposition of the huge Tharsis volcanic bulge, and the fact that the 2000 km wide impactor arrived at a low angle, thus excavating an elliptical depression.

This theory is supported by gravity data from the Mars Reconnaissance Orbiter, which show signs of a huge, but indistinct, impact basin. 8500 km across and 10,600 km long – the size of Asia, Europe and Australia combined – it seems to cover about 40% of the planet's surface, making it far larger than any other known impact basins in the Solar System.

Other studies have suggested that a number of separate impacts could have combined to reproduce the irregular dichotomy boundary, with evidence for numerous hidden craters (known as "quasi-circular depressions") provided by gravity and altimetry data. Likely subdued impact basins in the northern plains include Chryse Planitia, Acidalia Planitia and Utopia Planitia.

Tectonics

The Martian crust also bears scars of large-scale tectonic activity early in the planet's history, such as the Tharsis bulge and the Valles Marineris canyon system. However, there is little evidence for global plate tectonics, in which the motions of separate slabs of lithosphere drive the formation of fold mountains, volcanoes etc. (see Chapter 3).

Crustal spreading may have begun when Mars was very young, but then shut down quite rapidly when the interior cooled. Some of the oldest rocks are weakly magnetized, evidence that they cooled and solidified in the presence of a strong magnetic field that was generated by an internal dynamo. The banded pattern of strongly magnetized crust in the southern highlands bears some resemblance to the magnetic stripes imprinted on Earth's ocean floors by spreading on either side of a mid-ocean ridge.

The existence of a strong, magnetic field that reversed polarity is seen as good evidence that Mars once had a sufficiently high internal temperature to drive a dynamo in the iron-rich core. This would also be consistent with a vigorously convecting mantle, a thin lithosphere and – possibly – plate tectonics. However, Mars lost its internal heat very quickly, so that the lithosphere grew too thick and rigid to easily break up into tectonic plates.

The absence of widespread plate tectonics means that almost all of the familiar deformational features on Earth are rare or absent. Mountains on Mars are associated with large impact craters, rather than plate movements. One possible exception is the Thaumasia mountain range, located to the south east of the Tharsis volcanic province and south of the Valles Marineris. These curved uplands may have been created by powerful crustal forces associated with the formation of the major tectonic features.

The lack of lateral plate motion means that any hot spots in the mantle stay in a fixed position relative to the surface. This, along with the lower surface gravity, may account for the huge Tharsis bulge and its enormous volcanoes.

The Tharsis bulge dominates the tectonics of Mars. Although the origin of the huge dome is uncertain, the massive, widespread uplift of the crust has caused Tharsis to be the center of a vast array of radial fractures that spread across almost a third of the planet. To the east, several enormous interconnected canyons are aligned along these radial faults. The closely spaced, parallel graben (rift valleys) are evidence of tensional deformation where the crust has pulled apart.

Many of these fractures are buried by younger lava flows and are visible only on older, higher terrain. To the north of Tharsis, the fractures are diverted around the large volcano Alba Patera, forming a ring of faults. At least some of the faulting is associated with volcanic dikes, where magma rose to the surface through cracks in the crust. Chains of pits are also seen in places where the surface has collapsed along the faults, e.g. Tractus Catena.

Similar, curved faults occur at Acheron Fossae, north of Olympus Mons, where cracks in the crust formed when magma rising from deep in the mantle pushed up the overlying lithosphere (see Fig. 7.14). Eventually, the brittle crust broke along

Figure 7.12 A cross section showing the thickness of the crust – about 60 km in the south and 35 km in the north – based on altimetry data from Mars Global Surveyor. It also shows the difference in altitude above the mean datum level, with the southern highlands separated from the lowland plains of the north by a sharply delineated boundary near the equator. (NASA-JPL/MOLA)

Figure 7.13 A computer simulation of mantle convection, modeling the internal processes that may have produced the Tharsis bulge. Hot regions are red, cold regions are blue and green, with a difference of around 1000°C. Thermal expansion causes hot rock to have a lower density than cold rock, so hot mantle material rises toward the surface whilst colder material sinks, creating large-scale convectional circulation. The rising plume pushes up the crust, while the sinking mantle tends to pull the surface down. Deformation of the surface, shown in gray, is highly exaggerated: actual uplift in central Tharsis was about 8 km. This uplift also stretched the crust, forming fractures, grabens and Valles Marineris. The hot, rising mantle material melted as it approached the surface, producing volcanic activity. (Walter S. Kiefer, Amanda Kubala – LPI)

Figure 7.14 Acheron Fossae was once a region of intense tectonic activity and is part of a network of extensional fractures that radiates outward from the Tharsis bulge. This image from Mars Express shows curved depressions up to 1.7 km deep, created by faulting. When the crust dropped between parallel faults, a graben was formed. Fossa is the Latin word for trough. (ESA/DLR/FU-G. Neukum)

Figure 7.15 A Viking Orbiter mosaic of the north eastern Tharsis bulge showing an area of intense faulting to the west of the volcanoes Uranius Tholus (top) and Ceraunius Tholus. The presence of impact craters, particularly on Uranius Tholus, indicates that they are quite ancient and not active today. The summit crater of Ceraunius Tholus is about 25 km across. The true bases of the volcanoes are submerged beneath a flood of lava which produced the surrounding plain. The main field of radial fractures is known as Tractus Fossae, and the major double fault (left) is Tractus Catena. (NASA-JPL)

zones of weakness. When the crust dropped between parallel faults, it created deep grabens, whilst the remnants of the previous surface were left upstanding as block mountains or "horsts".

Circular fractures around large volcanoes such as Elysium Mons and Ascreus Mons probably formed in a similar way. On the other hand, curved faults around the Isidis and Hellas impact basins were created as the crust adjusted to the formation of these giant depressions and the sudden removal of huge amounts of crustal material.

There is also evidence of crustal compression, where linear or curved wrinkle ridges occur, e.g. in the ridged plains of Lunae Planum, to the east of Tharsis.

Magnetic Field

Mars does not possess a global magnetic field, although there is a patchy, localized surface field where the crust retains the magnetism implanted when Mars still had a global field. An induced magnetic field is also present on the dayside.

The patchwork magnetic field on the surface resembles scattered umbrellas that act as barriers to erosion of the atmosphere by the solar wind. They are produced by regions of highly magnetized rock up to 1000 km wide and 10 km deep. The surface fields are comparable to Earth's magnetic field – a few tenths of a gauss – and they can extend to an altitude of 1300 km. Where the anomalies are stronger, the ionosphere reaches to a higher altitude, an indication that the solar wind is being kept at bay.

The anomalies with the largest amplitude are in the southern highlands. Weaker anomalies occur on parts of the northern

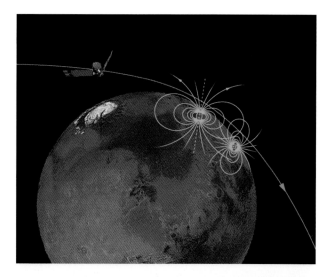

Figure 7.16 A schematic representation of localized magnetic sources in the Martian crust, buried beneath the surface, and revealed by observations of the magnetic field (blue) at the satellite's orbital altitude of 120 km. (NASA/JPL/GSFC)

lowlands, particularly near the boundary with the southern highlands. The major impact basins (Utopia, Hellas, Isidis and Argyre) have no apparent magnetic signatures, probably because the shock and heating that accompanied the massive impacts demagnetized the crust. This implies that Mars no longer possessed an active dynamo in its core by the time they were created: otherwise, the rocks would have been remagnetized once they cooled and solidified.

It seems that Mars was a magnetic world soon after its interior differentiated and its core formed in the pre-Noachian era. The strongest magnetic anomalies are associated with late Noachian and early Hesperian terrain, but surface magnetism is absent from younger volcanic or impact structures. Within the first billion years of Mars' existence, the core seems to have cooled considerably and largely solidified, shutting down the dynamo. When its global field died out billions of years ago, the solar wind was able to strip the atmosphere away.

The induced magnetic field is very different in origin. It is formed when the magnetic field that is carried by the solar wind is compressed at Mars and draped around the planet. In the absence of a strong global magnetic field, the solar wind interacts directly with the ionosphere. This interaction results in the formation of two boundaries: the bow shock and the magnetic pile-up boundary (MPB). At the bow shock, the solar wind is

East Longitude

$\Delta B_r/\Delta Lat$ (nT/deg)

-30 -10 -3 -1 -.3   +.3 1 3 10 30

Figure 7.17 The locations of magnetic anomalies, based on data from Mars Global Surveyor. The strongest crustal magnetization is largely confined to east-west "stripes" of alternating polarity in the ancient southern highlands. Regions of extensive volcanism (e.g. Olympus Mons, Tharsis Montes) are non-magnetic, as are regions surrounding the largest impact basins. (NASA-JPL/J. E. P. Connerney et al.)

decelerated from supersonic to subsonic speed as its flow is obstructed by Mars. Closer to the planet's surface is the MPB, which is the top of the induced magnetic field.

Volcanism

Superimposed on the dichotomy between the cratered southern uplands and northern, low-lying plains are two large crustal rises. The Tharsis bulge, centered on the equator at 100°W, is 5000 km across and 10 km high. The smaller Elysium bulge, centered at 30°N, 210°W, is 2000 km across and 4 km high.

Both regions display raised gravitational signatures, indicative of large masses of dense material close to the surface. They were probably formed through a combination of large scale, static, convective plumes in the mantle, and isostatic adjustment due to shifts in the mass of the crust, as well as intrusive (subsurface) and extrusive (surface) igneous activity. The likelihood of periodic volcanism in recent times suggests that the domes may still be supported by rising plumes of mantle material.

Both of these regions have been sites of volcanic activity throughout most of the planet's history. Although they seem to have originated in the Noachian era, there is evidence of episodic volcanic activity over successive eons. Low numbers of craters on the lava flows and evidence that Martian meteorites which have fallen to Earth have crystallized within the last 150 million years, suggest that the planet may still undergo periodic volcanic activity.

On Mars, volcanic landforms created by the eruption of fluid, basaltic lava are far more common than those formed through explosive activity and production of ash. The most notable examples are the large shield volcanoes, located primarily in Tharsis and Elysium, which have many similarities with those of Hawaii. They display broad domes with gently sloping flanks of 6° or less. Each has a summit caldera, formed by collapse, following eruptions on the volcano's flanks.

Alba Patera (40.5°N, 109.9°W) is unlike any of the other shield volcanoes. It is very large (over 1500 km across) and has a central caldera, but rises no more than 3 km above the surrounding plains. It appears to have been formed by numerous flows of non-viscous lava that were either much higher in output or much longer in duration than flows occurring on other volcanoes. Alba Patera is surrounded by a huge system of curved fractures, called "catenas."

Volcanoes located within the southern highlands have a very different morphology from the Tharsis or Elysium volcanoes. Two ancient, but very similar, examples are Tyrrhena Patera (21.9°S, 253.2°W) and Hadriaca Patera (30.6°S, 267.2°W).[3] Probably born around 2.5 billion to 3 billion years ago, each mountain is about 200 km across but has a gently sloping cone only a few hundred meters high. They represent the earliest central vent volcanism identified on Mars and may reflect a transition from the flood eruptions of very fluid lava which dominated early Martian history. Their slopes have been dissected by radial gullies,

Figure 7.18 Tyrrhena Patera is about 200 km across but has very shallow slopes, deeply dissected by broad channels, probably carved by running water, that radiate from the summit. The low relief and easily eroded flank materials indicate that most of the volcano is composed of pyroclastic (ash) deposits. (NASA-JPL/LPI)

hundreds of kilometers long, probably carved by streams of water flowing downhill and eroding weak layers of volcanic ash and lava. Massive clouds of steam billowing above the newborn volcano may have condensed into thick clouds from which heavy rain descended.

Not all Martian volcanoes are the result of quiet effusion of fluid lava from a single hot spot in the mantle. There are many other types of volcanic landforms, which range from small, steep-sided cones to plains covered in flows of solidified lava.

Volcanoes represent only a small fraction of the planet's volcanic activity. For example, surrounding the shield volcanoes in Tharsis and Elysium are extensive plains that were clearly built up from numerous lava flows superimposed on one another. These were probably fed by large fissures that are now buried beneath the flows.

Tharsis Giants

The most striking volcanic features on Mars are the three huge shield volcanoes on the Tharsis bulge – Arsia Mons just south of the Martian equator, Pavonis Mons on the equator, and Ascraeus

---

[3] Patera is the Latin term for shallow dish or saucer. It is used to describe a shield volcano of broad areal extent but little vertical relief.

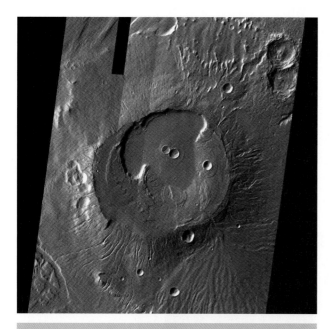

Figure 7.19 A THEMIS mosaic of Apollinaris Patera, an isolated, ancient volcano about 400 km across, located in the southern highlands (9°S, 186°W). To the west of the 80 km wide central caldera is a cliff, marked by landslides, and chaotic terrain, formed by the collapse of ice-rich materials. Extending south from a breach in the caldera wall is a fan of material that flowed out of the volcano after the main cone was built. To the north and east are the younger (ash?) deposits of the Medusae Fossae Formation, incised by numerous gullies. (NASA/JPL/Arizona State University)

Mons, 10 degrees north of the equator – and their larger, but isolated neighbor, Olympus Mons.

Each member of the Tharsis Montes chain is huge compared to terrestrial volcanoes – about 350 to 450 km across, towering to about 15 km above the surrounding plains.[4] Their location on the crest of a broad crustal dome means that they reach about the same elevation as the summit of the much taller Olympus Mons. The fractured region southeast of Pavonis Mons, named Noctis Labyrinthus, merges with the Valles Marineris canyon system to the east.

The flanks of the Tharsis volcanoes rise gradually, evidence that they formed from eruptions of fluid lavas that could flow for long distances before solidifying. Each has a large summit caldera, caused by a monumental collapse of the surface when the subsurface magma supply was withdrawn. The flanks have a striated pattern caused by long linear flows, some with central leveed channels (i.e. with natural embankments of solidified lava on either side). Each has vents on its north east and south west flanks which fed vast lava flows that extend far across the surrounding plains.

All three volcanoes were built from countless thin sheets of runny lava flowing from natural pipes or tubes that developed within the flows. These carried lava long distances from its source to the spreading flow fronts. The roofs of some tubes have collapsed in places, forming an open pit. Where a longer section of roof has collapsed, the pit becomes an oval. If the flow in the tube eroded the roof sufficiently, many pits and ovals merged to form an open channel.

The Tharsis trio is aligned in a south west – north east direction, suggesting that, like the Hawaiian islands, each one was formed by frequent, prolonged eruptions above a plume of rising magma. Studies of their surface features indicate that the oldest volcano, Arsia Mons, is in the south, and Ascraeus Mons is the youngest.

Since there is no evidence of plate movement, it seems likely that the plume moved beneath the stationary crust, instead of the other way round. Another possibility is a stationary plume that originated near Arsia Mons, then spread out toward Ascraeus as it neared the base of the crust.

Observations of surface roughness and crater counts indicate that they may currently be dormant, having erupted within the last few million years. It may be that future eruptions could contribute significant amounts of water and carbon dioxide to the Martian atmosphere.

Olympus Mons (18°N, 133°W) lies 1500 km north west of the Tharsis Montes, away from the main Tharsis rise. It is the largest volcano in our Solar System, with about 100 times the volume of Mauna Loa, the largest volcano on Earth. It is 25 km high (three times the elevation of Mount Everest above sea level) and about 620 km in diameter, or the size of the state of Arizona. The summit caldera alone is 80 km in diameter and 3 km deep.

Olympus Mons is so big that it would be impossible for an observer on the surface to see the entire edifice. The volcano is surrounded by a steep scarp, 6 km high in places, over which lava flows tumbled in the distant past. Extending several hundred kilometers from the base of the cliff is an aureole, consisting of several huge lobes of ridged terrain. The origin of the cliff and the aureole on the nearby plains is unknown – similar features are absent on the other Tharsis mountains.[5]

Plains

Much of Mars, particularly in the northern hemisphere, consists of low-lying plains – some of the smoothest surfaces in the Solar System. Like the plains of Mercury and the Moon, they have been constructed from innumerable thin lava flows. Laser altimeter measurements show that, apart from the craters, the plains typically vary in height by 50 meters or less. Shallow grooves and broad hollows are filled with wind-blown dust and dunes, separated by low rises. In places, the more resistant surfaces appear as higher land.

Wind is not the only agent of erosion and deposition to have modified the plains. There is ample evidence that large volumes of water carved deep channels and terminated in lowland basins

---

[4] The largest terrestrial shield volcano, Mauna Loa in Hawaii, is 120 km across and stands 9 km above the ocean floor.
[5] Aureole is Latin for "circle of light."

Figure 7.20 A THEMIS infrared mosaic of Arsia Mons – the southernmost, and possibly the oldest, of the Tharsis volcanoes. Indentations and collapsed lava tubes on the south west and north east sides align with Pavonis Mons and Ascreaus Mons to the north east – evidence that a large fracture/vent system fed the eruptions that formed all three volcanoes. Note the curved fractures around the caldera. (NASA/JPL/ASU)

Figure 7.22 An overhead view of the complex caldera at the summit of Olympus Mons. The striations are tectonic faults. After lava production ceased, the caldera collapsed over the emptied magma chamber, forming curved, extensional fractures. The level on which these fractures are observed represents the oldest caldera collapse. Later lava production and withdrawal produced new collapses, forming the other circular depressions. The HRSC image is 102 km across with a resolution of 12 meters per pixel. South is at the top. (ESA/DLR/FU Berlin – G. Neukum)

Figure 7.21 Olympus Mons is the largest volcano in the Solar System. The shield volcano is 624 km in diameter, 25 km high and rimmed by 6 km high cliffs. Note the 80 km wide central caldera. A massive aureole deposit was possibly produced by flank collapse. The vertical exaggeration is 10:1. (NASA/MOLA Science Team)

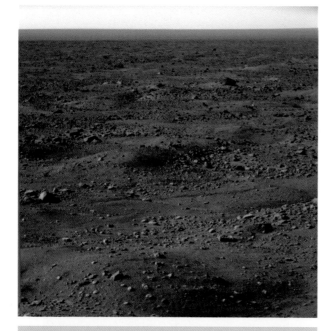

Figure 7.23 NASA's Phoenix spacecraft landed on the northern plains in a region covered by polygons, 1.5–2.5 meters across, that stretched to the horizon. Similar in appearance to surfaces in Earth's Arctic regions, the polygons are caused by continual expansion and contraction of ice wedges within the permanently frozen surface (permafrost). (NASA/JPL-Caltech/University of Arizona/Texas A&M University)

or spread out onto the plains, where sedimentary layers of rock and sand were deposited (see below).

Glaciation almost certainly modified the plains in the distant past, as ice sheets spread onto the plains from the polar ice caps and upland regions, even at low latitudes. The flowing ice sheets would have eroded and smoothed the terrain, eventually burying much of it under deep blankets of glacial moraine and outwash deposits (see below).

Even today, the plains contain a lot of water ice at shallow depths, as confirmed by the Phoenix spacecraft which landed in the Martian Arctic in 2008. Its panoramic images revealed vast plains corrugated by polygonal ice wedges.

Valles Marineris

Around the turn of the 20th century, some astronomers drew maps of linear channels – which they called "canali" – on the Martian surface. We now know that they were almost entirely optical illusions. One exception is a huge system of interconnected canyons, on the eastern flanks of the Tharsis bulge, which runs just south of the equator for about 4000 km.

This great rift in the crust was discovered in 1972 by the Mariner 9 spacecraft – hence its name, Valles Marineris (Mariner Valleys). It dwarfs all similar features in the Solar System, and would swallow up thousands of valleys like the Grand Canyon in Arizona.

Valles Marineris is formed out of several parallel, connecting troughs. Individual canyons are generally 50–100 km wide, reaching a depth of 8–10 km in places, six to seven times deeper than the Grand Canyon.

Figure 7.24 A mosaic of Valles Marineris obtained by the THEMIS camera on Mars Odyssey. The broad valley floor, parallel canyons and lateral landslides are clearly shown. Spatial resolution is about 100 meters. (NASA/JPL/ASU)

Their characteristics change from west to east. At the western end is a complex graben system known as Noctis Labyrinthus, with numerous, intersecting linear depressions. The depressions are generally aligned parallel to faults in the surrounding plateau.

To the east, the valleys become deeper, wider, and more continuous, forming canyons aligned roughly east-west. Three large canyons merge in the central section to form a huge depression up to 600 km across. Still further east, the individual canyons become shallower, and fluvial features become more common, both on their floors and on the surrounding plateau. At the eastern end, they lead into large areas of chaotic terrain.

The canyons probably formed during the Hesperian era, about 3.5 billion years ago, as the result of tectonic activity that accompanied the growth of the nearby Tharsis volcanoes (see above). The tectonic stress caused the surrounding crust to stretch. As the crust thinned and fractured, blocks of rock sank between parallel cracks. The result was a series of grabens – elongated, trench-like features bounded by parallel normal faults.

The faulting also opened paths for subsurface water to escape. As the surface broke apart, the water escaped, washing away material and causing the ground to collapse. The steep, newly exposed valley walls became unstable, causing numerous landslides, up to 100 km long, that widened the canyons further. The main extensional activity probably came to an end roughly two billion years ago, although landslides undoubtedly still occur.

Today, we can see where faults have opened into collapse pits, some of which then join together to form larger depressions. The process steadily ate away at the surface until the land between two neighboring faults was largely destroyed.

Massive floods seem to have escaped through several outflow channels at the valley's eastern end. The ground here was so saturated that when the water escaped, the surface collapsed almost completely, leaving only isolated mesas and hills which are termed "chaotic terrain."

In the north west of Valles Marineris, a similar torrent emerged from a depression called Echus Chasma, forming the Kasei Valles outflow channel. Again, the floodwaters eventually poured into the northern lowlands. It seems most likely that such floods occurred in several stages, rather than as single, catastrophic events.

Not all of the erosion within the canyons resulted from major floods. In places such as Louros Valles, numerous deep, tributary valleys have been cut into the canyon's southern rim, contributing to the overall widening of Ius Chasma. These minor tributaries were probably created by the release of local groundwater from springs.

As would be expected where flood water (and possibly glaciers) once existed, the canyons have also been shaped by large scale deposition of sediments. These have since been eroded and are now exposed as multiple layers, for example, in Candor Chasma (Fig. 7.26).

Figure 7.25 Noctis Labyrinthus, at the western end of the Valles Marineris, forms part of the graben system from which the canyons formed. The elongated, trench-like features, bounded by parallel faults, are some 5 km deep. They are greatly eroded, as seen from debris on the valley floors. Younger rock formations can be seen on the upper slopes. This view was based on High-Resolution Stereo Camera images from Mars Express. Vertical height is exaggerated. (ESA/DLR/FU Berlin-G. Neukum)

Figure 7.26 Layered sedimentary rock in western Candor Chasma. The area covered is 1.5 × 2.9 km. The MOC image reveals over 100 uniform beds on the canyon floor, each about 10 meters thick. Each layer has a relatively smooth upper surface and is hard enough to form steep cliffs at its margins. This suggests that the deposition was not continuous, but interrupted at regular intervals. The layers may have been deposited on a lake bed. (NASA-JPL/MSSS)

Polar Regions

At each pole, extending out to 80° latitude, is a permanent ice cap composed of finely layered deposits of ice and dust a few kilometers thick. In the north the cap rests on lowland plains, in the south on cratered uplands. The small number of superimposed impact craters suggests that they are relatively young, possibly only a few hundred million years old.

The residual northern cap covers an area bigger than Texas. Its volume of 1.2 million cubic km is less than half that of the Greenland ice sheet, and about 4% of the Antarctic ice sheet. In comparison, the residual southern ice cap is estimated to contain 1.6 million cubic km, making it the largest source of water on the planet's surface. Radar data indicate that it contains approximately 15% dust – significantly less than the northern cap.

Both polar caps possess a perennial core composed of frozen carbon dioxide mixed with some water ice. This is surrounded by a region of almost pure water ice, overlain by a thin veneer of dust. Extending for tens of kilometers around this is a vast area of dirty water ice, containing varying amounts of dust.

During autumn and winter, CO_2 condenses out in the polar regions to form a seasonal cap. In the northern hemisphere it can extend toward the equator as far as 60°N, while its southern counterpart extends to 50–60°S. The depth of the seasonal ice layer has been estimated to be no more than a meter or two, less at lower latitudes.

As summer approaches, the temperatures rise from the freezing point of CO_2 (−120°C) to around the freezing point of water (0°C). Not only does the CO_2 cap vaporize, exposing the underlying water ice, but the latter also begins to sublime, thus increasing the content of water vapor in the atmosphere.

There are some significant differences between the two ice caps. The northern CO_2 cap burns off completely in summer, leaving a residual water ice cap. At the south pole, the CO_2 cap does not completely disappear, since temperatures remain close to −120°C. Only a modest increase in water vapor is detected, although a permanent cap of water ice has been detected beneath the dry ice.

The southern cap is several kilometers higher – and therefore, cooler – than its northern counterpart, but the main reason for the colder summer conditions in the south may be the effect of dust in the atmosphere. Dust storms and dust devils are much more common during the southern summer, and the fine particles that are carried into the upper atmosphere block some of the incoming solar radiation, keeping the surface cooler than expected.

The north polar cap has a flat, pitted surface that has been likened to cottage cheese or the surface of a sponge. In contrast, the south polar cap has roughly circular depressions that are more like Swiss cheese, with flat floors and steep-sided walls (see Fig. 7.30). As the layered CO_2 ice is removed, it leaves a landscape of pits, troughs, flat-topped mesas and buttes.

Spacecraft images and laser altimeter data show that the polar caps are composed throughout of multiple, horizontal or near-horizontal, layers. These polar layered deposits are stacked on top of each other, reaching a maximum thickness of about 2.7 km in the north, and 3.1 km in the south. The layered deposits sit on top of a much older ice-sand deposit, known as the "basal unit."

Individual layers of ice are best seen in the walls of valleys cut into the sediments, such as Chasma Boreale. They are generally believed to be the result of variable amounts of dust and water ice being laid down over many climate cycles in the last four million years or so. The primary cause of the banding may be

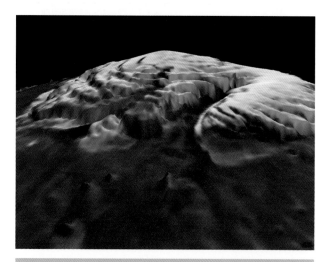

Figure 7.27 A 3-D view of the permanent northern polar ice cap and outliers seen from the south west. The cap is about 1200 km across, and nearly 3 km thick. Large areas are extremely smooth but the cap is dissected by canyons and troughs up to 1 km deep. Spatial resolution is 1 km and vertical accuracy 5–30 meters. Nearby mounds of ice, tens of kilometers across and around 1 km high, may be remnants of an older cap. Impact craters are filled with ice and dust. (MOLA science team/NASA/GSFC SVS)

Figure 7.28 The residual south polar ice cap lies on top of smooth layered deposits that overlie the cratered southern highlands. Nearby is a large area of dunes. (NASA/ MOLA Science Team)

Figure 7.29 Seasonal changes in the north polar cap imaged by the Hubble Space Telescope. In October 1996 (early spring), the almost circular cap extends to 60°N, nearly its maximum extent. Bluish areas on the edge of the cap are clouds. Some of the area was still in shadow. By January 1997, increased warming has sublimated the carbon dioxide ice and frost below 70°N. The faint, darker circle within the cap's boundary marks circumpolar sand dunes. These dark dunes are warmed more quickly than their brighter surroundings, so their frost coating sublimates earlier. By March 30, 1997 (early summer), the cap has retreated, leaving a small core of water ice, almost cut in half by the Chasma Borealis canyon. Outliers of ice persist south of the sand sea. Bright circular features are ice-filled craters. The edge of the maps is at 50°N latitude. 0° longitude is at the top. (Phil James/Univ. Toledo, Todd Clancy/Space Science Inst., Boulder, CO, Steve Lee/Univ. Colorado, and NASA)

Figure 7.30 (a) "Cottage cheese" terrain on the north polar residual ice cap. The flat surface has many closely-spaced pits up to two meters deep and 10–20 meters across. These form as water ice sublimes, growing slowly over thousands of years. This 1 × 1 km picture is illuminated from upper left. (b) The south polar cap is colder and coated with carbon dioxide ice. Large circular depressions resembling Swiss cheese are separated by flat-topped mesas and buttes up to four meters high. The area covered measures 3 × 9 km. (NASA-JPL/MSSS)

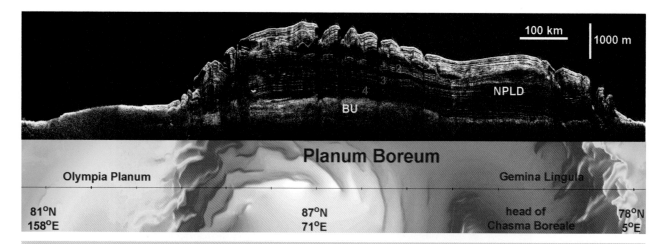

Figure 7.31   A cross section through the north polar ice cap (top), taken by the radar instrument on Mars Reconnaissance Orbiter, reveals layers of ice, sand and dust. The colored map shows the surface topography (red and white = higher ground, green and yellow = lower). The radar reveals four thick layers of ice and dust separated by layers of nearly pure ice. The boundary between the ice layers and the underlying surface is relatively flat, suggesting that the lithosphere is thick and rigid. NPLD = north polar layered deposits. BU = basal unit, an ice-sand deposit that lies beneath parts of the north polar layered deposits. (NASA/JPL-Caltech/University of Rome/SwRI)

periodic changes in the temperature and winds at the poles, induced by variations in the planet's orbit and rotation. These cycles affect the stability of water ice and CO_2 ice at the poles and the dynamics of the atmosphere.

The poles act as a cold trap for water. Any water entering the atmosphere as a result of geologic processes, such as volcanic eruptions or floods, is ultimately frozen out at the poles. They may also be a trap for dust, since dust nuclei can be taken out of the atmosphere as CO_2 freezes onto the pole in the fall and winter.

One of the most unusual features of both caps is the deep, curved valleys that dissect the residual summer ice. They are unlike any terrain on Earth, and various hypotheses for their formation have been put forward, including subsurface volcanic activity, glacial erosion, aeolian processes (such as local katabatic winds[6]) and sublimation, outflow of meltwater, or a combination of these processes.

The largest canyon in the northern cap is Chasma Boreale, which cuts into the layered polar terrain and crosses several smaller, curved troughs. It extends 500 km into the residual cap, has a width of up to 100 km and a depth of about 2 km. The head of Chasma Boreale ends in a steep, icy cliff more than 1 km high. The origin of the canyon is uncertain, but it seems to be associated with an old, linear depression that underlies the layered deposits. The depression was later modified by meltwater, katabatic winds or local variations in the rate of sublimation.

Both short and longer-term changes occur in the polar caps. Frequent avalanches have been observed on the scarps at the edges of the ice caps during spring. The falls appear to be composed

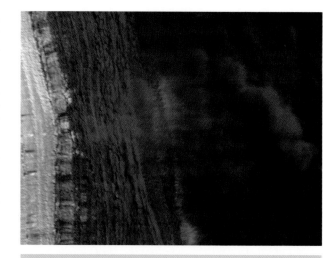

Figure 7.32   The HiRISE camera imaged an avalanche in progress on a steep scarp in the northern polar region. The 700 meter high cliff is made of layers of water ice with varying dust content. On top is white carbon dioxide ice and frost. The dust cloud is up to 50 meters high. (NASA/JPL/University of Arizona)

mostly of carbon dioxide frost, mixed with loose, dusty material. Frost builds up on the scarp during the long winter, then begins to sublimate when sunlight returns. This results in avalanches that create dust clouds tens of meters high.

---

[6]Katabatic winds occur when cold, dense air, slides downslope, becoming warmer and drier. Famous examples on Earth are the Chinook ("snow eater") and the Föhn.

Figure 7.33 Sand-laden jets shoot into the southern polar sky in this artist's view. These leave a delicate, spider-like pattern on top of the residual polar cap, after the seasonal carbon dioxide ice has disappeared. (Arizona State University/Ron Miller)

The carbon dioxide ice at the south pole is rapidly disappearing. Comparisons of images taken a few years apart show circular pits that have enlarged, shrunken mesas, and small buttes that have vanished. The average rate of retreat is about three meters per Martian year, though it varies from place to place. Studies suggest that an eight meter thick layer of CO_2 ice is being removed to reveal a permanent layer of water ice beneath. It seems that the planet was markedly colder only a few centuries to a few tens of thousands of years ago.

The residual southern cap also displays dark spots, typically 15–45 meters wide and spaced up to 100 meters apart. They appear every spring and last for three or four months before vanishing, only to reappear the next year, after a fresh layer of ice has been deposited during the winter. Most of the spots recur at the same locations year after year, and some of them display spider-like grooves that extend outward in all directions.

These features are created when powerful jets of trapped carbon dioxide erupt through weaknesses in the overlying ice as the temperature rises in spring. As the gas rushes toward the vents, it picks up loose particles of sand and erodes the ice, carving networks of spidery grooves that converge on the vents. The "geysers" climb high into the thin air, carrying fine, dark sand and spraying it for many meters around each vent.

Permafrost

On Earth, much of the ground surface adjacent to the polar ice caps is permanently frozen to a depth of many meters, a condition known as permafrost. On Mars, the area of permafrost is much more extensive. The subsurface ice is so cold that it resembles solid rock.

The precise depth of the Martian permafrost is uncertain, but it probably extends down more than a kilometer at the equator and several kilometers at the poles. At latitudes between 40° and 80°, water ice on the surface sublimes during summer, but it is very stable only a few meters below the surface. Only near the equator is ice unstable at all depths. This may enable liquid water to accumulate in porous aquifers deep underground. It may even exist close to the surface if the water contains a high concentration of salts which alters its freezing point.[7]

The extent of the ice, which is covered by a thin regolith, has been determined by measuring high energy neutrons and gamma ray emissions from the surface, which provide information on the hydrogen content. High concentrations of hydrogen detected in Mars' circumpolar regions are consistent with plentiful water ice in the shallow subsurface.

Visual support for the widespread presence of ground ice comes from the existence of numerous pedestal craters, each surrounded by what look like mud flows. These craters are created when a meteorite hits a region containing substantial subsurface ice (see below).

One of the key pieces of evidence for permafrost is the presence of a network of small polygonal features on flat, debris-covered plains (Fig. 7.23). Such polygons have been mapped from the edges of the polar caps to mid-latitudes in both hemispheres. In many cases, the pattern comprises large rectangular features, about 160 meters across, which are subdivided into less prominent sections, each separated into several dozen smaller polygons.

The polygons form when the ground contracts in the winter, creating small spaces. If they are infilled with dust, the ground ice has no room to expand again when the temperature rises in summer. As a result, the ground buckles, producing low mounds at the centers of polygons, and shallow troughs at the edges.

Pedestal Craters

The most distinctive impact structures on Mars are pedestal craters (also called rampart craters), which are particularly common at higher latitudes. Instead of linear rays of

---

[7] The presence of salt causes sea water to freeze at a lower temperature than pure water. This is also why common salt (sodium chloride) is used to de-ice roads and pavements in winter.

Lower-Limit of Water Mass Fraction on Mars

Figure 7.34   A map showing the estimated lower limit of the water content of the upper meter of Martian soil. The estimates are based on the hydrogen abundance measured by the neutron spectrometer on Mars Odyssey. The highest water mass fractions, exceeding 30% to well over 60%, are in the polar regions. Significant concentrations also occur in the uplands east of Valles Marineris (center), and to the south west of Olympus Mons (extreme left and right). (NASA/JPL/Los Alamos National Laboratory)

ejecta radiating from the impact site, they are notable for a raised platform (pedestal) that is piled up against the crater rim and lobes of debris. The unusual ejecta field is produced when water-rich material is blasted out of the ground and dropped onto the surrounding terrain. (see Fig. 7.35)

A thick, hot, slurry of mud and rock fragments flies away from the impact site at a steep angle and falls to the ground close to the newly formed crater. This inner ejecta apron forms a muddy blanket that flows sluggishly away from the crater rim.

Where the viscous mud comes to rest, about one crater diameter from the rim, it piles up to form a debris rampart composed of broken rocks and smaller fragments, mixed with groundwater and pieces of ice. A thinner, outer region of ejecta forms when the tall column of debris that rose over the impact site begins to collapse. At the surface, the debris column races outward until it loses its momentum and halts, roughly two crater diameters from the rim.

Only large pedestal craters are found near the equator, whereas much smaller examples can be found at higher latitudes. This makes sense if the permafrost is buried at great depth in equatorial regions, but close to the surface nearer the poles.

Glaciation

Since the temperature rarely rises above 0°C, water ice is more or less permanent on Mars. Any water ice that is warmed above its vaporization point sublimates, changing almost instantly from ice to gas. On the other hand, carbon dioxide ice is much less stable. CO_2 gas freezes out over much of the planet in winter, but the ice disappears through sublimation in summer (except in most of the residual south polar cap).

On Earth, regular glacial cycles have resulted in advances of ice sheets from the polar regions to lower latitudes. These cycles are

Figure 7.35 20 km wide Bacolor crater lies in Utopia Planitia (33.0°N, 118.6°E). The pedestal crater formed where the subsurface held a lot of water or ice. This false color view from Mars Odyssey shows a thick, inner layer of ejecta, surrounded by a thinner layer that extends off the image. The inner debris layer came to rest about one crater diameter from the rim, piling up a rampart 50–80 meters high. Note the two small craters, 14 km north east of Bacolor, which are surrounded by debris. (NASA/JPL/ASU)

Figure 7.36 This 5 km long, lobed feature on the inner wall of a large crater was imaged by Mars Global Surveyor. It appears to be a rock glacier, a mass of glacial ice overlain by rocky debris. The feature is located at 38°S, 113°E, a region famous for "softened terrain" which may involve flowing ice. (NASA/JPL/Malin Space Science Systems)

usually linked to changes in the Earth's axial tilt and orbit (see Chapter 3). Mars has also undergone numerous ice ages, but these are more extreme than on Earth, since the variations in insolation caused by periodic changes in its obliquity and orbit are more pronounced.

Under present conditions, there is little precipitation on Mars, so the formation of glaciers and ice sheets is unlikely. However, its climate has not always been so equable. During warmer periods, vaporization of water ice increases the atmospheric pressure and allows substantial amounts of water to circulate before eventually falling as snow. When the climate switches back, the glaciers become inactive or retreat. There is evidence that some glaciers still exist, protected from sublimation by a blanket of rocky debris.[8]

The walls of many mid-sized impact craters, for example in the Arabia Terra region of the northern hemisphere and the area east of the Hellas basin, display lobed features that resemble rock glaciers, with pitted surfaces and raised ridges, similar to moraines, on their lateral margins. Some major outflow channels, such as Kasei Valles, may also have been modified by subsequent glacial activity, which deepened and widened them to create U-shaped valleys.

Other terrain features between latitudes 30–45° may also be related to glaciation. Small arcuate (arc-shaped) ridges enclose depressions at the base of crater walls, often below gullies. These may be glacial moraines, remnants of unsorted glacial deposits from which the ice has since evaporated. Debris aprons – pitted and lineated deposits on crater floors – are similar to debris-covered glaciers or ice-rich landslides on Earth. Various types of narrow, sinuous channel are seen crossing crater floors, often breaching the crater walls and extending onto the terrain beyond. These channels may have been formed by flowing glacial meltwater.

An hourglass-shaped crater in Promethei Terra, near the Hellas basin, seems to contain debris-covered ice that flowed from one

[8] In theory, it is cold enough for glaciers to exist at the Martian equator today, but the prevailing low atmospheric pressure would cause ice to sublimate unless it was protected by a coating of surface dust.

side of a nearby massif. The "block glacier" almost filled a bowl-shaped impact crater, approximately 9 km wide, then flowed 500 meters downhill, through a breach in the crater rim, into a 17 km wide crater below. The lack of impact craters shows that the surface is only a few million years old.

Similar examples in the Hellas region and in the northern hemisphere have been documented by the SHARAD radar on Mars Reconnaissance Orbiter (see Fig. 7.37). The concealed glaciers extend for tens of kilometers from the edges of mountains or cliffs and are up to one kilometer thick. Such evidence supports the theory that snow and ice accumulated on higher topography within the last few million years and then flowed downhill.

Spacecraft images also indicate that glaciers covered the flanks of volcanoes and mountains only a few million years ago. Each of the Tharsis volcanoes has an unusual, fan-shaped deposit on its western or north western flank which may be formed by glacial processes.

Ancient Rivers

One of the most surprising discoveries of the 1970s was the presence of numerous dry channels, some of which closely resemble river networks on Earth. Crater counts show that the sinuous channels, hundreds or thousands of kilometers in length, date from the Hesperian era, around 3.5 billion years ago, a time of transition from a wet Mars to a dry Mars.

Careful study of the dry channels reveals features typical of erosion and deposition by water: terracing of channel walls, teardrop-shaped "islands," hanging channels that dead-end, and braided channels or oxbows that separate and then reconnect to the main branch. Lava is generally thought to be unable to create such a wide variety of features, although recent studies of a channel system on the flanks of Ascraeus Mons have shown that this is not always the case (Fig. 7.38).

Some channels begin abruptly, and are very deep and wide – sometimes tens of kilometers across. These outflow channels appear to be fed from a fairly localized source, often beginning in a box canyon or an area of collapsed or chaotic terrain that may be hundreds of kilometers across. Some in Elysium and around Hellas start adjacent to large volcanoes; others appear to originate at faults. Tributaries are few in number, and any that do exist are usually short and stubby. They typically have scoured floors, often with teardrop-shaped "islands," and convex walls.

The sudden emergence of full-sized channels and the lack of tributaries indicate that these features were carved by massive flash floods that eroded the surface over a short period of time. The floodwaters may have been fed by an abrupt release of meltwater, caused either by sudden heating of surface/near-surface ice – possibly caused by a large meteorite impact or volcanic activity – or the collapse of a barrier that was damming up a huge lake. Another possibility is that groundwater became trapped beneath a thick layer of permafrost and, as the pressure built up, it episodically burst onto the surface.

In the case of Mars, the water source is likely to have been an underground "reservoir" of meltwater, rather than a surface lake. When the softened ground collapsed, the torrent of water was released.

Some of the outflow channels dwarf anything on Earth. The most dramatic examples flow north from the Argyre basin and eastward from the Valles Marineris, cutting around and through impact craters, before disgorging onto the Chryse Planitia lowland.

Figure 7.37 Three craters in the eastern Hellas region which appear to contain concealed glaciers. On the left is how the surface looks today; on the right is an artist's concept, based on radar data, showing subsurface ice. The thickness of the buried ice ranges from about 250 meters in the upper crater, to 300 meters and 450 meters in the middle and lower craters respectively. Each image is 20 km across and 50 km deep. (NASA/Caltech/JPL/UTA/UA/MSSS/ESA/DLR)

Figure 7.38 Details of a channel (red) that meanders across the surface of Ascraeus Mons volcano. Some features resemble a dry river channel, but its origin is very different. The insets show close-ups of some of the structures that lava can create: (left) branched channels, (middle) a snaking channel, and (right) a roofed section that is probably a lava tube, with nearby pits and small mounds (rootless vents) where lava was forced out of the tube. Other rootless vents are marked by yellow spots on the main image. The length of channel shown is 270 km. (Jacob Bleacher)

The dimensions of these channels show that the volume of water must have been enormous, with discharges in the range of 100 million to one billion cubic meters per second, 1000 to 10,000 times greater than the Mississippi river. These discharges are so large that flow could have continued even under the present freezing climatic conditions. Thousands of cubic kilometers of material would have been removed by such floods.

However, such torrents could not have lasted for long, perhaps a matter of days or weeks. It is likely that the outflow channels were created by at least one catastrophic flash flood, possibly augmented by spasmodic, smaller events, rather than a prolonged, continuous discharge.

Although the temperature was frigid by Earthly standards, the water could have survived for weeks or months if it was briny (hence with a lower freezing point) and the air pressure was higher than it is today. The rate of evaporation would also have been reduced by a surface layer of ice, beneath which the river flowed.

One of the most dramatic examples of an outflow channel is Ares Vallis, which begins in a region of chaotic terrain known as Arram Chaos and flows 1600 km north to Chryse Planitia. The main channel is typically 2.5 km wide and varies in depth from 100 meters to one kilometer.

Ares Vallis is one of the few areas of Mars where a spacecraft has conducted surface measurements. Mars Pathfinder landed downstream from the channel mouth and sent back panoramic pictures of a dry, desert landscape littered with many dark, basaltic rocks separated by patches of sand and dust (see Fig. 7.39).

Channels such as Ares Vallis largely date back to Noachian or Hesperian times, but there are a few possible exceptions. One of these is the Athabasca Vallis/Marte Vallis system, which seems to cut through more recent, Amazonian lavas. It begins on the south eastern side of the Elysium volcanic region, in the Cerberus Fossae region, where there are numerous parallel fractures aligned from north west to south east.

Athabasca Valles resembles a flood-carved landscape, with branching channels, streamlined "islands" and dunes. However, its floor appears remarkably uneroded in high resolution images. What carved these relatively recent valleys? One theory is that they were eroded by a catastrophic flood originating from the Cerberus Fossae fractures or from a large lake in the Elysium Basin. The floodwaters which eventually spread out on the northern plains now take the form of broad, frozen "sea" known as Cerberus Palus. An alternative explanation is that Cerberus Palus was filled by lava flowing from Athabasca Valles.

Rivers, Lakes and Deltas

The dendritic (branching) valley networks that dissect much of the ancient cratered uplands are much more like terrestrial river valleys than the huge outflow channels. Not only are they narrower – up to a few hundred meters across – and much shallower, but many of them have tributaries (although these may be stunted) and increase in size downstream because they are fed by numerous, small channels.

The general consensus is that they formed as a result of slow erosion by water. Some of this water may have been derived directly from precipitation or from the melting of surface snow and ice, although searches have failed to find the tiny rivulets that would be expected if precipitation was the cause. However, any rivulets may have been blanketed and hidden by later deposits, and the dendritic networks remain the best evidence of surface runoff from rain or snowfall on young Mars. On the other hand, many of the networks have characteristics that suggest they were formed by seepage of groundwater, possibly from melting ice.

It is unlikely that these valleys could form under present climatic conditions, since even brine, which has a lower freezing point, is not stable for long in the low pressure environment. The

Figure 7.39   The Sojourner rover, which landed near the mouth of Ares Vallis, using its X-ray spectrometer to analyze the composition of a basaltic rock named Yogi. The rock's two-toned surface may be the result of accumulated wind-blown dust, or evidence of a break from a larger boulder as it was deposited in an ancient flood. Just beyond the ramp, on the left, is a silica-rich rock known as Barnacle Bill – a possible andesite. Patches of red or dark gray dust and sand lie between the rocks, with white material exposed by the rover's wheels. In the distance are the Twin Peaks, smooth-sloped hills that may have been shaped by fast-flowing water. (NASA-JPL)

Box 7.3   Mars Pathfinder

NASA's Mars Pathfinder was the first successful mission involving a roving vehicle on Mars.[9] Designed mainly as a technology demonstration mission, its primary objective was to demonstrate a new method of landing on Mars using a parachute and airbags.

Pathfinder consisted of a stationary lander and a surface rover, named Sojourner. The spacecraft was enclosed by three triangular solar panels which unfolded after touchdown. Once the airbags retracted, the six-wheeled rover extended to its full height and drove down a ramp onto the surface.

The lander carried a camera on a mast to survey its surroundings and atmospheric sensors on one of the solar panels. Sojourner was 0.28 meters high, 0.63 meters long, and 0.48 meters wide with a ground clearance of 0.13 meters, mounted on a rocker-bogie suspension. Power was provided by a deck of solar cells. Its targets were determined by a controller on Earth and the onboard computer navigated it around obstacles.

The spacecraft made a ballistic entry into the Martian atmosphere on July 4, 1997, taking measurements on the way down, and landed in southern Chryse Planitia (19.3°N, 33.5°W), 850 km ESE of the Viking 1 Lander. It bounced 16 times before coming to rest about 1 km from the initial impact site.

The landing site, at the mouth of a large outflow channel, Ares Vallis, was chosen in the hope of seeing evidence for the huge flood that probably inundated it billions of years ago.

Images revealed numerous boulders and dunes, with smooth hills on the horizon. The rover drove slowly around the lander for 83 Martian days, until a flat battery on the lander stopped communications on September 27. The rover's cameras took 550 images and 16 chemical analyses of rocks and soil were made by its X-ray spectrometer.

Pathfinder also detected the wind and pressure signatures of nearby dust devils, and imaged water ice clouds. The weather was similar to that encountered by Viking 1, with rapid pressure and temperature variations, downslope winds at night and light winds in general.

[9] The Soviet Mars 2 lander, which crashed in November 1971, carried a small 4.5 kg rover which was to have moved across the surface on "skis" while connected to the lander by a 15 meter cable. PROP-M rovers were also flown on the Mars 6 and 7 landers in 1973, but these missions also failed.

Figure 7.40 A Viking Orbiter 1 mosaic of part of Mangala Valles. The central area shows huge channels that appear to have been carved by floodwaters in the distant past. In places they have broken through crater rims and shaped elongated "islands." Areas of chaotic terrain, the remains of an ice-rich terrain that has largely collapsed, are visible in some of the channels. Resolution is 40 meters per pixel. (NASA-JPL)

networks probably formed when Mars was warmer and wetter, and possessed a thicker atmosphere. This is supported by the concentration of the channels on old, cratered uplands (e.g. in Terra Tyrrhena), although the presence of a few relatively young examples indicates that cyclical climatic changes may have allowed surface drainage in more recent times.

One of the most persuasive pieces of evidence for prolonged surface drainage is meandering channels, which are created where streams flow across a floodplain that is resistant to erosion. Erosion takes place on the outside of the bend, where the flow is faster, with deposition on the inside, where it is slower. Over time, the meander loops increase in size and migrate downstream, widening the flood plain. On Mars, meandering river sediments have been found in many places, suggesting that much of the planet experienced lengthy periods of surface run off. Low features within the bends, known as scroll bars, show where the meanders migrated over time.

Many of the meanders and dry channels are now seen as ridges rather than valleys. These examples of inverted relief are the result of wind erosion removing finer sediment from the surrounding flood plain (Fig. 7.43).

Rivers on Earth often enter lakes or the sea through deltas. The river splits into numerous distributaries as it crosses the flat, low-lying area which is created by deposition of a large sediment load when the rate of flow decreases. Deltas are also found on Mars, often at the margins of craters. In these cases, the channels probably disgorged into a lake which filled the impact basin.

small channel

Figure 7.41 (a) 2.5 km wide Nanedi Vallis is one of the valleys cutting through cratered plains in the Xanthe Terra region. The valley lacks lengthy tributary channels and has short, box-headed tributaries that suggest formation by collapse. (NASA-JPL) (b) Rocky outcrops occur along the upper walls and weathered debris lies on the lower slopes and the valley floor. A 200 meter wide channel in the valley bottom (upper right) is partially covered by dunes and debris. It suggests that a small river flowed in the valley after the main channel was created. The valley may also have been widened by slumps and landslides, wind and groundwater flow. This MOC picture covers an area 9.8 × 18.5 km. Spatial resolution is 12 meters. (NASA-JPL/MSSS)

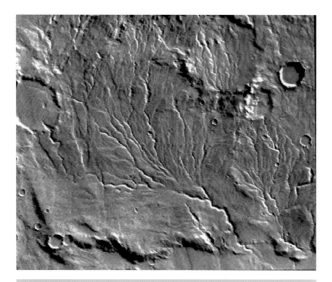

Figure 7.42 A dendritic valley network at 42°S, 92°W. The small, branching channels merge to form larger channels, although the network is less dense than typical drainage systems on Earth. They may have been eroded primarily by groundwater flow – perhaps beneath a layer of ice – rather than by runoff from rain. The area shown is about 200 km across. (NASA-JPL)

Figure 7.43 A fossil flood plain containing a meandering channel, indicated by red arrows. This is an example of inverted relief: the channel bed is now a ridge, rather than a valley, because wind erosion has removed finer sediment from the flood plain. The meanders were created by erosion of the bank on the outside of bends. A cut off may have occurred shortly before flow ceased above location "X," resulting in abandonment of the loop downstream. (NASA-JPL/MSSS)

The most famous example is a fossil delta in the Eberswalde crater at 24.0°S, 33.7°W. After its formation, the delta material was further buried by other materials – probably sediments – that are no longer present. The entire area became cemented

Figure 7.44 The Eberswalde delta lies in a crater in the southern highlands. It was formed when running water flowed over the surface and disgorged into a crater lake. Over billions of years, the delta sediments hardened to form rock, while the softer surrounding material was eroded by the wind. Today, the fossil delta is raised above the rest of the crater floor. This detailed MOC image shows meandering channels, a cut-off meander, and crisscrossing channels at different elevations. (NASA-JPL/MSSS)

and hardened to form rock. Subsequent erosion stripped away the overlying rock, re-exposing the delta. Today, the hardened delta sediments are preserved as ridges, an example of inverted relief.

Occasionally, fans of sediment are found where an ancient channel emptied into a large crater that was presumably filled by a lake. When the river slowed down upon entering the lake, its load of material was dropped, eventually building up into a wide fan crossed by a number of distributary channels. Many of these are fossil landforms, in which the channels through which sediment was transported are no longer present.

Not all polygons on Mars are associated with permafrost (see above). Networks of polygons up to 250 meters in diameter, which exist within large craters, may be desiccation cracks caused by the drying out of the floors of evaporating lakes (see Fig. 7.45).

Unexplained Gullies

No evidence of hot springs or geothermal activity has yet been detected on Mars, not even in volcanic fissures. However, many impact craters display fresh gullies on their steep walls, often with deposits at their lower ends. Similar gullies have also been found

Figure 7.45 Large-scale polygons on crater floors may be caused by a desiccation (drying out) process, with smaller polygons caused by thermal contraction inside. The central polygon is 160 meters in diameter, smaller ones are 10–15 meters wide, and the cracks are 5–10 meters across. (NASA/JPL)

on sand dunes. Clearly, some erosional process involving fluids or gases is active at these sites.

A typical gullied slope is smooth and fairly steep, but topped by a rocky outcrop. Many gullies originate from an indentation or "alcove" which is typically a few hundred meters across and located just beneath the upper rocky layer. On Earth, such alcoves form as water comes out of the ground and undermines the material from which it is seeping. The erosion of material causes rock and debris on the slope above to collapse and slide downhill, creating the alcove.

In the center, the gullies are fairly straight and follow the down slope. They are about 10 meters wide and hundreds of meters in length. At the bottom, some gullies fade away and disappear, whilst others end in a delta-shaped fan or apron of material – possibly a mixture of ice and loose debris. The aprons do not extend very far onto the crater floors, indicating that a limited amount of fluid reaches the bottom of the slope.

The most obvious explanation is that occasional flows of groundwater have eroded the crater wall and carried away loose material. However, it is very difficult for liquid water to penetrate the frozen crust and survive on the surface in the present Martian environment.

If the source is melting ice, then the gullies should be most frequent near the equator – but this is not the case. They are more common at mid-latitudes – particularly in the southern highlands – and almost extend to the fringes of the permanent ice caps. This may be explained by the fact that ground ice is deep below the surface in equatorial areas, whereas it is within a few meters of the surface in higher latitudes.

Another curiosity is the predominance of gullies on shaded, polar-facing slopes at mid-latitudes, though there is little preference nearer the poles. One possibility is that groundwater moving toward a cold, shaded slope freezes near the surface. Its passage is blocked until the barrier is warmed sufficiently for the dam to burst, enabling a sudden rush of water down the slope. The

Figure 7.46 Gullies in an impact crater in Newton Basin, Sirenum Terra (42.4°S, 158.2°W). Spanning a few kilometers, this image from Mars Global Surveyor shows gullies near the top of the crater wall, with dark sand dunes on its floor. Frost is visible near the top and on the dunes below. (NASA-JPL/MSSS)

trapped water could more easily remain in a fluid state if it contained dissolved salts – like sea water on Earth.

Other agents have been proposed, e.g. sublimating carbon dioxide ice which escapes, causing sand to become fluidized and flow downhill, eroding a gully. A similar scenario may apply to dunes that display gullies, e.g. in the Hellespontus and Noachis regions.

Several newly formed deposits have been observed in existing gullies (see Fig. 7.47). The lower end of the flow may split into several branches, and the material appears to be diverted around low obstacles. Such flows must have moved slowly, since they do not over-top some of the low obstacles in their path.

A Mars Ocean?

Did oceans ever exist on Mars? There is certainly a large area of flat lowland in the northern hemisphere onto which numerous channel networks converge from the southern highlands. They must have delivered large volumes of water to the lowlands, although the discharge was probably more episodic than continuous.

What happened to the floodwaters? Did they supply enough water to fill a Martian ocean? Studies of the polar caps, ground ice and numerous channel networks between the equator and

Figure 7.47 Images taken by Mars Global Surveyor in August 1999 (left) and September 2005 show new, light-toned deposits that appeared in an existing gully on the south east wall of a crater in the Centauri Montes region. The shallow, debris-laden flow was diverted around obstacles in its path and spread out into fingers at its lower end. (NASA-JPL/MSSS)

Figure 7.48 How Mars might have appeared some 3.5 billion years ago if an ocean covered the northern plains, fed by run off from the southern highlands. (University of Colorado)

mid-southern latitudes suggest that a plentiful supply of liquid water could have been available in the Noachian era, and favorable conditions may have recurred periodically in the Hesperian and Amazonian.

The fate of the water that flowed down the channels is uncertain. Any terminal lakes that pooled at the ends of the outflow channels may have quickly frozen, leaving thick bodies of ice that

may still be present. After freezing, the ice would have slowly sublimated into the atmosphere, unless the frozen lakes were quickly covered by insulating dust.

Such a frozen lake, covered with surface pack-ice, may have been identified in southern Elysium (5°N, 150°°E). The feature, presumably fed by water from the Cerberus Fossae fissures, measures about 800–900 km across and may be up to 45 meters deep

– similar in size and depth to the North Sea. (The "lake" has also been interpreted as a hardened lava flow – see above.)

Is there any evidence of a much larger body of water – a northern ocean? One clear indicator would be the presence of ancient shorelines: many former beaches can be seen on Earth. However, there is no reason to believe that such clearly demarcated shorelines would be visible on Mars. First, there are no large moons to create powerful ocean tides. Second, any Martian oceans may have been ice covered, which would prevent wave action.

One controversial analysis of Viking orbiter images supported the presence of two ancient shorelines. However, when Mars Global Surveyor mapped the topography to a resolution of 300 meters, it was found that the so-called Arabia "shoreline" varied in elevation by about 2.5 km, while the Deuteronilus shoreline varied by about 0.7 km, rising and falling like a wave with peaks several thousand kilometers apart. Since shoreline elevations, measured relative to sea level, are typically constant, the idea that Mars once had an ocean was widely rejected.

One attempt to overcome this objection suggests that the undulating shorelines can be explained by a surface displacement of Mars' poles by nearly 3000 km within the last two to three billion years. Because spinning objects bulge at their equator, this so-called "true polar wander" could have caused shoreline elevation shifts similar to those observed on Mars.[10]

It is argued that, since the solid lithosphere behaves elastically, its surface deforms differently than the sea surface, creating a non-uniform change in the topography. If a flood of water had filled the Arabia ocean to a depth of several kilometers, about three billion years ago, the additional mass at the pole might have been enough to shift the pole 50° to the south. Once the water disappeared, the pole could have moved back, then shifted again by 20° during the deluge that created the Deuteronilus shoreline.

An alternative scenario proposes that the two shorelines represent the receding edge of a large sea, rather than separate inundations, caused when the Arabia ocean retreated to the Deuteronilus shoreline, shifting the pole from 50–20°. Once the ocean disappeared entirely, the pole returned to its current position.

The presence of a northern ocean that covered one third of the planet's surface is also supported by detailed, global maps of channels and deltas. The regions that are most densely dissected by valley networks lie between the equator and mid-southern latitudes, consistent with past precipitation and the presence of an

Figure 7.49 Most valley networks are concentrated between the equator and mid-southern latitudes. These resemble river systems created by precipitation and surface run off. They end at two possible shorelines that marked the edge of northern oceans. This map shows the dissection density of valley networks, and a hypothesized northern ocean. Two possible sea levels are shown: contact 1 (pale blue) with mean elevation at −1680 meters and contact 2 (dark blue) with mean elevation of −3760 meters. (Wei Luo/Northern Illinois University, T. F. Stepinski, Lunar and Planetary Institute)

---

[10] Other possible causes of polar wander include a major shift of mass involving thermal convection within the mantle or a major asteroid impact.

ocean (Fig. 7.49). Rain would be mostly restricted to the area over the ocean and nearby land surfaces, whilst the southernmost regions, located far from the water, would be arid and without any valleys. This would also explain why the valleys become shallower from north to south.

The distribution of ancient deltas around the margins of the northern plains has also been claimed as a marker for a planetwide equipotential surface (i.e. a Martian sea level) along the margins of the northern lowlands. In other words, the levels of the deltaic landforms act as tracers of an ancient shoreline, where the Noachian and Hesperian rivers entered the ocean. Such data also support the existence of large seas in the Argyre and Hellas basins.

Chemical data have also been used to support the argument in favor of a northern ocean. The gamma-ray spectrometer on Mars Odyssey was used to detect elements buried 0.3 m underground by tracing their gamma-ray emissions. Concentrations of potassium, thorium and iron were found along what appeared to be two ancient shorelines – one that covered a third of Mars' surface, and another that bordered a younger, smaller ocean (Fig. 7.50). The elements were apparently leached out of the rocks on the neighboring uplands and carried by rivers into the oceans, where they were deposited.

To conclude, there is certainly plenty of circumstantial evidence that the northern, circumpolar plains may once have supported a large, shallow body of water. The proposed ocean would probably have resembled Earth's Arctic Ocean, with a surface layer of ice that waxed and waned with the seasons and helped to protect the underlying body of water from evaporation.

Hydrated Compounds

Apart from the persuasive physical evidence for running water and lakes, there is increasingly compelling chemical evidence that groundwater was widespread on ancient Mars. Thousands of outcrops of hydrated minerals, which have been chemically altered by exposure to water, have been found in the southern hemisphere, and some have recently been discovered in craters on the northern plains. However, there is still considerable debate over the amount in different localities.

Some of the debate centers around the presence (or absence) of carbonate compounds and a common mineral called olivine. Earth has enormous amounts of carbon locked up in its rocks in the form of carbonates. This occurred as the result of atmospheric carbon dioxide being dissolved in water and then reacting with other elements in the rocks, such as calcium.[11] These carbonates were washed into the oceans and eventually formed thick layers of chalk and limestone.

If Mars once possessed large lakes or oceans, similar carbonate beds might be expected to be widespread. However, these compounds have proved to be very elusive, even though many Martian meteorites contain small amounts of carbonate, as well as altered minerals, such as clays, salts and gypsum (see Chapter 13).

Carbonates originate in wet, near-neutral conditions, but are easily dissolved by acids, and many of the surface water and groundwater sources on the young Mars may have been sufficiently acidic to destroy any carbonates that formed.

However, recent observations have revealed some limited surface exposures of carbonates which have apparently been buried for billions of years. Mars Reconnaissance Orbiter has detected calcium or iron carbonate which was excavated from

Figure 7.50 This 3-D map superimposes gamma-ray data from Mars Odyssey onto topographic data from Mars Global Surveyor. The red arrow indicates the shield volcanoes of Elysium, seen obliquely to the south east. Blue-to-violet colors in Elysium and the highlands mark areas poor in potassium. Red-to-yellow mark potassium-rich sedimentary deposits in lowlands below the Mars Pathfinder landing site (PF) and Viking 1 landing site (V1). (University of Arizona)

Figure 7.51 A false color image of an outcrop, called "Comanche," in the Columbia Hills of Gusev crater. It has 10 times more carbonate than any previously identified Martian rock. Spectrometers on the Spirit rover found that about one-fourth of the rock is composed of magnesium iron carbonate, a mineral that formed when the environment was wet and non-acidic. (NASA/JPL-Caltech/Cornell University)

[11] Carbonates are also formed by living organisms, but there is no proof of their existence on Mars, so they will not be discussed here.

depths of up to 5 km by the giant impacts that created Huygens and Leighton craters.

Exposed magnesium carbonate ($MgCO_3$) has also been found in the Nili Fossae region, where it is closely associated with rocks that contain phyllosilicates (clays) and olivine. It probably formed during the Noachian or early Hesperian era from the alteration of olivine by hydrothermal fluids or near-surface water. The presence of carbonate with accompanying clays suggests that the waters were neutral to slightly alkaline at the time of its formation, so that acidic weathering, thought to be characteristic of Hesperian Mars, did not destroy the carbonates in this particular environment.

Small exposures of carbonate have also been found by the Spirit rover in the Columbia Hills of Gusev crater. Magnesium iron carbonate makes up about one quarter of the measured volume in the Comanche outcrop, a carbonate concentration 10 times higher than previously identified for any Martian rock. However, the presence of olivine indicates that neutral conditions may not have lasted very long.

Olivine, a common mineral in volcanic rocks, breaks down quickly when exposed to water. If olivine is still present today, then the implication is that it has not been exposed to wet conditions for any length of time. The mineral is widespread in areas such as Syrtis Major, but it has also been found in deep canyons, such as Ganges Chasma (Fig. 7.52).

In some places, such as Nili Fossae, clays appear beneath – and therefore were deposited earlier than – fresh, unweathered rock rich in olivine. It seems that such sites were initially wet, but since the volcanic rocks were laid down, they have remained dry for hundreds of millions, or even billions, of years.

Another key mineral is iron-rich hematite. On Earth, this is often associated with hot springs, where geothermally heated water dissolves iron-bearing minerals. After it evaporates, the iron compounds are left behind as colorful deposits. However, it can also be made in other ways, for example, from crystals that grow in volcanic lava.

Only a few exposures of hematite have been identified. The most abundant sites are Aram Chaos, near Valles Marineris, and

Figure 7.52 Almost 5 km deep, this section of Ganges Chasma reveals ancient rock layers in the canyon walls, as well as several enormous landslides. One landslide on the north wall flowed more than 17 km across the canyon floor, eventually coming to rest on top of an older landslide from the south wall. Outcrops of olivine-rich, basaltic bedrock appear as long, dark red ribbons at the base of the walls. They indicate that the region has remained relatively dry, because olivine is rapidly broken down in wet environments. The colors show night temperatures. Warmer, rocky areas are red and yellow. Cooler, dusty areas are green and blue. (NASA/JPL/ASU)

Figure 7.53 The Opportunity rover landed in a small hollow, named Eagle crater. Spectrometer data show the location of crystalline hematite, a product of past water activity. Red and orange indicate high levels of the iron-bearing mineral. Blue and green denote low concentrations. The crater was about 22 meters across, and the bedrock outcrop (mainly white) was 30–45 cm deep. The north eastern outcrop on the rim of the crater does not appear to contain much hematite. Also lacking hematite are the rover's airbag bounce marks (foreground). (NASA/JPL/Arizona State University/Cornell)

Figure 7.54 A near-true-color image of a rock in Eagle crater called "Berry Bowl." It shows numerous ball-shaped "blueberries." Rich in the iron compound hematite, the spherules are thought to have been deposited in water. The circular area (center) was brushed clean by the rock abrasion tool. (NASA/JPL/Cornell)

Figure 7.55 A shallow trench made by Spirit's wheel uncovered bright patches of almost pure, fine-grained silica – evidence for ancient, water-rich environments in Gusev crater. On Earth, such deposits can be created in warm, evaporating coastal waters or in hot, mineral-laden waters around geysers and hot springs. This false-color view was taken on May 17, 2007. (NASA/JPL-Caltech/ Cornell)

Meridiani Planum, the landing site of NASA's Opportunity rover. The rover came to rest in a small crater whose walls contained plenty of hematite in the form of small, spherical concretions which scientists dubbed "blueberries." Many more hematite/ blueberry deposits were later found, far beyond the landing site.

The blueberries – which are actually more gray than blue – were found in layers, which suggested that they had formed due to leaching of iron minerals in water, or the slow evaporation of iron-rich water in the rock's pore spaces – rather like stalactites and stalagmites in caves. Many blueberries had been released by weathering and wind erosion of softer rock, so that they gathered in small piles. Further support for an aqueous history was provided by cross bedding and possible concave patterns formed by underwater ridges within the rocks.

The Spirit rover made its own water-related discoveries, half way around the planet. Apart from the detection of carbonate in the Columbia Hills, one of the most surprising of these was a measurement of about 90% pure silica (SiO_2) in a patch of fine-grained soil that was churned up by a stuck wheel on the rover. Indeed, wherever the wheel dug into the surface, a white layer rich in sulfur or silica was revealed. The silica may have been laid down as the result of hot spring activity or when acidic, volcanic vapors interacted with water at the surface. High sulfate levels were also found in the Columbia Hills.

The picture that is now emerging is of a planet that was warm and wet more than four billion years ago. Atmospheric carbon dioxide interacted with ancient bodies of water, producing calcium- and iron-rich carbonates. Prolonged contact with water of low acidity – either on the surface or underground – meant that volcanic rock minerals such as olivine were rapidly hydrated, creating widespread phyllosilicate (clay) beds and magnesium carbonate.

As Mars began to cool and dry out, salt flats with intermittent pools between dunes became the norm. Planet-wide volcanic eruptions belched sulfur into the atmosphere, producing acid rain or highly acidic groundwater. Chemical reactions with this water again modified the surface rocks, creating different types of hydrated minerals, such as sulfates – especially where there was evaporation of salty, and sometimes acidic, water.

Hydrated silica, found in and around the margins of Valles Marineris and dry river channels, is evidence that some low temperature, acidic water remained on the surface until about two billion years ago. Finally, Mars became largely dry. During the second half of the planet's history, the rocks have been altered by slow weathering, punctuated only by occasional relapses to warmer, wetter conditions.

A Windblown Desert

Orange-red dust and sand cover almost the entire Martian surface. The color is the result of oxidation of iron minerals in the crust: essentially, the planet is "red" because its rocks have rusted. The iron component in the dust has been confirmed by spacecraft spectrometers and by magnets carried on surface landers and rovers, which have attracted particles of dust. The dust is dominated by silicate minerals.

Although the atmosphere is only about 1% as dense as Earth's at sea level, it is capable of picking up and transporting small, weathered debris on a massive scale. These shifting sands are largely responsible for the changing seasonal pattern of dark and light markings that have been recorded by astronomers for hundreds of years. (For several centuries they were thought to be associated with the expansion and dying off of Martian vegetation.)

In general, the classical bright features seen on pre-space-age maps are regions covered by a layer of dust up to 2 meters thick with particles in the size range 2–40 microns. The dark features

Figure 7.56 Many craters and canyons contain sand dunes. This Mars Express image shows a crater that contains a small field of dark dunes which measures 7 × 12 km. The crater, located at 43°S, 303°E, is about 45 km wide and 2 km deep. The dunes' asymmetric profile, with a gentle slope on the wind-facing side, suggests an easterly wind direction. The dark sands could be basaltic. (ESA/DLR/FU Berlin-G. Neukum)

are relatively dust-free, though still coated with particulate materials ranging from 0.1 mm to 1 cm across, rather than bedrock.

Tiny particles of dust – comparable to cigarette smoke – are also suspended high in the sparse atmosphere, producing an orange sky, as well as spectacular sunrises and sunsets. These can remain aloft almost indefinitely, although they are periodically replenished by dust storms and mini-tornados known as dust devils (see below).

The gradual build up of fine dust on exposed surfaces has been documented by the Mars Exploration Rovers, which experienced large drops in power when a coating of dust reduced the sunlight reaching their solar panels. However, Nature also intervened from time to time when gusts of wind removed much of the dust and boosted their power output.

As befits a desert world, Mars has many landforms which have been created by wind action, and dunes of all sizes are found. A huge dune field covers much of the north polar region, and there are many isolated sand seas, especially within old, degraded craters west of Hellas Planitia. In many ways these dunes resemble their counterparts on Earth. Some are dark, probably of basaltic composition, and others are bright.

Swarms of crescent-shaped barchans occur where the prevailing wind is mainly in one direction. Sand is moved up a gentle slope on the windward side. As the sand piles up at the crest, it collapses and forms a steep slope on the lee side. The entire dune is driven forward by this process. The horns of the crescent show the direction of movement and are created by sand moving around the dune's outer edge. Sometimes smaller, faster-moving dunes merge with larger, slower-moving dunes, resulting in single, large features over 500 meters wide.

Seif dunes are linear in shape, with their long axis parallel to the wind. These are often seen extending from sand sheets next to eroding mesas and rocky knobs, as well as from the horns of

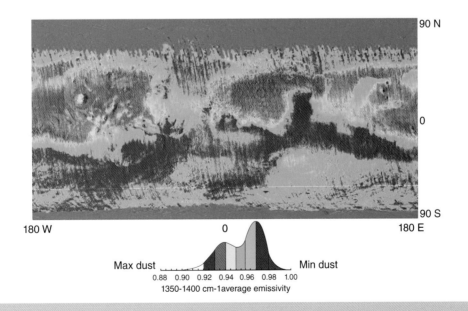

Figure 7.57 A global map of dust cover, based on data from Mars Global Surveyor's Thermal Emission Spectrometer. Orange, red, and white areas are probably dust covered, while blue and magenta areas are probably fairly dust-free. (NASA-JPL/ASU)

Figure 7.58 Barchan and linear dunes that formed on the floor of a crater and extended from an eroding mesa to the west of Hellas Planitia, near 41.8°S, 44.5°W. The prevailing wind blows from right to left. The location of the linear dunes coincides with areas where supply of loose material is greatest. The linear dunes break up into barchans with distance from the flow obstruction. (NASA-JPL/ University of Arizona)

Figure 7.59 A sea of sand ripples and wheel tracks imaged by Opportunity. The ripples in this part of Meridiani Planum are aligned generally north-south and the camera is pointing north. The distance between the wheel tracks is about one meter. (NASA/JPL – Caltech)

some barchans. They form when wind that encounters the dune obliquely is deflected on its lee side, causing it to flow along that side parallel to the crest line.

Most dunes on Mars are transverse, lying at right angles to the prevailing wind direction. Like barchans, they are asymmetrical, with a gently sloping windward side and a steeply sloping leeward side. However, in many places there is a juxtaposition of fresh and degraded dunes, including star dunes and transverse dunes with differing alignments, indicating multiple wind directions.

Images taken over many years show few changes: the larger dunes appear to be stationary, implying that many of them are "fossil" features left over from a different climatic regime. Only a few small, dome-shaped dunes have been seen to diminish in size or disappear completely, although streaks have been seen to appear on the downwind sides of some dunes, indicating that sand is moving. Calculations show that, on average, it takes about 4000 years for a Martian dune to shift by one meter. However, it is likely that dune change is limited to occasional dust and sand storms driven by strong winds.

Sand ripples that resemble small tranverse dunes are found almost everywhere. Over time, stationary ripples and small dunes develop a surface crust that is hard to identify. In March 2009, the Spirit rover became permanently stuck when its wheels broke through such a crusty surface and churned into soft sand hidden underneath. The rotating wheels revealed loose layers of basaltic sand, sulfate-rich sand and areas with silica-rich materials, possibly sorted by wind and cemented by chemical action.

The key to dune and ripple movement is saltation, a cascade effect which occurs when a sand particle starts to hop along the surface and dislodge other grains. The thin atmosphere means the wind has much less momentum and is much less likely to disturb a sand grain than on Earth. However, the lower gravity and air pressure enable a particle that has begun saltating to travel 10 times faster than it would on Earth. As a result, each Martian grain should eject about 10 times more sand upon impact than on Earth.

Other evidence that wind can move loose material around the surface is provided by "tails" of material on the leeside of craters which are probably accumulations of bright material shaped by wind currents.

Wind is an agent of erosion, as well as transportation and deposition. Perhaps the most obvious examples of aeolian erosion are yardangs, parallel ridges and grooves which are aligned in the direction of the prevailing wind (see Fig. 7.60). They are common in southern Amazonis and south of Elysium Planitia, sometimes reaching several kilometers in length, though they are often only a few tens of meters high.

Sand blasting of the lower parts of rocks and wind erosion of harder bedrock, such as lava flows, seems to be rare. This may be because wind speeds are generally low, and fine dust, rather than larger, more abrasive particles, is picked up and moved by the thin air.

Dust Storms

Astronomers have long been aware of regional dust storms that obscure the planet's surface. These are seasonal events, confined to spring and summer in either hemisphere, when the seasonal carbon dioxide ice cap sublimates in the spring sunshine. This

Figure 7.60 Parallel ridges and grooves known as yardangs are sculpted by wind-blown sand. They occur where the winds blow in the same direction for a long period. Weaker material is eroded faster than more resistant material. The Mars Express image was taken south of Olympus Mons at 6°N, 220°E. More resistant rock may occur in the three flat regions, which measure about 17 × 9 km. (ESA/DLR/FU Berlin-G. Neukum)

causes the atmospheric pressure to rise, thus enabling more dust to be suspended and remain aloft for a longer time.

Most are local or regional events, driven by strong surface winds, turbulence and rapid convection. They are very common near the poles in spring, powered by large temperature differences between the polar ice and the nearby regions of dark sand and rock which are heated and defrosted by the springtime Sun. The dust may obscure the ice caps as it is picked up and rapidly spread by the polar jet stream.

An even more important source region is the Hellas basin in the southern hemisphere, which is effectively a huge dust bowl. Because the basin is up to 8 km deep, air at the bottom is about 10°C warmer than at the top. This temperature gradient creates major updrafts and strong, turbulent winds, which can carry dust above its rim and out of the crater. On rare occasions, the dust from Hellas spreads far and wide, triggering other long-lived storms.

About 10 planet-encircling dust storms have been reported since 1877. The most famous example of recent times erupted in September 1971, whilst NASA's Mariner 9 and several Soviet spacecraft were en route to the planet. When they arrived in orbit, the planet was completely shrouded in dust. Only the south polar cap and four dark spots, eventually revealed as the tops of the Tharsis volcanoes, were visible. The dust sometimes rose to a height of 70 km, and the storm did not disappear until the end of the year.

The evolution of a similar storm was closely monitored in 2001, at the onset of the Martian southern spring. Studies showed a complex sequence of events which culminated in the dust eventually covering the entire planet.

It began as a small dust cloud inside the Hellas basin. At first there was little change. The storm only "exploded" after it crossed the equator on June 26. In less than 24 hours, additional dust clouds were being raised in Arabia, Nilosyrtis, and Hesperia, far from Hellas.

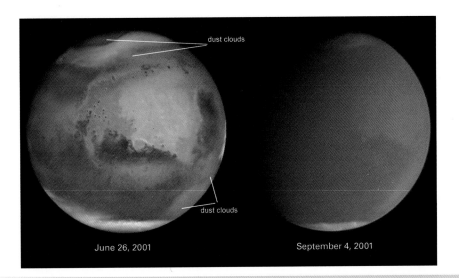

Figure 7.61 The evolution of a global dust storm. (Left) On June 26, 2001, one large dust storm was visible above the northern polar cap (top), with a smaller dust cloud nearby. Another large dust storm was spilling out of the Hellas impact basin in the southern hemisphere (lower right). White water ice clouds are seen near the polar caps. (Right) Over two months later, the entire planet was still blanketed by dust. (James Bell/Cornell Univ., Michael Wolff/Space Science Institute, Hubble Heritage Team)

Figure 7.62   The evolution of the June 2001 dust storm that began in the Hellas basin (near bottom), as measured by Mars Global Surveyor. The cloud grew until, by late July, it enveloped the whole planet. It then slowly subsided over the next few months. Blue represents a fairly clear atmosphere and red shows dense concentrations of dust. (NASA/JPL/Arizona State University)

During the following week, dust lifted into the stratosphere was carried eastward by the south circumpolar jet stream. At the same time, a turbulent wind front moved across Mars, favoring the creation of many other dust storms, sometimes thousands of kilometers from Hellas. By early July, dust had enveloped the entire planet. The storms largely subsided by late September, but the atmosphere remained hazy into November.

Dust storms play an important role in modifying the weather and climate of Mars. Smaller storms, especially near the poles, affect the rate at which seasonal frost evolves, and control local and regional weather patterns. The global storms temporarily alter the entire planet's heat balance, promote variations in seasonal frost formation and dissipation, and greatly affect the distribution of water vapor.

During the 2001 storm, the fine airborne dust blocked a significant amount of sunlight, causing a drop in the daytime average temperature on the surface of almost 3°C. This was hardly surprising, since images from the Viking landers showed a darkened sky during even local dust events (see Fig. 7.63), and this was confirmed by Spirit and Opportunity, which occasionally recorded a 99% drop in direct sunlight, enough to stop all surface operations.[12]

In contrast, the loss of heat from the surface at night was reduced by the dust blanket, so the nighttime average actually increased by 2–3°C. At the same time, the dust absorbed solar radiation and heated the upper atmosphere by some 30°C. The warmer atmosphere ballooned outward, causing increased aerodynamic drag on orbiting spacecraft.

[12] Before dust storms began blocking sunlight in June 2007, Opportunity's solar panels had been producing about 700 watt hours of electricity per day, enough to light a 100 watt bulb for seven hours. By July 18, the solar panel output had dropped to 128 watt hours.

Dust Devils

Loose surface material is also transported by much smaller atmospheric disturbances, known as dust devils, which form when the ground heats up during the day. Much larger than their counterparts on Earth, these column-shaped vortices resemble miniature tornadoes. As the warmer, lighter air nearest the ground begins to rise, it spins. When the spinning column moves across the surface it picks up dust, making the vortex visible.

The peak time for dust devil formation is around 2 p.m. local time, the warmest time of day in midsummer. On Earth, dust devils typically last for only a few minutes, but the Martian versions can survive much longer. Sometimes they grow to the size of terrestrial tornadoes, spiraling up to 9 km above the surface. (The height of a dust devil can be calculated from the length of its shadow and the height of the Sun above the horizon.)

The Mars Exploration Rovers observed hundreds of dust devils, often appearing in small swarms. Gusev crater alone probably experiences about 90,000 dust devils per sol, and these whirlwinds collectively hoover up and redeposit an estimated 4.5 million kg of sediment per sol. Clearly, these modest storms reshape the landscape over much of the planet and contribute to the build up of larger dust storms.

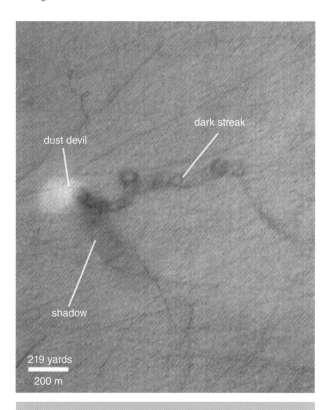

Figure 7.63 The impact of a dust storm at the Viking 1 landing site (22°N, 48°W) on sol 1742 of the mission, compared with illumination on other days. The relative levels of illumination show how atmospheric dust affects the brightness of the sky. The images have been computer enhanced. (NASA-JPL/Mary A. Dale-Bannister, Washington University, St. Louis)

Figure 7.65 A dust devil seen from above as it creates a curly, dark streak in Promethei Terra. Traveling from right to left (east to west), the rotating, dust-laden column throws a shadow. The dust devil is less than 100 meters wide. The picture covers an area of 1.5 × 1.7 km. (NASA-JPL/MSSS)

Figure 7.64 An image by the Spirit rover showing two dust devils crossing the dusty interior of Gusev crater. Small dust devils such as these appeared each afternoon during the summer months. (NASA-JPL)

Box 7.4   The Vikings

NASA's Viking 1 and 2, each consisting of an Orbiter and a Lander, were the first missions to obtain high resolution images of the Martian surface, characterize the structure and composition of the atmosphere and planetary surface, and conduct *in situ* biological tests.

Figure 7.66   Scientist Carl Sagan with a model of the Viking lander. The upright cylinder to the right of the dish-shaped high gain antenna is a TV camera. The meteorology boom is visible at right. (NASA-JPL)

Each 663 kg Lander was sterilized before launch and housed inside a protective, pressurized shell on the bottom of the Orbiter. It was 2.1 meters tall and three meters wide, with three legs and a three meter long robot arm to dig trenches and collect samples. Power came from two radioisotope thermoelectric generators, which used plutonium-238, and four batteries.

Each Orbiter was equipped with two vidicon cameras, an infrared spectrometer and an infrared radiometer. Their task was to image the entire planet at high resolution. The Landers carried two TV cameras, a biology lab with three experiments, a gas chromatograph mass spectrometer, and X-ray fluorescence spectrometer, pressure, temperature, and wind velocity sensors, a three-axis seismometer, a sampler arm and magnets. Other sensors measured air pressure, temperature and acceleration during the descent.

Viking 1 was launched on August 20, 1975, and arrived at Mars on June 19, 1976. The Viking 1 Lander touched down at Chryse Planitia (22.48°N, 49.97°W) on July 20, 1976. Viking 2 was launched on September 9, 1975, and entered Mars orbit on August 7, 1976. The Viking 2 Lander arrived on Utopia Planitia (47.97°N, 225.74°W) on September 3.

Originally designed to function for 90 days, the Vikings continued collecting data for more than six years. The Landers accumulated 4500 high resolution images of the Martian surface, whilst the Orbiters provided more than 50,000 images, mapping 97% of the planet.

Viking 1 landed on a rocky plain marked by occasional ridges and impact craters. The gently rolling landscape was yellowish-brown and strewn with rocks in the centimeter to meter size range. The rocks were volcanic and probably deposited as ejecta from distant large impacts. Between the rocks were drifts of fine-grained material, with bigger dunes in the distance. The drifts appeared to be formed from very fine-grained, wind-blown material. The surface was loosely cemented, probably by water-soluble salts such as sulfates, to form a crust.

Viking 2 also landed on a level, rock-strewn, but otherwise featureless plain. The source of the rocks was believed to be a 90 km diameter impact crater situated 170 km to the west. Although no chemical analyses were obtained from rocks at either Viking site, they were believed to be basaltic.

The Landers scooped up surface material in order to measure the chemical composition of the soils. They also performed a variety of chemical experiments on the soils, as part of their search for life.

Figure 7.67 The Viking 1 landing site on Chryse Planitia. Orange-red sand dunes and rocks covered most of the surface, interspersed by patches of darker material. The one meter high rock (extreme left) was named "Big Joe." Shallow trenches dug by the sampler arm (center) are visible in the foreground. Some of these were dug deeper to study soil which was not affected by solar radiation and weathering. The reddish sky was caused by dust suspended in the lower atmosphere. The boom holding the meteorology sensors is at left. (NASA-JPL)

Figure 7.68 Viking Lander 2 touched down on Utopia Planitia, further north than Lander 1. Although they were about 6400 km apart, the sites looked broadly similar, with a boulder-strewn surface reaching to the horizon nearly 3 km away. Fine particles of dust settled on the spacecraft. Three color calibration charts for the cameras are visible to the right of the Stars and Stripes. The circular high-gain antenna (top) was the main communication link with Earth and the Orbiters. (NASA-JPL)

Atmosphere

From the surface, the sky usually appears light brown or orange, due to light scattered from suspended dust particles, roughly 1.5 microns across. These can be raised to high altitude by dust devils and seasonal dust storms, remaining aloft for long periods before they slowly drift back to the surface.[13]

The atmosphere is about 150 times less dense than on Earth. It is dominated by carbon dioxide, which accounts for more than 95% of the atmosphere by volume. During the winter, when one of the poles is in continual darkness, the surface gets so cold that up to 25% of the atmospheric CO_2 freezes out at the polar cap. When the winter pole is again exposed to sunlight in summer, the CO_2 ice sublimates back into the atmosphere. There are also

[13]**The first color picture returned by Viking Lander 1 was hurriedly processed to produce a blue sky. Mission scientists soon realised their error, and all subsequent images showed the dust-laden, orange sky.**

Figure 7.69 The Opportunity rover took this image of vast plains of sand beneath an orange tinted, dusty sky, and wispy clouds of water ice. (NASA-JPL/Cornell University)

Table 7.3
Atmospheric composition of Mars (by volume)

Major gases	carbon dioxide (CO_2) – 95.32%; nitrogen (N_2) – 2.7%; argon (Ar) – 1.6%; oxygen (O_2) – 0.13%; carbon monoxide (CO) – 0.08%
Minor gases (ppm)	water (H_2O) – 210; nitrogen oxide (NO) – 100; neon (Ne) – 2.5; hydrogen-deuterium-oxygen (HDO) – 0.85; krypton (Kr) – 0.3; xenon (Xe) – 0.08

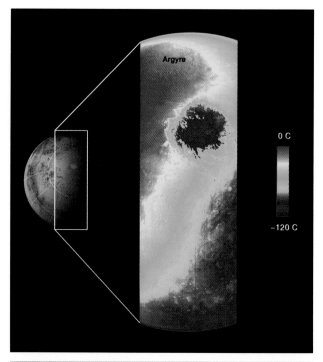

Figure 7.70 A thermal infrared image acquired by Mars Odyssey on October 30, 2001 – late spring in the southern hemisphere. It shows a temperature range of more than 120°C from the pole to the equator – a distance of over 6500 km. The near-circular blue feature is the southern polar cap of carbon dioxide ice, where the temperature is about −120°C. The cap was more than 900 km in diameter and shrinking as summer approached. Cool air blowing off the ice is shown in orange, to the left of the cap. The cold region at lower right is the night hemisphere. The warmest regions (red) occur near local noon. The 900 km diameter Argyre impact basin can be seen in the early afternoon. The thin blue crescent along the planet's upper limb is the atmosphere. (NASA/JPL/Arizona State University)

seasonal variations in the amount of water vapor in the atmosphere as the polar caps wax and wane.

The surface pressure ranges from about 14 mb (millibars) in the deepest parts of the canyons to 0.3 mb on top of the highest volcanoes. It also changes with the seasons, as a result of sublimation of ices in the summer hemisphere and freezing out of volatiles at the winter pole. This cycling of the polar ices causes the average atmospheric pressure to change by 20% as the seasons change. Variations of this magnitude are unknown on Earth, where even a hurricane causes a drop in pressure of only a few percent.

Since Mars is between 30% and 67% further from the Sun than Earth, it is correspondingly colder. The maximum solar irradiance on Mars is about 590 W/m², compared to about 1000 W/m² at Earth's surface – equivalent to the intensity on an autumnal day when the Sun is a mere 36° above the horizon.

Without a thick atmosphere to trap warmth by night or minimize the Sun's heat by day, Mars experiences extreme day-night temperatures. Local temperature variations are related to the reflectivity and thermal properties of surface materials.

Latitude also plays a part (see Chapter 3). The Spirit rover recorded greater temperature swings than Opportunity because it was farther from the equator, where there were greater variations in the Sun's height above the horizon and the length of the day.

Near-equatorial temperatures in summer typically range from −90°C at night to above 20°C at midday, but this variation increases if the surface consists of low density, fine-grained material. However, at depths of several centimeters below the surface, temperatures even out with a diurnal mean of about −50° to −60°C. The air temperature is also quite different only a few meters above ground level.

Conditions are even more extreme at the poles in winter, where temperatures drop below −120°C. At this point, carbon dioxide condenses out of the atmosphere to form a seasonal ice layer which covers the permanent polar cap and extends to lower latitudes.

Figure 7.71 Shade temperatures for the Spirit rover ranged from about 35°C in summer to −90°C in winter. The horizontal axis shows the number of Martian days (sols). The data were obtained by a sensor on the hazard avoidance camera, which was in shade beneath the solar panels. The background image shows a sunset. (NASA/JPL-Caltech/Cornell/NMMNH)

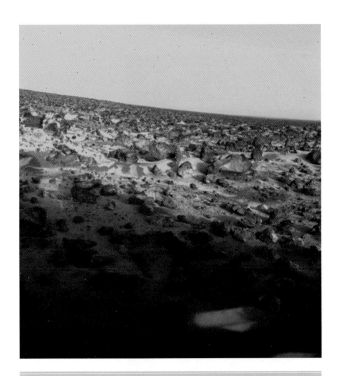

Figure 7.72 A frosty morning on Utopia Planitia, imaged by Viking Lander 2 on May 18, 1979. A similar coating of water ice on the rocks and soil appeared almost exactly one Martian year earlier, evidence of the planets' regular annual climate cycles. The frost remained at this mid-latitude site (48°N) for about 100 days. Dust particles in the atmosphere act as nuclei around which water and carbon dioxide freeze. The particles become heavy enough to fall to the ground. Warmed by the Sun, the carbon dioxide sublimates and returns to the atmosphere, leaving behind water ice and dust. The ice may be less than 1 mm thick. (NASA/JPL)

Clouds of water ice often gather above the slopes of the giant volcanoes and in the bottoms of the canyons and deep basins, such as Hellas, where they resemble blankets of fog or low-lying stratus. Wispy cirrus clouds also appear at high altitudes, both during the day and night. The daytime clouds have been found up to 100 km above the surface – the highest clouds ever detected on any terrestrial planet. Circular clouds are common in summer near the north pole, and may result from merging air currents.

Nocturnal clouds are five times thicker than their daytime counterparts and hover close to the ground. Despite their insubstantial nature, they do trap some heat and help to keep the surface warm when temperatures plummet after sunset. When the air becomes dustier, the upper atmosphere becomes warmer and the area of water ice clouds dwindles.

The Viking Landers sent back images of a frost-covered surface in winter, but snow was not thought to be likely. Nevertheless, the Mars Phoenix lander detected ice crystals falling from clouds at an altitude of about 4 km, so this form of precipitation may also be an important part of the water cycle. However, the total amount of water in the atmosphere is very small. If all the water in the Martian atmosphere simultaneously fell as rain, it would make a global ocean less than 0.01 cm deep.

On Earth, warm air rises in the tropics, travels poleward at high altitude, then cools and descends in the subtropics before returning toward the equator (see Chapter 3). Computer simulations suggest that a similar, though less complex, circulation also occurs on Mars, with a single, trans-equatorial Hadley cell which reaches an altitude of 50 km, compared with about 20 km on Earth. This cell is most prominent around the solstices.

The atmospheric flow associated with the Hadley cell transports surface air from the cold winter hemisphere to the summer hemisphere, e.g. at the northern winter solstice, the rising air is in the warmer southern hemisphere and the sinking air is in the northern hemisphere.

Figure 7.73 Thin cloud cover is widespread on a northern summer day in April 1999. This global view shows bluish-white water ice clouds, for example, above the Tharsis volcanoes (left). The bright white ice clouds (lower right) lie above the Hellas Basin. The equator runs across the center. The high southern latitudes (above 60°S) were in winter darkness. (NASA/JPL/Malin Space Science Systems)

Figure 7.74 Four images of an annular (circular) cloud that formed over the same terrain in the north polar region during successive summers. The first, taken in April 1999, was obtained by the Hubble Space Telescope. The others were taken by Mars Global Surveyor. Despite its apparent resemblance to a cyclone, the cloud does not rotate. It forms when different air currents merge in the morning; by afternoon, it usually dissipates or breaks up into smaller clouds. (NASA/JPL/Malin Space Science Systems)

Because the equator rotates faster than the rest of the planet, it creates a Coriolis force that deflects the zonal winds (see Chapter 3). The result is a trade wind-like pattern of easterlies in the winter hemisphere and westerlies in the summer hemisphere. A strong westerly jet also occurs near the winter pole, while an easterly jet occurs in the southern subtropics in summer.

Winds associated with the Hadley circulation can reach very high speeds: models indicate that mean winds in the subtropics during the southern midsummer exceed 160 km/h about 200 meters above the surface. Surface winds can also be very variable. At the Viking landing sites, typical wind velocities were from zero to 12–16 km/h in summer, with rare gusts in winter of 48–65 km/h. Wind patterns were also fairly predictable, with a repetitive pattern of wind direction and velocity for specific times of the day. However, during southern spring and summer, when dust storms are common, gusts of up to 144 km/h have been recorded.

Large scale variations in atmospheric pressure occur in the winter hemisphere, as high and low pressure systems travel eastwards – just as they do on Earth. Associated frontal systems – boundaries between cold and warm air masses – also occur on both planets.

The sparse atmosphere responds almost instantly to solar heating. This results in a daily pressure cycle, with minimum pressures near 4 a.m. and 6 p.m. local time, and maximum values near midnight and 10 a.m. It also causes strong winds across the day-night terminator. Winds are also modified by local topography, especially near the large volcanoes and in large basins such as Hellas, creating stationary eddies. Pathfinder found that winds at night came from the south, consistent with air flowing down Ares Vallis.

Despite its low density, the atmosphere does offer some protection from incoming interplanetary debris. The glow from at least one meteor burning up in the atmosphere was captured by the Spirit rover's panoramic camera, and meteor showers, similar to those experienced on Earth, should also be visible on Mars.

Over billions of years, most of the atmosphere has escaped into space or been frozen into the surface as permafrost. Much of the carbon dioxide may be tied up in the regolith or in the formation of carbonate rocks. Some of the water in the upper atmosphere has been broken up by solar ultraviolet light and the resultant hydrogen has escaped from the planet.

Solar Wind Interaction

Like Earth, Mars possesses an ionosphere – a layer of ionized (electrically charged) particles in its upper atmosphere. In this region, high energy solar radiation and particles in the solar wind split the atoms and molecules in the upper atmosphere, releasing free electrons.

There are two main layers, at altitudes of about 110–130 km. A third, non-continuous layer has been detected between 65–110 km. This lower layer may be created by the interaction of the atmosphere with fast, incoming meteors. This may involve an exchange of electrical charge between the atmospheric gases and magnesium and iron atoms present in meteors.

The ionosphere appears to vary in shape and composition over time scales of only a few minutes, and it is not always able to prevent the solar wind from penetrating. The solar particles then energize and ionize the upper atmosphere, causing the escape of

Figure 7.75 Four images, taken approximately two hours apart, show the evolution of a storm near the north pole on June 30, 1999 – late summer on Mars. The polar ice cap is at the center. White clouds are mainly water ice, orange/brown clouds contain dust. The "hook" in the clouds indicates a vortex associated with a storm front moving toward upper right. High surface winds raised dust and mixed it with water vapor over the polar cap to create this effect. Such storms are common in late summer – early autumn. (NASA/JPL/Malin Space Science Systems)

Box 7.5   The Methane Mystery

One of the most intriguing minor constituents of the atmosphere is methane (CH_4), the first organic molecule found on Mars. The global average abundance is tiny, around 10 ppb (parts per billion), although some measurements have claimed concentrations of up to 250 ppb. The methane levels change with the seasons: they are highest in autumn in the northern hemisphere and lowest in winter.

There are three regions in the northern hemisphere where methane concentrations are high: Tharsis and Elysium, the two main volcanic provinces, and Arabia Terrae, which has plentiful underground water ice.

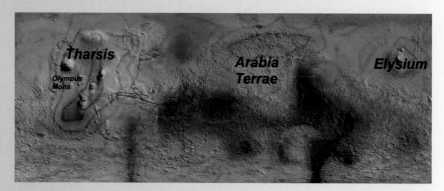

Figure 7.76   A map of methane distribution during the northern autumn, based on Mars Global Surveyor data. Highest concentrations (orange) occur in the Tharsis and Elysium volcanic regions, and Arabia Terra. Lowest concentrations are purple. (NASA-JPL, University of Salento)

Methane is an unstable gas which is soon broken down by natural processes, so, unless there is a source of continuous replenishment, it should have disappeared long ago from Mars.[14] How is the methane being generated? Geological sources are possible, such as hydrothermal and geothermal activity or crustal hot spots. The most exciting – and controversial – explanation is that the gas is produced by microscopic living organisms which survive far below the surface, where it is still warm enough for liquid water to exist.

hydrogen and oxygen ions and ionized molecular species, such as carbon dioxide.

One of the most surprising results from Mars Express was the discovery of an ionosphere above the night hemisphere. This can be explained if the solar wind flows around the planet and produces a population of free electrons in the upper atmosphere. The regions of high electron density above the night side are associated with strongly magnetized areas, especially south of the equator, where the local magnetic field lines are perpendicular to the surface.

The interaction between the solar wind and the ionosphere is responsible for erosion of Mars' upper atmosphere. This erosion is particularly marked during periods of high solar activity, when a strong solar wind and coronal mass ejections compress the ionosphere, pushing it to a lower altitude. This exposes more hydrogen and oxygen ions to the solar wind, which picks them up and carries them away.

Such erosion is thought to have increased around 3.5 billion years ago, when the planet's internal dynamo shut down. Erosion by the solar wind since that time could account for a loss of water

equivalent to a global Martian ocean about 10 meters deep. Today, the only regions to receive any protection from the solar wind lie above the stronger magnetic anomalies that still exist on the surface, where the altitude of the ionopause may exceed an altitude of 400 km.

Auroras

In the absence of a global magnetic field, auroras were not expected to occur on Mars. However, spacecraft have shown that they are very common. Hundreds have been recorded on the night side, particularly close to local midnight. Their locations coincide precisely with patches of strong magnetic field. Just as Earth's auroras occur where the magnetic field lines dive toward the surface at the north and south magnetic poles, Mars' auroras occur at the borders of magnetized areas where the field lines arc vertically toward the planet.

Earth's auroras light up when charged particles from the Sun are diverted down magnetic field lines and collide with atoms in

---

[14]The lifetime of CH_4 in the atmosphere of Mars is estimated to be 300–600 years, based on the rate of its destruction by photochemistry, or shorter if strong oxidants such as peroxides are present in the surface or on airborne dust grains.

Figure 7.77 The solar wind of electrically charged particles (ions and electrons) is slowed to subsonic speeds at a bow shock, upstream of the planet. Here, the magnetic field fluctuates wildly and the flow of the solar wind becomes chaotic. The absence of a global magnetic field allows the solar wind to interact directly with the ionosphere and upper atmosphere. Solar ultraviolet radiation ionizes gases such as oxygen and hydrogen, which can then be removed by the magnetized solar wind. Also shown is the near-polar orbit of Mars Global Surveyor. (NASA/JPL/GSFC)

the atmosphere to create an oval of light around each magnetic pole. The process on Mars is probably similar. Solar wind particles are funneled around to the night side of Mars where they interact with crustal field lines. The aurora is produced when the ions and electrons hit carbon dioxide molecules in the atmosphere.

Increased solar wind activity and solar storms tend to make the auroras brighter and stronger. The energetic electrons that interact with molecules in the atmosphere probably generate mainly invisible ultraviolet light, although the classical green emission line of oxygen might also be present.

Meteorites from Mars

Missions to bring samples of Martian material back to Earth for analysis have been under consideration for many years, but they have always been delayed by the huge cost and technological complexity. However, there are samples of Martian rock already available for study in terrestrial laboratories. These are meteorites that originated on the red planet, were ejected into space by impacts, and then fell to Earth after millions of years in interplanetary space.

At the time of writing, only 98 of the 40,000 or so meteorites discovered on Earth have been identified as originating from Mars.[15] All of them are made of igneous rock, and the vast majority are between 1.3 billion years and 165 million years old, much younger than typical meteorites, which are 4.5 billion years old (see Chapter 13). They also have higher contents of volatiles than igneous meteorites and oxygen isotope ratios that are distinctively different from terrestrial ratios.

The only plausible source that could have been volcanically active so recently is Mars. Conclusive evidence of their origin came from the measurement of gases trapped inside the EETA 79001 meteorite, which match those measured in the Martian atmosphere by the Viking Landers. Since the oxygen isotope content of the other SNC meteorites is the same, they must all come from the same parent object.

There are half a dozen sub-groups of these meteorites, representing rocks that formed in different locations on or beneath the planet's surface. Their varying mineral content and chemistry show that they cannot all have originated from a single impact event. Some contain shock-produced glass formed by partial melting as the result of a large impact event. Some have igneous minerals with little water, but many contain evidence of interaction with liquid water, especially carbonates, sulfates and clays.

The most unusual, and oldest, example is ALH 84001, which seems to have crystallized 4.5 billion years ago.[16] It displays evidence of a long, complex history of shock and thermal alteration, and also contains carbonates, evidence that it has been in contact with Martian water. However, the lack of hydrated minerals in ALH 84001 favors limited exposure to moisture in a region of restricted water flow, such as an evaporating saline pool. The meteorite made headlines around the world in 1996 when a team of US scientists announced that it may contain evidence of primitive life (see below).

Life on Mars?

Since life as we know it requires liquid water, the history of water on Mars and its current distribution are regarded as the key factors in determining whether the planet ever supported life and where it may have flourished.

It now seems certain that a denser atmosphere very early in Mars' history supported a warmer, wetter environment, creating conditions that could have been favorable for life. Hydrated mineral deposits suggests that the liquid water on or near the surface was widespread and of low acidity in the Noachian. However, even if life began within the first few hundred million years, its evolution may have been severely restricted.

The benign conditions changed in the Hesperian, to be replaced by a more acidic environment. Huge volcanic eruptions coincided with the shutdown of the global magnetic field and the depletion of the atmosphere as gases escaped into space. Apart from

---

[15] They are often called SNC meteorites after the names of three typical examples: Shergotty, Nakhla and Chassigny.
[16] The designation ALH indicates that it was found in the Allan Hills in Antarctica. The number 84001 indicates that it was the first meteorite found at that location in 1984.

Box 7.6 Mars Orbiters

The modern era of Mars exploration began on July 14, 1965, when the Mariner 4 spacecraft flew by the planet and sent back 22 pictures, with resolutions of several kilometers. Scientists had speculated that Mars might be more like Earth than the Moon, but the images revealed an ancient, cratered surface. Similiar views were sent back by the next two Mariner flybys.

Perceptions then changed dramatically in 1972, when Mariner 9 became the first Mars orbiter. Despite an initial delay caused by a global dust storm, the spacecraft mapped 70% of the planet and made the first detailed study of the atmosphere and the surface.

As the mission progressed, huge volcanoes, deep canyons, enormous dry river beds, and extensive dune fields came into view, and a complex geological history became apparent. It also sent back the first close-up images of Phobos and Deimos, and provided information on the shape of Mars and its gravity field.

Encouraged by the evidence that Mars was once much warmer and wetter, NASA sent two Viking spacecraft to survey the planet in much greater detail. By the end of the mission, almost all of the surface had been imaged from orbit at a resolution of about 250 m/pixel and smaller areas with resolutions as high as 10 m/pixel. The Viking Landers photographed their landing sites on the northern plains and successfully carried out a variety of experiments directed mostly toward detecting life and understanding the chemistry of the soil.

After the failure of the Vikings to detect any organic compounds, no more US spacecraft were dispatched to the planet for 16 years. Unfortunately, contact with Mars Observer was lost shortly before arrival at Mars, and further studies had to wait until Mars Global Surveyor entered orbit on September 12, 1997.

Propellant consumption was reduced by aerobraking, using friction with the planet's upper atmosphere to move into the operational polar orbit. MGS operated longer at Mars than any other spacecraft, and its Mars Orbiter Camera returned more than 240,000 images which revolutionized understanding of the planet's surface and weather. Discoveries included recent gully erosion, multi-layered sedimentary rocks and retreat of ice on the south polar cap.

Some of the most important breakthroughs came from the Mars Orbiter Laser Altimeter (MOLA), which provided the first accurate topographic maps. Other important results came from the magnetometer, which discovered local magnetic anomalies in the crust, and the Thermal Emission Spectrometer, which mapped surface minerals and discovered deposits of hematite – a clue to a watery past.

The next spacecraft to arrive was Mars Odyssey, which entered orbit on October 24, 2001. After aerobraking was completed, it entered a 400 km polar, nearly Sun-synchronous mapping orbit. Its gamma ray suite recorded significant quantities of hydrogen close to the surface – evidence of large deposits of underground water ice. Analysis of temperature data enabled scientists to distinguish between solid rock and a variety of loose surface materials. Odyssey was still operational in October 2011.

Figure 7.78 Mars Express was Europe's first planetary orbiter. Although its piggybacking Beagle 2 lander was lost during atmospheric entry, the orbiter continued to operate well beyond its design life. Among its instruments were a radar sounder, three spectrometers, a stereo camera and a particle instrument. (ESA)

Europe's first Mars orbiter was Mars Express, which comprised an orbiter and the Beagle 2 lander. Although Beagle 2 was lost during atmospheric entry on December 25, 2003, the orbiter was successfully placed in a near-polar orbit. Its radar experiment sent back the first subsurface soundings of Mars and the High Resolution Stereoscopic Camera provided a series of stunning 3-D images of the surface. Another breakthrough came with the discovery of phyllosilicates and other hydrated minerals by the OMEGA spectrometer, evidence of a warmer, wetter planet billions of years ago. Mars Express was still operational in October 2011.

Mars Reconnaissance Orbiter, the largest US spacecraft ever sent to the red planet, arrived on March 10, 2006. The spacecraft sends data to Earth at about 10 times the rate of any previous Mars mission. It carries five main instruments to characterize the surface, subsurface and atmosphere of Mars, particularly the history of water, past and present. The High Resolution Imaging Science Experiment (HiRISE) is most powerful camera ever flown on a planetary mission, capable of detecting features one meter across. The Italian-built Shallow Subsurface Radar probes to a depth of up to one kilometer to study layers of ice and rock, including the polar caps. By March 2010, it had returned 100 terabits of data, more than three times the amount sent back from all other deep space missions combined.

Figure 7.79 ALH 84001 is the oldest known Martian meteorite, which probably formed about 4.5 billion years ago. Since then, it has undergone shocks and metamorphism, probably associated with violent impacts on Mars. Carbonate minerals indicate that it was once in contact with ground water, possibly an evaporating pool of brine. (NASA-JSC)

occasional warmer spells and flash floods, the surface conditions have generally been far more extreme than anywhere on Earth for billions of years.

However, this does not necessarily exclude the survival of any primitive organisms that may have evolved long ago. Liquid water can exist deep below the surface, where temperatures are higher and there is no exposure to solar ultraviolet light. Furthermore, recent discoveries on Earth of extremophiles – organisms that can survive in conditions of extreme heat or cold, acidity or alkalinity, prolonged dryness, lack of sunlight and even high radiation – prove that life forms are remarkably resilient.

Today, although permafrost predominates at high latitudes, only a few centimeters below the surface, environments with liquid water probably exist a few kilometers below the surface, over most of the planet. This might enable organisms to survive in or beneath ice sheets, in subterranean aquifers, or in protected habitats such as lava tubes, caves and fissures. The potential for life could be enhanced in warmer, volcanic regions, such as Tharsis and Elysium.

Alternatively, life may have evolved to cope with alternating climatic cycles or different environments, so that dormant microbes could be present close to the surface whilst active variants survive in protected environments.

Since most potentially habitable environments are out of reach for the foreseeable future, the hunt for Martian organisms is currently focused on seeking evidence of past life, preserved as fossil biosignatures in surface rocks. This will require microscopic imagers capable of studying rocks in many wavelengths as well as seeing details only a hundredth of a millimeter across. Also needed are organic chemistry laboratories to analyze promising rocks.

The first attempt to identify life on Mars was made in the 1970s by the Viking Landers, which carried four experiments to search for indirect evidence of living organisms.

The biology lab on the Landers comprised three experiments to analyze gases released by soil samples under different conditions (see Fig. 7.80). The Pyrolytic Release Experiment was looking for organisms that could assimilate and reduce carbon dioxide or carbon monoxide (rather like the photosynthesis carried out by plants on Earth). Samples were tested in darkness and light, as well as in dry or humid conditions. After they were exposed to radioactively-tagged carbon dioxide, the "labeled" gas was removed and they were heated so that any gases that were released could be analyzed.

The fixation of gaseous carbon was detected, but similar results occurred even in darkness, and the reaction remained stable at a temperature of 90°C, suggesting a chemical, rather than biological cause.

The Labeled Release Experiment also used radioactive carbon-14 as a tracer, but in this case it was in the food supply – an inorganic nutrient that was added to the soil. It was hoped that any organisms would consume the nutrient and then exhale radioactive carbon dioxide or carbon monoxide.

Once again, there was an unexpected result. Part of the organic mixture was rapidly converted to CO_2, as would be expected in a biological reaction, but then the conversion largely stopped. A second injection of nutrient had no effect. This contrasted with the behavior of terrestrial organisms, which continue to take up nutrients until they are completely consumed. However, strong heating seemed to destroy whatever was converting the nutrient, a possible sign of organisms being killed during the sterilization process.

In the Gas Exchange Experiment, a very rich organic nutrient "broth" was added to the soil after it was initially dampened with water vapor. The atmosphere in the chamber was then monitored for changes that would indicate biological activity. Although a burst of oxygen was released after the sample was humidified, few changes occurred when the broth was added. This suggested that the soil contained a strong oxidizing substance, such as a peroxide or superoxide, rather than living organisms.

The landers also carried a Gas Chromatograph – Mass Spectrometer experiment. Several samples of material were heated in steps to 500°C, and the released gases were then analyzed for organic compounds. The experiment was capable of detecting simple carbon compounds in concentrations of about one part per million, as well as larger molecules in the parts per billion range.

Carbon dioxide and water absorbed in the soil were detected, as were some terrestrial contaminants, but no organic molecules were recorded, a remarkable result in light of the continuous delivery of carbon-based material by falling meteorites. After this negative result, it was generally concluded that any positive indications from the biology experiments were suspect. The argument against biological interpretations was reinforced by the apparent presence of a powerful inorganic oxidizing agent that decomposes organic compounds.

One of the principles behind the Viking experiments – the detection of life through the release of gases produced by metabolic processes – is also related to a more recent detection of localized methane sources on Mars (see Box 7.5). On Earth, more than 90% of atmospheric methane arises from current or past biological activity, but the question of whether the Martian methane is produced by living organisms or some inorganic process remains unresolved.

Other evidence is available from the small collection of Martian meteorites, most notably ALH 84001 (see above). In August 1996, NASA and Stanford University scientists announced that it contained probable evidence of past microscopic life.

Figure 7.80   The biology lab on the Viking landers comprised three separate experiments. In the Labeled Release Experiment, a radioactively-tagged liquid nutrient was added to soil and emitted gases were analyzed to detect any uptake by microbes. The Pyrolytic Release Experiment heated soil samples that had been exposed to radioactively-tagged carbon dioxide to see if it had been used by organisms to make organic compounds. The Gas Exchange Experiment analyzed gases released by soil to which a nutrient was added. (NASA-JPL)

Figure 7.81 Rounded globules in ALH 84001 are made of siderite (brownish iron carbonate) and magnesite (clear magnesium carbonate), while the dark, outer rinds are rich in iron oxide and sulfides. The globules formed after ALH 84001 solidified from lava billions of years ago, but their age is uncertain. Possible bacterial nanofossils were found in these globules. This thin-section view is about 0.5 mm across. (Allan Treiman/Lunar and Planetary Institute)

Figure 7.82 This segmented, worm-like object, seen in a scanning electron microscope image of ALH84001, has been interpreted as a possible microfossil bacterium. The image, one micron across, shows the surface of orange carbonates in the rock. (NASA-JSC)

The group cited four main arguments, all of which were associated with unusual orange globules of carbonate minerals, up to a few millimeters across:

1. Layered calcium, magnesium and iron carbonates were formed due to contact with non-acidic, liquid water on Mars.
2. Complex organic chemicals in the carbonates could be the decay products of microbes.
3. The carbonate globules contain tiny grains of magnetite, an iron oxide, and iron sulfides. These particles were similar to those associated with certain bacteria on Earth.
4. The globules contain segmented, tubular objects, less than one micron in length, similar to fossilized nanobacteria found on Earth.

Although each piece of evidence can be accounted for by inorganic processes, it was argued that biogenic activity is the simplest explanation to account for all of the observed features.

As a result of the furore, ALH 84001 became the most closely examined piece of rock in history. Although the meteorite was clearly very ancient, forming when liquid water was almost certainly present on its surface, and the carbonates almost certainly formed at that time, the other lines of evidence were not generally accepted.

Instead, it was argued that some of the features originally identified as possible fossil organisms may have been caused during sample preparation, while some others may be the result of weathering of its surface. The magnetite grains could be non-biogenic, while the organic compounds may be the result of contamination after the meteorite landed in the Antarctic.

Phobos and Deimos

After years of speculation that Mars may have some small moons, they were eventually found by American astronomer Asaph Hall during the opposition of 1877. He named the inner satellite Phobos (Fear) and its outer companion Deimos (Panic), after the companions of the Greek war god, Ares (the Roman god Mars).

The average distance of Deimos from the center of Mars is 23,459 km – only one tenth the Earth-Moon distance. It follows a near-circular path, inclined at an angle of 1.8° to the planet's equator. With an orbital period of 30 hours 21 minutes, Deimos takes a little over one Martian day to complete a single orbit. Consequently, an observer on Mars would see it track slowly westwards across the sky.

Phobos orbits only 9378 km from the planet's center – so close that it would be permanently below the horizon for anyone above latitude 69°. In order to avoid crashing into Mars, it sweeps around the planet in 7 hours 39 minutes (about one third of a Martian day). This means that an observer near the equator would see Phobos rise above the western horizon roughly every 11 hours. It would then cross the sky "backwards" (west to east) in 4½ hours, eventually setting below the eastern horizon. During this brief appearance, it would go through more than half its cycle of illuminated phases.

Both Phobos and Deimos are irregular in shape, with average diameters of about 22 and 12 km respectively. From Mars' surface, Phobos appears about one-third the diameter of the Moon, as seen from Earth, whereas Deimos appears as a bright star. Phobos is the denser of the two, though it still has less than one-thousandth the surface gravity of Earth.

Each satellite has been affected by tidal forces over billions of years, so that its long axis is aligned with the planet and the same

Box 7.7 The Mars Exploration Rovers and Phoenix

Two US Mars Exploration Rovers, Spirit and Opportunity, were launched in 2003 to search for evidence of liquid water, a possible precursor for the evolution of Martian life.

Both vehicles continued to operate far beyond their 90 day primary mission. At the time of writing, Spirit was no longer communicating, having become stuck in a sand trap, but Opportunity was still driving toward Endeavour crater, where clay minerals had been identified from orbit.

Figure 7.83 The rovers were equipped with navigation cameras, two high resolution stereo panoramic cameras and a spectrometer to study rocks at infrared wavelengths. The robotic arm carried a microscopic imager, for detailed views of rocks, and two spectrometers to determine surface composition. Also on the arm was an abrasion tool to remove the weathered outer crust from rocks. The six-wheeled rovers received power from a deck of solar panels. Eight radioisotope heaters enabled them to survive the cold nights and long winters. (NASA-JPL)

After considerable debate about where to send them, it was eventually decided to drop Spirit in Gusev crater, a 166 km wide impact structure on the edge of the southern highlands. Orbiter images indicated that it once contained a lake that was fed by an enormous outflow channel, Ma'adim Vallis. However, little evidence of the lake was found, perhaps because any sediments were subsequently buried by lava.

Opportunity's target was Meridiani Planum, a broad plain on the other side of the planet. This region was known to have deposits of gray hematite, an iron mineral that is usually produced in the presence of liquid water.

Spirit arrived on the surface, encased in airbags, on January 3, 2004. Its first images showed a flat, relatively featureless plain, strewn with small rocks. Some low hills were just visible 3 km away. Analysis of the soil showed that it was rich in silicon and iron, with significant levels of chlorine and sulfur. One unexpected finding was the presence of olivine, a mineral which is easily weathered. The surface material was found to be fine-grained basalt.

Opportunity touched down 10,000 km away on January 24, coming to rest inside a 20 meter diameter crater which featured a small outcrop of layered bedrock. There were thousands of gray, iron-rich spherules – dubbed "blueberries"– scattered across the surface and embedded in the exposed rock.

Over the coming months, both rovers found evidence that confirmed the presence of liquid water at their landing sites in the distant past. This was particularly true at Meridiani Planum, where the data indicated that finely layered rocks in Eagle crater and elsewhere were probably deposited on a salt flat, or playa, which was sometimes covered by shallow water and sometimes dry.

Eventually, the rovers began to explore further afield. Spirit passed by the 200 meter diameter Bonneville crater in mid-March 2004 and set off on a 2.3 km trek to the 120 meter high Columbia Hills, where it found layered rocks containing minerals that were either formed in, or altered by, water. Some of the rocks contained high concentrations of sulfates and carbonates.

Figure 7.84   The route followed by Spirit during the first 1391 sols, showing the main exploration sites. The traverse ended on the edge of a low plateau called Home Plate. The landing site is 3.8 km away to the north west. Sol numbers refer to when Spirit was at that location. Spirit reached the summit of Husband Hill on sol 619 (Sept. 29, 2005). (OSU Mapping and GSU Laboratory/NASA/JPL/Cornell University/MSSS)

During its descent from the summit, Spirit came across an unusual low plateau called "Home Plate," a possible site of ancient volcanic explosive activity. It continued to explore this area until April 2009, when its wheels broke through a surface crust and dug more than hub deep into sand. It was unable to escape from the sand trap. It ceased operating on March 22, 2010, during the Martian winter.

Figure 7.85   A wide-angle view of Burns Cliff inside Endurance crater, with an image of Opportunity superimposed. The approximately true-color rendering spans more than 180° side to side. The cliff walls appear to bulge outward. In reality, they form a gently curving surface. (NASA/JPL-Caltech/Cornell)

Opportunity initially examined sedimentary rock outcrops and sandy, windblown regolith within Eagle crater. It then spent six months inside Endurance crater, where it made a close inspection of rock layers exposed in "Burns Cliff," part of the crater wall. En route to its next target, Erebus crater, Opportunity found the first meteorite ever identified on another planet.

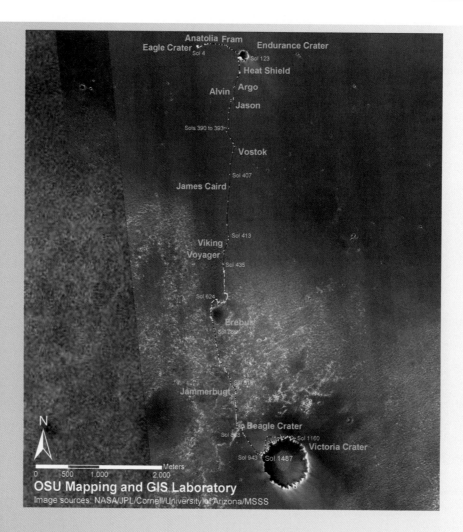

Figure 7.86 The route followed by Opportunity during the first 1487 sols, showing the main exploration sites prior to arriving at Victoria crater. After venturing inside Victoria crater, it set off south toward Endeavour crater. (OSU Mapping and GSU Laboratory/NASA/JPL/Cornell University/MSSS)

Despite becoming temporarily stuck in a sand dune, Opportunity arrived at the 300 meter wide Erebus crater in late September, where dunes and rifts lay between outcrops of rock. It continued south through etched terrain toward the 800 meter wide Victoria crater, where a substantially thicker sequence of layered rock was found.

After entering the crater on September 13, it experienced slopes of up to 25°, but was able to study the crater wall. After 11 months, Opportunity backed out the way it had come and headed for 22 km wide Endeavour crater, about 12 km away. It arrived at the crater rim on August 9, 2011. By then, Opportunity had driven 33.5 km, while Spirit had travelled over 7.7 km.

On May 25, 2008, **Phoenix** became the first lander to explore the near-polar regions when it touched down at 62.2°N on the plains of Vastitas Borealis. Surrounded by polygons in the permafrost, the spacecraft came to rest on a layer of hard water ice which restricted attempts to sample surface material. Nevertheless, heating of soil samples determined that the soil was alkaline (basic), and consisted of salts and aqueous minerals. Its five months of operation ended in November 2008.

hemisphere always faces Mars. Studies of their orbits show that Phobos is slowly accelerating and spiraling in towards Mars. About 100 million years from now, the moon will be destroyed when it collides with the planet.

Phobos has a heavily cratered, ancient surface which is dominated by a 9 km wide crater called Stickney, the maiden name of

Asaph Hall's wife. Also present are two 5 km wide craters – one of which is named after Hall himself.

Radiating away from Stickney are linear chains of small craters and parallel striations that are only 5–10 meters deep but up to 15 km in length. These have often been regarded as features related to internal fractures caused by the large impact that

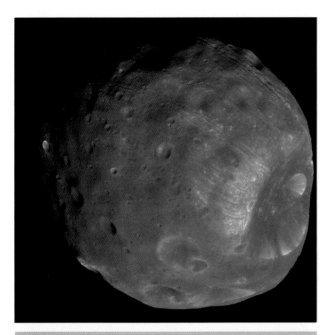

Figure 7.87 A false color image of Phobos, taken on October 23, 2007 by Mars Reconnaissance Orbiter, shows features 20 meters across. Ejecta around the rim of Stickney crater (right) appear pale blue. Linear grooves and crater chains radiate from Stickney. Landslides are visible on the internal walls of large craters, despite Phobos' weak gravity. (NASA/JPL-Caltech/University of Arizona)

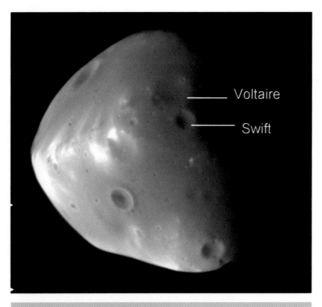

Figure 7.88 An enhanced-color view of the Mars-facing hemisphere of Deimos, taken with the HiRISE camera on February 21, 2009. At that time, only two craters (labeled) had been given names. North is up and the equator is roughly horizontal. (NASA/JPL/UA)

excavated Stickney (and almost destroyed Phobos). Also visible are linear grooves about 500 meters wide that may be evidence of lines of weakness produced by tidal forces and secondary impacts from material ejected from Stickney and other sizeable craters. Recent, multi-wavelength images show a coating of ejecta on the flank of Stickney and landslides on its inner walls.

Measurements made during close spacecraft encounters indicate that it has a lower density than expected. With 25–35% of its interior as empty space, it is now believed to be little more than a jumbled "rubble pile."

Deimos also has an irregular shape, with a several small craters, the two most prominent being Swift and Voltaire. However, its surface is much smoother than that of Phobos. There are no visible grooves or striations, suggesting that tidal forces and large impacts have played a lesser role in its evolution, and most of the craters appear to be partially filled or covered with loose regolith. Close range images also reveal a surface strewn with boulders that were presumably ejected during low-energy collisions with small meteorites.

Both objects are very dark, with albedos of only 5–7%. Early spectral measurements indicated that they resemble carbonaceous chondrite meteorites (see Chapter 13). This led to the belief that the moons are captured asteroids, although it was difficult to explain their near-circular, near-equatorial orbits.

More recent measurements have shown that they contain minerals that are common on Mars, including phyllosilicates that are created by interaction of silicate materials with liquid water. This has led to the theory that they formed relatively near their current orbits through re-accretion of material blasted into space by a catastrophic Martian impact. Such an origin would account for their rubble pile composition.

Questions

- Mars is often described as the most Earthlike of all the planets in the Solar System. Discuss the validity of this statement by discussing the major similarities and differences between the two planets.
- How is the Martian environment modified by variations in the planet's orbit and axis of rotation?
- Explain the importance of (a) dust, (b) water and (c) carbon dioxide in the Martian environment.
- Summarize the main phases in the geological evolution of Mars.
- Compare the magnetic environments of Mars, the Moon and Earth.
- Is Mars a likely location for the evolution of life? Explain your answer.
- Summarize how (a) orbiters and (b) landers have revolutionized our understanding of Mars in the past half century.
- Briefly summarize how and why the natural satellites of Mars differ from Earth's Moon.

EIGHT
Jupiter

Beyond the main asteroid belt is Jupiter, the largest planet in the Solar System, with more than twice the mass of all the other planets combined. Its gravitational field is so powerful that it can hold on to more than 60 satellites and thousands of Trojan asteroids, while removing asteroids from parts of the main belt. Its magnetic influence extends all the way to Saturn and the major moons are bathed in lethal doses of radiation. Yet, despite its huge size, Jupiter spins faster than any other planet, completing each rotation in less than 10 hours.

Jupiter's existence has been known since prehistoric times. This is hardly surprising, since it can reach magnitude −2.9 around the time of opposition, every 399 days or so. For about six months of each year, Jupiter is an unmistakable beacon, often dominating the night sky after midnight, when only the Moon (and very occasionally, Venus) rivals its brilliance.

Early observers were aware that Jupiter took almost 12 years to progress through the constellations of the zodiac – longer than any planet apart from Saturn. From its stately movement against the stellar background and prolonged illumination of the night sky, they deduced that the planet was further away than Mercury, Venus or Mars. Nevertheless, it was remarkably bright, so the Romans named it after Jupiter, the king of the Olympian gods.

Orbit and Physical Characteristics

Jupiter, the fifth planet from the Sun, lies beyond the outer edge of the main asteroid belt. Like most of the others, it follows an almost circular orbit which is very close to the ecliptic.

After the introduction of the telescope, it soon became obvious that Jupiter is the dominant body in the Solar System, after the Sun. With a volume equivalent to 1300 Earths, it is large enough to swallow all of the other planets. Although it comes no closer to the Earth than 599 million km at times of opposition, its visible disk dwarfs every planet apart from nearby Venus, offering a magnificent sight in telescopes of even modest aperture.

Jupiter has more than 2½ times the mass of all the other planets combined – or 318 times Earth's mass. As a consequence, its gravitational pull is enormous. To escape Jupiter's gravity, a spacecraft must travel away from the planet at a speed of 59.6 km/s, compared with 11.2 km/s for Earth. However, since its mean density is not much greater than that of water, it is clear that the bulk of the planet is made of gases and ices, rather than rock.

Despite an equatorial diameter 11 times that of Earth, Jupiter rotates once every 9 hours 50 minutes – faster than any other planet.[1] This means that the equatorial cloud deck is traveling from west to east at more than 45,000 km/h.

Such rapid rotation of a largely gaseous planet causes a very noticeable equatorial bulge. The outward centrifugal force of rotation opposes the effect of gravity. Since the speed of rotation is greatest at the equator and least at the poles, the planet expands into an oblate shape that is elongated along the equator. Jupiter's equatorial diameter is 9270 km larger than its polar diameter.

[1]The equatorial rotation period of about 9 hours 50 minutes is about 5 minutes shorter than the rest of the planet. Radio emissions from Jupiter's interior give a period of 9 hours 55 minutes 30 seconds.

Exploring the Solar System, First Edition. Peter Bond.
© 2012 by Peter Bond. Published 2012 by Blackwell Publishing Ltd.

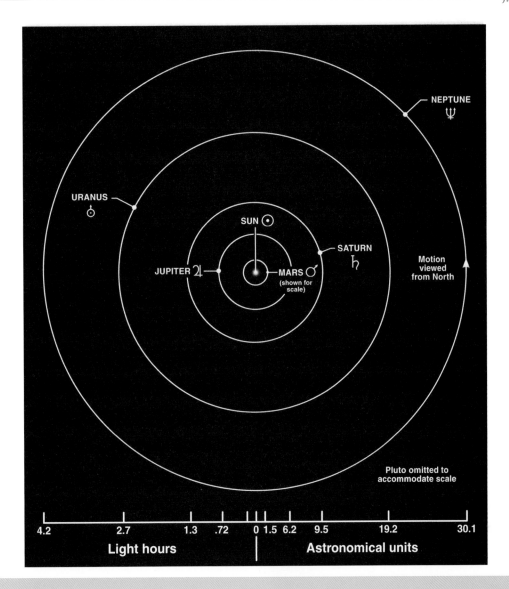

Figure 8.1 Jupiter, Saturn, Uranus, and Neptune are sometimes known as the Jovian (Jupiter-like) planets. This diagram shows their approximate distances from the Sun. Note the enormous gaps between their orbits, shown in astronomical units and light hours – the time light from the Sun takes to reach them. (LPI)

Zones and Belts

Jupiter is a dynamic, cloud-covered world with no visible solid surface. Most of the planet is dominated by reddish-brown "belts" and whitish "zones," linear cloud features that are parallel to the equator. These are associated with alternating east-west wind currents, with the fastest jets blowing from west to east in the equatorial region.

Although they are ever-present, the belts and zones are not unchanging. Amateur astronomers eagerly record the latest streaks, plumes and spots that form and evolve in the highly turbulent atmosphere. Some of the spots and plumes survive for several days, months or even decades. However, even Jupiter's turbulent atmosphere occasionally goes through a quiet spell, when the belts become fainter, or even disappear, and fewer cloud

variations are visible. A recent example was the 2010–2011 disappearance of the South Equatorial Belt.

The cloud patterns also reflect the vertical motion of the atmosphere in which the clouds form. The light-colored zones are regions of warmer, rising gas. When this reaches the cloud tops, it spreads out and sinks back in the neighboring dark belts. The direction in which the gas moves is also influenced by the extremely strong Coriolis force (see Chapter 3).

Above 50° latitude it is a different story. The stripes disappear, to be replaced by a mottled pattern of reddish-brown clouds in the form of numerous rotating cells. As is the case near the equator, the variations in color indicate cloud particles with different chemical compositions.

Infrared images show that the colors also coincide with altitude. Occasional bluish regions are warmest of all, indicating that

Figure 8.2 Jupiter has more than 2½ times the mass of all the other planets combined. Although 1300 Earths would fit inside it, gaseous Jupiter has only 318 times Earth's mass. Despite its enormous bulk, Jupiter rotates faster than any other planet, causing its equatorial region to bulge outward. The clouds are arranged in light and dark bands aligned parallel to the equator. The Great Red Spot (bottom centre) is a giant atmospheric storm. (Bob Kanefsky, NASA-Ames)

Figure 8.3 An image of Jupiter taken by the Cassini spacecraft on December 7, 2000 shows the shadow of the moon Europa on the planet's cloud tops. The arrows point in the direction of wind flow. Their length corresponds to wind velocity, which can reach 180 m/s in the jet near the equator. (NASA-JPL)

they are the deepest of the visible clouds and only visible through holes in the overlying blanket. Above these are brown clouds, followed by white. The highest clouds of all are reddish.

All of the major cloud layers predicted by models are white in color, so the precise composition of most of these compounds remains uncertain. Elemental sulfur, which can assume many different colors, depending on its temperature and molecular structure, is one possibility. Phosphine is another.

However, the clouds are certainly the products of active physical and chemical processes taking place in the atmosphere. Unstable compounds may form in a variety of ways: bombardment by charged particles; lightning discharges; rapid ascent and condensation; or solar ultraviolet light causing photodissociation of compounds such as methane.

Atmospheric Composition

Like all of the giant planets, Jupiter has an atmosphere dominated by the two lightest gases, hydrogen and helium. However, many more constituents have been found by spectroscopic observation.

The first major advances came in the 1930s, when dark lines in its spectrum were identified as ammonia (NH_3) and methane (CH_4). Around 1960, the presence of molecular hydrogen was confirmed, leading to the conclusion that Jupiter's overall composition must be very similar to that of the Sun – which is not

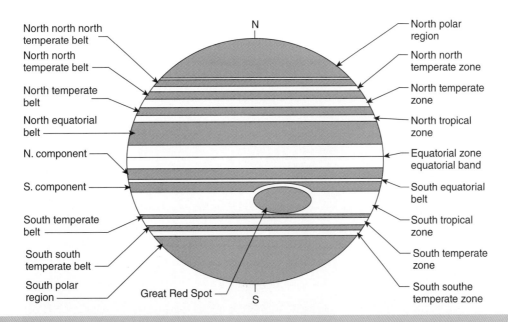

Figure 8.4 Jupiter's dark cloud bands are called belts, and the light bands are known as zones. The rest of each name is based on the corresponding latitude, for example, equatorial or tropical. North is at the top. (NASA)

Figure 8.5 Two cylindrical mosaics of Jupiter obtained during the flybys of Voyager 2 in July 1979 (**a**) and Cassini in December 2000 (**b**). Note how the parallel belts and zones are replaced by numerous rotating cells above 50° latitude. The Great Red Spot and numerous white ovals are visible in the southern hemisphere. (NASA-JPL)

Figure 8.6 False color, near-infrared images from the Galileo orbiter show clouds and hazes at various altitudes. The top left and right images, taken at 1.61 microns and 2.73 microns respectively, show the deep atmosphere, with clouds down to a level of about three Earth atmospheres. In the 2.17 micron image (top center), light is strongly absorbed by hydrogen. Only the Great Red Spot, the highest equatorial clouds, a small feature at mid-northern latitudes, and thin, high photochemical polar hazes can be seen. At 3.01 microns (lower left), deeper clouds are just visible, despite gaseous ammonia and methane absorption. The 4.99 micron image (lower middle), shows heat rising from the deep interior through thin cloud cover. In the image at lower right, red areas denote warmth from the deep atmosphere; green shows cool tropospheric clouds; blue shows the cold upper troposphere and lower stratosphere. Polar regions appear purplish, because small-particle hazes allow leakage and reflectivity. Yellowish regions at temperate latitudes may indicate tropospheric clouds with small particles which also allow heat to escape. The blue color of the Great Red Spot and equatorial region is caused by high- and low-altitude aerosols. (NASA-JPL)

too surprising if the planet formed from gas in the solar nebula. This opinion is still held today, although there are some small differences that are hard to explain (see Table 8.2).

The first detailed radio and infrared measurements of Jupiter's atmosphere were obtained during the Voyager flybys of 1979. This made it possible to produce pressure-temperature profiles to a depth of several hundred kilometers. Theoretical studies, along with the known gas abundances, indicate that there should be three cloud layers at varying depths.

At the top, where the atmosphere is coldest, the clouds are composed of ammonia ice crystals. A little lower should be a layer of ammonium hydrosulfide (NH_4SH) – a mixture of ammonia and hydrogen sulfide. Lower still, where the atmospheric pressure approaches 5 bars (five times sea level pressure on Earth), water ice clouds should form.

The only attempt to confirm conditions in these outer layers by *in situ* measurement took place in December 1995, when the Galileo probe made a suicidal entry into the atmosphere. Surprisingly, hazes predominated and no thick layers of cloud were detected. One cloud structure – presumably ammonia ice – was found near the 0.6 bar pressure level. Another well-defined cloud feature was found above the 1.6 bar level, presumably composed of ammonium hydrosulfide ice crystals. In both cases, the clouds seemed to be broken up and patchy.

Why were the models so wrong? Images of the probe's atmospheric entry point showed that – against all the odds – it had descended within a rare "hot spot". Such holes in the main cloud decks allow heat to escape from deep inside the planet, resulting in major changes to the normal atmospheric conditions.

The probe also found that the atmosphere is easily mixed in the vertical direction – a so-called "neutrally stratified atmosphere" – at depths below the 0.5 bar level. This confirmed that Jupiter's internal heat escapes principally by convective motion of the atmosphere at these levels.

Box 8.1   Mass and Weight

"Mass" and "weight" are often interchangeable in our daily speech, but they have very different scientific meanings. Scientists measure MASS in kilograms, whereas they measure WEIGHT in Newtons (N). So, instead of saying that someone weighs 100 kg, we should strictly say that their weight is 980 N.

The *mass* of a body is a measure of how much matter it contains. Mass is also a measure of an object's *inertia*, for example, how difficult it is to move from a resting position, or to slow down and stop if it is moving. Objects at rest typically want to remain at rest. Once they are in motion, they want to keep moving. This quality of matter is its inertia.

An object's weight is not only related to its mass, but also to how strongly gravity is pulling on that object. Every object in the Universe attracts every other object, but the amount of attraction depends on their masses and the distance between them. An apple has a very small mass, so it falls toward the center of the Earth, rather than the other way around.

The gravitational attraction between two apples is far too small to be measured. However, the pull between a very large object, such as Earth, and another object can be measured easily by using a weighing scale. Scales measure the amount of physical attraction between an object and Earth. This attraction is called *weight*.

Astronauts in orbit have to adapt to the fact that objects with mass can have no measurable weight. Everything on a spacecraft floats because it is (almost) weightless – a scale placed beneath an anvil in space would read zero. However, if you try to shake the floating anvil, you will soon note that it needs a hefty push to get it moving and another, equal force to stop its motion. This is because the anvil still has mass and inertia.

The pull of gravity is directly related to an object's mass. If its mass is doubled, its gravitational pull also doubles. Similarly, if the planet you are standing on is twice as massive as another planet of the same size, its surface gravity is twice as strong.

On the other hand, the pull of gravity decreases quite rapidly with distance from the center of a planet. The force drops off with the square of the distance. Hence, if a spacecraft doubles its distance from a planet, the pull of gravity is reduced to one fourth of its previous value. At 10 times the distance, the planet's gravitational attraction is 100 times weaker.

Sir Isaac Newton (1642–1727) was the first person to describe this relationship in his *law of gravitation*. This is shown by the equation: $F = GMm/r^2$. The F stands for the effective force of gravity between any two masses. G stands for the gravitational constant – a value which applies everywhere in the Solar System (6.67259×10^{-11} m³/kg/s²). The "M" and "m" are the masses of the planet and another nearby object and the "r" is the radius, that is, the distance of the object from the planet's center. Note that the force of gravity never becomes zero, no matter how far you travel from a planet.

When applied to giant planets such as Jupiter, the equation throws up some surprises. Since Jupiter has 318 times the mass of Earth, an astronaut standing on its visible surface might expect to weigh 318 times more than at home. This would be true if Jupiter and Earth were the same size, but Jupiter is much larger, so the astronaut would actually be over 11 times further from the center. This reduces the gravitational pull by a factor of 125.4, so Jupiter's surface gravity is "only" about 2.53 times stronger than on Earth. In other words, an astronaut "weighing" 10 kg on Earth would weigh 25.3 kg on the surface of Jupiter.

On smaller, less massive planets, the opposite is true. For example, surface gravity on the Moon is about one sixth that on Earth, so heavy lifting is not a problem for lunar explorers. A person with a mass of 100 kg weighs about 160 N on the Moon. Note that it is also possible for someone to weigh less on the surface of a very large planet whose mass is relatively low for its size. One example is Uranus, where the surface gravity is about 90% that on Earth (see Planets table in Appendices).

Table 8.1 Jupiter Summary	
Mass	1.9×10^{27} kg (317.9 Earth masses)
Equatorial Diameter	142,984 km
Polar Diameter	133,708 km
Axial Inclination	3.13 degrees
Semimajor Axis	778,330,000 km (5.2 AU)
Rotation Period (internal):	9.92 hours
Sidereal Orbit Period:	11.86 years
Mean Density	1.33 g/cm³
Number of known satellites:	65

Table 8.2
Atmospheric Compositions of Jupiter and Sun (%)

Molecule	Sun	Jupiter
Hydrogen (H)	84	86.4
Helium (He)	16	13.6
Water vapor (H_2O)	0.15	0.1 ?
Methane (CH_4)	0.07	0.21
Ammonia (NH_3)	0.02	0.07
Hydrogen sulfide (H_2S)	0.003	0.008

Figure 8.7 The Galileo probe detected clouds during its descent into Jupiter's atmosphere. One instrument observed variations in brightness and detected a cloud layer near the 0.6 bar pressure level. Another instrument observed laser light scattered by passing cloud particles. It found one well-defined, tenuous cloud structure with a base at the 1.6 bar level, though the higher cloud layer was not seen. This may have been because the upper cloud layer is patchy and the probe went through a clear area. Theoretical models (right) predict layers of ammonia (NH_3), ammonium hydrosulfide (NH_4SH), and water (H_2O) clouds. The explanation for their absence may be that the probe descended in an atmospheric "hot spot". (NASA-Ames)

The strong eastward winds were studied by the Doppler Wind Experiment, which measured changes in precisely tuned radio transmissions to the Galileo orbiter, 215,000 km above the probe. Winds speeds of 540 km/h were recorded just below the cloud tops, persisting and increasing with depth until they reached 720 km/h toward the end of the probe's transmissions. The presence of such strong winds at depth supported previous data that solar heating is supplemented by a powerful internal heat source.

The orbiter made its own contribution by detecting the first discrete cloud of pure ammonia ice found on the planet. The small, bright feature, perhaps 15 km thick, was named the Turbulent Wake Anomaly because it lay in the disturbed clouds downstream from the Great Red Spot, where powerful wind currents moving in opposite directions drew ammonia gas from below. The ammonia then cooled rapidly to form ice in the −120°C atmosphere.

The convective nature of the atmosphere, driven by solar radiation and heat from inside the planet, also accounts for the numerous thunderstorms on Jupiter. The Galileo orbiter identified many thunderheads towering above the main cloud deck to the north of the Great Red Spot.

One image sequence showed a dense, white convective storm, which measured 1000 km across and rose 25 km above most of the surrounding clouds. The cloud base nearby was shown to be at least 50 km below the general cloud level, so deep that it could only be made of condensed water.

Many lightning flashes, hundreds of times brighter than electric storms on Earth, have been seen on the planet. Interestingly, the Galileo probe did not observe any optical lightning flashes during its descent, although many discharges were observed at radio frequencies. The radio signals indicated that the discharges were many thousands of kilometers away. However, radio wave intensity suggested that lightning activity is 3–10 times less common than on Earth.

The Great Red Spot

All of the giant planets display discrete storm systems. In the case of Jupiter, these take the form of about 80 huge ovals – some more than half the diameter of Earth – that drift around the planet within the main jet streams. Most of these are distinguished by white clouds, although some take on a reddish hue – most notably the famous Great Red Spot (GRS). This enduring atmospheric feature may have existed for well over 300 years.[2] Its discovery is often credited to Englishman Robert Hooke, who sketched a spot on Jupiter in May 1664.

[2] There is an observational gap of 118 years after 1713, leading some to suggest that the current spot first appeared as a long, pale "hollow" in 1831, with the prominent red color first developing in the 1870s.

Figure 8.9 A Voyager 1 image showing lightning on Jupiter's night side, together with a faint aurora above the horizon. (NASA-JPL)

Figure 8.8 A false color mosaic of the northern hemisphere derived from near-infrared images taken by the Galileo orbiter. Atmospheric circulation between 10°N and 50°N is dominated by alternating eastward and westward jets. Light blue clouds are high, while thin, reddish clouds are deep, and white clouds are high and thick. The high clouds and haze over the whitish ovals extend into Jupiter's stratosphere. Dark purple probably represents haze overlying a clear atmosphere. Also visible are large white ovals, bright spots, dark spots, interacting vortices, and turbulent cloud systems. The north-south diameter of the two interacting white vortices is about 3500 km. (NASA-JPL)

Figure 8.10 A false color infrared mosaic of the Great Red Spot taken by the Galileo orbiter. High, thick clouds are white; high, thin clouds are pink; low altitude clouds are blue and brown. Clouds altitudes vary by more than 30 km, and the Great Red Spot rises about 8 km above the main cloud deck. The deepest clouds are in the collar surrounding the anticyclonic storm, and in the north west corner of the image. White cumulus clouds to the north appear to be thunderheads. The smallest clouds are tens of kilometers across. (Caltech)

The huge oval storm is located at about 22.4°S, within the South Tropical Zone. Although its distance from the equator is stable, the Spot is continually drifting in longitude. During the 20th century, it wandered a total of about 1200°, equivalent to more than three complete laps of Jupiter.

The Spot's reddish coloration first became noticeable in the 1870s. Since then, its color has frequently changed in intensity. It has also become smaller and reduced its rate of drift around the planet (though with considerable variation). Since 1940, the average time it takes to travel once around the planet has been 9h 55m 42s. The Spot now measures about 24,000 km by 12,000 km, little more than half its size 100 years ago. Although its dimensions have fluctuated, the Spot is still large enough to swallow several Earths.

No one knows precisely why it is red. However, observations show that it represents an enormous anticyclone which towers 8 km above its surroundings. This suggests that a powerful updraft is carrying material from deep beneath Jupiter's cloud tops. When it is exposed to solar ultraviolet radiation, the gas is altered by some unknown photochemical reaction, possibly involving phosphorus or sulfur, which results in the familiar brick color.

The movement of nearby cloud features shows that winds circulate around the GRS at up to 360 km/h. In Voyager images the clouds on the outskirts of the Spot took about seven days to complete one circuit, whereas there was little horizontal motion in the center of the storm. Amateur observations made in 2006 revealed that the rotation period had reduced to 4.5 days, indicating a substantial increase in wind speed.

The anticyclonic circulation around the GRS seems to be driven by powerful wind currents moving from east to west on its northern flank, and in the opposite direction on its southern edge. Like an island in a stream, the Spot acts as a barrier, causing considerable turbulence in the surrounding clouds, particularly on its western side. However, the clouds that enter its domain suffer various fates.

One bright feature imaged by Voyager circled the Spot for 60 days without any appreciable change. Some clouds swept around the northern edge and kept on going. Some were captured and consumed. Yet others split in two when they reached either end of the giant oval, and then traveled in opposite directions.

Cloud patterns around the Spot's edge also change fairly rapidly. In early 1979, Voyager 1 saw the GRS partially covered to the north by a veil of high ice cloud – probably ammonia. Four months later, a more marked ribbon of white cloud had developed around its northern edge, blocking the motion of small spots coming from the east and forcing them to reverse direction.

Merging Storms

The white ovals tend to survive for very long periods, but mergers between storms drifting at similar latitudes do occur. One remarkable sequence of mergers took place between 1997 and 2000, when three of them coalesced into one. Each of the ovals was a swirling, high-pressure vortex, marked by colder temperatures than its surroundings – evidence of upwelling at the center – and winds that rotated counterclockwise at about 470 km/h. The largest of them was about 9000 km across, the others slightly smaller.

One of the storms had been in existence for at least 90 years, while the youngest had first appeared in 1939, during a period of enhanced atmospheric activity. Over the following decades, they plowed inexorably around the planet, on the edge of the South Temperate Belt at 31–35°S (south of the Great Red Spot). Although they sometimes approached each other, no collisions took place. However, by early 1998, two of the ovals were approaching each other as Jupiter went into conjunction, behind the Sun. Between the ovals was a darker, warmer region of sinking gas – a low pressure system.

When the planet became visible once more, the pair had combined to form a single feature, while the pear-shaped, reddish storm had been weakened and disappeared. Curiously, the area of the newcomer did not match the combined size of its two components, suggesting that the new oval might have lost some material during the merger or grown vertically.

Figure 8.11 Near-infrared images from the Galileo orbiter show the "before" and "after" of a merger between two anticyclonic storms in February 1998. The top panel shows two white ovals separated by a darker, pear-shaped cyclonic region. Below, the low pressure system has disappeared and the ovals have coalesced to form a single, larger storm. Light blue clouds are high and thin, reddish clouds are deep, and white clouds are high and thick. The clouds and haze over the white ovals extend into Jupiter's stratosphere. High haze is largely absent above the cyclonic feature. Dark purple probably represents haze overlying a clear atmosphere. (NASA-JPL)

The second merger was not long in coming. By October 1999, the newly merged oval and the last of the original trio were approaching each other. Between them was a dark red storm, a low pressure system where clouds swirled clockwise. The darker oval was pushed out of the way and torn apart as the three pressure systems skirted past the Great Red Spot.

The disappearance of the cyclone cleared the way for the two white ovals to meet and merge. The process began in March 2000 and lasted about three weeks. At the cloud tops, the storms circled counterclockwise around each other, then consolidated into an enormous oval about 12,000 km across – becoming the second largest storm on the planet. (A similar merger centuries ago may have created the Great Red Spot itself – see below.)

Such events may be significant on a planetary scale, perhaps as part of a recurring climate cycle that causes Jupiter to lose most of its spots before the atmosphere is sufficiently destabilized to create new vortices.

Red Junior

The huge merged oval which was born in the year 2000 (see above) continued to drift around the planet over the next few

Figure 8.12 This Hubble Space Telescope image, acquired on April 8, 2006, shows Jupiter's two red spots. Oval BA (left) was drifting eastward while the Great Red Spot was moving westward. "Red Junior" passed to the south of the Great Red Spot in July 2006. Note a halo of high altitude cloud around Oval BA and a small area of pale cloud in its center. A large thunderstorm appears as a bright white cloud in the turbulent region west (left) of the Great Red Spot. (NASA-ESA, A. Simon-Miller/NASA-GSFC, Glenn Orton/NASA-JPL and Nancy Chanover/New Mexico State University)

Figure 8.13 A false color, near-infrared view of the two red spots brushing past each other in Jupiter's southern hemisphere in July 2006. White cloud features are at relatively high altitudes; blue indicates lower clouds and red represents still deeper clouds. Also prominent is an orange/red polar stratospheric haze, which is bright near the pole. Other white spots are regions of high cloud, such as thunderheads. (Travis Rector/University of Alaska Anchorage, Chad Trujillo and the Gemini ALTAIR adaptive optics team)

years. At first, the storm remained white in color, but by December 2005 it was slowly turning brown and by early 2006 its color was similar to the Great Red Spot.

Officially named Oval BA, but popularly nicknamed Red Junior, the changeling was about the same size as Earth – or about half the diameter of its famous rival. When viewed at near-infrared wavelengths which absorb methane gas, Oval BA was about as prominent as the GRS, suggesting that it, too, protruded above the main cloud deck. However, its darker appearance indicated that its top deck might be a little lower or less reflective. On the other hand, cloud studies showed that the highest wind speeds around Red Junior had increased from 430 km/h to 640 km/h – comparable to the GRS.

When the two storms passed close to each other in July 2006, there was little interaction, apart from a fairly normal stretching of Red Junior in an east-west direction. Since then, Oval BA has not only persisted, but it has retained its color. As long as the storm can maintain its high wind speeds and continue to tower above its surroundings, continually pulling up material from deep in the atmosphere, it seems likely to continue to thrive.

Interior

Jupiter's visible clouds are only the thin outer layer of an ocean of compressed gases thousands of kilometers deep. Until the 1930s, astronomers had very little idea about the nature of the planet's interior. Then Rupert Wildt (1905–1976) discovered methane and ammonia in the atmospheres of Jupiter and Saturn.

He went on to propose a model which included a solid, rocky core overlain by a layer of ices and topped by an atmosphere that was largely made up of hydrogen and helium, with small amounts of ammonia, methane and other gases.

Later researchers realized that the planet's low density probably indicated a hydrogen-rich composition all the way down to the core. However, under the pressure cooker conditions deep inside Jupiter, the highly compressed hydrogen ceases to behave like a gas and turns into a liquid.

Various models have been developed in recent years, based on theoretical calculations and laboratory experiments. It is now believed that hydrogen gas begins to change into a hot liquid at a pressure exceeding 100,000 bars. A further transition begins at a depth of perhaps 15,000 km, when the pressure reaches 1 million bars (1 megabar) and a temperature of about 6000 K – the same as the surface of the Sun.

The molecular and atomic bonds begin to break down under such extreme conditions, creating a new form of hydrogen which behaves like a metal. This liquid "metallic" hydrogen, which comprises ionized protons and electrons, makes up most of Jupiter. Since metallic hydrogen is an electrical conductor, slow-moving convection currents flowing through this layer would be expected to generate a magnetic field, and this is exactly what is found (see below).

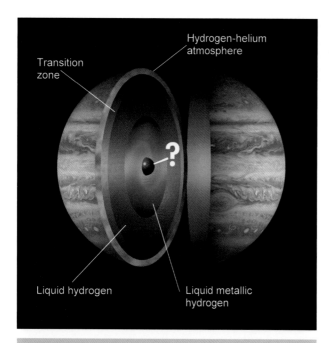

Figure 8.14 Jupiter's cloud layer extends to a depth of no more than 50 km. Beneath the cloud decks is a hydrogen–helium atmosphere about 1000 km deep. The gas then changes into liquid hydrogen, which extends to a depth of perhaps 15,000 km. When the pressure exceeds one million bars and the temperature exceeds 6000 K, the hydrogen starts to change phase, forming liquid metallic hydrogen. At the center there is probably a rock-ice core. (NASA)

The vast layer of metallic hydrogen is believed to have a depth approaching 50,000 km. At its base, the temperature reaches 20,000 K, with a pressure of around 20 megabars. Models indicate that there may be a gradual transition to a layer dominated by "ice," a liquid mixture of water, methane, and ammonia. Finally, at the center there may be a rock or rock-ice core of up to 10 Earth masses. If the core exists, it is probably molten, although it is possible that the tremendous pressure compresses it into a solid.

Some refinements to this overall picture have been proposed, based on the results sent back by the Galileo probe. Elements such as argon, krypton, xenon, carbon and nitrogen were shown to be 2–3 times more abundant on Jupiter than on the Sun. This enrichment can be largely explained by the addition of heavy elements from other planetesimals in the later stages of Jupiter's formation.

On the other hand, the abundance of neon was only 10% of the solar figure. This may be due to the removal of neon from the upper atmosphere by helium "rain" at a depth of about 10,000 to 13,000 km. According to this theory, helium does not mix with metallic hydrogen, but it does mix easily with neon. When helium droplets separate out and fall toward the interior, they carry neon with them.[3]

One other characteristic of Jupiter is an internal heat source which causes it to radiate more than twice as much heat as it receives from the Sun. The heat is thought to be generated by the slow gravitational compression of the planet. It flows outward from the core, driving the convectional circulation in the liquid hydrogen mantle and the outer atmosphere. Hot gas rises from the interior, then cools and flows back down.

Box 8.2   Metallic Hydrogen

Normally, hydrogen is a non-metallic element. Each hydrogen atom is composed of a single, negatively-charged electron in orbit around a single, positively-charged proton. However, as a result of the tremendous pressures deep inside Jupiter and Saturn, hydrogen is compressed so much that many atoms occupy the space normally filled by only a single atom.

Although there is still plenty of space in each hydrogen atom for the electron to circle the proton, high pressure laboratory experiments and theoretical calculations suggest that the molecular and atomic bonds begin to break at a pressure of about 1 million bars and a temperature of about 5700°C. As a result, electrons begin to migrate from atom to atom – the protons and electrons are said to be *ionized*. Some atoms contain so many electrons that the outermost one can easily be detached.

Under these conditions, hydrogen takes on the properties of a metal. As such, it becomes an electrical conductor and is opaque to visible radiation. Currents circulating within the metallic hydrogen are thought to be responsible for the powerful magnetic fields of Jupiter and Saturn.

In the case of Jupiter, the transformation from liquid hydrogen to metallic liquid hydrogen may occur at a depth of about 15,000 km. Saturn is considerably smaller, so the layer of metallic hydrogen is deeper and thinner. Internal pressures and temperatures are much lower for Uranus and Neptune, so metallic hydrogen is unlikely to exist.

Magnetic Field

The first planetary magnetic field to be discovered (other than Earth's), and by far the largest, was that of Jupiter. The breakthrough came in early 1955, when radio astronomers discovered a conspicuous radio source which shifted position from day to day. The signals coincided with the presence of Jupiter in the sky. It was the first radio emission detected from another planet.

At first it was thought that these decametric (tens of meters) wavelength radio emissions were caused by electrical disturbances in Jupiter's atmosphere. However, the regular variations in the intensity of radio waves eventually led to the conclusion that they were produced by electrons trapped in a strong magnetic field (i.e.

[3] Helium rain has also been proposed for the interior of Saturn (see Chapter 9). However, Saturn's helium rain was predicted because the planet is warmer than expected. The falling rain releases heat, accounting for the temperature difference. Jupiter's temperature is in accord with models of its cooling rate and age, so the hypothesis of helium rain was only introduced to account for neon depletion in the atmosphere.

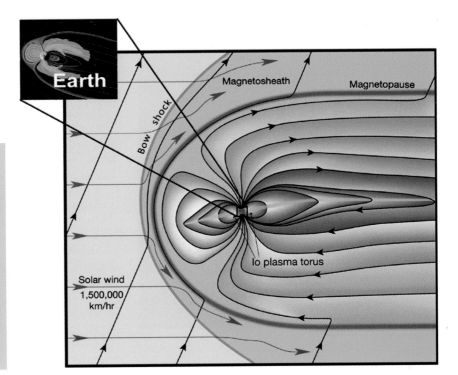

Figure 8.15 Jupiter is surrounded by an enormous magnetosphere which deflects the charged particles of the solar wind. The flow of solar particles shapes the magnetic bubble, creating a magnetotail that stretches all the way to Saturn's orbit. Trapped plasma is concentrated in a disk-like plasma sheet aligned with the planet's equator. The magnetic axis is inclined 9.6° to Jupiter's rotational axis. Earth's magnetic field (inset) is shown for size comparison. (SWRI)

radiation belts). This theory was confirmed in 1973, when the Pioneer 10 spacecraft flew past Jupiter and detected very intense radiation belts within an enormously powerful magnetic field.

As with most planets, the Jovian magnetosphere is shaped by the solar wind so that it resembles a giant tadpole, with a blunt head and a long tail. However, Jupiter's magnetic "bubble" is the largest single feature in the Solar System. If our eyes could see it, the magnetosphere would appear twice the width of a full Moon in the night sky.

Despite the pressure of the solar wind, the outer edge of the magnetosphere extends 3–5 million km toward the Sun.[4] On the planet's night side, the magnetotail stretches at least 740 million km (more than 100 Jupiter radii) – beyond the orbit of Saturn. All of Jupiter's major satellites are immersed in the magnetosphere – unlike the Moon, which dips occasionally into Earth's magnetotail.

Jupiter's magnetic field is also the strongest in the Solar System (except for fields associated with sunspots and other active regions on the Sun), with a dipole moment about 19,000 times that of Earth. However, the planet is so large that the magnetic field at its cloud tops is a mere 14 times stronger than at Earth's surface (about 4.2 gauss on the equator and 14 gauss at the poles).

Also like Earth, Jupiter's field is a dipole which generally resembles that generated by a bar magnet. The dipole's axis (which joins the magnetic poles) is offset from the planet's center and inclined by 9.6° to the rotation axis. (The offset is the result of the field being generated in the zone of liquid metallic hydrogen.) The polarity is opposite to Earth's, so a north-seeking compass would point south on Jupiter.

There are other differences. Jupiter's trapped plasma seems to co-rotate with the planet all the way out to the dayside boundary with the solar wind. At Earth, this co-rotation ceases well inside the magnetosphere. The Jovian plasma is composed not only of protons and electrons, but also of sulfur and sodium ions that originate in volcanic eruptions on the moon Io.

In the middle magnetosphere, from 20–60 Jupiter radii, ionized particles form a sheet of electric current. This "ring" current produces a magnetic field which is actually stronger than the planet's magnetic field. The current is concentrated near the equator, creating a "magnetodisk" where the magnetic field is much more flattened than Earth's.

The major satellites sweep high-energy particles from the radiation belts, so that total radiation near Jupiter is reduced by as much as 100 times. By far the largest number of particles is removed by Io, the innermost of the large moons. When Io passes into Jupiter's shadow, a faint auroral glow appears in the satellite's sparse atmosphere due to high speed collisions with ions from the magnetosphere. The "sponge effect" makes it possible to fly spacecraft close to the planet without major radiation damage.

Jupiter's Auroras and Io

Jupiter displays brilliant auroras which are caused when electrically charged particles trapped in the magnetic field spiral inward at high energies toward the north and south magnetic poles. When these particles collide with the rarefied upper atmosphere, they excite atoms and molecules, causing them to glow (similar to fluorescent street lights).

---

[4] The magnetosphere expands and contracts under the fluctuating pressure of the solar wind.

Some of these particles come from solar eruptions, but many originate from the large satellites. As the plasma trapped in Jupiter's rotating magnetic field passes by, it sweeps charged particles off the surface of Io and the other moons. Io also contributes about one tonne of material per second through its continuous volcanic eruptions.

This enormous amount of material is stripped of its electrons (ionized), so that the particles become trapped in Jupiter's magnetic field. They are then carried around Jupiter once every 10 hours, creating a donut-shaped ring or "torus" – predominantly ionized oxygen and sulfur – that occupies Io's orbit.

A cloud of neutral sodium gas also surrounds Io and trails around its orbit, as well as chlorine in the plasma torus. The source of these elements, gaseous sodium chloride, was confirmed by radio astronomers in 2002. Although the salt constitutes only about 0.3% of the moon's tenuous atmosphere, it may be present in much larger concentrations near the active volcanoes.

As Io sweeps through Jupiter's magnetic field during each orbit of the planet, the moon acts as an electrical generator, producing about 400,000 volts of electricity and electrical currents of 3 million amps. This enormous electrical current flows along the magnetic field lines to Jupiter's ionosphere. When the energetic ions hit the ionosphere, at the end of the flux tube, they generate a narrow, linear auroral trail about 250 km above the main cloud decks. Similar auroral footprints are visible for the weaker flux tubes of Ganymede and Europa.

The Galilean Moons

In January 1610, Galileo Galilei found four bright "stars" near Jupiter (see Chapter 1). Observations over the following weeks led

Figure 8.16 A composite, ultraviolet image of Jupiter's polar auroras, taken with the Hubble Space Telescope in September 1997. The bright ovals are created by highly energetic particles which have been accelerated by the planet's powerful magnetic field. These electrons and ions flow along magnetic field lines toward the magnetic poles, generating auroras when they collide with the upper atmosphere. Narrow, curved trails outside the main ovals show where the powerful electric current from Io's flux tube enters the atmosphere. (NASA-ESA)

Figure 8.17 Io's yellow cloud of neutral sodium gas, imaged by the 0.6 meter telescope at Table Mountain Observatory, California. The images of Jupiter and Io's orbit were superimposed. The sodium is ejected from Io's volcanoes. Io's size and location are shown by the orange dot within the crossbars. (NASA-JPL/Bruce Goldberg and Glenn Garneau)

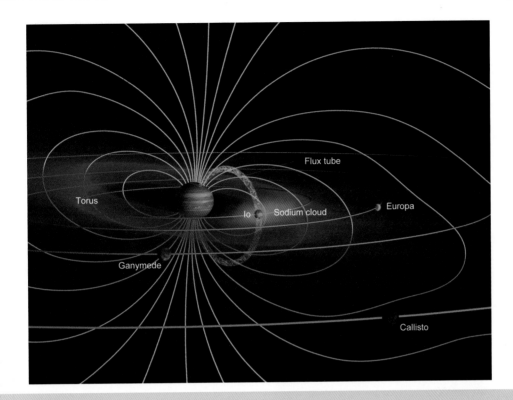

Figure 8.18 The main features of the inner Jovian magnetosphere. The plasma torus (red) of ionized sulfur and oxygen is trapped in the rotating, tilted magnetic field, so it co-rotates with the planet. The neutral sodium cloud (yellow), also derived from Io, is mainly concentrated close to the satellite. The Io flux tube (green and brown) is a flow of electrical current connecting Io with Jupiter's ionosphere (upper atmosphere). (John Spencer, Lowell Observatory)

Figure 8.19 A Hubble Space Telescope ultraviolet image of the Jovian aurora, which is centered on the magnetic north pole, as well as more diffuse emissions across the polar region. Also visible are the magnetic "footprints" of three of Jupiter's largest moons. These emissions, produced by electric currents generated by the satellites, flow along Jupiter's magnetic field lines. (NASA-ESA/John Clarke, University of Michigan)

him to conclude that they were actually satellites, with orbital periods that varied from about 42 hours to 17 days. The so-called Galilean moons – visible to anyone with a modest pair of binoculars – are named Io, Europa, Ganymede and Callisto.

These satellites provide a rich variety of visual phenomena – eclipses, when they disappear and reappear from Jupiter's shadow; transits, when they cross the planet's disk; occultations, when they disappear and reappear from behind the planet; and transits of their shadows across the cloudy atmosphere. On rare occasions, the moons pass in front of each other or are eclipsed by the shadow of a neighbor.

Until the Space Age, little was known about the satellites, although it was noticed that a resonant configuration exists among the inner three Galileans, so that innermost Io completes four orbits to every two of Europa, and every one of Ganymede.

Their relative sizes and densities were determined by direct observation and calculations based on their motions and mutual gravitational perturbations. Ganymede, Callisto and Io are all larger than the Moon, while Europa is only a little smaller. Indeed, Ganymede is the largest satellite in the Solar System, and noticeably larger than the planet Mercury. Callisto is only fractionally smaller than Mercury.

Despite their size, the moons have lower densities than the rocky terrestrial planets, and only Io has a higher bulk density than the Moon. This implies that they are a mixture of rock and ices. Their densities decrease with distance from Jupiter, probably a reflection of conditions within the nebula of dust and gas that surrounded the planet, their birthplace more than 4 billion years ago. Like the solar nebula, temperatures would have been much higher near the center, so that fewer volatiles condensed out. Further away from Jupiter, the colder conditions would have allowed more ices to condense.

Like our Moon, all of them have been influenced by tidal interactions with their planet, so they keep the same face towards Jupiter. Furthermore, since the satellites are embedded within its radiation belts, they are bombarded by energetic particles that cause chemical reactions and physical erosion of their surfaces (see above).

Io

Until the Voyager flybys of 1979, Io remained an enigma. Radio astronomers knew that it was linked to the bursts of radio noise from Jupiter and that it was surrounded by a cloud of sodium. Visual observations showed a yellowish object with reddish polar caps. When the moon emerged from Jupiter's shadow, it was frequently (but not always) a few percent brighter than usual for about 15 minutes. The mystery deepened in 1973, when Pioneer 10 detected a layer of ionized particles about 100 km above Io's surface. Clearly, Io had a tenuous atmosphere.

Voyager 1 sent back pictures of a "pizza" world, marked by irregular patches of white, yellow, black and orange. The absence

Figure 8.20 This false color, infrared image by the Hubble Space Telescope shows Io and Ganymede. Also visible are their shadows and the shadow of Callisto, which is out of view. (NASA, ESA, Erich Karkoschka/Univ. of Arizona)

Figure 8.22 A dramatic Voyager 1 view of an umbrella-shaped plume rising 300 km above Pele, one of Io's largest volcanoes. Pele's vent is marked by a dark spot just north of the triangular-shaped plateau. The plume fallout covered an area the size of Alaska, but the cloud was so sparse that it can hardly be seen against the surface. Note the slab-like mountains near the limb (right). (NASA-USGS)

Figure 8.21 The four largest moons of Jupiter, shown to scale. From left to right, they are Ganymede, Callisto, Io and Europa. Ganymede, the largest satellite in the Solar System, is bigger than Mercury. Io is larger than Earth's Moon. (NASA-JPL/DLR)

of any impact craters indicated that Io's surface was extremely young and global resurfacing was still ongoing.

Then an optical navigation engineer named Linda Morabito noticed a strange, crescent-shaped cloud extending beyond the satellite's limb. After eliminating all obvious possibilities, she concluded that it must be a cloud of material rising hundreds of kilometers above the surface.

Morabito had stumbled upon Io's secret. The satellite is the most volcanically active world in the Solar System. Further observations showed at least 8 active volcanoes, with mushroom-shaped plumes rising up to 250 km into space. Six of these eruptions were still taking place when Voyager 2 arrived four months later, although the largest plume – from a volcano named Pele (after the Hawaiian goddess of the volcano) – had ceased.[5]

Independent confirmation of the nature of Io's volcanoes and surface material comes from measurements of thermal emission and spectral data. The volcanic centers are hundreds of degrees hotter than their surroundings. The plumes of sulfur dioxide (SO_2) gas are the source of the sulfur and oxygen ions in Jupiter's magnetosphere.

The presence of sulfur had been suspected even before the Voyager flybys, since its color changes dramatically according to temperature. Although the "normal" color of sulfur is yellow, the element changes to orange, red and black when exposed to heat.

The roughly circular volcanic centers on Io are indicated by black markings a few kilometers across, where sulfur has melted and then formed a dark crust. These are volcanic calderas filled with lakes of lava. Extensive lava flows spread out from these basins, either as meandering rivers or broader fans and lobate

features. Extensive white patches are attributed to deposits of sulfur dioxide frost or snow.

Io's surface is continually changing. In the 17 year interval between the Voyager and Galileo missions, a dozen areas the size of Connecticut had been resurfaced. At least 120 volcanic hotspots were recorded, compared with about 600 on the much larger Earth. Although Io's larger centers of activity could remain active for many months, or even years, some of the smaller volcanoes were found to turn off and on over a period of a few weeks.

During repeated encounters, the orbiter's cameras recorded towering plumes associated with the growth of new dark spots and surface deposits. On one occasion, it flew through a plume of microscopic particles that soared to a height of 500 km above Pillan Patera – the most powerful eruption ever recorded on Io.

In many respects, Io's plumes resemble high temperature geysers on Earth, such as Old Faithful in Yellowstone National Park. However, the low gravity of Io (about 1/6 that of Earth) and the extremely low atmospheric pressure allow the sulfur dioxide to soar to much greater heights. When superheated, liquid sulfur dioxide rises quickly through fractures in the surface, it escapes as a high-velocity plume of gas. In the extreme cold of Io, where the typical temperature is about −146°C, the gas immediately starts condensing into snowflakes which fall onto the frigid surface.

Galileo also sent back the first direct images of surface volcanism, capturing a 60 km long curtain of lava erupting along the edge of a 200 km wide caldera named Tvashtar Catena. In another center, known as Loki, a gigantic block of solidified sulfur dioxide was floating on the dark crust that covered a lake filled with

Figure 8.23 Volcanoes and the previous day's sunshine warm the nighttime surface of Io in a thermal image from Galileo (left). Several volcanoes are identified: L-K is Lei-Kung Fluctus, L is Loki, Pi is Pillan, M is Marduk, and Pe is Pele. Much of the heat comes from a few discrete volcanoes. The brightest is Loki, which radiates roughly 15% of Io's total volcanic heat. Second brightest is Pillan, where the heat is radiated by extensive cooling lava flows erupted in 1997. Although Pele produces Io's largest plume, its small crater emits very little heat. Blue indicates the coldest temperatures, near 90 K (−183°C). Oranges and yellows show the hottest areas, above 170 K (−103°C). Small volcanic areas exceed 1500 K (1227°C), but they are not visible in this low-resolution view. Io's overall temperature varies little with latitude. (Right) A visible light image of the same night hemisphere. A thin crescent marks the area in sunlight. (NASA-JPL)

---

[5] Place names for Io are taken from various mythological associations with fire and volcanoes.

Figure 8.24 Galileo images showing major changes caused by volcanic activity between April 4 (left) and September 19, 1997 (right). The appearance of a new, 400 km wide dark spot around Pillan Patera was caused by its eruption of a 120 km high plume. The pinkish deposits of sulfur around Pele were partially covered and other subtle changes are visible. Some apparent differences are due to variations in illumination, emission and phase angles, notably at the dark volcano named Babbar Patera. (NASA-JPL/Univ. of Arizona)

Figure 8.25 Two images of volcanic eruptions at Tvashtar Catena, a chain of calderas centered at 60°N 120°W. The view on November 25, 1999 (left), is a drawing produced after the camera was saturated by extremely hot lava. It shows an orange-red lava curtain (a chain of lava fountains) and surface flows. The lava fountains are up to 1.5 km high. Several other lava flows can be seen on the caldera floors. The darkest flows are probably the most recent. In the February 22, 2000, image (right), the November outburst appears as a dark, L-shaped feature. White and orange areas are fresh, hot lava. Two small, bright spots show where molten rock is exposed at the ends of lava flows. The orange and yellow ribbon is a cooling lava flow over 60 km long. Dark, diffuse deposits around the active lava flows were also recent. The main caldera is almost surrounded by a mesa about 1 km high. In places the mesa's margins are scalloped, due to an erosional process called sapping. (NASA-JPL)

molten rock. Along one shoreline, a ribbon of glowing lava had broken through the overlying crust.

In another volcanic center, Prometheus, the active vent had migrated 70 km since 1979. Images indicated the presence of a subsurface lava flow which moved west before eventually breaking through the overlying crust and interacting with sulfur dioxide frost to produce a new umbrella-shaped plume.

The largest known active lava field in the Solar System lies in the Amirani region on Io, where 23 distinct new flows were visible (see Fig. 8.26). The area spans more than 300 km, and individual

Figure 8.26 The Amirani lava field, 500 km long and 180 km wide, is a patchwork of separate, overlapping flows. The most recent lavas are the darkest because they are too hot to be covered by deposits from sulfur dioxide plumes. Fresh lava is leaking from at least 5 areas at the northern end and 3 places in the middle. The lava probably originates in the south, then travels north beneath a solid crust of older lava before breaking out onto the surface hundreds of kilometers from the vent. Diffuse, white halos around the darkest flows are probably sulfur dioxide snows and frosts. (Moses Milazzo, Lunar and Planetary Lab., University of Arizona)

Figure 8.27 Like many of the mountains on Io, these raised blocks alongside Hi'iaka Patera (the irregular, dark depression at the center) seem to be in the process of collapsing. Huge landslides have left piles of debris at the bases of the mountains. The ridges parallel to their margins also indicate that material is sliding downhill due to gravity. By measuring the length of the shadows, it has been estimated that the sharp peak in the north is about 11 km high, while the elongated plateaus are about 3.5 km high. The two blocks have similar shapes which appear to fit together – possible evidence of horizontal rifting which left a depression that has been filled by lava flows. (Moses Milazzo, Lunar and Planetary Lab, University of Arizona)

flows are several kilometers long. Galileo images showed that new lava flows at Amirani covered about 620 km² in less than five months.[6] Despite the huge amount of lava disgorged – over 100 tons every second – the eruptions are relatively tranquil, with few explosive outbursts.

The extreme vulcanicity of Io was reinforced once more in February 2001, when the Keck II telescope in Hawaii detected the largest eruption ever observed in the Solar System. The high-temperature event involved incandescent lava fountains which were kilometers high, propelled by expanding gases and accompanied by extensive lava flows. Its thermal output almost matched the entire energy output from all of the other volcanoes on Io.

For many years, the volcanoes on Io were thought to be driven by a sulfur cycle, but this theory was disproved when infrared data indicated temperatures of 1700°C, making it the hottest lava ever recorded. Since liquid sulfur would vaporize at this temperature, it seems certain that silicate magma is associated with such eruptions – although this does not eliminate the possibility that some lava flows on Io are composed primarily of sulfur.

The arguments favoring silicate volcanism are further supported by the presence of plateaus and mountains, some more than twice as high as Everest. The mountains and deep, steep-sided calderas must be underlain by a crust of considerable rigidity and strength, something which cannot be explained by a sulfurous surface.

About two percent of the surface is occupied by block- or slab-shaped mountains. Io does not have tectonic plates such as those on Earth, but vertical movements seem quite common, as well as occasional rifting. Uplift and thrust faulting have caused some

mountains, such as Euboea Montes, to reach heights of up to 13 km.

How these mountains form is not clear. One theory suggests that crustal recycling may be responsible. Since Io is being resurfaced continuously at a rate of about 1 cm per year, the surface may be buried by up to one kilometer of material over a period of 100,000 years. Under the weight of these deposits, the crust may sink and merge with the molten mantle. The resulting compression may force large crustal blocks upward along deep faults, at the same time enabling magma to seep to the surface. Another possibility is that some of the mountains have been pushed up by intrusions of rising magma.

The remarkable volcanoes and tectonic activity are sustained by heat generated by tidal flexing, similar to the tides on Earth, but on a huge scale. Io's shape is periodically distorted by the gravitational pull of Jupiter, Europa and Ganymede.[7] This tug of war creates tidal bulges up to 100 meters high in Io's solid crust. This squeezing and stretching strongly heats the interior, with the heat being released at the surface by global volcanism.

Unlike the other Galilean satellites, Io seems to be lacking in water. Its extremely dry surface (and, presumably, interior) can be explained by the high rate of volcanic activity and resurfacing throughout much of its history. The silicate-sulfur volcanism, possibly involving magnesium and iron silicates, indicates a rocky lithosphere at least 30 km thick.

The uniform, global distribution of volcanic centers suggests the presence of a convective, partially molten asthenosphere beneath the solid crust, with magma rising and sinking in cells several hundred kilometers apart. Models of the interior suggest

[6] Amirani is huge even when compared to other Ionian lava flows. The Prometheus lava flow field covered only about 60 km² during the same period.
[7] The other Galilean moons are also affected by tidal heating, but to a much lesser extent, since they are further from Jupiter.

a silicate mantle overlying a core of molten iron, or a mixture of iron and iron sulfide, which fills 35–60% of Io's radius.

Europa

Beyond Io is Europa, an icy world which is lacking in volcanoes, mountains or craters, but covered by enormous, linear fractures – rather like a cracked egg shell. Instead of closely packed impact craters, Europa displays a young, smooth surface – evidence that resurfacing is still taking place.

Its most striking characteristic is a global network of narrow, criss-crossing, linear features. Many of these are straight, but some are curved or irregular. Ranging from a few hundred kilometers to 3000 km in length, and varying from several kilometers to 70 km in width, they appear to be evidence of tectonic processes.[8]

Although the dark "streaks" are the most obvious features, closer examination shows a comparable web of smaller, bright lines. No more than about 10 km across, these curved features can be seen when the Sun is low above Europa's horizon. Largely

Figure 8.28 An enhanced color image showing the region around Pwyll, one of the few impact craters on Europa. The bright rays – made of fine water ice ejecta – indicate that the 26 km diameter crater is geologically young. Its central peak, not visible here, stands about 600 meter above the floor and is much higher than its rim. The rays extend for over 1000 km in all directions. Also visible are dark linear features known as "triple bands." Pwyll is on the trailing hemisphere, at 11°S 276°W. The area shown is about 1240 km across. (NASA-JPL, PIRL/University of Arizona)

invisible when the Sun is overhead, their vertical height is estimated to be only a few hundred meters. However, the pattern of curves tends to repeat regularly every 100–300 km.

Most of the satellite is classified as smooth or ridged plains. Smooth plains appear to be among the youngest surfaces on Europa. One possibility is that they have been created by a form of ice volcanism, in which fluid erupted from below floods a low-lying area and freezes.

Europa's ridged plains are notably older and lack the bland, ice rink appearance of the smooth plains. They generally contain multiple, overlying ridges that have formed at different times. The linear ridges range in complexity from simple parallel pairs to intricate, braided triple bands where two ridges, about 10 km wide, lie either side of a steep-sided, narrow valley. Color images show that dark material – perhaps in the form of mineral-rich liquid water or warm ice – appears to have welled up, filling the gaps created by fracturing along these ridges.

More difficult to explain are the numerous dark patches, surrounded by older ridged and fractured plains. Typically 50–100 km across, they are named "lenticulae," Latin for "freckles." Although many of the features are domes, others are small depressions. Some are smooth, others are rough.

Comparisons of their irregular shapes and the alignment of nearby ridges and fractures indicate that these regions are not controlled by pre-existing cracks. One model for their formation is complete melt-through from an ocean below. Another possibility is that warm ice welled up from below, resulting in partial melting and disruption of the surface.

Close-up images of chaos regions such as Conamara show that they are made up of a disrupted matrix which resembles a mixture of frozen slush interspersed with rafts of pre-existing crust. Some blocks of ice appear to be shifted, rotated, tipped or partially submerged. This chaotic pattern is thought to be the result of blobs of warm ice moving up through colder, near-surface ice. Rough terrain was created if the surface was broken up and partially melted. Smooth lenticulae were formed if meltwater spread onto the surface and then quickly froze.

There has been considerable debate about the composition of the dark materials. Both sulfuric acid and compounds such as sulfates and carbonates – presumably originating in a subsurface ocean – have been suggested. This theory was supported by Galileo's discovery of marked fluctuations in the magnetic field around Europa. The moon has a weak magnetic field induced by its interaction with Jupiter's powerful magnetic field. Its rapid, systematic variations favor an electrically conducting fluid, such as salty water, rather than a dynamo in a molten, iron core.

The thickness of Europa's crust remains uncertain, as does the presence of liquid water. Estimates for the thickness of the crust are in the range 3–40 km. However, images of the few sizeable impact craters, such as Pwyll, show central peaks and sharp topography that can only be explained by a fairly thick, strong crust. The impacts that formed them clearly did not break through to the moon's postulated ocean.

One scenario favors an ocean up to 100 km deep – 10 times deeper than any ocean on Earth. If so, it would contain twice as

---

[8] The markings only appear dark in comparison with the brightness of their surroundings: the difference in albedo is only about 10%.

Figure 8.29 Galileo images of a double ridge that cuts across older plains and a darker, wedge-shaped band. (North is to the right.) The parallel ridges run east-west and are separated by a valley about 1.5 km wide. A perspective view (upper right) shows that bright material, probably pure water ice, lies on the ridge crests and slopes while most of the dark material (perhaps ice mixed with silicates or hydrated salts) is confined to lower areas. The northernmost slope is darker than the south-facing slopes. A color-coded relief model (lower right) shows that the ridges (red) rise more than 300 meters above the surrounding furrowed plains (blue and purple). The crust seems to have pulled apart, allowing dark material to well up from below. (DLR, NASA-JPL)

Figure 8.30 An enhanced color mosaic of Thera and Thrace Macula, two dark, reddish "chaos" regions found on the ridged plains of the southern hemisphere. Thera (left) measures about 70 × 85 km and appears to lie slightly below the level of the icy plains. Curved fractures along its boundaries suggest that collapse may have been involved in its formation. Thrace (right) is hummocky and appears to lie slightly above the plains. On its southern edge, material from Thrace overlies a gray band known as Libya Linea. The image covers an area of 525 × 300 km. (NASA-JPL, PIRL/University of Arizona)

Figure 8.31 An enhanced color view of thin, disrupted, ice crust in the Conamara region (9°N 274°W) that resembles Arctic ice floes on Earth. White and blue areas are blanketed by fine ice particles from the Pwyll impact crater about 1000 km to the south. A few secondary craters, less than 500 meter across, can be seen. The reddish-brown color is due to mineral contaminants released from below the crust when it was disrupted. The image covers an area of 70 × 30 km. (NASA-JPL, PIRL/University of Arizona)

much water as Earth's oceans. However, whereas Earth's oceans contain sodium chloride brought down by rivers and concentrated by evaporation, Europa's ocean would more likely contain salts such as magnesium sulfate which have been derived from warm water circulating through its rocky mantle. Alternatively, Europa's outer layer may simply consist of warm, convecting ice.

If Europa has a mineral-rich ocean, there is the tantalizing possibility that it may be able to support some form of life. Protected from harmful radiation by the overlying ice, it is possible to envisage ecosystems flourishing alongside hydrothermal vents, where the water is warm, there is an abundance of chemical compounds, and chemical reactions provide the energy required for each organism to function.

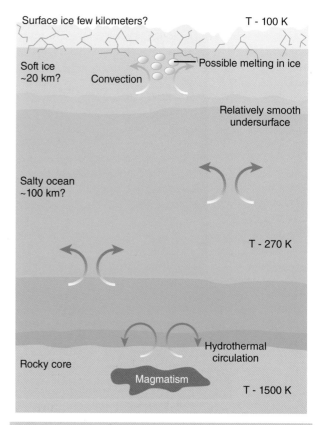

Figure 8.32 Two different models of Europa's internal structure. (Top) A warm, "slushy" convecting ice layer located several kilometers below a cold, brittle ice crust. (Bottom) A layer of liquid water, possibly more than 100 km deep, beneath the icy crust. Most of Europa is made up of a rocky mantle with an iron or iron-iron-sulfide core. (NASA-JPL, SETI Institute)

Figure 8.33 Europa has an icy surface ice which is extensively cracked and deformed. Magnetometer and other data suggest that there may be a briny ocean at a depth of perhaps 10 km – a possible environment suitable for life. Tidal heating may allow liquid water to rise into the crust, possibly causing occasional eruptions or surface flows. (Science)

Europa has an extremely thin atmosphere which is created by "sputtering," when fast-moving particles in Jupiter's magnetosphere hit the moon's surface and dislodge water molecules.[9] There may also be some sublimation of ice due to solar heating. However, its low gravity means that Europa cannot hold onto an atmosphere for very long.

Water molecules lost from the surface are quickly separated into neutral oxygen and hydrogen. The molecules can also be quickly ionized by ultraviolet radiation and charged particles in the vicinity. As a result, Europa has a neutral atmosphere as well as an ionosphere. The relatively lightweight hydrogen gas escapes into space, while the heavier oxygen molecules accumulate to form an atmosphere which may extend 200 km above the surface. The oxygen slowly leaks into space and must be replenished continuously.

This "leakage" means that, like Io, Europa is the source of a huge torus, a donut-shaped cloud of hydrogen and oxygen atoms that stretches for millions of kilometers around Jupiter. However, there is no evidence of explosive geysers (or volcanoes) sending clouds of gas into space.

Ganymede

With a diameter of 5262 km, the third Galilean satellite, Ganymede, is the largest moon in our Solar System. Pre-Voyager studies indicated a heavily cratered world, but the spacecraft revealed a satellite which had undergone considerable modification and resurfacing during its 4.5 billion year history.

About one third of its surface is covered by dark, heavily cratered terrain. The most notable example is a roughly circular region named Galileo Regio, which measures about 3200 km across. Located in the northern hemisphere on the side facing

---

[9] Europa's surface pressure is barely one hundred billionth that of the Earth.

away from Jupiter, the region is pockmarked by lighter impact features and partially blanketed by a thin layer of frost. Its circular outline suggests it was created by an enormous impact early in the satellite's history. Superimposed on it is a series of slightly curved, concentric markings – again presumably the result of an ancient impact, although the center of the former basin has since been erased by resurfacing.

Like other regions of ancient, icy crust, Galileo Regio displays a variety of impact craters and furrows. More than 4000 craters larger than 10 km in diameter have been identified in this region alone. The oldest impact features take the form of lighter colored palimpsests or "ghosts" which show no present-day topography.[10] This may be because they formed when the crust was warmer and malleable.

In contrast, the younger, usually smaller, craters display more typical impact features, such as deep basins and uplifted rims. The largest preserved impact basin, named Gilgamesh, is about 800 km across. It was created after the formation of nearby grooved terrain, suggesting that the crust eventually became rigid enough to preserve the impact structures. Other evidence for relatively recent impacts comes in the form of a number of bright ray craters.

Much of Ganymede shows evidence of a complex geological history, with a mottled pattern of light and dark regions. Some of the bright regions have a lower density of impact craters, indicating a more recent origin; others appear (based on the crater count) to be only a little older than the dark terrain.

Bright, grooved regions, such as Uruk Sulcus, which marks the western boundary of Galileo Regio, cover more than half of Ganymede. These regions of linear ridges and troughs have a vertical relief of a few hundred meters and extend for thousands of kilometers, but most of them seem to have a different origin from the grooved terrain on Europa. Instead of new material welling up from below, they seem to have been created by crustal tension in the dark terrain.

Once the icy surface was fractured and pulled apart, it was broken into many parallel ridges and valleys. This tectonic process was associated with parallel faulting, which resulted either in the formation of graben separated by raised blocks of crust, or by tilted blocks that resemble toppled dominoes.

In some cases, the surface is seriously disrupted by multiple fractures and grooves running in different directions – presumably created at different times. This creates a broken landscape of separate, small hills known as "reticulate terrain."

The brightness of the grooved terrain may be accounted for by cryovolcanism which blanketed the surface with fresh ice. This theory is supported by a number of impacts which have punched through the bright surface layer to dredge up dark material from beneath. Also present are peculiar, caldera-like depressions which seem to have been formed by volcanic flows of liquid water and slush.

Although there was a lot of tectonic activity during the first billion years of Ganymede's history, crater counts show that relatively little has changed since then. It was, therefore, something of

a surprise when Galileo's instruments showed that the moon generates a magnetic field about 1% the strength of Earth's.

The satellite could have a molten, metallic core, although this might be too small to generate the required dynamo effect. An alternative solution would be the presence of a salty, subsurface ocean – similar to that postulated for Europa, but much deeper. Natural radioactivity, possibly combined with modest tidal heating, might provide enough warmth to maintain a stable layer of liquid water several kilometers thick between two layers of ice at a depth of 150–200 km.[11] Further evidence was provided by infrared spectrometer data which showed the presence of hydrated salt minerals similar to those found on Europa.

These observations, plus Ganymede's fairly low mean density, suggest that the moon has a crust of water ice lying on top of a thick ice mantle, with a core of silicate rock and iron. However, this generalized model of uniform layers hides a more complex reality, as exemplified by Galileo tracking data which revealed the presence of irregular "lumps" beneath the surface. These masses may be rock formations which have been supported by Ganymede's icy shell for billions of years.

Like Europa, Ganymede has an extremely sparse exosphere of atomic hydrogen and molecular oxygen which is formed when charged particles trapped in Jupiter's magnetic field rain down onto the icy surface. As the particles penetrate the surface, molecules of water are disrupted, leading to production of ozone (O_3). This gas is then broken down to form oxygen.

Callisto

Of the four Galilean moons, only the outermost, Callisto, lived up to expectations by revealing a heavily cratered, ancient surface. Callisto is the only body with a diameter greater than 1000 km that shows no signs of undergoing any extensive resurfacing since the end of the period of heavy bombardment, about four billion years ago.

Its surface seems saturated with 100 km diameter craters, making it the most heavily cratered object in the Solar System. However, only a few huge impact structures are visible, suggesting that the remainder were erased by the subsequent bombardment of smaller impactors.

The largest of these structures is Valhalla, a bright, circular feature about 600 km across. The flat central region is surrounded by numerous concentric rings which extend about 1300 km from the center. The absence of high ridges, ring mountains or a large central depression suggests that the impact resulted in melting of the surface and the outward spread of shock waves. Refreezing took place in time to preserve the concentric shock rings.

Some minor resurfacing of Callisto's surface is taking place, since there are many fewer small impact craters than expected. Instead of deep, saucer-shaped basins, the craters are very shallow, suggesting that they have been gradually filled and partially obliterated over billions of years.

---

[10] The term "palimpsest" comes from reused parchment on which previous writing was poorly erased.

[11] At the present time, Jupiter and the other Galilean moons cause only a small amount of tidal heating on Ganymede. However, tidal effects were probably much more pronounced when Ganymede's orbit was more elliptical, perhaps three billion years ago.

Box 8.3   Voyagers at Jupiter

The Voyager missions to the outer planets were made possible by a planetary alignment that occurs once every 176 years. On such rare occasions, the four gas giants are suitably placed in their orbits for spacecraft to use gravity assists from Jupiter and Saturn in order to reach Uranus and Neptune.

Figure 8.34   The Voyagers carried instruments to investigate atmospheres, satellites and magnetospheres. The TV cameras were located on a movable scan platform alongside other experiments that required tracking. Color images were obtained by combining pictures taken through blue, orange and green filters. The Voyagers were powered by three radioisotope thermoelectric generators. (NASA-JPL)

Voyager 2 was the first to leave Earth, blasting off from Cape Canaveral on August 20, 1977, followed 16 days later by the identical Voyager 1. Leaving Earth faster than any previous spacecraft, it crossed the Moon's orbit in less than 10 hours and overtook Voyager 2 soon after entering the main asteroid belt.

During the cruise to Jupiter, the scan platform on Voyager 1, which carried the cameras and other tracking instruments, temporarily stalled. More serious was the failure of Voyager 2's primary radio receiver, followed by a malfunction in the back-up receiver. Concerned about a total breakdown in communications from Earth, the mission team uploaded a command sequence that would ensure a minimum science return at Jupiter.

Voyager 1 flew past Jupiter on March 5, 1979, followed by its sister craft on July 9. The dual flybys made it possible to arrange complementary observations of the Jovian system. Voyager 1 was able to study five of Jupiter's 13 known satellites, obtaining remarkably detailed images of Io, Ganymede Callisto and Amalthea after it passed by the planet. Voyager 2 encountered the satellites on the inward leg, enabling it to image the anti-Jupiter hemispheres of Callisto, Ganymede and Europa, with a long-range reconnaissance of Io.

Each spacecraft took more than 15,000 pictures of Jupiter and its moons, which were used to create movies of the planet's cloud motions and to map the surfaces of the major satellites. The glows of auroras and lightning were detected for the first time, as was a dark ring system. Three small satellites were also discovered close to Jupiter. Perhaps the most exciting finds were the sulfur-laden volcanic eruptions and flux tube of Io, together with the "egg shell" surface of Europa.

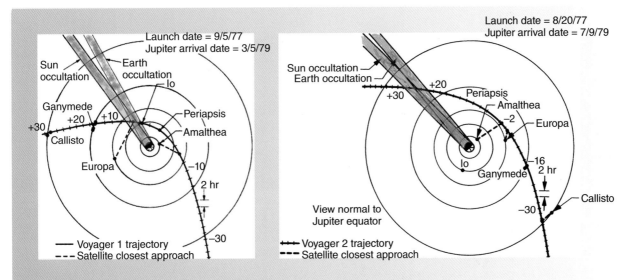

Figure 8.35 The Voyager flybys of Jupiter took place four months apart. In both cases, they flew very close to the planet's equatorial plane. Voyager 1 imaged the major satellites on its outward leg, while Voyager 2 observed the opposite hemispheres on the inward leg. Voyager 1 was able to study Io at close quarters, while Voyager 2 obtained the most detailed images of Europa. (NASA-JPL)

Figure 8.36 A near-global view of Ganymede taken by the Galileo orbiter. The large, circular, dark region is Galileo Regio, centered at 36°N, 138°W. On its western boundary is a curved region of parallel ridges and troughs known as Uruk Sulcus. The dark areas are older and more heavily cratered, whereas the light areas are younger, tectonically deformed regions. The brownish-gray color is due to mixtures of rocky materials and ice. Bright spots are geologically recent impact craters and their ejecta. (NASA-JPL)

Figure 8.37 A Voyager 2 view of the western edge of Galileo Regio. The dark, fairly ancient terrain contains numerous small impact craters, many with central peaks. The larger, bright circular features – known as "palimpsests" – show little relief and are the remnants of old craters which have been filled by the flow of icy material. Slightly curved lineaments in the dark region (center) suggest the presence of a large impact basin to the south west which was later hidden by the formation of the grooved terrain of Uruk Sulcus (bottom left). (NASA-JPL)

Figure 8.38 Topographic detail of Galileo Regio is seen in this 3-D view. Note the deep furrows and impact craters. Bright walls show where cleaner ice may be exposed, and deposits of dark material fill the lower terrain. (NASA-JPL)

Figure 8.39 An unusual, caldera-like depression in the Sippar Sulcus region, possibly the result of water ice volcanism. This depression, marked by scalloped walls and internal terraces, is about 55 km long and 17–20 km wide. On the floor of the inner depression is a lobate, flow-like deposit 7–10 km wide, with ridges that are curved outward (and apparently downslope) toward a junction of smooth terrain and a grooved band. The morphology suggests that volcanic eruptions created a channel and eroded the surface. The mosaic, centered at 31°S 189°E, covers an area of 91 × 62 km. (Brown University, NASA-JPL)

Figure 8.40 Not all bright terrain on Ganymede shows parallel ridges and troughs. In this Galileo view, an 85 km wide grooved feature, known as Erech Sulcus, runs north–south between two blocks of the older, dark terrain of Marius Regio. The numerous grooves probably formed when tectonic forces pulled apart the icy surface. In the south it meets a much smoother region known as Sippar Sulcus, which may have been flooded by ice volcanism. (LPI/NASA-JPL)

major differences. Callisto's bulk density (1.86 g/cm³) is the lowest of the Galileans, indicative of a world which is composed of a mixture of about 40% ice and 60% rock and iron.

There is very little tidal heating of Callisto at the present time and it seems that the satellite has never undergone the strong heating experienced by its neighbors. Tracking data for the Galileo orbiter suggest that little, if any, internal differentiation has taken place. Models indicate the absence of a rock-metal core, with most of the interior filled with a mixture of ice and rock. The surface crust may be composed of ice and rock less than 350 km thick.

In addition to the expected water ice and hydrated minerals, four unusual spectral features have been detected. One of these appears to show carbon dioxide trapped in the surface. Two others may represent sulfur derived from Io's volcanic eruptions. The fourth corresponds to carbon-nitrogen compounds – possibly complex organic molecules known as tholins.

Callisto may also possess a subsurface ocean. As was the case with Europa and Ganymede, the first evidence came from measurement of magnetic fluctuations around the satellite. However, subsequent studies showed that, unlike Ganymede, but like Europa, Callisto is not generating its own magnetic field. Instead, an electrically conducting layer beneath the surface makes the moon act like an enormous electromagnet. The magnetic field is induced by its interaction with Jupiter's magnetosphere.

How can Callisto be hot enough to support an ocean without melting its interior and producing separate layers of ice and rock? Some feeble heating of the interior may still occur through radioactive decay of elements. The very cold, rigid crust also acts as a lid, keeping in the heat. This may be sufficient to melt the ice at a depth of at least 150 km below the surface, creating a subterranean ocean. An ocean at least 10 km thick would have the

The entire satellite seems to be coated with a blanket of fine, dark material of uncertain origin – so Callisto is the least reflective of the Galilean moons.[12] On steep slopes, this dark coating slides downhill, leaving bright crater rims where underlying water ice is exposed. Intercrater plains typically appear quite dark and smooth.

Callisto is the third largest satellite in the Solar System. Although it is only a little smaller than Ganymede, the two moons display

[12] Callisto's surface is the darkest of the Galileans, but it is still twice as reflective as the Moon.

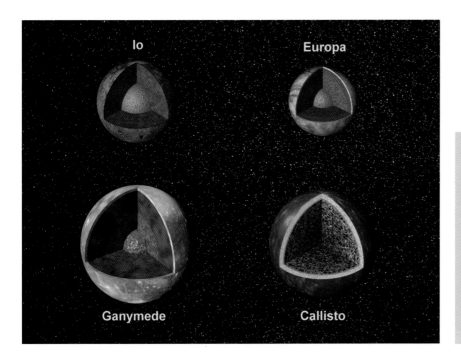

Figure 8.41 The possible internal structures of the Galilean satellites, shown to scale. Three of the satellites have iron-nickel cores (gray) and rocky mantles (brown). Io's rock (silicate) shell extends to the surface, while the mantles of Ganymede and Europa are overlain by water ice or liquid oceans (white/blue). Callisto has a relatively uniform mixture of ice and rock, perhaps with a shallow ocean beneath the icy crust. (NASA-JPL)

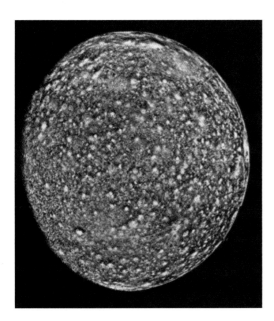

Figure 8.42 This Voyager 2 image of Callisto shows a uniform distribution of medium-sized impact craters on the hemisphere facing away from Jupiter. Typically about 100 km across, many of these craters have bright rims, evidence of exposed water ice. Some relatively fresh ray craters are visible, as well as several larger impact structures. One of these, Asgard, is visible near the limb (top) and has about 15 concentric rings around a bright central spot. The smooth limb confirms the absence of high topography. (NASA-JPL)

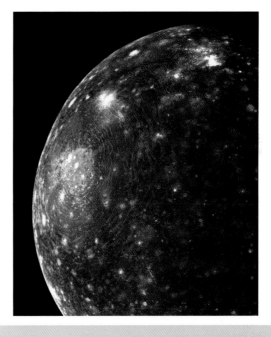

Figure 8.43 The largest impact feature on Callisto is the multi-ringed Valhalla (left). Unlike craters on rocky planets, Valhalla is quite flat, with no encircling mountains or central peak. The bright central plains were possibly created by the excavation and ejection of "cleaner" ice from beneath the surface, with fluid impact melt filling the crater. The bright central region is around 600 km across, while the surrounding rings extend 1500 km from the center. Valhalla is on the hemisphere facing Jupiter. (NASA-JPL)

Figure 8.44 Four Galileo views show how increasing resolution modifies interpretation of Callisto's surface. In the broad view (top left) there are many small, bright spots around the Valhalla impact structure. The regional view (top right) shows the spots are large craters with bright rims. The local view (bottom right) shows some smaller craters and bright material on crater rims amid a smooth, dark layer that covers much of the surface. In the close-up frame (bottom left), where spatial resolution is 30 meters, the surface is very smooth between the craters. (DLR/NASA-JPL)

required conductance if the water was comparable in salinity to Earth's oceans.

Callisto's sparse exosphere seems to be mostly carbon dioxide, presumably derived from dry ice on the surface. A few other constituents, notably sulfur dioxide, have been tentatively identified. Hydrogen atoms escaping from Callisto are derived from water ice on its surface. Solar ultraviolet radiation striking the rock-hard ice is probably the primary mechanism for separating the hydrogen and oxygen atoms. Sputtering is less likely, since Callisto orbits outside the radiation belts and does not interact as strongly as Ganymede with the charged particles in Jupiter's magnetosphere.

Inner Satellites

For almost three centuries, the only known Jovian satellites were the four Galileans. Then, in 1892, American astronomer Edward Barnard (1857–1923) discovered Amalthea. It was the last planetary moon to be found without the aid of photography.

Amalthea orbits only 181 300 km (2.5 Jupiter radii) from the center of the planet. At this distance, within Jupiter's rings, the satellite is bombarded by intense radiation and micrometeorites.

It has an irregular shape, pockmarked with numerous impact craters. Two of these are quite large. Pan, located in the northern hemisphere, is about 90 km in diameter and at least 8 km deep. Gaea, which straddles the south pole, is about 70 km across but probably twice as deep. Overall, the potato-shaped moon measures $270 \times 165 \times 150$ km, and, like most satellites of the giant planets, it is in a synchronous rotation, with its long axis always pointing towards Jupiter.

One of Amalthea's most surprising characteristics is its dark red color – in fact, the satellite is one of the reddest objects in the Solar System. This is thought to be the result of its surface being sprayed by sulfur and sodium derived from Io's volcanoes.

Precise tracking of the Galileo orbiter indicated that Amalthea's overall density is well below that of water ice. Instead of being a solid chunk of rock and ice, the satellite seems to be a loosely packed mixture of rubble in which the voids take up more of its volume than the solid material. It seems that Amalthea was once shattered by a collision and the pieces are now held together through their mutual gravitational attraction.

The density is so low that even its solid rubble is apparently less dense than Io. This contradicts models suggesting that moons forming close to Jupiter would be made of denser material than those farther out (see Box).

Box 8.4   Birth of the Jovian Satellites

It is generally accepted that the Galilean satellites were created in orbit around Jupiter. Most models assume that they formed within a gaseous disk, in which temperatures were highest near the young, rapidly growing planet, and lowest on the outskirts. This interpretation is supported by the bulk densities of the moons, which decrease with distance from the planet. This trend indicates an increasing proportion of low-density ice and a decreasing proportion of rock and iron, so that the outer moons, Ganymede and Callisto, contain up to 50% ice by mass. The combined mass of the Galilean satellites indicates that the nebula was equal to about 2% of Jupiter's mass.

This view has been challenged by researchers who argue that temperatures in such a massive, gas-rich disk would be too high to retain ices, even in its outer reaches. In addition, the gravitational interaction of the satellites with the disk material would have caused their orbits to decay rapidly inward, probably leading to their destruction.

The alternative model does not require that all of the mass needed to form the satellites was present in the disk from an early stage. Instead, it suggests that the disk grew slowly over a long period of time. As the inflow of gas and dust from solar orbit fed the growth of Jupiter, it also supplied material for the disk. In this "gas-starved" disk, temperatures remained low enough for ice to exist in the region of Ganymede and Callisto. According to this model, the satellites formed fairly slowly, over 100,000 to 1 million years.

The model accounts for the moons' bulk compositions, the unusual internal structure of Callisto, and the ability of the satellites to avoid rapid inward migration of their orbits due to interaction with the disk.

It also predicts that the satellites would migrate inward during their formation, each at a rate proportional to its mass. This inward migration could lead to the establishment of an orbital resonance, so that Io completes four orbits to every two of Europa, and every one of Ganymede. The existence of the resonance leads indirectly to internal heating of the satellites, causing Io's extensive volcanism, for example.

However, the discovery that Jupiter's fifth largest moon, Amalthea, which lies inside the orbit of Io, has a low density and probably contains a lot of water ice, is not explained by any of the disk models. Temperatures would have been too high at Amalthea's current position for an icy moon to form. This suggests that Amalthea was born in a colder environment. One possibility is that it formed later than the major moons. Another is that it formed beyond the orbit of Europa or in a cold region of the solar nebula. It would then have migrated or been captured in its current orbit around Jupiter.

The multitude of tiny satellites that lie close to Jupiter or follow eccentric orbits far from the planet are likely to be asteroids that were captured by the giant planet's gravity.

Figure 8.45   The left image shows escape velocities color-coded on a shape model of Amalthea. Blue areas have the lowest escape velocity, barely 1 m/s near the anti-Jupiter end; red shows the tiny regions of much higher escape velocity, nearly 90 m/s. The weak gravity is due to its low density and rapid rotation. The middle image also shows Amalthea from the anti-Jupiter side. The visible area is about 150 km across. Since the Sun is overhead, there are no shadows, but brighter markings on the ends of a ridge are prominent. The image at right shows bright ridges on the satellite's leading side and a 40 km wide crater with two nearby ridges. (Cornell University, NASA-JPL)

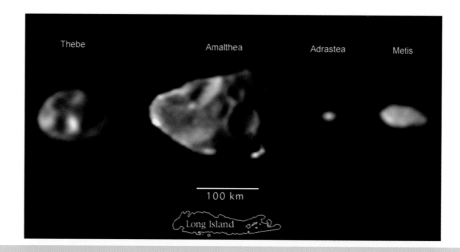

Figure 8.46   Jupiter's inner satellites compared in size to Long Island, which is 190 km E-W. These small satellites have very low surface gravities: a person weighing 68 kg on Earth would weigh about 0.2 kg on Amalthea, and approximately 0.01 kg on Adrastea. Impact craters 35–90 km across are visible on the larger satellites. (NASA-JPL)

Three more moons lie close to the planet. The orbits of Adrastea and Metis are less than 1000 km apart, and within 130,000 km of the center of Jupiter. Both lie deep within the Jovian rings. Thebe, the largest of the trio with a maximum diameter of 110 km, orbits at the outer edge of the rings (see below). All three are irregular in shape and images show Thebe has a number of sizeable impact craters.

Outer Satellites

Jupiter has 65 known satellites – more than any other planet. 57 of these are small objects that orbit far beyond Callisto.[13] All of these follow eccentric, highly inclined orbits. These orbits are stable over the age of the Solar System, although they are strongly influenced by solar and planetary perturbations.

There are five distinct groups, each dominated by one relatively large body. In general terms, the innermost irregular satellites follow prograde orbits. They are named the Himalia and Themisto groups after the largest object in each. Most of the more remote moons travel in the opposite, retrograde direction. There appear to be at least three groups (named Pasiphae, Ananke and Carme).

The groups were probably formed through collisional shattering of larger precursor objects after they were captured by Jupiter. At present, it is very difficult for Jupiter to capture satellites permanently because no efficient means of changing an object's orbital momentum and energy exists. However, satellite capture could have occurred more easily in the past, for example due to gas drag from the planet's bloated atmosphere and/or a higher probability of collision with nearby small bodies. This would suggest that most of the outer satellites were captured towards the end of Jupiter's formation, around 4.5 billion years ago.

The sizes of the outer satellites can be estimated by assuming geometric albedos (reflectivity) of 4%. This makes Himalia the largest irregular satellite of Jupiter with a diameter of about 160 km. Each retrograde group contains one or two large objects a few tens of kilometers in diameter, but the remainder are mere fragments, typically 2–4 km across.

Jupiter's Rings

Like the other giant planets of the Solar System, Jupiter is circled by a ring system. It was discovered when Voyager 1 crossed the planet's equatorial plane and its narrow angle camera captured light reflected from a ring of tiny particles as it looked back toward the Sun.

Unlike the spectacular disk that surrounds Saturn, the Jovian rings are extremely dark, with an albedo of about 0.05 (5% of the sunlight falling on the rings is reflected). They would probably be invisible from the Galilean moons, except when backlit by the Sun. The rings are made of billions of microscopic specks of dust, similar in size to particles of cigarette smoke. Unlike the rings of Saturn, they contain little or no ice.

Four distinct regions have been identified. The innermost of these is a faint torus of material, about 10,000 km thick, that extends down to the planet's cloud tops. Its particles are very small – perhaps 100 times smaller than the width of a human hair. Particles this size should survive for only a short time, so they

---

[13] The rapid rate of discovery in recent years, made possible by the introduction of large format charge-coupled devices (CCDs), indicates that many more have yet to be found. One estimate suggests that there may be about 100 irregular satellites larger than 1 km in diameter.

Box 8.5 The Galileo Mission

NASA's Galileo mission, named in honor of Galileo Galilei, comprised two spacecraft: an orbiter and an atmospheric probe. The orbiter was designed to study Jupiter's atmosphere, satellites, and surrounding magnetosphere, while the probe would make the first *in situ* measurements of a giant planet's atmosphere.

The spacecraft was carried into Earth orbit by Shuttle Atlantis on October 18, 1989. After release, a two-stage booster sent it toward Jupiter along a complex trajectory which involved one flyby of Venus (February 10, 1990) and two flybys of Earth (December 8, 1990 and December 8, 1992). It also made the first flybys of main belt asteroids, discovered a satellite around Ida, and observed the impacts of fragments of comet Shoemaker-Levy 9 with Jupiter.

Figure 8.47 The Galileo spacecraft weighed 2223 kg at launch and measured 5.3 meters from the top of the high gain antenna to the bottom of the probe. It was the first dual-spin planetary spacecraft. The upper, spinning section rotated at about 3 rpm and contained the power supply, propulsion module and most of the computers and control electronics, as well as instruments to study magnetic fields and particles. The de-spun section was counter-rotated to provide a fixed, stable orientation for cameras and other remote sensors. Power was provided by two radioisotope thermoelectric generators. (NASA-JPL)

The piggybacking probe was released from the mother craft on July 12, 1995. The 339 kg probe slammed into Jupiter's atmosphere on December 7, 1995 at a speed of 170,000 km/h. For the next 58 minutes it sent back data on lighting levels, heat flux, pressure, temperature, winds, lightning and atmospheric composition. By the time high temperatures silenced its transmitters, it had reached a pressure 22 times the sea level pressure on Earth.

The main craft also arrived on December 7, 1995, becoming the first spacecraft to orbit an outer planet. Over the next eight years it completed 34 orbits of Jupiter and 35 flybys of the major satellites – 7 of Io, 8 of Callisto, 8 of Ganymede, 11 of Europa and one of Amalthea.

The orbiter carried 11 scientific instruments, seven of which were located on a section that provided pointing stability by spinning. The four remote sensing instruments, which sent back images and data on composition and temperature, were on a de-spun section. Its radio link to Earth and the probe-to-orbiter radio link were also used to conduct scientific investigations, such as mapping of gravity fields.

Figure 8.48 The Galileo probe, with its heat shield (below) and parachute (above), entering Jupiter's upper atmosphere at a speed of about 48 km/s. Data from its six instruments were relayed to Earth via the orbiter before it was destroyed by extreme temperature and pressure. (NASA/JPL)

Figure 8.49 The probe measured the variation of temperature and atmospheric density with pressure and altitude to a depth of about 20 km. The probe stopped transmitting at a pressure of 22 bars – equivalent to the pressure at an ocean depth of 230 meters on Earth – and a temperature of 153°C. (NASA-Ames)

Unfortunately, the amount of data sent back from the mission was severely curtailed when the high-gain antenna failed to deploy. Although far fewer images could be returned than originally anticipated, the mission team estimated that 70% of the original science objectives were accomplished. In particular, Galileo provided major advances in our understanding of Jupiter's magnetosphere, and the nature of its large moons, including evidence for saltwater oceans on Europa, Ganymede and Callisto.

In December 2000, Galileo participated in a unique dual examination of the Jovian system which coincided with the distant flyby of the Cassini spacecraft en route to Saturn.

must be replenished by material derived from other parts of the ring system.

Outside this is the brighter, denser, main ring, which begins at a distance of about 122,500 km and ends at about 129,000 km, very close to the orbit of Adrastea. The brightness of this ring drops markedly at about 127,850 km, near the orbit of another moon, Metis. The outer ring segment, which is very faint, is known as the gossamer ring, and extends all the way out to the orbit of Thebe. It is divided into two distinct sections.

Jupiter's four inner satellites, Metis, Adrastea, Amalthea and Thebe, affect the structure of these tenuous rings. The main ring is supplied by fine particles knocked off Adrastea and Metis by small meteoroid impacts. Similarly, Amalthea is the source of material for the inner gossamer ring, while Thebe feeds the outer gossamer ring.

Instead of tailing off to nothing, the gossamer ring is unusually thick, with a sharp outer edge. The depth of this faint, outer ring is accounted for by the inclined orbits of Thebe and Amalthea, which spread out the ring debris as they travel above and below Jupiter's equator.

Figure 8.50 A mosaic of Jupiter's ring system taken when the Galileo spacecraft was in Jupiter's shadow and looking toward the Sun. This reveals tiny, dust-sized particles in the ring and upper atmosphere of Jupiter. Only the main ring and a hint of the surrounding halo can be seen in this image. (Cornell University, NASA-JPL)

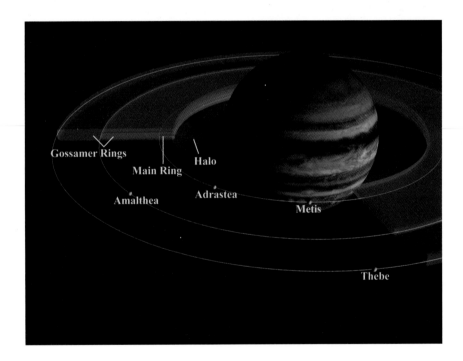

Figure 8.51 The geometry of Jupiter's rings in relation to the four inner satellites, the sources of the ring material. The innermost and thickest ring (gray) is the halo that ends at the main ring. The thin, narrow main ring (red) is bounded by Adrastea and shows a marked decrease in brightness near the orbit of Metis. It is composed of fine particles ejected by meteoroid impacts on Adrastea and Metis. (The orbits of Adrastea and Metis are about 1000 km apart.) Thebe and Amalthea, which follow inclined orbits, supply dust to the thicker gossamer rings (yellow and green). (Cornell University, NASA-JPL)

Questions

- Why is Jupiter sometimes described as a failed star?
- Compare and contrast the atmospheric features imaged during the Voyager 2 and Cassini flybys, shown in Figure 8.5.
- Make a list of the similarities and differences between the atmospheric conditions on Earth and Jupiter.
- Why does Jupiter display belts and zones?
- Describe and explain the main features of the red spots and white ovals on Jupiter.
- The Galilean satellites are comparable in size to Mercury. Describe the major differences and similarities between each of these satellites and the innermost planet.
- Briefly explain the scientific rationale for believing that at least three of the Galilean satellites may have subsurface oceans.
- Why is Europa often cited as a high priority target for future space missions?
- Describe the relationship between the Jovian satellites and the ring system.
- Compare and contrast the magnetic fields and magnetospheres of Jupiter and Earth.

NINE
Saturn

Saturn is the outermost of the naked eye planets. As such, it was long regarded as the furthest outpost of the Solar System. In many ways, Saturn may be regarded as a smaller version of Jupiter, with a hydrogen-helium composition and parallel bands of cloud driven by high-speed jets. What make the Saturnian system unique are the planet's enormous rings and its many-faceted satellites. Most notable of all is Titan, the only satellite with a dense atmosphere and a world that has been likened to a primitive Earth in deep freeze.

The oldest written records to mention Saturn date back to the Assyrians, around 700 BC. A few centuries later, the Greeks named it in honor of Kronos, their god of agriculture and the father of Zeus. The Roman equivalent was Saturnus. This name turned out to be very appropriate, since Saturn was later found to be a giant, second only to Jupiter among the Sun's family.

By studying its movement against the "fixed" stars, early astronomers discovered that Saturn took 29½ years to complete one circuit of the Sun. However, every 13 months or so, it would brighten as it neared opposition. From its slow motion through the constellations and modest brightness, it was clear that Saturn was the furthest of the known planets. We now know that Saturn never comes closer to Earth than about 1.2 billion km, or 8 AU. Its average distance from the Sun is about 1.4 billion km (9.5 AU).

With a diameter more than nine times that of our world, 764 Earths would fit inside Saturn. Despite its impressive size, it is only 95 times more massive than Earth. Clearly, Saturn is a largely gaseous object, with an average density 0.7 times that of water. If a sufficiently large ocean could be found, it would easily float. This combination of great size and low density means that the pull of gravity at its visible surface is lower than on Earth.

The planet has the most oblate ("squashed") shape of all the planets, with a marked outward bulge at the equator: the equatorial diameter is almost 13,000 km greater than the polar diameter – a difference of 11%. This is explained by its rapid rotation and gaseous composition. By observing occasional long-lived cloud features, it is possible to calculate that Saturn rotates once every 10 hours 14 minutes. This is a little longer than Jupiter's rotation period, but shorter than the all the other planets.[1]

Saturn's orbit is inclined about 2.5° to the ecliptic and its axis of rotation is inclined almost 27° to the orbital plane. As a result, different hemispheres are tilted towards us during Saturn's journey around the Sun. (Our views of the rings also vary dramatically – see below.) As on Earth, the marked axial tilt produces noticeable seasonal variations, exacerbated by the growth and decline of the shadow cast by the broad rings.

Winds and Clouds

At first glance, Saturn's atmosphere seems to be a pale imitation of Jupiter's. It appears as a fairly bland, tangerine disk, marked only by muted bands and occasional spots, rather than turbulent, colorful ribbons of cloud and huge, long-lived storm systems.

Saturn's more quiescent version of Jupiter is probably explained by the fact that it receives only 25% as much heat from the more

---

[1] The period of rotation at the visible surface was shown to about 20 minutes longer at higher latitudes – the result of different wind speeds at cloud level. Similar variations occur on Jupiter. The period of internal rotation, based on regular variations in the magnetic field, is approximately 10 hours 47 minutes.

Exploring the Solar System, First Edition. Peter Bond.
© 2012 by Peter Bond. Published 2012 by Blackwell Publishing Ltd.

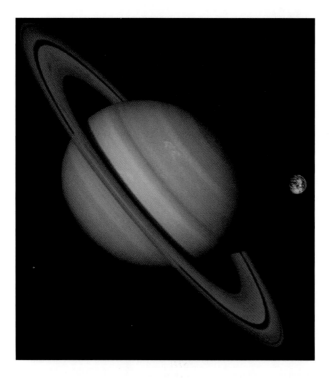

Figure 9.1 Saturn is the second largest planet in the Solar System, with an equatorial diameter of 120,000 km. If the planet and its main ring system were placed between Earth and Moon, there would not be much room to spare. 764 Earths would fit inside Saturn, but the gas giant's bulk density is well below that of water, so its mass is only 95 Earth masses. (NASA)

Figure 9.2 When Cassini arrived at Saturn in 2004, its northern hemisphere appeared pale blue, much like the deep, clear atmospheres of Uranus and Neptune, rather than the usual tan color. Since it was winter in the northern hemisphere, there was less sunlight reaching the atmosphere – particularly since it was largely covered by the ring shadow. The cooler conditions seem to have caused the tan clouds to sink to depths where they are no longer visible. The upper atmosphere became relatively haze- and cloud-free, so the blue color was probably caused by Rayleigh scattering of sunlight (like the blue skies on Earth). The thin rings are barely visible (center), but they cast broad, dark shadows. (NASA/JPL/Space Science Institute)

Table 9.1
Saturn Summary

Mass	5.68×10^{26} kg (95.16 Earth masses)
Equatorial Diameter	120,536 km
Polar Diameter	108,728 km
Axial Inclination	26.73°
Semimajor Axis	1,433,530,000 km (9.58 AU)
Rotation Period (internal)	10.78 h
Sidereal Orbit Period	29.46 years
Mean Density	0.69 g/cm³
Number of known satellites	62

distant Sun, resulting in lower cloud condensation levels. Its lower gravity also leads to a smaller pressure gradient, with greater potential for the scattering of light by a deep layer of high-level haze above the main tropospheric cloud deck.

Curiously, the pattern of belts and zones seems to have little connection with the wind pattern. Bands of cloud running parallel to the equator are surprisingly common, with 24 counted in the southern hemisphere alone. A number of undulating, ribbon-like streamers are also seen within unstable, high speed jet streams. The major seasonal variations caused by Saturn's pronounced axial tilt and the umbrella effect of the huge ring system may account for this apparent anomaly.

Observations of occasional long-lived cloud features enable astronomers to recognize differences in rotation rates and wind speeds at various latitudes. As on all other planets with dense atmospheres, the winds are modified by the Coriolis force and arranged symmetrically on either side of the equator (see Chapter 3). However, the overall pattern is more Earthlike and less complex than on Jupiter. Only above 40° latitude does Saturn's atmospheric circulation mimic that of its larger neighbor.

On Saturn, almost all of the zonal winds blow from west to east, in the same direction as the planet's rotation.[2] Most notable is a broad equatorial jet located between 35° N and S, where winds speed around the planet at up to 1720 km/h. This is more than four times the velocity of Jupiter's equatorial jet stream and much

---

[2] Meteorologists refer to these as "westerlies" because they come from the west.

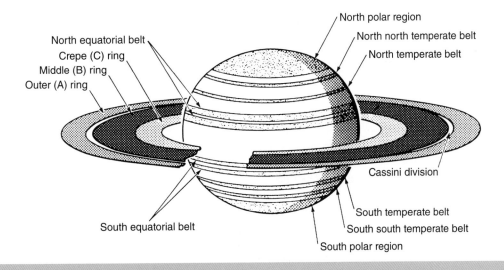

North polar region
North north temperate belt
North temperate belt
North equatorial belt
Crepe (C) ring
Middle (B) ring
Outer (A) ring
Cassini division
South temperate belt
South south temperate belt
South equatorial belt
South polar region

Figure 9.3 Saturn has a regular pattern of belts and zones aligned parallel to its equator. These are less prominent and more difficult to observe than the clouds on Jupiter because they are obscured by a thick, high level haze. The bands disappear at high latitudes. (NASA)

faster than any wind measured elsewhere in the Solar System. However, Cassini data show that the winds at the equator vary by ~200 m/s, which is 10 times larger than the variation of Earth's jet streams. Whether this is a variation with time or a variation with altitude is still an open question.

Recent analyses of Hubble Space Telescope and Cassini images show a slowdown in Saturn's equatorial jet at the visible cloud level since the Voyager flybys of the early 1980s. Since Cassini data suggest that winds slow down with altitude in the equatorial region, the apparent reduction of the equatorial wind speed may be due to the fact that the "tracer" cloud features are now at higher altitudes, rather than an actual decrease in velocity at a given level.

Only four easterly wind currents were identified from Voyager images, all of them very weak. Despite the planet's large axial tilt, there are no major seasonal differences in local weather between the northern and southern hemispheres. Wind speeds tail off dramatically at high latitudes, causing the parallel cloud bands to disappear.

How are such remarkably rapid jet streams created? The primary sources of heat energy are the Sun and the planet's interior. Observations indicate that Saturn emits a similar amount of heat to that received from the distant Sun. The resultant convection produces an alternating pattern of belts and zones, aligned parallel to the equator, where gas is rising or sinking. The parallel pattern is the result of the Coriolis effect created by the planet's rapid rotation.

As with Jupiter, it has generally been assumed that the bright cloud bands (zones) are regions where warm gas rises and the dark bands (belts) are where cooler gas sinks. However, detailed observations show that towering plumes of white cloud associated with thunderstorms, generated by condensation of water and powerful updrafts, occur only in the dark belts. Further evidence that the belts are regions of convection is provided by the interaction between small eddies (rotating storms) and the atmospheric jets.

Figure 9.4 Saturn's wind speeds were calculated by studying Voyager images showing the movement of long-lived spots and eddies. The planet's huge equatorial jet propels the clouds toward the east at up to 480 m/s (1720 km/h) – the fastest winds on any planet in the Solar System. (NASA-JPL)

Clearly, the atmosphere cannot be rising everywhere. One possibility is that the link between zones/belts and rising/sinking gas might not apply on Saturn. Alternatively, it may be that any storm clouds in the belts are local phenomena, surrounded by other

Figure 9.5 Small-scale cloud features are associated with turbulent eddies and wind shear close to one of Saturn's westerly jet streams. The jet, located at 27.5 ° S, is indicated by the horizontal arrow. Winds in this jet blow continuously at speeds close to 320 km/h. By tracking the movements of the cloud features for about 10 hours (roughly one Saturn day), it was shown that eddies on either side of the jet transfer energy and momentum to help sustain the jet. The small arrows indicate the direction in which this transfer takes place. The strong winds that accomplish this combine to stretch out the eddies into bright, tilted streaks parallel to the arrows. (NASA/JPL/Space Science Institute)

Figure 9.6 As on Jupiter, Saturn's upper atmosphere is thought to be made up of three main cloud decks: water ice (H_2O), ammonium hydrosulfide (NH_4HS) and ammonia (NH_3). In the troposphere, temperature increases steadily with depth at the so-called "adiabatic lapse rate". In the stratosphere, above the tropopause, the temperature rises with altitude, probably due to absorption of solar radiation by gases and haze particles. Zero altitude marks the level where atmospheric pressure is 1 bar. (After Open University)

regions in the belts that are sinking. Upwelling may still be occurring in the zones, though obscured by clouds.

Boundaries between eastward- and westward-flowing jet streams create turbulent, eddy-filled regions that pump energy into the never-ending gales. The eddies apparently power the horizontal jet streams, rather than the other way round. For example, Cassini pictures have confirmed that eddies on either side of a jet at 27.5°S supply its energy and momentum. Fed by heat that rises from deep within Saturn, the spinning storms and occasional thunderstorms help to maintain the powerful eastward and westward currents in much the same way as rotating gears power a conveyor belt.

Cloud Decks

Like Jupiter, Saturn is completely covered by dense cloud, although its turbulent activity is better hidden by high level hazes. On Saturn, the cloud decks occur at lower levels and higher pressures than on Jupiter, since the low temperatures required for cloud formation occur at greater depths.

Calculations based on the known composition of the atmosphere indicate that there are at least three main cloud layers. The predicted altitude of these cloud decks is based upon the temperature at which a particular gas will condense into droplets. The outer atmosphere consists of about 97% hydrogen and 3% helium, with minor amounts of methane, ammonia, water, acetylene, ethane and phosphine. However, since hydrogen and helium will not condense at Saturn's temperatures, it is the minor constituents that are the key to cloud formation in the troposphere.

As with Jupiter, the lowest cloud layer is probably made up of water ice. Some 10 km thick, it occurs about 250 km below the tropopause (the boundary between the troposphere and stratosphere) where the temperature is about 250 K (−23°C) and pressure is greater than 10 bars. (Deeper layers of cloud presumably exist, but they can not be seen.) Some 70 km above this is a cloud layer composed of ammonium hydrosulfide ice (NH_4SH), where the temperature is about 180 K and pressure is about 5 bars. 70–80 km higher still are the upper clouds, composed of white crystals of ammonia (NH_3). At this level, about 100 km below the tropopause, the temperature is approximately 120 K and the pressure is between 0.5 and 1 bar.

Above these are found occasional clouds that are propelled upward by convection, with a stratospheric haze layer, probably composed of ammonia crystals and photochemical products – hydrocarbons produced by the effect of sunlight on methane.

Saturn Storms

As on Jupiter, cyclonic and anticyclonic storms sometimes appear through the haze, particularly away from the equator, where wind speeds are comparatively modest. Relatively warm, rising gas from the interior creates sizeable storm systems, with numerous white or brown spots and peculiarly shaped convective clouds that are

Box 9.1 How Fast Does Saturn Spin?

For many centuries, astronomers studied the motions of major cloud features in order to determine the length of Saturn's day. The results indicated that there were different zones, each with its own periodicity. At the equator, clouds complete one circuit of the planet every 10 hours 14 minutes – the result of a high speed westerly jet stream. This was known as System I. At higher latitudes, the typical rotation period (System II) is around 10 hours 38 minutes. But what about the rotation of the planet's interior (System III)?

Saturn is a source of radio emissions, although these are considerably weaker than Jupiter's. It has a broad band of emission in the frequency range 20 KHz to 1 MHz, with maximum intensity between 100 and 500 KHz. Saturn's kilometric radio emissions show a regular variation in intensity, measured by the Voyagers to occur every 10 hours 39 minutes 24 (±7) seconds. This was taken to be the planet's internal rotation period, although the cause of the emissions was hard to understand, since no asymmetry of the internal magnetic field has been detected. (In the cases of Jupiter, Uranus and Neptune, the internally generated magnetic field is asymmetric about the spin axis, so the direction of the field seems to nod up and down as the planet spins. This means that variations in the intensity of the observed radio emissions are directly linked to the rotation period.)

The question of Saturn's rotation rate has been thrown wide open by recent data, which show that the periodicity of the radio signals is surprisingly variable over long and short time scales. The Cassini orbiter found that Saturn apparently rotates in 10 hours 45 minutes 45 seconds, ±36 seconds. It is inconceivable that a planet could slow its rotation by six minutes in a few decades. Furthermore, Cassini's observations of radio emissions have shown that they vary by as much as one percent in a week, and that their timing changes at different rates in different hemispheres.

Since the planet's spin cannot account for the irregular radio signals, some other process must be the cause. One possibility is that gusts in the solar wind cause the emission to vary when they impact the magnetic field. Studies show that there is a characteristic variation in the behavior of the short-period radio emission every 25 days. This period happens to coincide with the Sun's rotation rate, as seen from Saturn.

Another influence on the radio emissions is Saturn's icy moon Enceladus. The geysers on Enceladus may be more active now than a few decades ago. Large numbers of gas molecules ejected into space by the satellite become electrically charged, and are captured by Saturn's magnetic field, forming a torus – a disk of plasma, which "weighs down" the magnetic field. As a result, the rate of rotation of the plasma disk slows down slightly.

Meanwhile, scientists are still struggling to deduce Saturn's true internal rotation rate. The most reliable indicator may be features (e.g. auroras and charged particles) associated with the magnetic field, which suggests an internal rotation rate of 10 hours 48 minutes 6 seconds (±40 seconds).

---

associated with gigantic thunderstorms. Their color is clearly related to cloud composition and altitude.

The smaller eddies last only a few days, although some of the larger features found by Voyager 1 were recovered in Voyager 2 images, demonstrating that they could survive for nine months or more. The most impressive storms resemble smaller versions of Jupiter's white ovals. Like their Jovian counterparts, they are anticyclonic in nature, rotating in a clockwise direction in the northern hemisphere. Nothing comparable to Jupiter's Great Red Spot has been seen, although Voyager 1 did discover a much smaller red feature, about 16,000 km across, at latitude 55°S.

Cassini has monitored Saturn's cloud patterns more or less continuously since it entered orbit around the planet in June 2004. Some of the most interesting features are occasional white clouds, often irregular in shape, which appear in a "storm alley" at about 30–35°S. This area lies between two jets moving at different speeds. The resultant wind shear seems to generate tremendous turbulence, allied to strong updrafts of warmer gas from the interior.

One of the best documented of these active systems was the so-called "Dragon Storm," a complex feature which arose in the same region that had earlier produced large, bright convective storms (see Fig. 9.7). This suggested that it was a long-lived feature that originated deep in the atmosphere and periodically flared up to produce dramatic white plumes.

During July and September 2004, the storm was shown to be associated with powerful radio emissions that resembled short bursts of static generated by lightning on Earth. The radio bursts started while the storm was below the horizon on the night side and ended when the storm was on the day side. This led to the conclusion that it was a giant thunderstorm whose precipitation generates electricity as it does on Earth.

Images taken over many months showed that the Dragon Storm spawned three small, dark ovals that broke away. Two of these subsequently merged while the wind current to the north carried the third oval off to the west. Such small storms are generally stretched out until they merge with opposing currents to the north and south.

Every few decades, the equatorial region hosts even more dramatic atmospheric eruptions, usually in the form of a huge white spot that may spread right around the planet (see Fig. 9.8). Notable disturbances which lasted for several months were recorded in 1876, 1903, 1933 and 1960 – suggesting an upwelling of warmer gas roughly every 30 years. The link between the genesis of these storms and mid-summer in the northern hemisphere was confirmed when another equatorial eruption appeared in 1990, and over the next few weeks it spread around the planet before eventually fading. The fundamental trigger for these eruptions remains a mystery.

Figure 9.7 A false color, near-infrared, image of the "Dragon Storm," a large, complex convective feature that appeared in the southern hemisphere in 2004. The image was taken through filters that reveal different amounts of methane gas. Regions with an abundance of methane above the clouds are red, indicating clouds that are deep in the atmosphere. High clouds are gray and brown clouds are at intermediate altitudes. Powerful radio emissions, similar to bursts of static generated by lightning, indicated that it was a giant thunderstorm. (NASA/JPL/Space Science Institute)

Polar Hexagons and Hot Spots

Saturn's considerable distance from the Sun means that the visible cloud layers are cold, well below the freezing point of water.[3] The average temperature of the cloud tops is about −140°C. The absorption of sunlight by high altitude hazes means that temperatures increase in the stratosphere, rising to −130°C at a height of 200 km above the tropopause. Unlike Jupiter, no large hot spots pierce the all-embracing cloud blanket.

An axial tilt of almost 27° and huge shadows cast by the rings either side of the summer or winter solstices inevitably cause some differences in atmospheric temperature, although seasonal effects are quite small. The immense atmosphere is so massive that it only responds sluggishly to the modest changes resulting from summer heating and winter cooling.

Nevertheless, ground-based measurements taken in 2005 during the southern midsummer showed that there was a general warming towards the south pole, which had been exposed to continuous, direct insolation for many years. In general, the southern atmosphere was up to 10°C warmer than the north.

Less expected was a temperature increase of about 2°C in the upper troposphere between 69° and 74°S, together with another marked rise of some 2.5°C between 87° and 90°S. These elevated temperatures correlated with regions of lower visible reflectivity. Most notable was a dark spot surrounded by high walls of clouds at the south pole that coincided with the highest temperatures measured in the atmosphere.[4]

Surprisingly, Cassini infrared measurements obtained in 2007 showed that the north pole possessed a hot spot similar to the southern vortex, even though it had been in winter darkness for more than a decade. Temperatures at the polar tropopause, the top of the atmosphere's convective layer, peaked at −189°C.

Both hot spots coincide with large cyclonic circulations in the atmosphere. The centers of both vortices are depleted in phosphine gas, an imbalance probably caused by a downward motion of the atmosphere. It seems that the hot spots are the result of gas from equatorial regions moving poleward, then being compressed and heated as it descends over the poles.

The main difference between the polar regions is the presence of a hexagonal feature encircling the north pole (see Fig. 9.10). First imaged by the Voyagers, it has persisted to the present day, and is visible in thermal images taken at night by Cassini. (Jupiter's northern polar region also exhibits a quasi-hexagonal feature that extends vertically down from the stratosphere to the top of the troposphere.)

The infrared images show that the hexagon is largely a clearing in the thick clouds that extends some 100 km below the cloud tops. Although Saturn's northern aurora lies nearly overhead, this considerable depth indicates that the hexagon is not linked to Saturn's radio emissions or to auroral activity (see Fig. 9.16). The hexagon appears to have remained fixed with Saturn's rotation rate and axis since first glimpsed by Voyager 26 years ago, suggesting that it may be an unusually strong planetary wave.[5]

Hubble also imaged a smaller, arrowhead-shaped storm that appeared near the planet's equator in September 1994. The east-west extent of this storm was about 13,000 km, similar to Earth's diameter. The most powerful eastward flowing jet created a dark "wedge" in the western side of the bright central cloud, as well as a secondary disturbance that generated faint white clouds to the east of the storm center. The clouds were probably composed of ammonia ice crystals that formed in the towering thunderhead.

The most recent planet-wide eruption was something of a surprise, since it began in December 2010, only 20 years after its predecessor. Analysis indicated that it was caused by a rapid upwelling of warm, moist gas at 38° N. This penetrated a cold layer in the upper atmosphere which included particles of ammonia ice and water. Over the next 55 days, these were carried rapidly around the planet by strong zonal winds. The storm was also associated with electrical activity, consistent with suggestions that these events are moist convective storms driven by condensation of ammonia and water, at the 1–2 bar and 10–12 bar levels, respectively.

[3] On average, Saturn receives about 1% as much solar radiation as Earth.
[4] Neptune has a similar south polar hot spot.
[5] A planetary wave (or Rossby wave) is a large, long-lived wave in a planet's atmosphere (like a meander in a jet stream) caused by temperature differences and modified by the Coriolis effect produced by the planet's rotation.

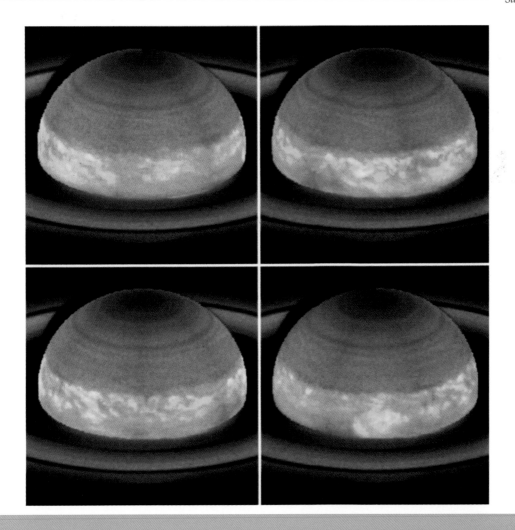

Figure 9.8 Saturn's fairly bland appearance is occasionally altered by the growth of huge, bright storm systems that appear in the equatorial region during the northern summer. They are generated by large-scale upwelling of warmer gas. These Hubble Space Telescope images were taken on November 17, 1990, two months after the eruption was first noticed by amateur observers. It eventually extended right around the planet and lasted for several months. (STScI/NASA)

Interior

The exact nature of Saturn remained uncertain for a long time, though it was thought to be largely gaseous with a fluid interior. Then came the introduction of the spectroscope, which led to the 1932 discovery of methane and ammonia – simple compounds that combine hydrogen with carbon and nitrogen.

This led Yale University's Rupert Wildt to suggest that Saturn had formed by pulling in a huge amount of hydrogen, helium and other gases from the original solar nebula. He envisaged an Earth-like, rocky core surrounded by a thick layer of water, ammonia and methane ices, topped by a hydrogen-rich atmosphere. More recently, it has been suggested that the hydrogen atmosphere is compressed so much that it changes into metallic hydrogen at a depth of about 30,000 km (see Chapter 8).

Today, our knowledge of Saturn's interior is still based upon its size and bulk density, together with measurements of its atmospheric composition compared with the elemental composition of the Sun. The most obvious characteristic is its remarkably low density, approximately $0.7 \, \text{g/cm}^3$, or 70% the density of water. When combined with the fact that 97% of the atmosphere (by molecular abundance) is hydrogen, with almost all of the remainder accounted for by helium, it is clear that Saturn is made almost entirely of the lightest gases in the Universe.

Additional constituents include ammonia (NH_4) and methane (CH_4), a small amount of water vapor, and trace quantities of compounds such as carbon monoxide (CO), phosphine (PH_3), ethane (C_2H_6), acetylene (C_2H_2), methyl acetylene ($CH_3C{\equiv}CH$) and propane (C_3H_8). In the upper atmosphere, most of these are probably produced by photochemical reactions in which solar ultraviolet radiation breaks down methane. In the cloudy troposphere, lightning discharges may also contribute to these chemical processes. Lower down, a water/methane reaction may result in the formation of carbon monoxide.

Jupiter and Saturn are believed to have similar internal structures, with a thick layer of liquid hydrogen beneath their thick, gaseous atmospheres. As pressure increases with depth, this turns into liquid metallic hydrogen at a depth of about 30,000 km. This

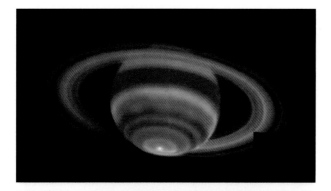

Figure 9.9   An infrared mosaic obtained by the Keck I telescope on February 4, 2004, showing temperatures at Saturn's tropopause, well above the main cloud decks. Lighter shades show warmer temperatures. The bright hot spot (bottom) is located above the south pole. Temperature increases slightly toward the pole, rising from 88 K (−185°C) near 70°S to 91 K (−182°C) at the pole. Note the cooler region in the ring shadow. The rings (orbiting clockwise in this image) also show temperature variations. Ring particles are coldest just after leaving Saturn's shadow (lower left), and become warmer until they pass behind Saturn again (lower right). (NASA/JPL)

SATURN

Figure 9.11   Saturn is composed mainly of hydrogen, with a substantial amount of helium. Beneath the 1000 km thick atmosphere is a layer of liquid hydrogen that changes to liquid metallic hydrogen as temperature and pressure increase. The liquid hydrogen layer is probably thicker than that of Jupiter, while the metallic hydrogen layer is thinner. There may be a core composed of rock and ice. (LPI)

Figure 9.10   A false color thermal image from Cassini, showing night temperatures from 30°N to the north pole in the center. Data relate to temperatures near the 100 millibar tropopause, the top of the convective layer. At the north pole is a relatively hot, cyclonic vortex very similar to that found at the much sunnier south pole. Colors show temperatures from 72–84 K (−201—189°C). The polar hexagon is evident in the warm belt between 75 and 80° N. (NASA/JPL/GSFC/Oxford University)

highly compressed, conductive form of hydrogen behaves like a metal and is probably the source of Saturn's magnetic field.

Pressures and temperatures are inevitably lower in the smaller of the two gas giants, so Saturn's metallic hydrogen zone is relatively thin and located at much greater depth. On the other hand, its liquid hydrogen layer is likely much thicker.

At the center, there may be a fairly small core – perhaps with a radius of about 16,000 km – made of water, methane and ammonia ices overlying rocky materials such as silicates and iron. (Another model has only a rocky core with the ices dispersed through the overlying hydrogen-helium layers.) The core temperature is in the region of 13,000°C, with pressure of about 18 megabars.

Infrared data show that Saturn has a strong internal heat source. It actually radiates 1.8 times more heat than it receives from the Sun, compared with 1.7 times for Jupiter. Since this internal powerhouse cannot be explained by emission of heat left over from its formation and subsequent contraction (Saturn should have completed this long ago), some other process must be at work.

Supporting evidence is provided by the depletion of helium in Saturn's atmosphere, compared with Jupiter and the Sun. The favored theory is that metallic (ionized) hydrogen and neutral helium atoms – once thoroughly mixed – are separating deep inside the planet. Some two billion years ago, the helium began to form tiny liquid droplets which are raining toward the center of the planet. The process of differentiation provides a source of internal heat.

Magnetic Field

Although Saturn's radio emissions indicated the presence of a magnetic field, the existence of a magnetosphere was not

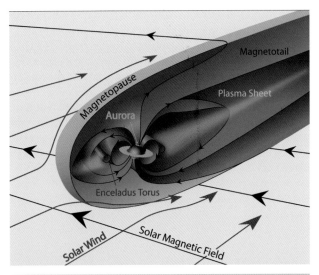

Figure 9.12 A false color mosaic taken above the northern, unlit side of the rings over a period of 13 hours. On the night side (right), Saturn's internal thermal radiation is shown (wavelength 5.1 microns). The speckled and banded patterns are created by variable, thick clouds, which block some of the heat deep in the atmosphere. These clouds are likely made of ammonium hydrosulfide and cannot be seen in reflected light on the dayside (left), since they are hidden by overlying hazes and ammonia clouds. Tiny, high altitude particles are only half as numerous in the northern hemisphere, so it is twice as bright as the southern hemisphere. The icy ring particles are highly reflective at 2.3 microns (blue). (NASA-JPL/University of Arizona)

Figure 9.13 Saturn is surrounded by a "normal" tadpole-shaped magnetosphere, with many similarities to those of Earth and Jupiter. The solar wind is slowed at the bow shock and diverted around the magnetopause, the outer edge of the region dominated by Saturn's magnetic field. Saturn is the only planet whose magnetic field is almost exactly aligned with its rotation axis. (Fran Bagenal/ University of Colorado, Boulder)

confirmed until the flyby of Pioneer 11 in September 1979. The data indicated that Saturn's magnetosphere is, in many ways, a smaller version of Jupiter's, with a blunt sunward side and a long downwind tail that is shaped by the solar wind. Since then, further details have been revealed by the Voyagers and the Cassini orbiter.

The first interaction between the solar wind and Saturn's magnetosphere takes place at the bow shock, where the incoming particles are dramatically slowed. The average distance from the planet of this outer boundary is about 1.8 million km (30 Saturn radii). However, changes in solar wind pressure mean that the bow shock oscillates between 20 and 35 Saturn radii. During its insertion into orbit around Saturn in 2004, Cassini crossed the boundary 17 times.

After crossing the bow shock, the solar wind changes direction and is forced around the magnetosphere. The magnetopause typically lies about 1.2 million km above Saturn's sunlit hemisphere, although the distance varies with the level of solar wind activity. The magnetotail likely stretches hundreds of Saturn radii downwind.

Of the planets in our Solar System, Saturn's magnetic field strength is second only to that of Jupiter. Its magnetic moment is 600 times Earth's, but only about 3% that of Jupiter. Although its magnetosphere dwarfs that of Earth, Saturn is so large that the strength of the magnetic field at its cloud tops is actually lower than on Earth's surface: 0.2 gauss, compared to 0.5 gauss. At

present, the polarity of the field is opposite to that of Earth, but like our planet and Jupiter, the field is dipolar.

Like Jupiter, the key factors in the generation of Saturn's field are its rapid rotation, an internal heat source and the presence of metallic hydrogen, a conductive material. These combine to drive a dynamo involving moving currents deep in the interior. Curiously, Saturn is the only planet whose magnetic field axis is almost exactly aligned with its rotation axis, even though the center of the field is displaced northwards along the axis by about 2400 km. Until this discovery, theories for the generation of planetary magnetic fields required the field to be inclined to the planet's rotation axis.

Saturn's magnetosphere is populated by numerous heavy ions (charged particles), largely derived from geysers on Enceladus, sputtering by charged particles on the surfaces of Saturn's icy satellites and meteoroid bombardment of ring material. Smaller contributions come from the planet itself and Titan's nitrogen-rich atmosphere. Disks of neutral and ionized hydrogen and nitrogen surround Saturn near the orbit of Titan, and plumes of heavy ions may also extend beyond the satellite's orbit.

Geysers on Enceladus are thought to deliver as much as 300 kg/s of water to Saturn's inner magnetosphere. The neutral water molecules are split by sunlight and turned into ions, which Saturn's magnetic field captures and shapes into a torus of energized plasma. The rotating plasma disk slightly slows down the rate of rotation of the planet's magnetic field.

Saturn's plasma torus, similar to the one around Jupiter, is the biggest plasma structure associated with any planet. It extends out to 15–20 Saturn radii. The torus contains approximately 3000 particles per cm³.

The rotating plasma torus is also associated with an electrical current (the "equatorial ring current") of about 10 million amps. Ring currents are caused when plasma becomes trapped between magnetic field lines and gradually drifts around the planet. The aggregate motion of the hot ions distributed around the equator generates an electrical current. The ring current occupies a region of the equatorial plane between 540,000 and 1,080,000 km from the center of Saturn.

Whereas Earth's ring current is associated with hydrogen and appears during solar flares, Saturn's is largely associated with ionized oxygen and is always present. Unlike Earth's fairly stationary ring current, Saturn's rotates with the planet and is asymmetric, with the side facing the solar wind showing evidence of squeezing.

Auroras and Radiation Belts

Saturn's magnetosphere traps ionized particles in radiation belts where the number of particles is similar to the terrestrial magnetosphere. The most heavily populated belts occur between 139,000 km and 362,000 km from Saturn's center (i.e. outside the main rings), and contain particles with energies up to tens of MeV. The belts are largely composed of oxygen and water-derived ions produced by the bombardment of the rings and icy moons by solar ultraviolet light and particles trapped in the magnetic field.

Until recently, the inner edge of the radiation belts was thought to be marked by the main rings, which absorb any particles that

Figure 9.14 Saturn's main radiation belts are revealed by Cassini data of the energetic ion and electron population. Ion intensity is shown above the horizontal line as energy increasing upward. Electron intensity is shown below the horizontal line as energy increasing downward. Red indicates high particle intensity, blue is low intensity. The vertical energy scales run from 30 keV to several MeV. The data are shown for regions between 783,000 km and 78,000 km from Saturn's center. The region above the rings was devoid of ions and electrons. The region between the D ring and Saturn was not sampled. (NASA/JPL/APL)

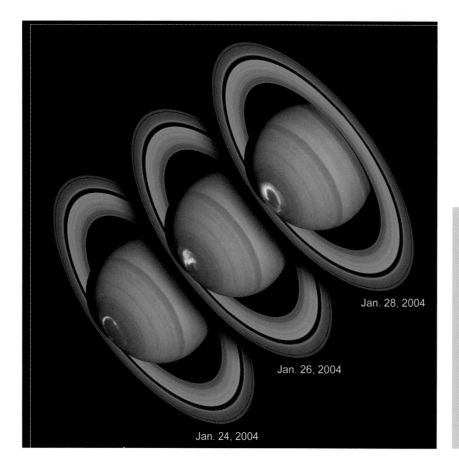

Jan. 28, 2004

Jan. 26, 2004

Jan. 24, 2004

Figure 9.15 A sequence of images of Saturn's southern hemisphere taken by the Hubble Space Telescope, January 24, 26 and 28, 2004. The ultraviolet aurora is superimposed on visible images. The brightening on January 28 (right) corresponds with the arrival of a large disturbance in the solar wind. When Saturn's auroras become brighter (and more powerful), the ring of light encircling the pole shrinks in diameter. (NASA, ESA, J. Clarke/Boston University, and Z. Levay/STScI) encounter them. However, Cassini found another radiation belt that stretches between Saturn's cloud tops and the inner edge of the D ring (the closest ring to the planet). The new belt is less than 6000 km thick and is not known to contain particles with total energy above about 150 keV.

The belts are more intense on the night side of the planet, and there are marked gaps created as the trapped ions collide with small, embedded moons, the tenuous E ring, and gas.

Another phenomenon associated with energetic particles trapped in the magnetosphere is the aurora. On Earth, auroral storms may develop in about 10 minutes and last for a few hours, whereas Saturn's auroras always appear bright and may last for several days. The auroral zones typically measure about 9000 km across and rise more than 1600 km above the cloud tops, but their extent and brightness vary over time. These changes indicate that the aurora is primarily shaped and powered by a continual interaction between Saturn's magnetic field and the solar wind.

As on Earth, the auroras occur when charged particles from the Sun spiral down the magnetic field lines, making the molecules in the upper atmosphere glow. This results in circular or oval auroras, shaped by the magnetic field lines that converge like narrowing funnels towards the magnetic poles. Unlike Earth, Saturn's auroras are only "visible" in ultraviolet light.

Studies of interplanetary shock waves that strike Saturn's magnetosphere show that they result in a brightening of auroral emissions, with accompanying increases in radio emissions. This suggests that there is a direct correlation between auroral intensity and the state of the solar wind – similar to the situation on Earth. (Jupiter's auroras are largely controlled by processes taking place inside its huge magnetosphere, rather than the conditions in the solar wind – see Chapter 8.)

However, further study indicates that the main factor controlling Saturn's auroras appears to be compression of the magnetopause as a result of increased solar wind dynamic pressure. Unlike Earth, the orientation of the interplanetary magnetic field (i.e. its north-south polarity) and the resultant magnetic reconnection that links the solar and planetary magnetic fields is less important. This may be because, during Saturn's southern summer, the solar magnetic field is generally aligned at right angles to the planet's field, making it less easy for magnetic reconnection to take place.

Discovering Saturn's Rings

The rings of Saturn were discovered in 1609 by Galileo Galilei. Using a homemade "spyglass" that could magnify a mere 20 times, he noticed that Saturn seemed to be sandwiched between two identical, but smaller planets. When he returned to the scene two years later, the companions had vanished and Saturn was now in splendid isolation. Subsequent observers produced drawings of detached *ansae* ("handles") attached to either side of the planet.

Confusion reigned for more than 40 years, until a Dutch astronomer named Christiaan Huygens announced that he had discovered a satellite – now known as Titan – and found an

Figure 9.16 A Cassini infrared view of Saturn's north pole at night reveals heat rising from a depth of at least 75 km below the upper clouds. Surrounding the north pole is a nearly stationary hexagonal feature, possibly an unusually strong planetary wave that extends deep into the atmosphere. The blue color shows auroral emissions caused by charged particles entering the atmosphere. The aurora seems to coincide with the hexagon, though it lies at a much higher altitude. (University of Arizona–LPL, NASA-JPL)

explanation for Saturn's *ansae*. Curiously, he decided to disguise the answer in the form of an anagram and waited three years to unveil the secret in his book, *Systema Saturnium* (The Saturn System).[6]

Huygens declared that the mystery appendages were actually a thin, solid ring that circled above its equator. Since the planet was inclined to the plane of its orbit, the ring, too, was tilted by the same amount. This meant that the ring's appearance changed during each 29½ year-long orbit of Saturn. When the ring was edge-on to Earth, during Saturn's spring and autumn, it seemed to virtually disappear. During summer or winter, when the poles were tilted toward the Sun and Earth, the ring opened out and appeared in all its glory. Then the entire cycle operated in reverse until the disk was viewed edge-on once more (see Fig. 9.19).

But it soon became clear that the ring was not solid. In 1675, Giovanni Domenico Cassini noticed a continuous dark region, about two thirds of the way out from Saturn, which he correctly surmised to be a gap – now known as the Cassini Division – that separates an outer A ring from an inner B ring. Cassini also stated that the ring was composed of a large number of small satellites orbiting Saturn.

As telescopes improved, further structure was observed in the ring system. The faint, diaphanous, C or "Crepe" ring was

---

[6] The solution to the anagram (translated into English) was: "It is surrounded by a thin flat ring, nowhere touching, and inclined to the ecliptic."

Box 9.2   Voyagers at Saturn

NASA's Voyager spacecraft were not the first visitors to Saturn. A gravity assist from Jupiter enabled the much simpler Pioneer 11 to fly past Saturn on September 1, 1979. Pioneer's TV camera showed very little cloud detail on the planet but there was clear evidence of structure in the main rings. Its major discoveries included the thin F ring, about 3600 km beyond the A ring, and the presence of a small satellite between the new ring and the orbit of Mimas. Pioneer also made the first survey of Saturn's magnetosphere and particle environment.

The Voyager spacecraft were designed to carry out a comprehensive exploration of Saturn's rings and satellites. Furthermore, the second spacecraft could be reprogrammed to take advantage of any discoveries by its predecessor.

Titan was the key to the Voyager 1 encounter. In order to observe this mysterious moon, while obtaining a good view of the rings and images of some smaller satellites, the spacecraft was targeted to arrive from above the ring plane. It then swung over Saturn's southern hemisphere at a height of 124 200 km and returned northward, using the planet's gravity to head toward interstellar space.

Figure 9.17   The flight paths of Voyager 1 and 2 through the Saturn system, as seen from above the planet's north pole. Voyager 1 was targeted to fly close to Titan. Saturn's gravity then diverted it above the ecliptic to send it out of the Solar System. This successful flyby meant that Voyager 2 could be redirected toward Uranus. (NASA-JPL)

During the Saturn approach, Voyager 1 skimmed past Titan on November 11, 1980, just 4520 km above the orange clouds. It then sent back close-ups of Tethys before plunging beneath the rings. The camera turned toward Mimas, Enceladus and Dione before Voyager recrossed the ring plane and obtained detailed images of Rhea. On the outward leg, as seen from the Earth, the spacecraft disappeared behind the planet and the main ring system.

Although the flight from Titan to Rhea lasted little more than 24 hours, Voyager's instruments monitored the Saturn system continuously for almost four months. The rings were found to contain hundreds of ringlets, some elliptical, while the F ring comprised several braided strands.

Another puzzle involved shadowy "spokes" in the B ring that survived despite different rotation rates within the ring material. The existence of the D and E rings was confirmed, and a seventh main ring (G) was discovered. The theory that some narrow rings were confined by nearby satellites was confirmed by the discovery of two tiny "shepherds" on either side of the F ring and another just outside the A ring.

Although Voyager's cameras were unable to pierce the smog that shrouded Titan, new data were returned about the temperature, pressure and composition of its atmosphere, and the high-altitude haze layers. Particularly intriguing was the dense, nitrogen-rich atmosphere in which complex organic compounds were created.

Voyager 2 arrived more than nine months later. Equipped with a more sensitive camera and an approach trajectory well above the ring plane, it was ideally placed to study the rings. The northern side of the rings was now receiving more direct sunlight and so appeared much brighter.

From early June to late September 1981, Voyager 2 sent back more than 18,500 images of Saturn, its rings and satellites. Closest approach occurred on 25 August, 101,000 km above the cloud tops.

Thousands of ringlets were detected, though a search for small satellites within them was unsuccessful. Saturn's bland clouds came alive with images of brown and white spots, and undulating, ribbon-like jet streams. Titan was only viewed from afar, with Hyperion, Enceladus, Tethys and Iapetus the satellites that gained most attention. Voyager 2 crossed the equator only once, providing an opportunity to look back at the shaded side of the rings.

Two thirds of the Enceladus pictures were lost through jamming of the scan platform, which carried the cameras, 45 minutes after the ring passage. However, the platform was operational again nine days later, when Voyager made a distant pass of Phoebe, the outermost satellite then known.

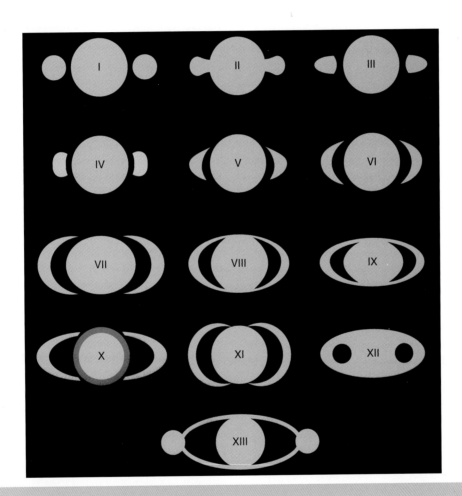

Figure 9.18 Sketches of Saturn made by various observers during the first half of the 17th century. Galileo's observation of 1610 is labeled I. Not until 1659 did Christiaan Huygens reveal that the "handles" were actually a ring above the planet's equator. (NASA)

discovered in 1850, extending inward from the B ring. A sizeable gap, now known as the Encke Division, was also noticed near the outer edge of the A ring.

Meanwhile, the true nature of the rings was revealed when James Clerk Maxwell proved mathematically that a very thin, solid ring circling the planet would be destroyed by gravitational forces. His prize-winning paper noted that the rings must be composed of small particles in separate orbits, with the inner ring particles obeying Kepler's laws and traveling faster than those further out.

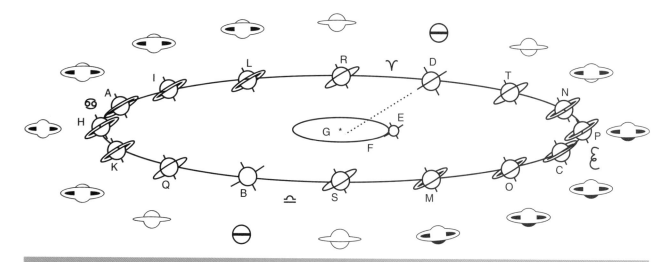

Figure 9.19   In this diagram from his book, *Systema Saturnium*, Christiaan Huygens showed how its large axial tilt resulted in seasons on Saturn and caused different amounts of its ring to be visible from Earth. (NASA)

For many years, it was thought that the rings were formed at the same time as the planets, coalescing out of swirling clouds of interstellar gas, 4.5 billion years ago. An alternative explanation was provided by French mathematician Edouard Roche, who calculated that an object would be torn apart by tidal forces if it approached too close to Saturn. The distance at which this would occur – the so-called Roche limit – more or less corresponded to the position of the rings.

The Roche limit is an imaginary boundary around a planet within which a "rubble pile" satellite will be torn apart by gravitational forces. At the same time, particles in this region cannot gravitationally coalesce to form a satellite because of the disruptive effect of planetary tidal forces.

Today, the origin of the main rings is still unknown. The possibility remains that a comet or small satellite approached too close to Saturn and was broken apart by tidal forces. Alternatively, they may have formed after an icy moon about the size of Mimas was broken up by an impact with a comet or asteroid. The thinner, dustier rings are produced by collisions between small objects within the rings, or by meteorite erosion of these bodies. Geysers on Enceladus provide tiny grains of water ice that generate the huge E ring, while an even larger outer ring is fed by material removed from the moon, Phoebe.

Multiple Rings

Today, there are seven recognized major rings, which are labeled in the order in which they were discovered. From the planet outward, they are D, C, B, A, F, G and E. Other faint rings or ring arcs are associated with Janus, Epimetheus, Anthe, Methone and Phoebe.

From outer edge to outer edge, the main system measures 374,000 km, so it would almost span the gap between Earth and

the Moon. Despite their vast width, they are generally only a few tens of meters thick – the reason why they "disappear" when they lie edge on to the Earth. The rings are best envisaged as thousands of ringlets, each comprising innumerable particles and small chunks of icy debris in separate orbits. The particles are composed primarily of water ice mixed with silicates, and range in size from microns to tens of meters. Above this size range, their number drops sharply. A typical ring particle is more like a fluffy snowball than a hard chunk of ice.

Subtle color variations reveal differences in the composition and particle sizes within the main rings, with the largest objects furthest from Saturn. The cleanest ice occurs in the A ring, whereas the C ring is noticeably dirtier. The Cassini Division and other gaps also contain thin ringlets composed of an unidentified "dirty" substance which appears remarkably similar to the surface material seen on Saturn's remote moon, Phoebe.

The main rings are thought to be made up of boulder-size snowballs, but the size of the ice crystals on their surfaces ranges from very small, like powdery snow, to larger grains, similar to more granular snow.

Until recently, it was assumed that the rings were ephemeral, and would disperse over time. However, support is mounting that they may be a permanent feature. Cassini observations show that the A ring is primarily empty space, with ever-changing clumps of material that range in size from 2 to 13 meters. The gaps between the clumps are larger than the clumps themselves.

The individual clusters are largest near the middle of the ring and became smaller toward the edges of the ring. These clusters are torn apart and reassembled by Saturn's gravity. Even when they fragment, they tend to come back together, thereby maintaining the overall ring structure.

Similarly, the B ring contains numerous clumps that are 30–50 meters wide and very flat, like big sheets. Once again, there is a constant destruction process as Saturn's gravity pulls them apart,

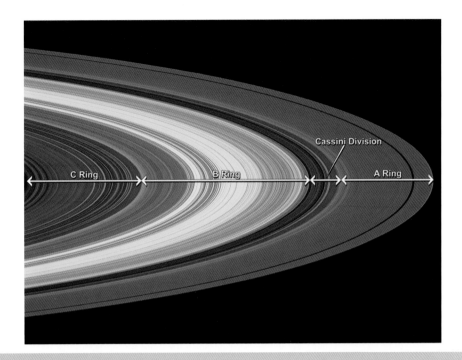

Figure 9.20 A Cassini false-color image showing the size and density of ring material, based on observations of the changing brightness of stars as they pass behind the rings. The B ring is densely packed with clumps, separated by nearly empty gaps. These clumps, 30–50 meters across, are constantly colliding. Particles in the central, yellow regions are too densely packed for starlight to pass through. (NASA/JPL/University of Colorado)

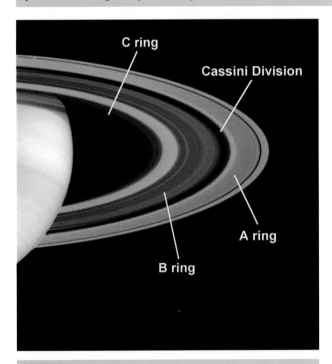

Figure 9.21 The temperatures of the unlit side of Saturn's rings are shown in this false-color image from Cassini. Red is about −173° C, green −183° C and blue −203°C. Opaque regions are cooler, for example, the outer A ring and the middle B ring. More transparent sections are relatively warm, for example, the Cassini Division or the inner C ring. (NASA/JPL/GSFC/Ames)

followed by reformation. The particles move from clump to clump as they are destroyed and new ones reformed. Such recycling may enable the rings to persist almost indefinitely.

An atmosphere of molecular oxygen and neutral hydrogen extends 60,000 km above and below the main rings and some distance beyond the A ring. Water ice is the likely source for this cloud. Water molecules are driven off the ring particles by solar ultraviolet light. They are then split into hydrogen, as well as molecular and atomic oxygen – a process known as photodissociation. Much of the hydrogen gas escapes to space, while the atomic oxygen and any remaining water are frozen back into the ring material due to the low temperatures. This leaves behind a concentration of oxygen molecules.

A and B Rings

Both the A and B rings are very bright and easily observed from Earth. However, the B ring is much denser and more opaque than the other rings, hence its dark appearance when illuminated from behind.

The A ring, the outermost of the main rings, is subdivided by two "empty" regions. Each of these gaps is cleared of particles by a small satellite embedded in the ring. The 42 km wide Keeler Gap is swept clear by 8 km Daphnis, while the 330 km Encke Gap is occupied by 26 km diameter Pan.

The A and B rings are separated by the Cassini Division, the largest gap in the rings. The Division is cleared by a 2:1 gravitational resonance with the moon Mimas. Particles orbiting at this distance from Saturn travel twice around the planet for every single orbit of Mimas. This regular gravitational interaction pushes them into different orbits, opening up a gap.

Once thought to be empty, spacecraft images have shown that the Division, which measures about 5000 km across, actually contains five broad rings that display additional structure. The effective particle diameter in these rings, derived from radio occultation measurements, is about eight meters.

At least four faint, dusty ringlets have also been observed in the Encke Gap. One of them occupies almost the same orbit as Pan, suggesting that the satellite is the source of the material. The presence of the other narrow ringlets indicates that there may be other unseen moonlets in the Encke Gap. The ringlets vary in brightness and appear to move in and out along their length, resulting in notable "kinks" which are similar in appearance to those in the F ring (see below).

Both Daphnis and Pan perturb the particles along the borders of the gaps they inhabit. When the satellite passes, its gravity generates a series of visible stripes or "wakes" in the ring material on either side of it. These move diagonally away from the gap's edge. In the faster-moving ring material closer to Saturn, they drift ahead of the moon as leading wakes. Similar trailing wakes are created as Pan overtakes the slower-moving ring material beyond its orbit. Since Pan is larger than Daphnis, it creates correspondingly larger wakes.

Cassini images of the middle A ring also show "propeller-shaped" disturbances that are thought to be caused by embedded objects approximately 40–120 meters in diameter. These moonlets are not large enough to create a continuous gap in the ring, but they can make a partial clearing which is shaped like an airplane propeller. Their gravity perturbs the surrounding ring material, creating wavelike features in their wakes. Since four "propellers" were found in two images covering only a tiny fraction of the ring, it is likely that millions of these objects exist within the A ring alone.

Cassini also detected numerous wavy features called "density waves" in ring A. Many of these were in its outer region, and generated by repeated gravitational interactions (resonances) with the small moons orbiting just outside the ring. Spiral bending waves due to Mimas produce vertical corrugations in the ring with amplitudes as large as 1 km.

The B ring has a very complex structure. It contains rings that are hundreds of kilometers wide and vary greatly in the amount of material they contain. A thick, 5000 km wide core contains several bands of material that are nearly four times as dense as ring A and nearly 20 times denser than ring C. Organized into uncountable ringlets, many of which are eccentric (not circular), the B ring is a dynamic, constantly changing region. Once again, the ringlets may be formed by the presence of numerous unseen moonlets.

Since there are so few gaps, most of the radial structure in the B ring must be due to variations in its optical thickness due to density waves, gravitational instabilities or dynamical instabilities. Density waves are generated by the gravitational effects of Saturn's satellites, propagating out from the resonant orbits (e.g. at the 2:1 resonance with Janus), where particles travel twice around the planet for every single orbit of the satellite.

The outer edge of the B ring is elliptical, with a difference of some 140 km in its maximum and minimum diameters. This eccentricity is accounted for by the 2:1 resonant interaction with Mimas.

Figure 9.22 An artist's concept of Saturn's ring material. The mini-satellites are composed mostly of ice, but are not uniform. They clump together to form elongated, curved aggregates, continually forming and dispersing. The space between the clumps is mostly empty. The largest individual boulders shown are a few meters across. (NASA-JPL/ University of Colorado)

Figure 9.23 The 8 km wide Daphis occupies an inclined orbit within the Keeler Gap in Saturn's A ring. The moon's gravity perturbs the orbits of the particles forming the gap's edge, creating horizontal and vertical waves. Here, the vertical waves cast dark shadows on the rings. Measurements show that they are 0.5–1.5 km high, up to 150 times taller than the thickness of the ring. Daphnis also casts a shadow onto the nearby ring. (NASA-JPL)

Figure 9.24 Propeller-shaped disturbances in the middle of the A ring. These indicate the presence of at least four small, embedded moons. This was the first evidence of the objects intermediate in size between the ring moons Pan and Daphnis and much smaller particles that comprise the bulk of the rings. The moonlets may be remnants of the body whose break-up produced the rings. (NASA/JPL/Space Science Institute)

Figure 9.25 A cross section of the main rings showing how they differ in structure, color and density. The natural color mosaic was acquired as the Cassini spacecraft soared above the shaded side of the rings. Major gaps are named at the top. Scale in the radial (horizontal) direction is about 6 km per pixel. (NASA/JPL/Space Science Institute)

Spokes

The B ring is also characterized by sporadic radial markings or "spokes." In Voyager images these faint, wedge-shaped features typically extended outward 10,000 to 20,000 km across the B ring. When seen at low phase angles, they appeared dark; at high phase angles they appeared bright – indicative of tiny dust particles.[7] They formed at a distance from Saturn where the rotational speed of the ring particles roughly coincides with that of the planet's magnetic field lines.

The dark spokes appeared very quickly after the ring particles entered sunlight and then dissipated gradually as they moved toward the planet's shadow. Broader, less radial spokes that showed normal orbital behavior were also visible, probably older

remnants that had survived. In some cases, new spokes appeared to be superimposed on older ones.

The importance of electrostatic charging effects in spoke formation and dynamics was supported by the detection at radio wavelengths of electrostatic discharges from the rings. These discharges loosely correlated with Saturn's rotation.

The spokes are made up of particles less than a micron across that collect electrostatic charges in the plasma environment of the rings and become subject to electrical and magnetic forces. Under the right conditions they gain an extra electron, allowing them to briefly levitate en masse from the ring's surface to a height of perhaps 80 km above the B ring.

The spokes were clearly visible to the Voyagers when Saturn was near its equinox, but in 1998 they faded from view and were not

[7] The D, E, F and G rings all show similar forward scattering properties to the B ring spokes, evidence that they are made of very small particles.

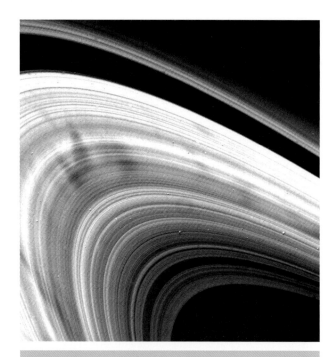

Figure 9.26 A Voyager 2 image of numerous, dark "spokes" in the B-ring. Often more than 10,000 km long, they could appear within a matter of minutes. The spokes survived for several hours, even though the particles in the B ring orbit at different rates. They are probably created by electromagnetic forces acting upon micron-sized ice particles. (NASA-JPL)

seen by Cassini when it arrived during Saturn's southern summer in 2004. It was not until 2005 that the first spokes reappeared.

Their visibility is now believed to depend on the elevation of the Sun above the rings and the changing light-scattering geometry: the less sunlight, the more likely they are to form. It seems that the spokes are visible for a period of about eight years, when the rings are nearly edge on to the Sun, and then disappear for six or seven years.

C and D Rings

The C ring is fainter and more transparent than the A and B rings – hence its other title, the "Crepe Ring." The fine dust particles that make up the ring scatter light forward. It contains a number of dense ringlets whose locations are unrelated to orbital resonances with the larger satellites.

Within the C ring is the 270 km wide Maxwell Gap, which contains an eccentric ringlet whose width varies from about 30 to 100 km. A more diffuse ring may indicate the presence of unseen moonlets in the gap.

The still fainter D ring extends from the C ring to within 7000 km of Saturn's cloud tops. Details are difficult to observe, except with long exposures taken at favorable viewing angles. Cassini discovered that, although some ringlets within the D ring

had stayed in the same place since the Voyager flybys, one of its brighter ringlets has dimmed considerably and moved 200 km closer to Saturn during that 25-year period.

Another example of dynamism in the D ring is a series of vertical or corrugated ripples, possibly generated by a recent collision with a comet or asteroid.

F Ring

Beyond the A ring is a gap known as the Roche Division, which contains two recently discovered, diffuse rings. R/2004 S1 is a 300 km wide feature that lies along the orbit of Atlas, while R/2004 S2 is very similar and located immediately interior to the orbit of Prometheus. The moon's orbit is elliptical, swinging inward to the outer edge of R/2004 S2 and outward to the inner boundary of the F ring. The region occupied by its orbit has been cleared of material through the gravitational sweeping action of the 102 km wide satellite.

The association of R/2004 S1 with Atlas and the main Encke Gap ringlet with Pan (plus two other diffuse rings either side of the G ring) suggests that all of these ringlets are fed by material derived from their orbital companions. The weak gravity of these small moons means they cannot retain any loose material on their surfaces. When they are struck by meteoroids, this material is blasted into space, creating diffuse rings. Collisions among several moonlets, or clumps of boulder-sized rubble, may also lead to debris trails. Many of the particles in these ringlets may be micron-sized dust, implying that non-gravitational forces also affect the ringlets' dynamics.

On the outer edge of the Roche Division is the F ring, the weirdest member of the entire system. The narrow ring is shepherded on either side by two tiny moons, Prometheus and Pandora. Discovered by Pioneer 11 in 1979, the F ring displays kinks, clumps and strands, as well as braids.

Voyager images showed one bright "core" strand and four fainter ones, each about 70–100 km across. During a stellar occultation, the core was seen to be subdivided into many narrower strands about 3 km wide, possibly as a result of larger particles in the ring. Clumps of material in the ring were shown to be long-lived, and fairly uniformly distributed at roughly 9000 km intervals around the ring.

The clumps and some of the other ring patterns are undoubtedly formed as a result of gravitational tugs from the nearby shepherding satellites. Since encounters between a moon and a ring particle recur during each orbit of the planet, the overall effect is a pattern that roughly repeats, with a spacing proportional to the difference in orbital speeds between the ring particle and the moon. The length scale of most F-ring structures indicates that Prometheus, the larger and closer of the two moons, is responsible for most of the braiding.

However, perturbations by the shepherds cannot explain many of the ring's other features. Voyager images showed that the brightest clumps are not regularly spaced. Some of them, called "bursts", appear suddenly and then spread out over periods of days to weeks. Others evolve more slowly but fade away after a few months. One hypothesis is that the most rapid "bursts" are dust clouds arising when a meteoroid hits the ring; the other

Box 9.3   Cassini

NASA's Cassini is one of the largest and heaviest scientific spacecraft ever launched. Together with the European Space Agency's piggybacking Huygens probe, the spacecraft weighed about 5.7 tonnes at lift off on October 15, 1997. Despite the use of a powerful Titan IV-B/Centaur launch vehicle, Cassini had to rely upon gravity assists from Earth, Venus and Jupiter before it reached Saturn on June 30, 2004.

During the Saturn orbit insertion, the spacecraft passed through the gap between the F and G rings, looped above the ring plane, then descended between the F and G rings. The closest approach to Saturn throughout the entire mission occurred shortly before burn completion, some 20,000 km above the cloud tops. To protect the spacecraft from small ice or dust particles, Cassini was turned so that its high-gain antenna acted as a shield.

During its four–year prime mission, Cassini completed 76 orbits of Saturn, including 45 close flybys of the Titan and numerous flybys of other moons. The mission has since been extended until at least 2014. Titan's gravity is used to change Cassini's course and navigate around the Saturn system.

The 12 remote sensing experiments include cameras, various spectrometers, an imaging radar, a magnetometer, a magnetospheric imaging instrument, a dust counter and a radio science experiment.

Many of the most surprising results have come from investigations of the satellites. The first of these to be seen at close quarters was Phoebe, which proved to have a very dark, ancient surface. However, the main focus of attention is Titan, whose orange clouds, dense atmosphere and hidden surface are being studied. The radar instrument, in particular, has revealed smooth plains, icy uplands, impact craters, dry river beds and methane-filled lakes.

Perhaps the greatest surprise has been Enceladus. Cassini discovered a tortured surface of "tiger stripes" where tidal forces have fractured and buckled the icy crust. An underground heat source at the south pole feeds geysers that propel a plume of water vapor, ice and dust particles thousands of kilometers into space. These vents are the source of the minute particles that make up Saturn's enormous E ring.

Figure 9.27   The nuclear-powered Cassini orbiter carries 12 remote sensing instruments to study the Saturn system. The European Huygens probe piggybacked on the ride to Saturn. (NASA-JPL)

The most bizarre, cratered landscape was found on Hyperion, an irregularly shaped rubble pile whose sponge-like surface displays closely packed and deeply etched pits. The cratered landscapes of Rhea and Dione also show signs of past geological activity in the form of fractures and ice cliffs.

Cassini's instruments have also studied the smaller moons and the rings, e.g. the ring shepherd, Prometheus, which creates knots and kinks in the F ring. Studies of clumps and waves in the rings show an ever-changing population of ice-rich particles.

Saturn's cloudy atmosphere has also been scrutinized, showing high speed jet streams, long-lived oval storms and thunderclouds, as well as a hexagonal cloud feature at the north pole.

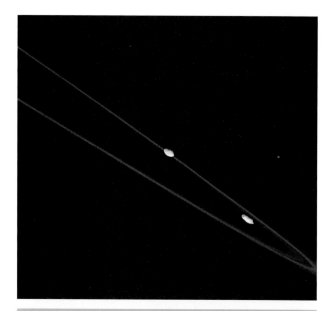

Figure 9.28 Two shepherd moons, Prometheus (right) and Pandora (left), are seen on either side of the F ring. Pandora is closer to the spacecraft in this view. Prometheus is slightly closer to Saturn, so it travels faster around its orbit, periodically overtaking Pandora. The moons' gravitational interactions disturb the ring particles, creating complex braids and other structures. (NASA/JPL/Space Science Institute)

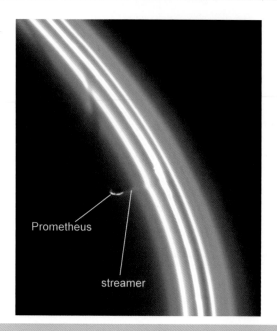

Prometheus

streamer

Figure 9.29 Prometheus creates complex structures in the F ring, shown here as five distinct strands. The inner strands are disrupted during the moon's regular passes on the inside of the ring. Prometheus is pulling material from the ring, as shown by the faint streamer. Discontinuities or "kinks" in the ringlets are seen, as well as gaps in the diffuse inner strands. (NASA/JPL/Space Science Institute)

is that all the clumps arise from mutual impacts among ring bodies

The ghostly ringlets flanking the ring's core are arranged into a spiral structure that wraps around the planet three times. Comparable structures in the main rings are associated with density and bending waves initiated by the gravitational influence of a large moon. However, the F ring's spiral structure contains very little mass and appears to originate from material that is somehow ejected periodically from the core strand and then sheared out due to the different orbital speeds followed by the constituent particles.

The spiral may be caused by one or more tiny moons crossing the F ring and spreading particles around, or by a rare, large impact with the ring. Cassini found at least two moonlets or sizeable clumps of material that appear to cross the F ring periodically. One of them, S/2004 S6, can intersect the F ring at very high speeds and may be responsible for disrupting the ringlets and forming the spiral. S/2004 S6 may be a particularly large shard left over from the proposed impact.

How was the F ring formed, and why it is so strange? First, it orbits at the edge of the Roche limit, where some re-accretion is possible, so ring bodies can continuously break up and join back together. Second, it is faint and narrow, so that small injections of new dust are quite noticeable. The ring and its shepherds all follow eccentric orbits, which means that the ring is perturbed quite markedly as Prometheus and Pandora approach and recede during each orbit.

On the other hand, it seems to have sufficient mass to maintain a fixed eccentricity, overcoming the tendency of ring particles to precess at different speeds. The overall picture favors a ring that arose from the disruption of a small moon – perhaps the size of Prometheus – but subsequently suffered too many external disturbances to ever settle down into a uniform, circular feature.

Outer Rings

Beyond the F ring are three much wider, but less noticeable, rings, labeled G, E and Phoebe, plus a number of faint rings that are co-orbital with small satellites. All of these are many times more tenuous and transparent than the inner rings.

The Janus-Epimetheus ring is visible between the G ring and the bright main rings and is about 5000 km wide. Another recent discovery, the Pallene ring, is about 2500 km across, and located between the G ring and the E ring (and between the moons Mimas and Enceladus). The satellites are the source of material for the rings. Impacts on Janus, Epimetheus and Pallene knock particles off their surfaces and inject them into Saturn orbit.

The G ring, first imaged by Voyager, lies more than 15,000 km from the nearest sizeable satellite and about 170,000 km from the center of Saturn. Like the D ring, it contains very little material, and is largely made of tiny, icy particles.

Its existence at this location was a long-standing mystery. However, Cassini observations show that it contains a bright, long-lived, arc of material. The arc is very tightly confined to the inner edge of the ring. It extends one-sixth of the way around

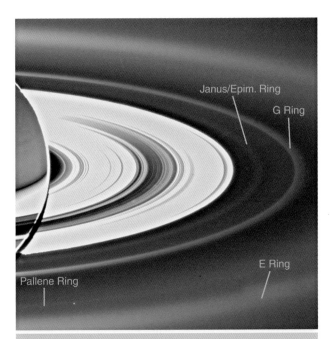

Figure 9.30 This view, acquired on the shaded side of Saturn, reveals two previously unknown faint rings. One of these is a broad, diffuse ring coincident with the orbits of Janus and Epimetheus. It lies between the F and G rings and is about 5 000 km wide. The other, coincident with the orbit of Pallene, is about 2 500 km across and located between the E and G rings. Pallene orbits Saturn between Mimas and Enceladus. The bright dot in the Pallene ring is a background star. (NASA/JPL/Space Science Institute)

Saturn and is about 250 km wide, much narrower than the full 7000 km width of the ring.

Cassini also discovered a moon, called Aegaeon, which is estimated to be 0.5 km in diameter, orbiting within the G ring arc. This seems to be the major source of the ring material. Similar to the other dusty rings, debris escaping from the moon forms the arc, then spreads to form the rest of the ring.

Encircling the entire Saturn system is the diffuse E ring, which extends about 1 million kilometers from the orbit of Mimas as far out as Rhea. Discovered by ground-based telescopes in 1967, the thick disk is made of ice and dust particles only one micron across. Five of Saturn's seven largest moons are embedded within it: Mimas, Enceladus, Tethys, Dione, and Rhea.

The E ring has a very different origin from its neighbors. Its material is largely derived from geysers on the surface of the geologically active moon Enceladus (see below). The vents eject clouds of water ice particles which are spread into a broad torus around the planet by the motions of the moons and Saturn's magnetic field.

An even larger ring was discovered by the Spitzer Space Telescope in February 2009. Made of ice and dust particles, the sparse ring lies in the plane of Saturn's orbit. Its inner edge is known to start about 7.7 million km from the planet and extends out to at least 12.5 million km, but it may be even larger. Its remarkable depth of about 2.4 million km is accounted for by the presence of the heavily cratered moon Phoebe within the ring.

Continuous impacts with Phoebe produce the particles that feed the ring. However, Phoebe has an elliptical orbit whose plane is inclined to Saturn's orbit. Bodies that are ejected into Phoebe's orbit share this inclination, so they travel above and below the

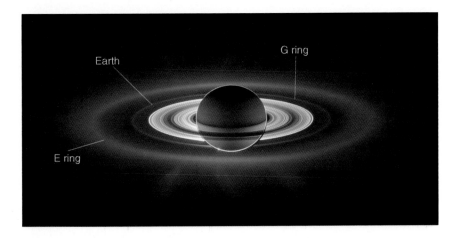

Figure 9.31 The entire inner ring system seen when Cassini was in Saturn's shadow so that microscopic ring particles appear brighter. The mosaic shows one faint ring along the shared orbit of Janus and Epimetheus, and another along Pallene's orbit. Color contrast is greatly exaggerated. Reddish, radial "flares" are artifacts caused by scattered light within the camera's optics. (NASA/JPL/Space Science Institute)

main ring plane. Over time, their orbits precess at different rates, depending upon their average distance from Saturn, so the ring particles spread out over a huge area.

Moreover, the particles are affected by re-radiation of absorbed sunlight, which exerts an asymmetric force that causes them to spiral in towards Saturn. Most material from 10 microns to centimeters in size ultimately collides with Iapetus, with smaller amounts striking Hyperion and Titan (see below).

Satellites

Saturn has 62 known satellites (not counting the numerous moonlets in the rings), more than any planet other than Jupiter. One of these, Titan, is a planet-sized object, comparable to Jupiter's four Galilean moons. The family also includes seven intermediate-sized worlds. Mostly between 400 and 1500 km across, they have very low bulk densities, evidence that they are mostly composed of water and other ices. All of these "regular" satellites move along nearly circular, prograde orbits in the planet's orbital plane.

However, by far the majority of the satellites are much smaller. They can be subdivided into two main groups: the inner moons associated with the ring system (which also follow regular orbits) and a retinue of captured objects that occupy eccentric, often retrograde, orbits far from Saturn.

The first to be discovered was the largest and brightest – Titan. Found by Christiaan Huygens in 1656, it was only the sixth known planetary satellite. Iapetus, Rhea, Dione and Tethys were subsequently found by Giovanni Domenico Cassini between 1671 and 1684.

More than a century passed before William Herschel discovered Enceladus and Mimas in 1789. Saturn's eighth satellite (Hyperion) was discovered by William Lassell and William Bond in 1848. Phoebe, found by William Pickering in 1898, was unusual, since it followed a retrograde, eccentric path with an average distance of 13 million km from the planet – far beyond the other known satellites.

More recent additions to the satellite list have come from spacecraft observations and dedicated searches carried out using ground-based telescopes. The most favored times to search were once considered to be when the rings were edge-on, but the latest discoveries have been made far from the rings and the ring plane. Discoveries of these small, outer satellites require some of the world's largest telescopes in combination with a sensitive, wide-field digital camera.

Orange Titan

Long before Voyager 1 flew past Titan in November 1980, ground-based observations had shown that the satellite was unique, with

Figure 9.32   In February 2009, the Spitzer Space Telescope discovered a huge ring between 7.7 million and 12.5 million km from Saturn. It is tilted about 27° to the main ring plane and encompasses the orbit of Phoebe. Both the ring and Phoebe orbit in the opposite direction to the other rings and most of Saturn's moons. Material blasted off Phoebe by small impacts remains in its neighborhood for millions to billions of years. Some of it spirals in toward Saturn and is swept up by Iapetus, Titan, and Hyperion. (NASA-JPL)

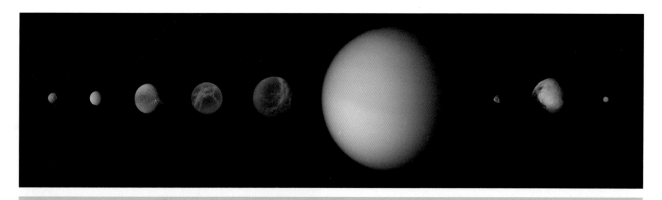

Figure 9.33 The relative sizes of Saturn's nine major satellites in order of distance from the planet (left to right: Mimas, Enceladus, Tethys, Dione, Rhea, Titan, Hyperion, Iapetus, and Phoebe). Most of them are between 400 and 1500 km across and have very low bulk densities, evidence that they are mostly composed of water and other ices. Titan is similar in size and density to Jupiter's largest satellites, Ganymede and Callisto. (Paul Schenk, Lunar and Planetary Institute)

an intriguing orange hue and spectroscopic evidence of methane and hydrogen gas. Clearly, Titan was large enough to have a significant atmosphere, and speculation was rife that, beneath its clouds, it might have a nitrogen-rich atmosphere and an icy surface coated with tarry organics or oceans of liquid methane.[8]

Voyager's cameras revealed a satellite surrounded by an impenetrable cloud blanket, relieved only by a dark polar hood and high level hazes. Information on atmospheric density and temperature was provided by studying radio signals transmitted to Earth, revealing the profile all the way to the surface.

Stellar occultations and direct measurements had indicated that Titan is the largest satellite in the Solar System, but more accurate radio occultation data obtained using Voyager revealed that it is actually a little smaller than Ganymede, with a diameter of 5150 km. The reason for the error was that Titan's atmosphere was unexpectedly thick – some 10 times deeper than Earth's.

With a bulk density almost twice that of water, the moon is made up of equal amounts of ice and rock. In that sense, it is a near twin of Ganymede and Callisto, with a rocky, silicate core that extends about two-thirds of the way to the surface. Recent models suggest that the overlying layers comprise various high-pressure forms of ice. These are topped by an ocean of liquid water mixed with ammonia (which acts as an antifreeze) and a surface crust of rock-hard "ordinary" ice a few tens of kilometers thick. Despite the possible presence of a subsurface ocean, no evidence of a magnetic field has been found.

The methane that plays such a key role in Titan's atmospheric and surface evolution is locked in the crust of water ice. This may be in the form of methane clathrate – methane trapped in the crystal structure of the ice or in pores within the crust. The gas would then be released by seasonal warming or localized volcanism.

Like many satellites, Titan is tidally locked to Saturn, and hence always presents the same face toward the planet. To view all hemispheres of Titan from Earth therefore requires observations during almost one entire orbital period, 16 days.

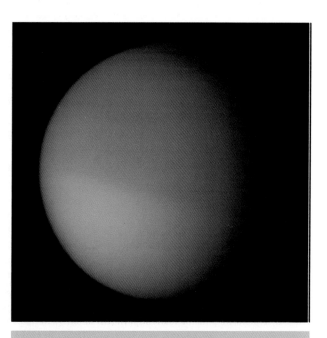

Figure 9.34 A Voyager 2 image showing Titan completely shrouded in a high level, orange smog made up of complex hydrocarbons. The main features visible are a relatively bright southern hemisphere and a dark northern polar hood. The image was taken on August 22, 1981, from a range of 4.5 million km. Spring was just beginning in its northern hemisphere. (NASA-JPL)

Since its rotation axis is tilted at an angle of 26.7° to the Sun, Titan has seasons. With a year on Titan lasting almost 30 Earth years, each hemisphere experiences winters and summers lasting more than 7 years. The southern summer ended with the

---

[8]Titan's average distance from Saturn is 20.3 Saturn radii, so it is sometimes inside, and sometimes outside, the magnetosphere.

autumnal equinox in late 2009, after which the northern hemisphere started to tilt towards the Sun.

Today, Titan's atmosphere has a surface pressure of about 1.5 bars, 50% greater than on Earth. It is likely to have been much more massive in the past. Why does Titan have such a thick atmosphere when Ganymede and Callisto have essentially none?

The low temperature of the nebula that surrounded the young Saturn allowed Titan to acquire and hold onto the moderately volatile compounds methane (CH_4) and ammonia (NH_3), in addition to water. As sunlight split the ammonia, it was converted to nitrogen and hydrogen. While most of the hydrogen escaped into space, the heavier nitrogen remained. The higher temperatures of the Jovian moons, which were closer to the Sun, prevented them from retaining such an atmosphere.

Today, about 95% of the satellite's atmosphere is nitrogen, with most of the remainder in the form of methane. In fact, Earth and Titan are the only places in our Solar System with a dense atmosphere of molecular nitrogen.

The dominance of nitrogen on both worlds has led to descriptions of Titan as an early Earth in deep freeze. Titan's present environment could be similar to that on Earth billions of years ago, before life forms began pumping oxygen into the atmosphere. However, a surface temperature of −180°C suggests that the emergence of life – at least on the surface – is most unlikely.

Another intriguing aspect of the atmosphere is the substantial amount of methane. At the temperatures present on Titan, methane can take the form of a gas, liquid or solid – just like water on Earth. This means that Titan has a methane cycle, with ethane-methane lakes that slowly evaporate, eventually resulting in cloud condensation that leads to methane rain or snow.

However, the entire methane content of the atmosphere would be destroyed in only 10 million years by the action of solar ultraviolet light. In an irreversible process, the liberated hydrogen escapes to space while organic photolysis products such as ethane drizzle down to the surface.

So how can the gas have survived in large quantities to the present day? The only explanation is that some source – perhaps an underground ocean, a large aquifer or cryovolcanism – is continually replenishing the methane supply. (An unlikely alternative is that biology may also play a role in creating Titan's methane.) Outgassing from the interior may have pumped methane into Titan's atmosphere over billions of years.

Chemical Factory

Methane and nitrogen are the raw materials for many chemical reactions that lead to the creation of complex organic molecules in Titan's atmosphere and on the surface.

Much of the chemistry takes place at high altitude. Several tenuous haze layers occur up to 700 km above the hidden surface. These and the main brownish haze layer at an altitude of 200 km were created by interactions between the nitrogen-methane atmosphere, ultraviolet light from the Sun, cosmic rays and charged particles trapped in Saturn's magnetic field. The hydrocarbons produced by this methane destruction form a

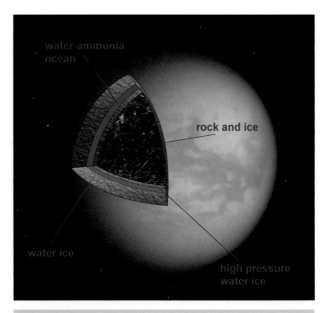

Figure 9.35 A cross section of Titan showing a possible ocean of liquid water and ammonia, sandwiched between layers of water ice. At the center is an undifferentiated "core" of rock and ice. (NASA-JPL)

photochemical smog similar to that found over large cities, only much thicker.

Many minor constituents have been detected in the smog layer, including ethane (C_2H_6), acetylene (C_2H_2) and propane (C_3H_8). Methane interacting with nitrogen atoms also forms "nitriles" such as hydrogen cyanide (HCN). Lower down is a denser layer of tiny, frozen condensate particles, possibly nitriles and hydrocarbons such as ethane.

The photochemical reactions result in large amounts of liquid and solid hydrocarbons raining onto the frigid surface. The heaviest rainstorms come from low-level methane-ethane clouds created by localized convection or gas rising over mountains. However, the solid smog particles, including acetylene and other compounds, also fall very slowly through the atmosphere, taking about a year to reach the surface. There they accumulate over millions of years, possibly forming a layer of hydrocarbon sludge hundreds of meters thick.

Methane Rain

As is the case with water on Earth, Titan has a methane cycle: the gas evaporates from the surface, then cools and condenses as it rises, releasing latent heat and causing cloud formation. Occasional rainstorms then return liquid methane to the surface, feeding rivers and lakes.

Titan's climate changes with the seasons. One of the most obvious clues is the north-south asymmetry, as shown by the development of a polar hood. At the time of the Voyager flybys, the northern hemisphere – just emerging from a long winter –

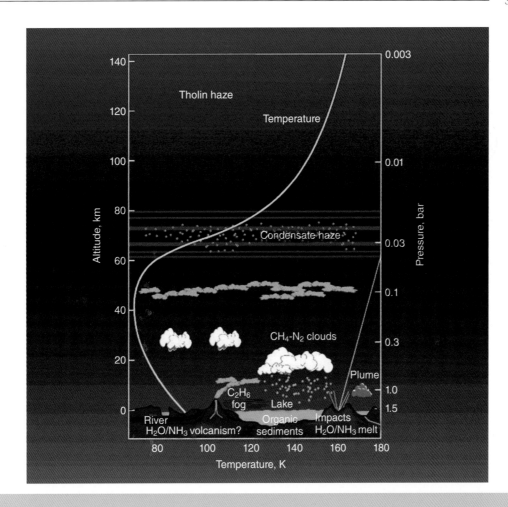

Figure 9.36 A cross section through Titan's atmosphere. High altitude haze is made of complex organic molecules known as "tholins." Lower down is a denser layer of tiny frozen condensates, including hydrocarbons such as ethane. In the troposphere are localized clouds of liquid methane which occasionally produce heavy rainstorms. The methane cycle is completed by evaporation from surface lakes and cryovolcanism. (NASA-JPL)

was shrouded in a dark hood. As the seasons change, the appearance is reversed.

Ground-based infrared observations and images from Cassini reveal clouds in three distinct latitude regions of Titan. Widespread ethane clouds at altitudes of 30–50 km blanket the north pole, due to a constant influx of ethane and aerosols from the stratosphere. There is also patchy, convectional cloud at the south pole and in a narrow belt around 40°S.

At the height of the southern summer, when the south polar region is in permanent sunlight, large, convectional clouds occur within 30° of the pole.[9] When the southern fall arrives, these clouds disappear.

The southern polar clouds have lifetimes of several weeks and result from heating of the surface, possibly combined with rising gas where the summer Hadley cell meets cooler polar air.

Short-lived (typically one terrestrial day), often elongated, clouds also appear regularly in mid-southern latitudes (40°S)

during the summer. Solar heating alone may be the cause, although other explanations have been put forward, including a seasonal shift in global winds or enhanced surface activity which disturbs the atmosphere.

Clouds are rare in Titan's equatorial regions, unlike on Earth. At Titan's equator, the atmosphere is so thick that the temperature at night is the same as during the day. This is because the thermal response of the atmosphere at such a great distance from the Sun is much longer than the Titan day (16 Earth days). In the equatorial regions of Titan, the temperature has been largely unchanged in 20 years.

Significant changes took place between July 2004 (early southern summer) and April 2010, the start of northern spring. Whereas both polar regions had been heavily overcast during the late southern summer, the clouds at the south pole completely disappeared just before the equinox in August 2009 and the clouds in the north began to thin out. The cloud patterns are

[9]Titan's southern summer solstice took place in October 2002.

Figure 9.37 This dark feature in Titan's south polar region marks a former or present lake of liquid hydrocarbons. The brightest features are methane clouds. A red cross below center marks the pole. This is a composite of three Cassini infrared images taken on June 6, 2005. Resolution is approximately 3 km per pixel. (NASA/JPL/Space Science Institute)

expected to reverse with the changing seasons as the regions of ascending air and cloud formation switch to around 40° and 70° N for Titan's northern summer.

Severe convective storms, accompanied by intense precipitation, sometimes occur. The strongest storms grow when the methane relative humidity in the middle troposphere is above 80 percent, producing updrafts with maximum velocities of 20 m/s that reach as high as 30 km before dissipating in five to eight hours. Raindrops 1–5 mm across produce torrential downpours.

Less spectacular is the bank of high, layered cloud that causes a widespread, persistent morning drizzle or mist of methane over the western foothills of Xanadu, the main 'continent' on Titan. This drizzle may be formed as moisture-laden winds travel upslope, resulting in condensation. The drizzle or mist seems to dissipate after about three Earth days, when the Sun is fairly high in the sky.

Such observations support Huygens data indicating widespread clouds of frozen methane at a height of 25–35 km, plus liquid methane clouds below 20 km, with rain or drizzle at lower levels. On Titan, methane droplets are predicted to be at least one millimeter across, 1000 times larger than water droplets in terrestrial clouds.

Observations of the northern polar regions were difficult during their long, dark winter. However, in September 2006, Cassini imaged a huge cloud formation at a height of 40 km that extended almost to latitude 60°N. This cloud, over 2400 km in diameter, is probably composed of frozen ethane, rather than methane droplets, since the particles were only 1–3 microns across. This may produce rain (or snow) as strong downdrafts drive organic particles towards the surface. This precipitation would feed the numerous methane/ethane lakes found on the surface.

Figure 9.38 The percentage of cloud cover on Titan, July 2004 to April 2010, based on data from Cassini. Black is cloud free, yellow shows complete cloud cover. During the northern winter, clouds of ethane formed in the northern troposphere. In the warmer southern hemisphere, convection caused methane to condense, forming mid- and high-latitude clouds. (NASA/JPL/University of Arizona/University of Nantes/University of Paris Diderot)

Rivers and Lakes

The Voyager flybys led to speculation that large expanses of liquid ethane and other hydrocarbons might exist on Titan. Infrared and radar observations made by the Hubble Space Telescope and ground-based instruments seemed to confirm this by revealing dark and bright patches – possibly oceans separating large, icy continents.

The Cassini orbiter and Europe's Huygens probe have discovered a surface that bears some remarkable similarities to Earth. Bright, icy uplands are drained by branching river systems which flow towards much darker plains that likely flood from time to time, although they are largely dry at present. The plains seem to be covered with carbon-rich sediment that has probably been removed from nearby uplands by rainstorms or seasonal springs, collected by streams and rivers, then transported downstream to lowland basins.

The evidence for fluvial erosion, transportation and deposition was supported by images of the Huygens landing site. The surface had the consistency of soft sand and was covered with numerous rounded stones and pebbles made from a mixture of water ice and hydrocarbons. The largest boulder measured about 15 cm across. Hollows at the base of these ice pebbles resemble those caused by erosion on river beds.

Cassini's radar instrument also found numerous extremely smooth, dark patches around Titan's north pole.[10] These were assumed to be lakes – at least some of which contained liquid – but confirmation had to wait until Cassini obtained an image of sunlight reflected from a liquid surface (see Fig. 9.41).

There are dozens of these features. Some are less than a kilometer wide, but others are many times larger. Ontario Lacus, the largest lake in the southern hemisphere, is about 235 km long, although it may be only partly wet. However, it is dwarfed by Kraken Mare, which lies near the north pole. The largest hydrocarbon lake on Titan covers nearly 400,000 sq. km – more than the Caspian Sea on Earth.

The lakes, probably filled with liquid methane or a combination of methane and ethane, are the source of the hydrocarbon smog. In some cases, rims can be seen, suggesting deposits that form as liquid evaporates. Nearby channels seem to have been carved by liquid. The surface is also covered with alluvial features, suggesting that that methane rain sometimes feeds surface flow that results in widespread deposition.

Figure 9.39 A mosaic of Titan's surface taken from an altitude of 16 km during the descent of ESA's Huygens probe on January 14, 2005. The bright ridge is marked by numerous black, branching channels that appear to have been formed by flowing liquid – probably methane loaded with organic sediment. The darkness of the channels and the nearby lowland suggests that hydrocarbon dust has drifted from above onto the surface, been washed into the dry river beds during a rainstorm and transported to the low-lying terrain – possibly an evaporated lake. (ESA/NASA/JPL/University of Arizona)

Figure 9.40 The Huygens landing site on Titan was imaged after touchdown by the probe on January 14, 2005. The surface was darker than expected, and covered with numerous pebble-sized stones made from a mixture of water ice and hydrocarbons. The largest boulder, 85 cm from the camera, was about 15 cm across. The ice pebbles rest in small hollows like those caused by erosion on river beds. The hazy image suggested the presence of methane or ethane mist. (ESA/NASA/University of Arizona)

---

[10] Dark regions in radar images generally mean smoother terrain, while bright regions mean a rougher surface.

Figure 9.41 Numerous dark patches resembling lakes are visible in radar images of Titan's north polar region. With a surface temperature of about −180°C, the lakes probably contain liquid methane or a mixture of methane and ethane. In this false color view, regions of low backscatter are blue. Radar-bright regions are tan. The image, acquired by Cassini on July 22, 2006, is centered near 80°N, 35°W. Image width is about 140 km and the smallest features are about 500 meters across. (NASA/JPL/USGS)

Figure 9.42 A Cassini synthetic aperture radar image of long, dark ridges – longitudinal dunes created by prevailing winds. They may be composed of solid organic particles or ice coated with organic material. Up to 3 km apart, the dunes curve around bright features that may represent high topography. The image is centered at 8°N, 44°W, and covers approximately 160 × 325 km. Spatial resolution is about 500 m. (NASA/JPL)

Dunes, Mountains, Volcanoes and Craters

Radar also reveals dune fields, mainly concentrated within 30° of the equator. They are thought to be composed of small hydrocarbon or water ice particles – probably about 250 microns in diameter, similar to sand grains on Earth. The "sand" may have formed when methane rainstorms produced flash floods that eroded particles from the icy bedrock. Alternatively, it may come from organic solids that are produced by photochemical reactions in the atmosphere and then fall onto the surface.

The huge longitudinal dunes are 100 m high and run parallel to each other for hundreds of kilometers. Up to 10 km apart, they form and grow through the action of fairly strong (1 m/s or more) west-to-east surface winds.

Unlike most satellites, impact features are rare on Titan. Only about half a dozen certain impact craters have been found at the time of writing, with 44 other likely candidates. This is partly due to comets and meteoroids breaking up and incinerating during descent through the thick atmosphere, but it likely also indicates relaxation of the icy surface over time, as well as weathering by methane rain, fluvial erosion and burial by photochemical sediments.

Cassini's radar has imaged one large, possibly multi-ringed crater – known as Menrva – with an outer diameter of 440 km. The feature lacks a central peak, suggesting that it has been eroded or otherwise modified since formation.

Another, known as Selk, measures roughly 90 km in diameter, and is surrounded by bright material that is probably an ejecta blanket covering the surrounding plains (see Fig. 9.43). The inner part of the crater is dark and may represent smooth deposits. There is also evidence of a central peak. The smallest yet observed is 30 km wide Ksa, which seems quite fresh, with a central peak and a broad ejecta blanket.

Cassini has also discovered several mountain ranges near Titan's equator. The highest peaks rise to a height of about 2 km, comparable to the Appalachian Mountains of the USA. The main range, which runs south east to north west, is about 150 km long and 30 km wide. Together with smaller mountainous areas to the west and east, it probably results from cooling of the interior and resultant shrinkage which causes the outer ice crust to compress and fold.

Cryovolcanism (ice volcanism) may be fairly widespread on Titan, involving emissions of a slurry of ice and liquid hydrocarbons, instead of rocky lava. However, proof is hard to find. Radar images of apparent domes, cryovolcanic flows and changes in surface brightness can be interpreted in various ways. The most likely candidate found to date is Sotra Facula, where there are three possible volcanic cones more than 1000 meters high, with deep craters and lobed flows that have a different composition from their surroundings (see Fig. 9.44).

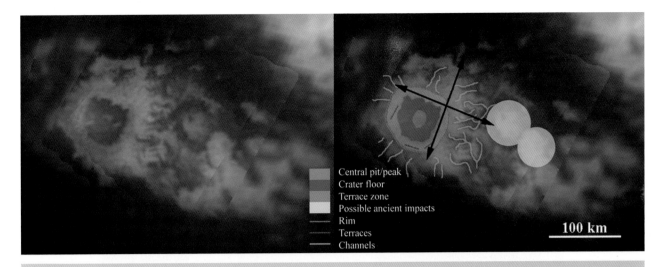

Central pit/peak
Crater floor
Terrace zone
Possible ancient impacts
Rim
Terraces
Channels

100 km

Figure 9.43 A Cassini image of 90 km diameter Selk impact crater (left) and a simple geology map of the area (right). The crater's rim is in the middle of the optically bright area and the interior walls are terraced. The crater rim is polygonal, indicating some possible planes of weakness in the crust (black arrows). The exterior is cut by drainage channels (blue lines), and the crater may be superposed on ancient, degraded craters (yellow circles). (NASA-JPL/UA)

Figure 9.44 A false color, 3-D radar image of Sotra Facula, the best known candidate for a cryovolcano on Titan. There are two peaks more than 1000 meters high and multiple craters up to 1500 meters deep, as well as lobe-shaped flows. Topography is vertically exaggerated by a factor of 10. Dunes appear brown-blue. Possible lava flows appear bright yellowish-white, like the mountain and caldera. (NASA/JPL-Caltech/USGS/University of Arizona)

The cryovolcanoes may eject "lava" composed of water and ammonia. Since liquid water is denser than ice, the process that enables it to rise to the surface is unclear, although the presence of ammonia could substantially lower its density and prevent it from freezing. Methane gas is also a likely ingredient, helping to explain the persistence of that gas in Titan's atmosphere.

Tiger Stripes on Enceladus

Dynamic processes were expected on planet-sized Titan, but there was little reason to anticipate anything other than ancient, heavily cratered surfaces on the smaller, satellites. Then, in 2005, the Cassini magnetometer discovered evidence of an extended, asymmetric atmosphere around the 504 km diameter moon, Enceladus. The sparse gas was acting as an obstacle to the planet's magnetic field and plasma.

Subsequent flybys detected dust particles around Enceladus, and a huge cloud of gas over its south pole. Water vapor comprised about 65% of the cloud, with molecular hydrogen at about 20%. The rest was mostly carbon dioxide and other simple, carbon-based molecules.

Close-up images revealed a landscape near the south pole almost entirely free of impact craters and littered with house-sized ice boulders. The most noticeable features were continuous, ridged, slightly curved and roughly parallel faults nicknamed "tiger stripes" due to their distinct appearance when viewed in false color. These unique tectonic fractures are approximately 150 km long and about 40 km apart. Compounds found in the tiger stripes include carbon dioxide and hydrocarbons such as methane, ethane and ethylene.

Cassini's infrared spectrometer also showed the south pole to be much warmer than expected. Enceladus is one of the coldest places in the Saturn system because its surface reflects more than 80% of the incoming sunlight. The temperature near the equator of Enceladus is −193°C, similar to predictions. However, although the poles should be even colder because the Sun shines so obliquely there, the average figure near the southern fractures is a remarkable −93°C. Clearly, the satellite has a strong internal heat source – unprecedented for an object of Enceladus' size.

Box 9.4   Huygens

In June 1982, less than a year after Voyager 2 flew past Saturn, a joint USA-Europe working group recommended development of a Saturn orbiter and a Titan probe. Named in honor of Christiaan Huygens, the probe was eventually selected as part of ESA's Horizon 2000 long-term science plan.

The probe was carried to Saturn by the Cassini spacecraft, which would relay data from Huygens during its descent through Titan's atmosphere. After launch, a design flaw was discovered in the Huygens communications system, so the mission plan had to be revised. It was decided to shorten Cassini's first two orbits around Saturn and insert an extra orbit. This meant that Cassini would fly over Titan's cloud tops at an altitude of about 65,000 km, more than 50 times higher than originally anticipated.

Figure 9.45   After a journey lasting almost seven years, Cassini-Huygens entered orbit around Saturn on June 30, 2004. The piggybacking Huygens probe was released on December 25, and entered Titan's atmosphere on January 14, 2005. This artist's view shows the descent sequence of the 2.7 meter diameter probe after separation from Cassini (top left). Saturn was not visible through the dense clouds. (ESA/David Ducros)

After a four billion km piggyback ride, Huygens was released toward Titan on December 25, 2004. It slammed into the upper atmosphere at a speed of 6 km/s and headed for a near-equatorial landing site on Titan's sunlit hemisphere.

The 319 kg, wok-shaped probe was equipped with six instruments and 39 sensors that would enable it to investigate Titan's haze layers and atmospheric composition, search for evidence of lightning and measure wind speed, air pressure and temperature. If it survived the landing, Huygens carried a Surface Science Package that its creators hoped would send back the first *in situ* measurements of an alien sea.

Protected by an ablative heat shield, Huygens' descent rapidly slowed, enabling it to deploy a 2.6 meter pilot chute at an altitude of about 160 km. After 2.5 seconds, the main 8.3 meter parachute was released to stabilize its swinging motion. Once the front shield was released, its instruments began to sample the atmosphere, take pictures and study the strange environment. Microphones recorded the sound of wind rushing by the probe.

The descent lasted for 2 hours 28 minutes, but transmissions from the surface continued long after Cassini dropped below the horizon and was no longer able to receive the data. More than 474 megabits of data were received, including some 350 pictures taken during the descent and on the ground. The atmosphere was probed and sampled from an altitude of 160 km all the way to the surface, revealing a uniform mix of methane with nitrogen in the stratosphere.

Methane concentration increased steadily in the troposphere, with clouds of methane at about 20 km altitude and methane or ethane mist near the surface. There were some indications of rain falling from clouds. The measurement of atmospheric argon–40 indicated that Titan may have experienced cryovolcanic activity associated with water ice and ammonia.

Although some Doppler data and images were lost, the probe's signal was monitored by a global network of radio telescopes on Earth, making it possible to analyze variations in Titan's winds by reconstructing the descent trajectory with a remarkable accuracy of 1 km.

Results showed that the winds generally blow west to east above 10 km altitude, but in the opposite direction nearer the ground. They were weak near the surface, less than 5 m/s, increasing slowly up to an altitude of about 60 km. Higher in the atmosphere, Huygens was given a bumpy ride, probably due to significant vertical wind shear. The maximum wind speed of roughly 430 km/h was measured at an altitude of about 120 km. This powerful, high latitude wind is an example of superrotation, which is also seen in Venus' atmosphere.

Panoramic images returned during the descent showed a complex network of narrow, black, steep-sided drainage channels running from brighter highlands to lower, flatter, dark regions. The branching channels merged into rivers that seemed to run into lake beds featuring offshore "islands" and "shoals." Although any surface liquid seemed to have evaporated or seeped underground, leaving the rivers and lakes dry, it was thought rain may have occurred in the recent past.

Bright hills rose about 200 meters above the dark plains. The uplands were made of water ice (which is rock-hard at Titan's temperature of about −180°C), covered by organic deposits that drift down from above. Methane rains apparently wash the hydrocarbon dust off the hillsides into rivers, eventually depositing it on the lowlands.

Huygens landed at 10°S, 192°W. A penetrometer on the probe recorded the moment of landing at a speed of 4.5 m/s. The surface resembled wet sand or clay, composed of a mixture of water ice and hydrocarbons, and overlain by a thin crust. Bursts of methane gas boiled out of the damp surface as the result of heat generated by Huygens.

Bathed in subdued orange light, the landing site resembled a dry river bed, overlain by small, rounded pebbles between 3 mm and 15 cm across. The absence of sharp edges and hollows at the bases of some rocks suggested that the area was sometimes inundated by flash floods of liquid methane (see Fig. 9.40).

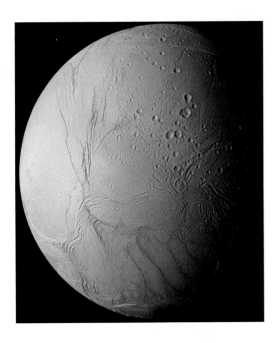

Figure 9.46 The anti-Saturn hemisphere of Enceladus imaged on July 14, 2005. The false-color view includes images taken at wavelengths from ultraviolet to infrared. Softened craters and complex, fractured terrains are visible. Many long fractures in the south (shown here in blue) are markedly different in color from the surrounding terrain. These walls of these "tiger stripes" may expose coarse-grained ice, whereas powdery material covers the flatter surfaces. (NASA/JPL/Space Science Institute)

Then, on November 27, 2005, Cassini observed the night side of Enceladus. Clearly illuminated above the limb was a series of at least eight tenuous plumes erupting from geysers close to the south pole. The jets of fine particles reached hundreds of kilometers into space. All of the plumes were found to be associated with the "tiger stripes," or sulci.

The geysers are venting water vapor and fine, salt-rich, water ice particles that have crystallized on Enceladus' surface, as well as carbon dioxide, carbon monoxide and organic compounds. One theory suggests that they are driven by liquid salt water under pressure below the surface, possibly containing methane that has been locked up inside the satellite and is now escaping through the vents.

Studies suggest that the large number of ice particles and the steady rate at which these are produced require temperatures close to the melting point of water ice. This may lead to the creation of a subsurface lake, similar to Lake Vostok beneath Antarctica. Vapor evaporating from the water expands and cools as it rises to the surface through cracks in the ice crust. The resultant ice grains are carried along by the vapor plumes as they blast upwards at speeds of 300–500 m/s. On reaching the narrow vents at the surface, the faster-moving water vapor shoots high above Enceladus, becoming trapped in Saturn's magnetosphere. However, most of the condensed ice particles fail to reach the escape velocity of 240 m/s and fall back to the surface. Only about 10% escape to form Saturn's E-ring.

With hindsight, these findings were not really so surprising. Enceladus has long been known to have an extremely reflective surface, suggesting that it is coated with almost pure water ice. Ground-based observers also noted that Enceladus appears brighter when its south pole is visible. If the moon had frosted over, it must have happened recently or radiation would have changed the chemistry of the ice and darkened it.

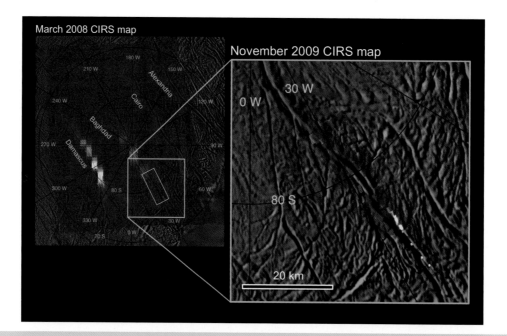

Figure 9.47   A false color, infrared view of Enceladus' south polar region (left) shows heat radiating from the 150 km long "tiger stripe" fractures in March 2008. Temperatures of at least 180 K (−93° C) were recorded along Damascus Sulcus. The higher resolution infrared image (right), obtained on November 21, 2009, shows the warm regions are less than 1 km across within the Baghdad Sulcus fracture. The intensity of heat radiation increases from violet to red, orange and yellow. (NASA/JPL/GSFC/SwRI/SSI)

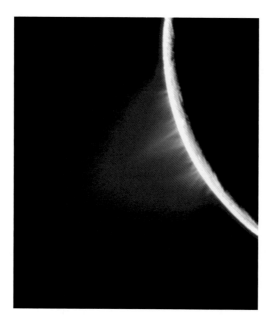

Figure 9.48   A false color Cassini image shows numerous gaseous plumes erupting into space from the south polar region of Enceladus. This view was processed to enhance the individual jets. Some artifacts due to the processing occur in the image. (NASA/JPL/Space Science Institute)

The Voyagers showed that Enceladus has one of the youngest surfaces in the Saturn system, with sizeable expanses of relatively crater-free terrain. Some areas were moderately cratered with hollows up to 35 km across, some had numerous small craters, while others were smooth, grooved plains. Many of the craters were softened in outline, suggesting some relaxation of the surface, possibly due to warming of the ice. There was also evidence of fairly recent geological activity in the form of sinuous mountain ridges 1–2 km high and linear fractures where the crust had apparently shifted. No giant craters were seen – evidence that the entire satellite had been resurfaced at least once.

How can a satellite which should have frozen solid long ago retain enough internal heat to energize such powerful geysers? And why are they only found near the south pole? Enceladus is unlikely to contain enough internal rocky material for radioactive heating or enough ammonia to lower its melting temperature significantly.

Tidal heating may contribute, even though the satellite's orbit is almost circular. Perhaps the moon is stretched between Saturn's powerful gravity and the opposing pulls of Tethys and Dione. In particular, Enceladus is currently near the 2:1 mean motion resonance with Dione, which is the main cause of its small orbital eccentricity. The high heat flow observed at present may vary over time as Enceladus' orbital relationship to Saturn and Dione changes, allowing periods of more intensive tidal heating, separated by more quiescent periods. The heating would then be due to localized tidal dissipation within either the ice shell or the underlying silicate core. This may be sufficient to cause geological

Figure 9.49 Various models have been proposed for the plumes of water vapor and ice particles on Enceladus. This model shows a subsurface lake or layer of water, possibly containing ammonia "anti-freeze." Vapor evaporating from the water expands and cools as it rises through cracks in the ice crust to the surface. The resultant ice grains are carried along by the vapor plumes as they blast upwards at speeds of 300–500 m/s. Possible heat sources are internal radioactive decay of elements and tidal heating. (NASA/JPL/Space Science Institute)

activity, resulting in shear heating caused by movement along polar faults.

In a more exotic version of this theory, the current polar location of the hotspot can be explained by reorientation of the satellite's rotation axis. A plume of relatively warm ice at any point within its icy shell could have caused Enceladus to reorient itself so that the warm region becomes the new south pole. Because the warmer ice expands and becomes less dense, the resulting uneven distribution of mass shifts the rotation axis.

Other Major Moons

All of the other medium-sized moons have surfaces that are heavily cratered and display varying degrees of tectonic activity in the past.

Mimas has a flattened or oblate shape with an equatorial diameter of 397 km, nearly 10% larger than its polar diameter. Its low density suggests that it is primarily composed of ice. It has one of the most heavily cratered surfaces in the Saturn system.

Its most distinctive feature is the 130 km diameter Herschel impact crater, which lies near the equator on the leading hemisphere. Herschel has a large central peak caused by crustal rebound during the impact (see Chapter 4). The collision which created the ancient crater may have almost destroyed Mimas, spreading ejecta over much of the surface.

Most of the surface displays numerous, ancient, impact craters of various shapes. These often overlap – evidence of the intensity of the bombardment. Many of the larger craters appear to be filled in with landslides that may have been triggered by subsequent impacts elsewhere.

Mimas also has a series of grooves and canyons, some more than 1 km deep and over 100 km in length, that are generally aligned away from Herschel. These features may be related to the giant impact that created Herschel. Alternatively, they may indicate that the moon once had a warm, active interior.

Beyond Enceladus is Tethys, something of a Mimas look-alike, though much larger, with a diameter of more than 1000 km. Once again, the leading hemisphere is dominated by a huge impact feature. Known as Odysseus, the 450 km wide, multi-ringed basin is remarkably well-preserved, even though a superimposed peppering of craters indicates its ancient origin.

Another striking feature is Ithaca Chasma, a huge, ancient rift, marked by parallel canyons running north-south, on the Saturn-facing hemisphere. It is over 1000 km long, averages 100 km in width and up to 4 km deep.

Dione is slightly larger than Tethys, and, after Titan, the densest of Saturn's moons – implying that it has a fairly large rock core under the layers of ice. However, its terrain is surprisingly varied. Although much of the surface is heavily cratered, other regions display only smaller craters, while the trailing hemisphere displays a broad area of lightly cratered, smooth terrain which may have been resurfaced by cryovolcanism.

Voyager images showed broad, bright wisps spreading out across this smoother region. These are now known to be numerous, roughly-parallel, cliffs and grooves. Cutting across older

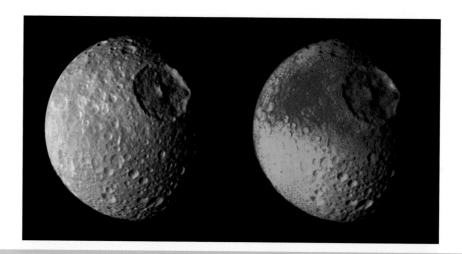

Figure 9.50 The clear-filter image of Mimas (left) has been processed to enhance the brightness and sharpness of visible features. The false color, multi-spectral image (right) shows subtle variations in the composition or granular texture of its surface. Shades of blue and violet reveal materials that are bluer in color and have a weaker infrared brightness than more typical materials, shown in green. Herschel crater is visible at upper right. The bluer color may represent material that was excavated from depth when the Herschel impact occurred. (NASA/JPL/Space Science Institute)

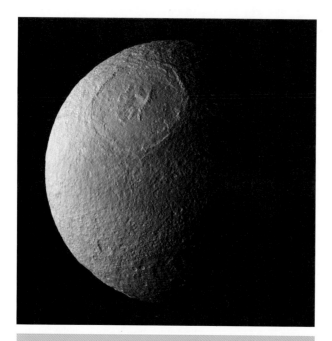

Figure 9.51 Multi-ringed Odysseus crater, 450 km in diameter, covers much of northern Tethys in this view of the leading hemisphere. (NASA/JPL/Space Science Institute)

Figure 9.52 Dione's bright cliffs diverge across the smooth plains on its trailing hemisphere. At lower right is a feature called Cassandra, which is surrounded by thin, linear rays. (NASA/JPL/Space Science Institute)

craters, many of the cliffs are sharp and fresh looking. Other fractures in the southern polar region and some craters have a softened appearance, suggesting that they are older. Dione has clearly been tectonically active for much of its existence, presumably driven by internal heating.

Cassini has sent back other indirect evidence that both Dione and Tethys may still be geologically active. Streams of plasma are apparently coming from the moons and becoming trapped in Saturn's magnetosphere. A cloud of methane and water ice has also been detected around Dione, suggesting possible cryovolcanism, similar to that on Enceladus, but on a much smaller scale.

Just as tidal heating involving Dione may be the cause of Enceladus' activity, so the reverse may also be true. The satellites may have been hot early in their lives and remained warm for billions of years through continued heating caused by Saturn's strong gravitational pull. The 1:2 orbital resonance between Dione and Enceladus may also have played a role, so that they are gravitationally squeezed. Since Dione is further from Saturn, it experiences less severe flexing than Enceladus. Radioactive decay of rocky elements may also have played a part.

Beyond Dione is Rhea, second in size and mass only to Titan among Saturn's satellites. Rhea is heavily cratered, with two large impact scars on its leading hemisphere. One of them, Tirawa, is about 360 km across and slightly elliptical in outline. Its shape and elongated central peak complex infer an oblique impact. Tirawa overlaps a larger and more degraded basin to its southwest. Both are covered in smaller craters, indicating that they are quite ancient.

Another striking feature is a bright ray crater on the leading hemisphere. About 50 km across, the relatively recent impact feature has sprayed fresh ice for hundreds of kilometers across the older terrain.

The Voyagers revealed some intriguing bright streaks that run for hundreds of kilometers across Rhea's trailing hemisphere – in a region that is generally darker and less cratered than its surroundings. Cassini showed that these resemble the bright, icy canyons seen on Dione, and they are probably due to a combination of crustal extension and shearing. Models of Rhea's interior suggest that it is almost undifferentiated, with a very large, uniform, core surrounded by a relatively thin ice shell.

Cassini's Magnetospheric Imaging Instrument found indirect evidence for a broad debris disk, possibly extending up to 4800 km from the moon's center, and several rings around Rhea. The debris

appeared to be shielding Cassini from the usual rain of electrons. Brief, sudden drops in electrons on both sides of the moon also suggested the presence of three rings or extended arcs of material within the disk. However, no visual evidence for the disk or rings has been found.

Spongy Hyperion

The satellites orbiting beyond Titan also possess unique characteristics. Hyperion measures about 270 km × 201 km × 336 km and is one of the largest irregularly shaped moons in the Solar System. Its axis of rotation tumbles unpredictably. Its slightly eccentric orbit makes it subject to varying gravitational forces from Saturn, so that its rotational period differs from one orbit to the next. The 4:3 orbital resonance between Titan and Hyperion may also contribute to the random motion.

It has one major, degraded crater and several others that are 120 km in diameter and 10 km deep. However, most of the satellite is peppered with medium-sized, well-preserved craters, typically 2–10 km across and 400 meters deep. They are unlike any craters seen on other bodies, creating a strange, sponge-like appearance (see Fig. 9.54). Hyperion has a reddish surface, coated with water and carbon dioxide ices, as well as dark material that fits the spectral profile of hydrocarbons – particularly on the crater floors.[11]

The crucial factor in explaining the weird landscape seems to be Hyperion's low density – only about 55% as dense as water. This can only be explained if the moon is a rubble pile with large empty spaces. This means that impactors striking the porous outer layers form craters more by compressing the surface than by blasting out material. Furthermore, its weak gravity allows much of the ejected material to escape completely, rather than re-impact the surface. As a result, the craters are sharper and less blanketed by debris than on other bodies.

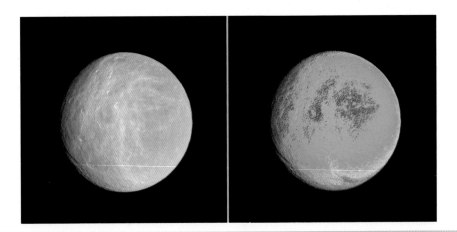

Figure 9.53 A natural color view (left) of Rhea's trailing hemisphere show bright, wispy terrain that resembles the heavily fractured landscape on Dione. The false color image (right) shows that the wisps cut across an older, cratered surface. A series of thin, north-south lineaments (also likely fractures) is visible at left. The false colors reveal subtle differences in surface composition or grain size. Rhea's diameter is 1528 km. (NASA/JPL/Space Science Institute)

[11]The dark hydrocarbon material on the crater floors seems to be similar to material on the dark side of Iapetus and the surface of Phoebe.

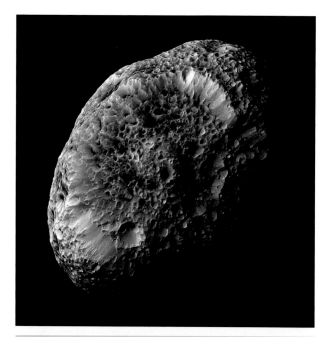

Figure 9.54 A high-resolution, false color mosaic of Hyperion. At the center is a possible circular impact feature upon which numerous smaller craters have been superimposed. Its walls display fresh ice that has been exposed quite recently. Dark organic material covers many crater floors. Numerous bright-rimmed craters indicate that the dark material may be only tens of meters thick with brighter material beneath. Downslope movement causes craters to fill with debris. (NASA/JPL/Space Science Institute)

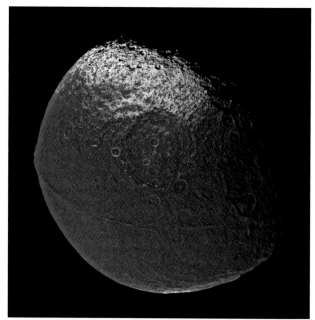

Figure 9.55 Iapetus has a dark leading hemisphere and a bright, icy, trailing hemisphere. The dark, heavily cratered area, called Cassini Regio, is centered on the equator at roughly 90° W. Within Cassini Regio, deposits with an albedo of only 4% coat nearly everything. Above about 40° N the surface gradually becomes much brighter. An ancient, 400 km wide impact basin is visible near the center. Also shown near the equator is a mountain ridge, at least 1300 km long and 13 km high. (NASA/JPL/Space Science Institute)

Two-Toned Iapetus

Iapetus, with a diameter of 1436 km, is the outermost of Saturn's large, icy satellites. Like most of Saturn's moons, it always keeps the same hemisphere toward the planet. Its two-toned nature has been recognized since 1671, when Giovanni Domenico Cassini noted that it appeared bright to one side of Saturn and much dimmer on the opposite side. The great observer deduced that the moon must have a lighter, trailing hemisphere abutting a much darker leading surface.

His deductions have since been confirmed by spacecraft, revealing one coal-black hemisphere, known as Cassini Regio, alongside a region of clean water ice more than five times as reflective. The dark material is thought to be composed of nitrogen-bearing organic compounds called cyanides, hydrated minerals and other carbonaceous minerals.

Iapetus displays at least nine large impact basins, including several on the bright hemisphere. Cassini Regio also displays a 600 km wide feature alongside a 400 km diameter basin, both overlain by smaller, more recent, craters. Around their rims are steep scarps up to 15 km high and altered in places by major landslides. Many of the scarps appear bright, probably due to exposed outcrops of relatively clean ice. The scarps and walls of nearby craters are brightest on north-facing slopes, while the opposite scarps are stained with darker material.

Even though the Sun is over 1.4 billion km away, its warmth is able to vaporize ice from sunward-facing surfaces, leaving behind dark material that is mixed with the ice. Subsequent downslope motion slowly transports this to the crater floors and other low lying terrain.

Although the surface is heavily cratered and ancient, the carbon-rich, dark coating is obviously much more recent. However, the presence of very small craters that reveal bright ice beneath indicates that the dark layer is generally thin. The material is almost certainly dust which spirals in from Phoebe and is deposited on the forward-facing hemisphere.

Dark surfaces absorb more sunlight and warm up faster than bright, reflective regions. On Iapetus, infrared observations show that the dark material is warm enough (−146°C) to enable a very slow release of water vapor from water ice. The gas then condenses on the nearest cold spot, such as the moon's poles and icy areas at lower latitudes on the trailing side of the moon. According to this model, the dark hemisphere loses its surface ice and becomes even darker, while the brightness of the trailing face is enhanced by fresh deposits of ice.

The most surprising feature is an elongated mountain ridge that stretches at least 1300 km along the satellite's equator, bisecting the entire dark hemisphere. No more than 20 km wide, the ridge rises at least 13 km above the surrounding terrain, higher than Mount Everest. The ridge is pitted with craters, indicating that it is old, but its origin remains a mystery. Is it a mountain belt that has somehow been folded upward, or does it mark a fracture through which viscous material erupted onto the surface and then piled up, rather than spreading outward?

Iapetus' rotation period and shape, and its 15° inclination to Saturn's equator, indicate strong early heating. As Saturn's gravity slowed the moon's spin rate to match its 79 day orbit, gravitational tides within the satellite would dissipate the rotational energy (see Chapter 4). In addition, Iapetus' early rotation may have been sufficiently rapid – once every 17 hours or less – and its surface warm enough to be easily deformed at the equator. As the moon cooled, the bulge "froze" in place.

Phoebe

For many years, Phoebe was thought to be outermost moon of Saturn. Unlike the other sizeable satellites, it follows a highly eccentric and retrograde path at an average distance of 13 million km from the planet. The orbit is inclined approximately 27° to Saturn's equator.

Phoebe is almost four times farther from Saturn than its nearest major neighbor, Iapetus, and substantially larger than Saturn's other outer moons. It rotates every nine hours and 16 minutes and completes one orbit of Saturn in about 18 months. Phoebe is also unusually dark, reflecting only 6% of the sunlight it receives.

Images reveal an irregular chunk of ice, about 220 km across and heavily potholed with craters, large and small. Also visible are bright streaks in the ramparts of the largest craters, bright rays emanating from smaller craters, and long, uninterrupted grooves. The plethora of 50 km craters suggests that debris from past impacts created many of the tiny outer moons that follow eccentric orbits much like Phoebe's.

The moon's low mean density indicates that it is mainly composed of water ice and carbon dioxide ice with a little rock, overlain by a layer of darker material – probably carbon-rich compounds – perhaps 300–500 meters thick. This primordial mixture suggests that Phoebe may have formed as a Kuiper Belt object in the outer Solar System before it was nudged inward and captured by Saturn (see Chapter 12). Material ejected from its surface feeds a huge ring and coats some of the inner satellites, most notably Iapetus (see above).

Small Satellites

As with Jupiter, there is a mixed collection of small, irregularly shaped moons orbiting quite close to the planet. As mentioned above, they are closely linked with the ring system.

Pan and tiny Daphnis orbit within the main rings and are responsible for clearing gaps in the ring particles. Pan keeps the 330 km Encke Gap open while Daphnis is responsible for the narrow Keeler Gap within the A ring. Atlas lies just beyond the A ring and is probably responsible for confining the ring particles and creating a well-defined edge to the ring.

It is believed that these moons began as leftover shards from larger bodies that broke apart. These acted as cores that have grown through the accretion of ring material. The result is a

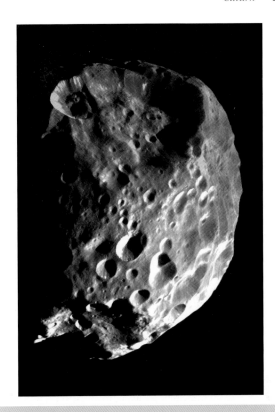

Figure 9.56 Cassini's view of Phoebe, taken from a distance of 32 500 km. 220 km across, Phoebe is an ice-rich body coated with a thin layer of dark material. Small, bright impact craters reveal fresh ice and are probably fairly young. On some crater walls, dark material has slid downwards, exposing brighter ice. The largest crater, about 100 km across, is named Jason (top). Some areas that appear particularly bright – especially lower right – are over-exposed. (NASA-JPL/Space Science Institute)

Figure 9.57 The highest resolution images of Pan and Atlas show distinctive "flying saucer" shapes, caused by equatorial ridges. From left to right: a view of Atlas' trailing hemisphere; Atlas seen from mid-southern latitudes, with the sub-Saturn hemisphere at the top and leading hemisphere to the left; Pan's trailing hemisphere seen from low southern latitudes; an equatorial view of Pan's anti-Saturn hemisphere. Pan measures 33 × 21 km. The dimensions of Atlas are 39 × 18 km. (NASA/JPL/Space Science Institute)

moon with a small, dense ice core that is covered by a thick shell of porous, icy ring material.

The equatorial ridges of Pan and Atlas (see Fig. 9.57) may have been created during a secondary stage of accretion that occurred after their main growth was completed and after the rings became extremely thin and flat. This would have allowed the rings to form small accretion disks above the satellites' equators.

The multi-stranded F ring is shepherded on either side by Prometheus and Pandora (see above). Pandora keeps the narrow ring from spreading outwards, while Prometheus shapes its inner edge. Observations show that they behave unpredictably – Pandora is far ahead of where it should be in its orbit, while Prometheus is lagging far behind. Their motions can be explained by chaotic gravitational interactions between them each time Prometheus overtakes Pandora, about once every 28 days. Since their orbits are elliptical, the distance of closest approach – and the strength of the gravitational push – varies on each occasion.

Other pairs of small satellites closely interact in the Saturn system. Epimetheus and Janus follow almost identical paths between the F and G rings. Every four years the inner moon overtakes the outer one and swaps positions – one of them moves closer to Saturn and the other moves farther away. Janus was the innermost of the pair 2006–2010, then it switched positions.

Saturn is the only planet known to have "Trojan" moons, so called because they resemble Jupiter's Trojan asteroids by orbiting near stable "Lagrange points." In Saturn's case, both Tethys and Dione are accompanied by pairs of co-orbital satellites with positions centered roughly 60° ahead and behind them. All four Trojans are very small and irregular in shape.

Telesto is located 60° ahead of Tethys, at the L4 Lagrange point, whilst Calypso orbits behind Tethys at L5. Helene and Polydeuces are co-orbital with Dione. Polydeuces is much more of a wanderer than Helene, coming as close as 39° to Dione and then drifting up to 92° away. It takes over two years to complete its journey around the L5 point.

The vast majority of Saturn's small moons orbit millions of kilometers beyond Phoebe. Most of them have been discovered in recent years during dedicated searches by Scott Sheppard, David Jewitt and Jan Kleyna, among others, using some of the world's biggest ground-based telescopes and large, sensitive CCD detectors.

The satellites generally follow eccentric, inclined, and retrograde orbits, although a minority occupy prograde orbits with inclinations of less than 50°. These remote moons are probably icy remnants of much larger objects that were destroyed during cataclysmic impacts and then gravitationally captured by Saturn.

Questions

- Both Jupiter and Saturn are described as "gas giants." Summarize the main similarities and differences between them.
- Why has it been difficult to determine Saturn's sidereal rotation period?
- Describe and explain the periodic global storms that occur on Saturn.
- How were the different rings of Saturn probably formed?
- What are the main characteristics of these rings:
  1. A and B;
  2. C and D;
  3. F;
  4. E and Phoebe?
- Titan is comparable in size to Jupiter's four Galilean moons, but otherwise very different in nature. Describe and briefly explain the main differences between these planet-sized satellites.
- Summarize the evidence for geyser/plume activity on Enceladus. What factors contribute to making this activity possible on such a small satellite?
- Briefly describe the following features of Saturn's system:
  1. magnetosphere;
  2. auroras;
  3. Trojan moons;
  4. Titan's methane cycle.
- Summarize the main features of Saturn's major moons (other than Titan and Enceladus) and explain their similarities and differences.

TEN
Uranus

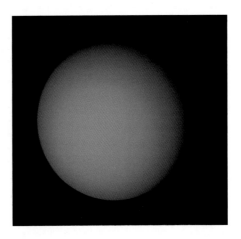

The third of the Solar System's giant planets is Uranus, a blue-green world which seems to be mainly composed of liquid water, methane and ammonia. Orbiting twice as far from the Sun as Saturn, Uranus is a frigid world where the feeble solar radiation is the major driver of occasional storms in the deep atmosphere. Many mysteries remain, not least the reason for the planet's 98° inclination, which causes the most unusual seasons in the Solar System.

Discovery

Until the late 18th century, studies of the Solar System were restricted to the Sun, Moon, six planets (including Earth) and a number of planetary satellites. No one was prepared for the major upheaval that occurred on March 13, 1781 – the first recorded discovery of a planet.

The revolution began with a routine sky survey by a German-English musician who also happened to be an enthusiastic amateur astronomer and telescope-maker. Working from his house in Bath, with the willing assistance of his sister, Caroline, William Herschel (1738–1822), was using a home-made 6.2 inch (15.7 cm) reflecting telescope when he noted in his journal:

> *"In the quartile near Zeta Tauri the lower of the two is a curious either nebulous star or perhaps a comet."*

It soon became clear that the object had no tail and was not following a comet-like orbit. The mystery was eventually solved when calculations showed that it was following an almost circular path around the Sun.

The seventh planet turned out to be a giant, though smaller than either Jupiter or Saturn. Located more than twice the distance of Saturn from the Sun, it was obviously a dark, cold world. At a stroke, Herschel had doubled the size of the Solar System.

In gratitude for an appointment as King George III's personal astronomer (and a royal pension of 200 pounds a year) Herschel suggested that the new planet be named Georgium Sidus (George's Star). Not surprisingly, this title was completely unacceptable to other nations, and many years passed before the name Uranus was officially recognized.[1]

The Seventh Planet

Herschel was not the first person to see Uranus. John Flamsteed had noted it on six separate occasions between 1690 and 1715 without recognizing its true nature. In fact, the planet is just visible with the unaided eye if you know exactly where to look. Nevertheless, with a disk measuring only four arc seconds in diameter, it was difficult to find out more than the basic information about the planet's size and orbit.

With a semimajor axis of 2871 million km and an average orbital velocity of 6.8 km/s, Uranus takes 84 years to complete one revolution. Since its discovery, the planet has only completed two full circuits of the Sun. The orbit is only slightly elliptical, ranging between 3005 million km and 2734 million km, so its apparent size and brightness vary little. Uranus last reached aphelion on February 27, 2009.

For many years, there was uncertainty about its size, density and rotation period. Today Uranus is known to be the third largest planet in the Solar System, with an equatorial diameter of 51,118 km – four times greater than Earth. 64 Earths would fit

---

[1] In Roman mythology, Uranus was the father of Saturn and grandfather of Jupiter.

Exploring the Solar System, First Edition. Peter Bond.
© 2012 by Peter Bond. Published 2012 by Blackwell Publishing Ltd.

Table 10.1
Uranus Summary

Mass	86.8×10^{24} kg (14.5 Earth masses)
Equatorial Diameter	51 118 km
Polar Diameter	49 946 km
Axial Inclination	97.77 degrees
Semimajor Axis	2 872.46 million km (19.2 AU)
Rotation Period (internal):	17.24 hours
Sidereal Orbit Period	84.011 years
Mean Density	1.27 g/cm³
Number of known satellites	27

inside Uranus. Like the other giant planets, its rapid rotation causes the disk to appear noticeably flattened at the poles – evidence of a deep atmosphere and the absence of a solid surface. But where are the poles?

A key clue to the axial inclination of Uranus came in 1787, when Herschel discovered two satellites (Titania and Oberon) whose orbits were tilted almost 90° to the plane of the ecliptic. If the moons were orbiting above the planet's equator and in the same direction as the planet's rotation – like most of the larger moons – then Uranus must have an extreme axial inclination. Further confirmation came in the 19th century when spectral observations of the planet's upper atmosphere showed that it appears to be rotating in a retrograde direction – the opposite of most of the other planets.

We now know that Uranus spins like a top that has been knocked over onto its side, with the rotation axis lying in almost the same plane as the orbit. The actual axial inclination is 97.77°,

so what would normally be regarded as the north pole actually lies below the ecliptic plane.[2]

This unusual arrangement means that the poles endure extreme seasonal variations. For about 21 years, the south pole is in permanent sunlight, while the north pole endures continuous darkness – a situation that reverses when Uranus has moved half way around its orbit. In between, the equatorial regions have more sunlight and conditions are less extreme.

The south pole was last aligned toward the Sun in 1985, since when the planetary aspect has been changing. In 2007, the rings and equator were aligned with the Sun and Earth (as they were in 1964 and 1922), and by 2028 the north pole will be facing the Sun.

Even with this knowledge of its obliquity, the bland appearance of the bluish disk made it extremely difficult to determine the rotation period. Before the Voyager flyby of 1986, the best estimates, based on observations of occasional, isolated clouds, indicated a period of about 16 hours. Today, the accepted rotation period – based on radio signals generated by particles trapped in the magnetic field – is 17 hours 14 minutes, the slowest of the giant planets.

Why is Uranus spinning on its side? The most likely explanation is that at least one large planetary embryo was involved in an off-centre collision with the young Uranus. The positioning of the planet's rings and inner moons above the equator suggests that this took place more than four billion years ago, perhaps when Uranus was still migrating out to its present orbit (see Chapter 1).

Atmosphere

Since Uranus lies more than 19 AU from the Sun, it receives 360 times less light and heat from the Sun than Earth. As a result, its atmosphere is extremely cold, with a temperature of about −214°C at the 1 bar pressure level (equivalent to the air pressure at sea level on Earth).

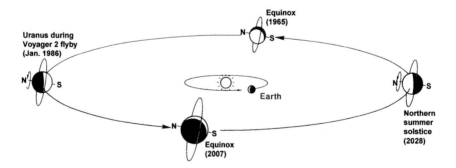

Figure 10.1 The axis of Uranus is tilted almost parallel to the plane of the planet's orbit. As a result, one pole is in permanent sunlight for 21 years, while the opposite pole is in permanent darkness. In between, when the equator receives more sunlight, conditions are less extreme. At such times, the rings and the orbits of the satellites are seen almost edge-on. (NASA / Peter Bond)

[2]There is some confusion over the naming of Uranus' poles. The International Astronomical Union has declared that all poles above the ecliptic should be regarded as north poles. Under this system, the south pole of Uranus was in sunlight during the Voyager 2 flyby in 1986. However, the Voyager scientists preferred to refer to it as the north pole.

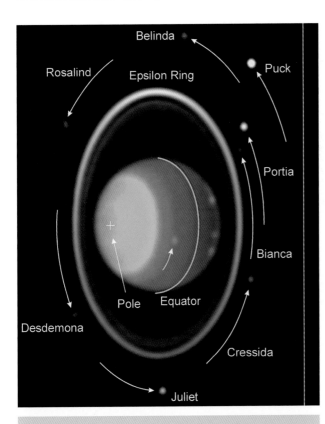

Figure 10.2 From the late 1960s Uranus' south pole was aligned with the Sun and Earth. However, when this Hubble Space Telescope image was taken in November 1997, the planet had moved far enough around its orbit to make part of its northern hemisphere visible for the first time since the early 20th century. (Erich Karkoschka/ NASA)

Voyager 2 found that temperatures were almost identical everywhere on Uranus, evidence that the weak heat received from the Sun is rapidly spread by winds around the planet. However, Voyager's instruments did detect a slightly colder band between 15 and 40°S, where temperatures were 2–3°C lower. The minimum near the tropopause (the troposphere – stratosphere boundary) was −221°C at the 0.1 bar pressure level. The temperature rose with increasing altitude, reaching −123°C in the rarified upper atmosphere. Below the tropopause, temperatures also increased steadily. When extrapolated, these show that the temperature soars to thousands of degrees in the interior.

The pale blue color of Uranus indicates that its atmosphere differs from those of Jupiter and Saturn. For many years, the reason for this coloration was unclear. Then, in 1932, Rupert Wildt announced that dark lines in the spectra of all four giant planets could be attributed to absorption of sunlight by methane gas (CH_4). Subsequent measurements show that Uranus has 10

times more methane in its upper atmosphere than Jupiter or Saturn. The gas acts as a filter, scattering blue wavelengths but absorbing red light. As a result of this filtering effect and the presence of a widespread photochemical haze at high level, the underlying cloud features appear washed out and fairly featureless.

Spectroscopic data also confirmed that molecular hydrogen (H_2) is the most abundant gas on Uranus, implying that helium must also be a major atmospheric constituent.[3] Confirmation came from the Voyager 2 spacecraft, which measured 83% hydrogen, 15% helium and 2% methane.

Other minor constituents include hydrogen sulfide (H_2S), deuterated hydrogen (HD), ethane (C_2H_6) and acetylene (C_2H_2). The last two are created by photochemistry – the action of sunlight on methane gas. Despite their scarcity, these minor ingredients do have a subtle influence on their surroundings: they condense to form high level hazes which absorb and reflect sunlight, raising the temperature of the upper atmosphere while lowering the temperature of the deeper atmosphere.

Telescopes often showed no features at all, although some ground-based observers recorded two faint belts on either side of a bright equatorial zone and dark hoods over the poles. This absence of cloud features persisted throughout the Voyager 2 flyby of 1986, when even the most extreme computer processing of images revealed only 8 discrete clouds and a high layer of photochemical haze around the sunlit pole.

Only since 1994 have individual cloud structures become visible from the Hubble Space Telescope and modern ground-based telescopes equipped with adaptive optics. As the seasons changed and the northern hemisphere emerged from its prolonged winter darkness, the banded structure near the south pole slowly evolved, while bands of bright, high-altitude clouds were seen for the first time in the northern hemisphere.

Images taken at infrared wavelengths (not possible with Voyager) show a global blanket of methane ice crystals with bands and spots caused by convectional storms – comparable to continent-sized thunderheads on Earth – developing at the top of huge columns of warmer, rising gas. (Since there is very little temperature contrast, only small inputs of energy are required to drive the weather on Uranus.)

Some cloud features evolve rapidly, others persist over a number of years. They vary greatly in size. Some are dim and diffuse, others bright and sharp. One large, bright feature near 34°S was first recorded in 1986 and was still visible in 2004. During that period its latitude oscillated between 32°S and 36.5°S over a period of about 1000 days. It also varied its longitudinal drift rate between −20° per day and −31° per day as it drifted in the strong zonal wind, behavior similar to that of the DS2 feature observed on Neptune by Voyager 2. This sizeable storm was thought to be the result of vigorous convection created as the southern hemisphere shifted to day-night illumination instead of permanent sunlight.

Like the other giant planets, the winds on Uranus blow parallel to its equator, resulting in faint cloud belts that are pale imitations

---

[3] Helium is impossible to detect spectroscopically since it does not absorb light at any detectable wavelengths. Estimates of its abundance can be obtained by studying the effects of collisions between hydrogen and helium molecules on the hydrogen spectrum, and by using radio occultation data to determine the molecular weight of the gases in the atmosphere.

Figure 10.3   These two near-infrared images, taken with the Keck II telescope on July 11–12, 2004, show both hemispheres of Uranus, with its north pole at 4 o'clock. The rings appear red because of processing to reveal methane absorption in the atmosphere. 31 cloud features are visible (13 in the left image and 18 at right), more than the total observed up to the year 2000. The bright white feature (right) reached a high altitude and was very variable. The isolated bright green cloud (lower left limb in the right hand image) is a long-lived feature that oscillates in latitude. Note the line of bright, convective clouds in the northern hemisphere, which was coming out of a long winter. The 28 800 km long band of clouds dissipated completely within one month. (Lawrence Sromovsky, UW-Madison Space Science and Engineering Center)

of those on Jupiter and Saturn. The zonal pattern is apparently controlled by the planet's rapid spin rather than by direct heating from the Sun.

Voyager 2 enabled scientists to track the rare clouds and determine the wind speeds in the southern hemisphere. The images showed that mid-latitude winds blow in the direction of the planet's rotation at 40–160 m/s (145–575 km/h). Radio science experiments found winds of about 100 m/s blowing in the opposite direction at the equator. The atmospheric circulation in the dark northern hemisphere remained a mystery.

With the shifting of the seasons, it is now possible to observe the cloud motions to the north of the equator. Observations made in October 2003 showed wind speeds of 107–111 m/s, reaching a peak of 218 m/s (785 km/h), the highest wind velocity seen yet on Uranus. At northern mid-latitudes (20° to 40°), the winds appeared to have accelerated when compared to earlier HST and Keck observations. However, the comparable wind speeds at southern mid-latitudes had not changed since the Voyager measurements in 1986.

The equatorial region exhibits a subtle wave structure, indicated by diffuse patches roughly every 30° in longitude. The largest discrete cloud features show complex structure and nearby streaks extending over tens of degrees, reminiscent of activity seen around Neptune's Great Dark Spot in 1989. There has been no sign of a northern "polar collar" resembling that seen in the south,

though it has been suggested that a number of discrete features at high latitudes might mark the early development of such a collar.

Current models suggest that the atmosphere becomes warmer and denser with depth. Far below the visible clouds of methane ice, models suggest that there may be thicker layers of ice cloud, similar to those found at depth on the other giant planets. Methane freezes at the lowest temperature, so it forms the highest layer. The uppermost of these hidden layers would be hydrogen sulfide (H_2S), overlying ammonium hydrosulfide (NH_4SH) and water ice (see Chapters 8 and 9).

Interior

Even the most basic physical data indicate that Uranus must be very different in make-up from the larger gas giants. With a volume less than 5% that of Jupiter and 8% that of Saturn, Uranus is clearly dwarfed by its neighbors. At the same time, its overall density is not much higher than water – twice that of Saturn, but very similar to that of Jupiter. Uranus' relatively modest size and bulk density indicate that pressure levels in the planet's interior are low compared with its larger cousins.

Low-density elements must account for a substantial fraction of the planet's mass. However, to be consistent with the bulk

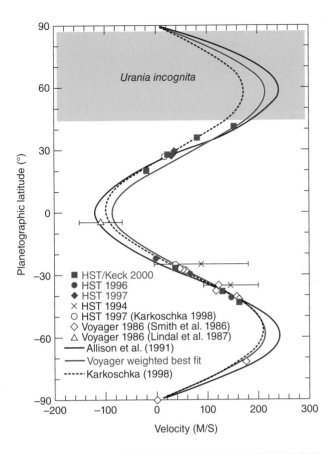

Figure 10.4 This analysis of recent data from various sources shows a symmetrical pattern with two eastward blowing jets at about 60° N and S, and a westward jet at the equator. Most storm activity occurs around 30°N and S. (Heidi Hammel)

density, Uranus cannot be composed largely of hydrogen and helium. Instead, roughly 80–90% of the total mass of Uranus must be ice and rock (compared with 10% for Jupiter). The high temperatures and pressures in the planet's interior mean that the "ices" are actually in the form of a compressed liquid. In other words, Uranus is a fluid planet, whose gaseous atmosphere gradually merges with the liquid interior.

The formation of Uranus (and Neptune) is not well understood. One reason for its different composition is probably the planet's slow growth in the outer regions of the solar nebula. The relatively slow orbital motion of the proto-Uranus would have limited its rate of accretion from the dusty disk. By the time a significant rocky core had grown, much of the gas in its neighborhood may have been cleared by a powerful outflow of radiation from the young Sun. The smaller core and limited supply of gas meant that the planet captured a less massive atmosphere.

Various models for the internal structure of Uranus have been proposed to account for these properties, and even today a number of questions remain. In 1934, Rupert Wildt suggested that it has a rocky core overlain by a thick layer of ice and a deep atmosphere of hydrogen, methane and ammonia. An alternative model put forward in 1951 by W. Ramsey preferred a world largely composed of ammonia, methane and water.

By the early 1980s, this was refined to a three-layered planet with a fairly small rocky core made of a mixture of silicate, magnesium and iron-nickel, overlain by a liquid mantle of water, methane and ammonia, and topped by a thick hydrogen-helium atmosphere.

Current models suggest that the hydrogen-helium atmosphere extends about 7500 km below the visible cloud tops. There is then a layered mantle of rock and "ices" – water, ammonia and methane in solar proportions – with traces of other substances. In this ice layer, pressure and temperature increase from 20 GPa and 1700°C at the upper boundary to 600 GPa and 7700°C at the mantle-core boundary.

Figure 10.5 The overall density of Uranus is only a little higher than that of water, less than all the planets except Saturn. Current models of the interior favor a hot, rocky core, surrounded by a liquid mantle of water, methane and ammonia and a thick atmosphere of hydrogen, helium and methane. (LPI)

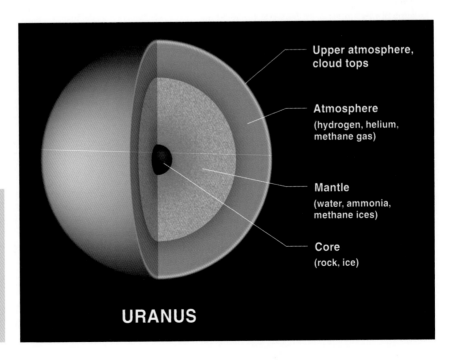

Shock-wave experiments indicate that, under such extreme conditions, methane dissociates – possibly into molecular hydrogen and carbon. Some scientists have suggested that the high pressures and temperatures result in the carbon taking the form of tiny, solid diamonds. At the very center of the planet may be an Earth-sized rocky core with a radius of around 6000 km and a temperature of about 8000°C.

Missing Heat

All of the planets should have started as hot objects, when gravitational energy and kinetic energy from impacts were transformed into heat during planetary accretion. Over the age of the Solar System, the smaller planets have lost most of this heat, but the massive giant planets store heat well and radiate poorly from their cold surfaces. Consequently, they should have retained much of the heat produced during their formation and this should still be escaping today.

The total heat output of a planet is determined from its infrared emission, while the heat input is determined from the fraction of incident sunlight that is absorbed, i.e. not scattered back into space. The fraction that is scattered is called the albedo, which for Uranus is about 0.3 (30%).

By studying the far-infrared emissions and albedo of Uranus, its internal heat output can be estimated. The surprising conclusion is that Uranus is the only giant planet that does not radiate a substantial amount of heat from its interior. Voyager data confirmed that the internal heat source – if it exists – produces no more than 6% of the energy it receives from the Sun, though the degree of uncertainty meant that there could be no excess at all. (The equivalent figures for the other giant planets exceed 70%.)

Since Uranus and Neptune appear to be virtual twins, the reasons for this energy balance are not readily apparent, though the planet's relatively bland appearance seems to be one obvious consequence. The lack of vertical motion and storm activity is the result of limited interaction between the escaping heat and the atmosphere.

One possibility is that some form of layering within the mantle is preventing large-scale convection. If layers of differing composition, density and temperature exist, they may confine the regions where convective motions occur. Instead of being allowed to rise rapidly to the surface by convection, Uranus' heat flow may occur thru a much slower diffusive process in regions in the upper mantle which have a high molecular weight gradient. Such an interpretation is supported by the presence of a highly unusual magnetic field (see below).

An alternative hypothesis is that the high obliquity of the rotation axis may have been responsible for a higher rate of heat loss, resulting in the current temperature being much lower than expected. The possible massive impact with a planetary embryo soon after its formation may also have caused it to lose most of its primordial heat, leaving it with a low core temperature.

Magnetic Surprise

Although its fluid interior and rapid rotation indicated that Uranus possesses a magnetic field, it was not possible to directly measure it from the Earth. The best indirect evidence came in the early 1980s when the International Ultraviolet Explorer spacecraft detected "aurora-like" emissions from hydrogen in the planet's upper atmosphere.

Everything changed in 1986, when Voyager 2 flew past the planet. Radio emissions detected several days before closest approach provided the first conclusive evidence that Uranus possesses a magnetosphere. The magnetic field turned out to be 50 times stronger than Earth's (though at Uranus' visible surface the field is generally weaker than on our planet).[4]

The alignment of the magnetic field was a major surprise. The field is not only tilted at an angle of 58.6°to the rotational axis (compared with 11° for Earth), but it is also offset from the planet's center by about 8000 km, or one third of the distance to the cloud tops. This means that the magnetic poles are nowhere near the geographical poles. Instead they lie at 15.2°S and 44.2°N on the visible surface. This unusual offset is thought to be caused by the field being generated by an electrically conductive fluid layer far above the core.

The source of the magnetic field remains uncertain, although it has been suggested that it may arise from nearly complete ionization of water, the most abundant component of the icy mantle.

One important clue is the "roughness" of the magnetic field. Instead of being predominantly dipolar (like a bar magnet), the field also has a quadrupole component (as though produced by a

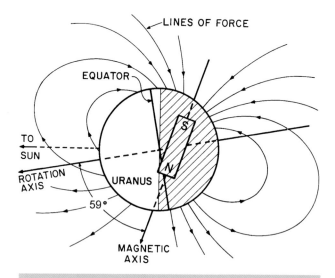

Figure 10.6 The magnetic field of Uranus, shown as a dipole tilted about 59° from the rotation axis and offset 8000 km toward the dark north polar region. (NASA-JPL)

---

[4]**The intensity of the magnetic field at Uranus' visible surface varies considerably from place to place because of its large offset from the planet's center. Earth's field strength is about 0.3 gauss everywhere on the surface, whereas on Uranus it varies from 0.1 to 1.1 gauss.**

combination of two bar magnets, with two north and two south poles). However, small-scale components of a field die out rapidly above the electrically conducting region. When combined with the unusual tilt and offset location from the center, this evidence suggests that the field is generated at shallow depths within the planet.

Since water and ammonia dissociate into positive and negative ions at relatively low pressures and temperatures, the field could easily be generated close to the surface where these moderate conditions exist. According to current models, the dynamo that creates the magnetic field occurs in a thin shell of convecting, ionized fluid that lies above a non-convecting, ionized "ocean."

The planet's highly inclined rotation produces an unusual corkscrew motion in the field lines as the cylindrical magnetotail, shaped by the movement of the solar wind past the planet, rotates in space. During the late 20th century, the magnetic dipole axis appeared to rotate in the anti-clockwise direction as seen from the Sun, with the solar wind approaching Uranus more or less from above the south pole. With the changing seasons, this alignment has now changed, so that the solar particles arrive at latitudes close to the equator. This latter configuration causes the magnetic poles to gyrate toward and away from the Sun every 17 hours (Uranus' rotation period).

The magnetosphere has many of the characteristics of those around other planets. The field repels the stream of charged particles in the solar wind, creating a huge magnetospheric cavity, filled with trapped charged particles, that surrounds the planet and extends downstream. The magnetotail extends at least 10 million km (400 radii) beyond Uranus's night side.

On the upstream (sunward) side, the boundary between the magnetosphere and the solar wind extends to approximately 25 planetary radii. A bow shock at about 33 planetary radii deflects the supersonic flow of the solar wind in front of the magnetosphere.

The magnetosphere of Uranus is fairly empty, since it encompasses only a few, relatively small, icy satellites and solar wind-driven convection moves particles through the region in a matter of days. There is no evidence of helium, which might originate from the solar wind, and no evidence of heavier ions (e.g. oxygen) that might be sputtered from the surfaces of the moons.

The rings of Uranus extend to 1.95 planetary radii and sweep away much of the material there. This means there is only a small plasmasphere containing trapped particles, and a lot less plasma can be forced from the magnetosphere down into the atmosphere to create auroras.

Some auroras seem to be produced by current systems which connect the moon Miranda to the magnetic poles of Uranus. However, from what little we know about the magnetosphere, many of the auroras are apparently created from the diffuse precipitation of particles into the atmosphere, creating much less dynamic curtains of light than those found at the poles of Earth or Jupiter.

Voyager 2 found radiation belts of similar intensity to those at Saturn, although they differ in composition. The radiation belts at Uranus appear to be dominated by hydrogen ions, probably derived from the planet's upper atmosphere and ionosphere. The escaping hydrogen atoms glow in ultraviolet light on the sunlit side, close to the magnetic poles.

The radiation belts are so intense that irradiation would darken any methane trapped in the icy surfaces of the inner moons and ring particles within 100,000 years. This process may have contributed to the darkened surfaces of the moons and ring particles.

Major Satellites

Prior to the Voyager 2 flyby, only five medium-sized moons were known to orbit Uranus. The two largest, Oberon and Titania, were discovered by William Herschel six years after he first observed Uranus. Ariel and Umbriel were found by William Lassell in 1851, while the smallest, Miranda, was discovered in photographs taken by Gerard Kuiper in 1948. All of these were located well outside the rings, following normal orbits of low eccentricity and inclination (except Miranda, whose orbit is inclined about 4° to the planet's equator).

The satellites were generally much darker than the moons of Saturn and all but Miranda appeared to be partially covered by spectrally neutral material similar to charcoal or carbonaceous chondrite meteorites. (The spectral reflectance of this dark material also closely resembled that of the rings). Their bulk densities were higher than those of Saturn's icy satellites, but lower than the Galilean giants of Jupiter – suggesting a fairly even mixture of water ice and silicate. Since the moons were considered to be heavily cratered, ancient relics left over from the formation of the planet, no evidence of geologic activity was anticipated.

Like most satellites in the Solar System, the moons of Uranus are in synchronous rotation, always keeping the same face toward the planet. Normally, this would enable partial imaging of both the northern and southern hemispheres. However, in the case of Uranus' satellites the Voyager 2 spacecraft passed above their fully illuminated south poles, so imaging was limited to their southern hemispheres.

The other giant planets all possess at least one satellite comparable in size to Earth's Moon, but the main moons of Uranus are all quite modest in size. Their mode of formation is unclear. Unlike Jupiter and Saturn, the young Uranus almost certainly was unable to draw in a large cloud of dust and gas. Hence, the satellites could not have formed during the infall of material from the solar nebula. However, the accretion process would have caused the planet to be hotter and larger than it is today. As it cooled and shrank, it may have left behind a "spinout disk" of material that eventually condensed and stuck together. In the classical accretion process, the debris could have grown into the satellites we see today.[5]

The larger satellites, Oberon and Titania, are also the most distant from Uranus. They are remarkably similar in size, density, color and albedo, although their surfaces are quite different. Oberon has numerous large craters between 50 and 100 km in diameter. Spatial resolution in Voyager images was only about 12 km, but it was possible to identify a large mountain, probably

---

[5] Another possibility is that they were formed from a cloud of material that was ejected by a colossal impact that knocked Uranus on its side.

Box 10.1   Voyager 2

Most of what we know about Uranus comes from the brief Voyager 2 flyby in January 1986. Although the Voyagers were not optimized to operate beyond Saturn, the mission team hoped that it would be funded to include visits to Uranus and Neptune – completing the "Grand Tour" of the outer Solar System. A rare alignment of the outer planets made it possible to redirect the spacecraft by a gravitational "slingshot" from each planet.

Since Uranus orbits about twice as far from the Sun as Saturn, the spacecraft's cameras had to cope with light levels that were four times lower. When combined with the high speed of the flyby, this increased the likelihood of image smear, particularly if the spacecraft vibrated while the shutters were open. Imaging was further complicated by the darkness of Uranus' rings and moons.

Voyager 2 was already suffering from a number of malfunctions, most notably the loss of its primary radio receiver and limited communication via the back-up receiver. In addition, output from its nuclear power source had dropped to about 400 watts – not enough to operate all spacecraft subsystems simultaneously.

In order to obtain the maximum data return, the spacecraft's stability was improved by modifications to the attitude control system. To avoid image smear, the entire spacecraft was programmed to turn at the correct rate, like a camera panning on a fast-moving vehicle – even though this caused the main antenna to point away from Earth.

Greater distance also meant reduced rates of data return. In order to minimize the loss of information, the spacecraft was programmed to compress the image data, transmitting only the change in brightness between neighboring picture elements (pixels) rather than the absolute brightness values. The ground facilities were also upgraded by electronically combining the signals received by several antennas.

The encounter began on November 4, 1985 – 81 days before closest approach, with a long range survey of the planet, its rings and satellites. The bland, blue sphere of Uranus showed only a dark polar hood and the occasional high altitude cloud. By studying the rare methane clouds, it was found that mid-latitude winds were circling the planet in the same direction as its rotation at speeds of 40–160 m/s.

Figure 10.7   Two views of Uranus returned by Voyager 2's narrow angle camera on January 17, 1986. The spacecraft was 9.1 million km from the planet, several days from closest approach. The picture at left – a composite of images taken thru blue, green, and orange filters – shows the southern hemisphere in true color. False color and extreme contrast enhancement reveal a polar hood of high altitude methane haze (right). (NASA-JPL)

Uranus was found to be extremely cold, with an average temperature of –210°C. Temperatures were fairly uniform at similar levels in the atmosphere, with the exception of a slightly colder band between 15–40°S. Even the sunlit south pole was similar in temperature to the north pole, which had not seen the Sun for two decades.

As expected, the main constituents of the atmosphere were found to be hydrogen and helium, although the 15% helium abundance was much lower than some Earth-based studies had suggested. However, this figure was comparable to the amount of helium thought to be present in the original solar nebula.

The mystery of the planet's internal rotation was solved by studying fluctuations in the magnetic field and associated radio signals. A day on Uranus was found to last for 17 hours 14 minutes, 1½ hours longer than anticipated. There was no significant internal heat source.

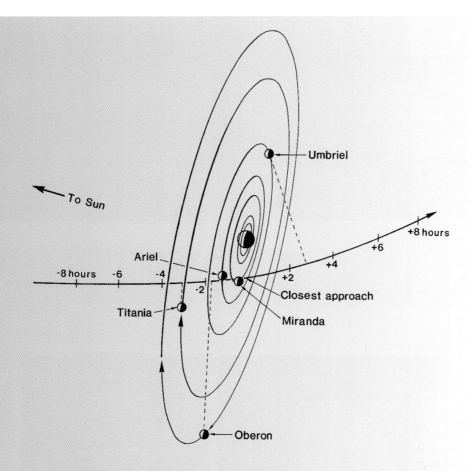

Figure 10.8 Voyager 2 flew to within 81,600 km of Uranus' cloud tops on January 24, 1986. Since the planet, its rings and the satellite orbits were arranged like concentric rings on a target, the best photo opportunities were limited to a period of less than six hours around the time of closest approach. The trajectory was also influenced by the need for a gravitational assist toward Neptune. This meant that images of the larger moons were taken from fairly long range, while Miranda received the closest scrutiny. (NASA-JPL)

Spacecraft measurements showed that the magnetic tail extended at least 10 million km behind the planet. The strength of the field at Uranus' surface turned out to be roughly comparable to that on Earth. What was not anticipated was the unique magnetic field, which was tilted 59° to the spin axis and offset from the center of the planet by about 8000 km.

Since the magnetic poles were not expected to be so close to the equator, most scans by Voyager's instruments were aimed toward the wrong places. However, the ultraviolet experiment did find some evidence of an aurora on the night side, along with an "electroglow" caused by energized electrons above the dayside.

Voyager 2 also obtained brief glimpses of the individual rings and moons. One priority was a search for "shepherd" satellites that might be confining the narrow rings (only two were found). Altogether, 10 new satellites were discovered, nine of them orbiting between the epsilon ring and the orbit of Miranda. Of the previously known moons, Miranda was the star of the show, with a bizarre surface marked by three regions of parallel ridges, one of which resembled a chevron.

the central peak of a large impact crater, rising 11 km above the limb of Oberon. Also visible were several large craters with dark floors and bright rays of ejecta. The largest of these (Hamlet) had two dark spots on either side of a bright central peak. Perhaps isolated, icy volcanism, had released a dark fluid in the distant past. Further evidence of past surface activity included linear and curved features that appear to be enormous faults.

Titania revealed a number of large impact basins (100–200 km in diameter) and far fewer craters than Oberon in the 50–100 km size range, with an abundance of 20 km diameter craters. The size-frequency distributions of these smaller craters suggested that they were mainly secondary features excavated by debris from larger impacts.

Several patches of smoother terrain indicated a prolonged early period of resurfacing. There was also an extensive network of branching, intersecting faults, with a number of inward-facing scarps on either side of grabens. Calculations based on length of shadows near the terminator showed that the scarps rise to

Figure 10.9   A "family portrait" of Uranus' five largest moons. From right to left, in order of decreasing distance from the planet are Oberon, Titania, Umbriel, Ariel, and Miranda. The image shows correct relative sizes and brightness. The two largest, Oberon and Titania, are about half the size of Earth's Moon, or roughly 1600 km in diameter. All five satellites are generally gray, with only slight color variations on their surfaces. (NASA-JPL)

Figure 10.10   Oberon (left) and Titania (right) are the largest moons of Uranus. They are very similar in many respects, although Titania shows more small craters and seems to have undergone global extension of its crust after the main cratering phase ended. (NASA-JPL)

heights of 2–5 km, while the grabens are 20–100 km across. The longest graben extends for at least 1500 km and its stepped floor is several kilometers deep. The absence of superimposed craters suggests the faults are some of the youngest features on the satellite.

Closer to Uranus are Umbriel and Ariel. Again, they are almost identical in size and density, but with significant surface differences. Umbriel is the darkest of all the major Uranian satellites and has a weaker spectral signature for water ice. The black surface shows little variation in brightness, the most conspicuous feature being a ring, about 30 km wide, on the floor of a sizeable crater called Wunda. The population of large craters indicates that, along with Oberon, it has the oldest surface of Uranus' major

satellites. The reasons why Umbriel is darker than its neighbors (particularly Ariel) and apparently unaffected by internal melting remain unknown.

On Ariel's relatively bright, complex surface, most of the larger, ancient craters have been erased, either by viscous relaxation of the icy surface over billions of years or by extrusion of fluid material (or both). The largest remaining crater appears flattened, with a gently domed floor partly surrounded by a shallow trough.

Voyager also revealed extensive faults and flat-floored rift valleys over 10 km deep and hundreds of kilometers long. Lacking craters, the spectacular canyons have been filled with relatively recent volcanic flows. The resurfacing appears to be too extensive to be accounted for solely by radiogenic heating (heat from the

Figure 10.11 Umbriel (left) and Ariel (right) are similar in size and density. Umbriel has a dark, ancient surface covered in impact craters. Only a few brighter features are visible, most notably a ring on the floor of Wunda crater (top). Much of Ariel has been resurfaced, leaving few large craters. Also visible are numerous linear faults and smooth-floored grabens filled with volcanic flows. (NASA-JPL)

decay of radioactive elements in the moon's interior). Instead, tidal heating may have been involved. Although it has no orbital resonances with other satellites today, couplings with Umbriel, and possibly Titania, may have occurred in the past, causing repeated gravitational tugs that heated Ariel's interior. As a result, a cryogenic mix of water ice, methane and small amounts of ammonia could have risen to the surface and filled many of the valleys and impact craters.

In addition to the ubiquitous water ice, carbon dioxide ice has been discovered on the surfaces of Ariel, Umbriel and Titania.[6] It may be primordial (from the solar nebula) or delivered by impacting comets. Alternatively, the dry ice could be the result of greater plasma irradiation on the trailing side as protons and electrons trapped in Uranus' magnetosphere overtake the moons from behind and impact their surfaces. The CO_2 ice is only found on the trailing side of the satellites, suggesting that the ice may have been eroded or buried by meteorite impacts on the more exposed forward-facing hemisphere.

Miranda

Voyager 2 imaged almost the entire southern hemisphere of Miranda at fairly high resolution, and the smallest of the major satellites turned out to be the most intriguing. Its unique landforms, superimposed on more heavily cratered terrain, included a "chevron" and two huge polygonal features with rounded corners that were nicknamed "circi maximi" after Roman chariot-racing tracks.[7] All three features displayed parallel faults and

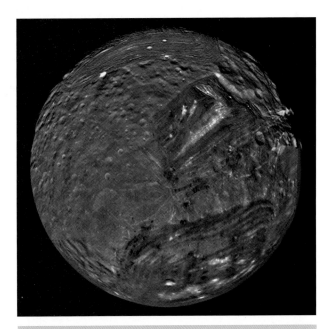

Figure 10.12 Although it is only 472 km across, Miranda displays a bizarre variety of landforms. About half of the surface is heavily cratered, but the most obvious features are the three coronae. The V-shaped "chevron" is known as Inverness Corona. The ridged polygon at the bottom is Arden Corona, while the similar feature on the opposite limb is Elsinore Corona. Also visible (right limb) is the 20 km cliff known as Verona Rupes. (NASA-JPL / USGS)

[6] Carbon dioxide ice has also been observed in the reflectance spectra of Jupiter's satellites Europa, Ganymede and Callisto, Saturn's satellites Phoebe and Iapetus, and Neptune's moon Triton.

[7] These polygonal features are known as "coronae," although they bear little resemblance to the ovoid coronae on Venus.

ridges with a scattering of fresh impact craters. Clearly they had been formed in relatively recent geological times.

On the day-night terminator, close to the chevron (Inverness Corona), could be seen a 20 km high cliff (twice the height of Mount Everest), now known as Verona Rupes. In the weak gravity

of Miranda, an object would take almost 10 minutes to fall to the valley floor below.

Various theories have been put forward to explain this tortured terrain. One suggestion was that the satellite slowly reassembled itself after being shattered by a huge impact. Each corona was

Box 10.2   Discovering Uranus' Rings

On March 10, 1977, several teams of astronomers observed a rare stellar occultation by Uranus. Their intention was to time very precisely the moments when a star known as SAO 158687 was extinguished by the intervening bulk of the planet and when it reappeared. Such measurements taken from many different locations would be used to refine the planet's size and orbit.

Since the precise time of the occultation was unknown, the teams' observations began well ahead of the predicted commencement. The resulting light curves caused a sensation. A series of blips in the light levels were visible, both before and after the star was eclipsed. Eventually, James Elliot, leader of the American team using the Kuiper Airborne Observatory over the Indian Ocean, realized that five sizable dips in the starlight appeared as a mirror image on either side of the planet. They had discovered a set of narrow rings around Uranus.

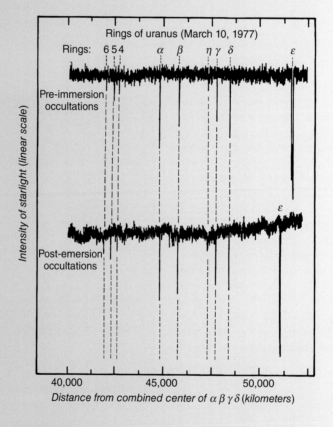

Figure 10.13   On March 10, 1977, astronomers on board the Kuiper Airborne Observatory (KAO) detected a number of dips in the starlight before and after Uranus passed in front of the 9th magnitude star SAO 158687. Analysis of data from the KAO, Perth, and other locations showed that Uranus had a system of at least nine rings. (James L. Elliot, Annual Reviews of Astronomy and Astrophysics, 1979)

Similar dips in the starlight were recorded by groups in India, Japan and South Africa. One team at Perth Observatory in Australia saw them, even though Uranus passed to the north of the star.

When the data from all of the sites were taken into account, it became clear that at least eight separate rings had been detected. The three rings closest to Uranus and first recognized by the Perth group, led by Robert Millis, were known as 6, 5 and 4. The five rings announced by Elliot's team were given Greek letters whose alphabetical order corresponded to their distance from the planet: alpha (α), beta (β), gamma (γ), delta (δ) and epsilon (ε). A less obvious ring that was identified at a later date was labeled eta (η). This complex, rather haphazard system of nomenclature has survived to the present day.

thought to represent a discrete fragment. However, this is difficult to reconcile with the much older terrain between the coronae.

Alternatively, the coronae may have formed during past episodes of tidal heating. Multiple eruptions of viscous fluid material, possibly associated with stretching of the crust, could have created the broad lanes of parallel ridges. Bright patches on the surfaces of Arden and Inverness Coronae could be deposits of material erupted from explosive vents. The development of the coronae may have ceased prematurely when the satellite cooled before its internal differentiation into ice and rock was complete.

Rings

For many centuries, Saturn was assumed to be the only planet to possess a set of rings. Then, in March 1977, during a stellar occultation by Uranus, astronomers noticed brief dips in the starlight, both before and after the planet passed in front of the star (see Box 10.2). By combining observations from different locations, nine separate rings were recognized.

Since then, four more rings have been discovered, bringing the total to 13. Two of them – lambda (λ) and zeta (ζ) (formerly 1986U2R) – were discovered in Voyager 2 images. Another pair, mu (μ) and nu (ν), (originally known as R/2003 U1 and R/2003 U2) was discovered in Hubble Space Telescope images in 2005.

Figure 10.14 In this false color view, taken by Voyager 2, the nine original rings are shown. The fainter, pastel lines between them are the result of computer enhancement. Six 15-second narrow-angle images were used to image the extremely dark, faint rings. The brightest, or epsilon, ring (top) is neutral (gray) in color, while the eight fainter rings show slight color differences. The lambda ring is too faint to be visible. Spatial resolution is about 40 km. (NASA – JPL)

Even before the Voyager 2 flyby, it was possible to gather some basic information about the rings from ground-based observations and analysis of data from various occultations. The nine known rings were shown to begin about 16,000 km above the planet, with their outer limit about 25,600 km from Uranus' cloud tops. Unlike most of the rings around Saturn, they had very sharp edges and were very dark, reflecting less light than coal dust.[8]

From the very brief blips in the light curves, it was obvious that the rings are very closely spaced – often only a few hundred kilometers apart. They are also remarkably narrow, and, with the exception of the outer, epsilon ring, generally no more than 10 km across. Even more surprising was the fact that some of the rings are slightly elliptical, rather than circular. In particular, the distance of the epsilon ring from Uranus was found to vary by about 800 km.

The rings reflect only a few percent of what little sunlight reaches them, so imaging the charcoal black rings against the dark background is extremely difficult. Occultation data at radio and optical wavelengths showed that particles smaller than tens of centimeters were absent from the epsilon ring, and the rings as a whole are remarkably free of dust. However, Voyager images taken looking back towards the Sun did reveal that most of the region inward of the lambda ring is filled with faint, broad bands and ringlets of micron-sized particles.

Since the effects of sunlight and drag from the tenuous hydrogen gas of Uranus' outer atmosphere would be expected to remove the smoke-sized particles within one million years, the particles must be slowly replenished – presumably by material derived from collisions involving ring boulders or larger moons. As it is, the amount of dust in the rings is extremely small. If it could all be swept into a single pile, it would only make a ball about 100 meters in diameter.

Voyager also detected a sparse but broad band of dust 90,400 km above the cloud tops, between the epsilon ring and Miranda. 30 hits per second were recorded from particles striking the spacecraft during this crossing of the ring plane – very modest compared with 600 hits per second during the ring plane crossing at Saturn.

At least six of the rings were inclined to the planet's equator, though only by a few hundredths of a degree. Several seemed to wobble and "pulsate" in and out, probably as the result of Uranus' slightly flattened shape, along with the gravitational influence of nearby moons.

The epsilon ring, the broadest and brightest of the family, was one of the few rings orbiting above the equator, but its orbit was also the most eccentric. Not only did its distance from Uranus vary, but its width changed proportionally – at its closest to the planet, it was only 20 km across, whereas at its furthest the ring expanded to 100 km.

Voyager observed some fine structure when the epsilon ring passed in front of distant stars. Some camera images revealed a bright outer component about 40 km wide, a darker central region of comparable width, and a bright inner strip about 15 km across. Studies of the way light passed through the ring suggested that it was mainly made of boulders up to one meter across.

---

[8] Since the black rings are the same color as the nearby small moons, it seems likely that both are being darkened by the same process – the bombardment of organic molecules on their surfaces by high energy particles.

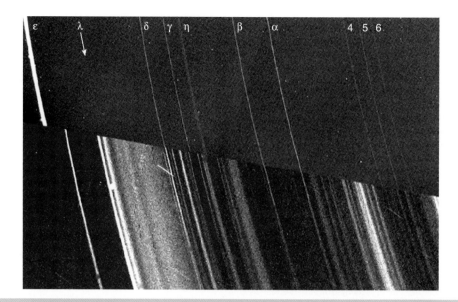

Figure 10.15   Two Voyager images of the main Uranian ring system compared. The top image, obtained one day before closest approach, shows 10 narrow rings from their sunlit side. The lower image, taken looking almost directly towards the Sun (phase angle 172.5°), reveals a sheet of micron-sized dust particles inward of the lambda ring. The differing distance of the elliptical epsilon ring is clearly shown. (J. A. Burns/Cambridge University Press)

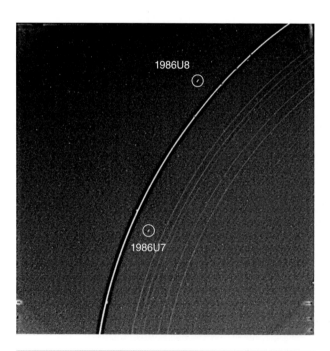

Figure 10.16   Two small moons, both about 40 km in diameter, were found by Voyager 2 on either side of the epsilon ring. Cordelia (1986 U7) and Ophelia (1986 U8) – the tiny specks inside the circles – are the only known "shepherd moons" in Uranus' ring system. Their gravitational influence confines the ring's boulder-sized debris. (NASA-JPL)

The most important discovery about the epsilon ring was the presence of two shepherding satellites, Cordelia and Ophelia, on either flank of the epsilon ring – comparable to the moons that accompany Saturn's narrow F ring. However, no other shepherds were found.

Orbiting only 4000 km apart, the moons are able to force wayward particles and boulders back into the epsilon ring. Ophelia, the outer shepherd, completes 13 orbits in the time a particle at the outer edge of the epsilon ring completes 14 orbits. As they slowly overtake Ophelia, the particles are attracted by the moon and lose energy. This causes them to "fall" closer to Uranus.

In contrast, particles at the inner edge of the ring complete 24 orbits in the same time that Cordelia completes 25 orbits. This is known as a 24:25 resonance. As Cordelia overtakes the ring particles, it gives them increased energy and forces them into higher orbits. The combined result is to give sharp edges at both the inner and outer boundaries of the ring.

The almost circular orbit of Cordelia and the slightly elliptical orbit of Ophelia complicate the gravitational interaction. This may explain the fact that the average eccentricity of the epsilon ring is between that of the orbits of the nearby shepherds.

Resonances with these companions of the outer ring may also modify the behavior of some of the other rings, though other unseen shepherds only a few kilometers across could also exist between the lanes of particles.

Two more rings located well outside the main system were found with the Hubble Space Telescope in 2003–4. Both of these faint rings are much wider than the previously known rings. The outer component, now known as mu (μ), was discovered orbiting between 86,000 and 103,000 km from the planet's center, well

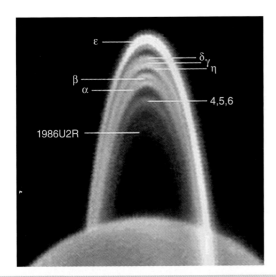

Figure 10.17 At 2.2 microns, methane and hydrogen in Uranus' atmosphere absorb most of the light falling on the planet, so little is reflected back into space. Imaging at this wavelength results in a dark Uranus, making the faint ring system clearly visible. The region interior to rings 4, 5, and 6 is a broad, dusty ring known as zeta (1986U2R). This near-infrared image was taken using adaptive optics with the Keck 2 telescope in July 2004. (Imke de Pater, Seran Gibbard, and Heidi Hammel)

beyond the 11 recognized rings.[9] The ring shares its orbit with a small satellite called Mab. Since it is brightest along Mab's orbit, it is probably replenished by material ejected during impacts with the moon. This is supported by its blue color, evidence that the ring is made of tiny water ice particles, mostly less than a tenth of a micrometer across.[10]

The other recently discovered ring, now designated nu (ν), is almost 4000 km wide and squeezed between the orbits of Portia and Rosalind, which probably confine its limits. The ring is centered about 67,700 km from Uranus, in a region with no known satellites to provide source material. This ring may be comprised of debris from the collisional disruption of a moon.

Their considerable variability in width and opacity suggests that the rings are likely to be younger than the planet. Since the rings all lie inside the Roche limit, one possibility is the collisional or gravitational break-up of a small moon (see Chapter 9). If so, it must have been quite small – the entire ring system could be compressed into an icy body only 30 km in diameter.

Chaotic Inner Moons

By the time the data from the Voyager 2 encounter were analyzed, the satellite count had trebled, with the discovery of 10 satellites between Uranus and Miranda. An eleventh moon, Perdita (S/1986 U10), was found 13 years later in a re-analysis of the Voyager images.

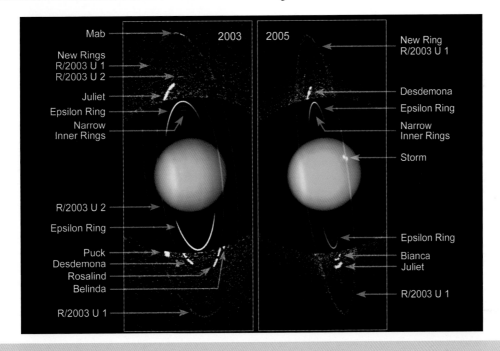

Figure 10.18 These Hubble Space Telescope images taken in 2003 and 2005 show the most recently discovered rings of Uranus. The outer mu (μ) ring (R2003 U1) lies almost twice as far from the planet as the outer edge of main ring system. Hubble also detected two small satellites, named Mab and Cupid. Mab shares an orbit with the outer ring and is probably the source of dust that replenishes it. The nu (ν) ring (R2003 U2) lies about 67 700 km from Uranus, in a region with no known satellites. (NASA, ESA and M. Showalter-SETI Institute)

[9] Although they were first noticed in 2004, the rings were subsequently found in HST images taken the previous year – hence their designation.
[10] The only other known blue ring is Saturn's E-ring, which surrounds Enceladus. The ring is fed by jets of water vapor spewing out of the moon.

Since then, improvements in ground-based instruments have led to the discoveries of two more satellites inside the orbit of Miranda. One of them, Cupid, comes to within 800 km of Belinda. It lies in the midst of the so-called "Portia group", a family of nine moons (collectively named after their largest member), which occupy the relatively confined space between the orbits of Bianca (59,166 km) and Perdita (76,417 km).

The other recent discovery, Mab, is the only moon known to orbit between Puck and Miranda. It is also the only satellite known to be the primary source of material for one of Uranus' rings. Like the previously known moons, the newcomers follow nearly circular orbits above the planet's equator and are assumed to have a synchronous rotation.

Puck was the first and largest of the moons discovered by Voyager. Orbiting at a distance of 86,000 km from the planet's center, it takes a little over 18 hours to complete one circuit. It is about 160 km across and roughly spherical, with a large impact crater visible on the limb. It reflects only 7% of the incoming sunlight – much less than the larger moons, but more than the rings.

Images of the other inner moons show only smudges, so few details of their physical nature, including their masses or densities, are known. However, estimates of their sizes can be derived from their brightness and distance. Assuming that they reflect as much sunlight as Puck, their diameters have been calculated to be between 20 and 135 km. Their dark surfaces suggest a mixture of water ice and organic (carbon-based) compounds that have been modified by interactions with charged particles trapped in the magnetosphere.

Most of the inner moons are more or less uniformly distributed between the rings and Puck, their orbits typically separated by a few thousand kilometers. Showalter and Lissauer have measured numerous changes to the orbits of these moons since 1994. Their simulations reveal that there is a continual exchange of energy and angular momentum between the satellites. As a result, their orbits are chaotic and dynamically unstable over time scales of only one million to 100 million years.

Probably the most unstable moon of all is Cupid, which approaches the considerably larger Belinda. Calculations predict that that the pair will collide within a few million years. A similar process in the past could have led to the birth of Cupid and Perdita, which may be pieces that broke away when a comet smashed into nearby Belinda.

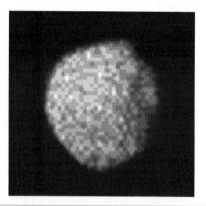

Figure 10.20 Puck is roughly spherical and about 160 km across, making it the sixth largest of Uranus' moons. Its surface is almost as black as the planet's rings. A 45 km wide impact crater is visible near the limb (right), with another large indentation above it. (NASA-JPL)

Figure 10.19 The crowded region within the orbit of Miranda contains 11 known rings and 13 small satellites. (Left) The mu (μ) and nu (ν) rings lie between the orbit of Miranda and the narrow rings detected in 1977. (Right) An illustration of the region near the outer rings, showing the satellites' orbits. The satellites are shown to their correct relative sizes, but exaggerated by a factor of ~16 compared with the width of their orbits. (Carl D. Murray)

Box 10.3   Naming the Moons of Uranus

The International Astronomical Union (IAU) has the task of assigning names to newly discovered objects in the Solar System. In the case of planetary satellites, the IAU usually follows the tradition of naming them after characters in Greek or Roman mythology. However, in the case of Uranus, the first four moons had been named in the mid-19th century after characters from the literary works of Shakespeare (Oberon, Ariel and Titania) and Alexander Pope (Umbriel). Miranda and the more recent discoveries have also been named after Shakespearean characters, with the exception of Belinda (1986 U5), which was named after a character from Pope's "The Rape of the Lock."

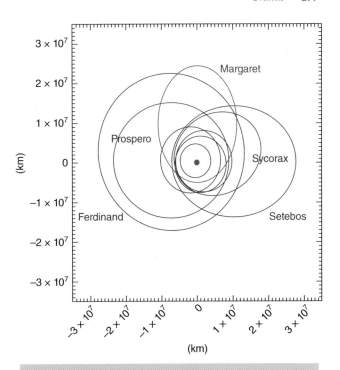

Figure 10.21   The orbits of the outer satellites. All are highly elliptical and steeply inclined to the planet's equator. They also travel in a retrograde direction (with the exception of Margaret, which is prograde). The satellites were probably captured by Uranus. (Adapted from Scott S. Sheppard)

Remote Moons

Uranus has at least nine small satellites beyond Oberon, bringing the total number to 27. All but one of these outer moons follow irregular, retrograde orbits. The orbits are also highly elliptical and inclined to the planet's equator – further evidence that they have been captured by Uranus.

The process of capture could have taken place in two different ways. The moons could have simply been trapped by Uranus' gravity as they came close to the planet, or the young Uranus could have been surrounded by a gaseous nebula that would have caused a drag on objects passing nearby.

Almost all of the outer satellites have been discovered since 1997 by ground-based teams of observers. Since they are very faint (below magnitude 20), large modern telescopes with adaptive optics are required to find them. The first to be recognized were Caliban and Sycorax. Both were noticeably red, resembling the Kuiper Belt objects that exist beyond Neptune.

Caliban, the smaller of the two moons, has a diameter of about 70 km and orbits Uranus at an average distance of about 7.2 million km, taking 1.6 years to complete one revolution. Sycorax appears to be about twice the diameter and has an orbital period of 3.5 years, with a mean distance of about 12.2 million km. However, Sycorax has a much more elliptical orbit than Caliban, bringing it as close as 6 million km to the planet and as far away as 18 million km.

In 1999, three more irregular moons (Stephano, Prospero and Setebos) were found with the Canada-France-Hawaii Telescope. Considerably smaller than the moons discovered two years earlier, these are about 20 times fainter and orbit Uranus at average distances of 8 million, 16.2 million and 17.4 million km. The orbit of Setebos is even more eccentric than that of Sycorax.

Francisco and Trinculo were found in 2001, followed by Margaret and Ferdinand in 2003. Francisco is the innermost irregular moon so far discovered, with a mean distance of about 4.3 million km from Uranus – more than seven times further out than Oberon. Ferdinand is the outermost member of the group, taking almost eight years to complete one orbit at an average distance of

20.9 million km. The odd one out is Margaret, which travels in a "normal" prograde direction, though it has the most eccentric orbit of all the Uranian satellites.

No missions to Uranus are planned for the foreseeable future.

Questions

- Uranus was the seventh planet to be discovered orbiting the Sun. Why did it take so long to find?
- Summarize the major differences between Uranus, Jupiter and Saturn.
- Why does Uranus rotate on its side? Is this likely to have been an ancient or relatively recent event? Explain your answer.
- How does Uranus' high obliquity influence conditions on the planet?
- Why were scientists surprised by the magnetic field of Uranus? Briefly explain how this field is thought to be generated.
- Compare and contrast the rings of Uranus with those of Saturn and Jupiter.
- What are the major similarities and differences between the satellites of Uranus and Saturn?

ELEVEN

Neptune

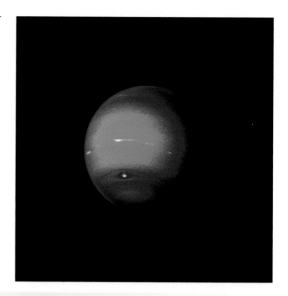

Another blue-green ice giant, 1600 million kilometers beyond Uranus is its near-twin, Neptune. Despite their many similarities, the planets reveal a number of unexplained differences, notably their contrasting heat output and storm activity. Although most of what we know about this remote world was returned by the Voyager 2 spacecraft in 1989, its changing atmosphere and elusive ring system continue to be examined by the Hubble Space Telescope and state-of-the-art ground-based instruments. These observations make it clear that Neptune is one of the most dynamic members of the Sun's family.

With the discovery of Uranus in 1781, the Solar System once again seemed complete (see Chapter 10). However, when astronomers began to refine their calculations of its orbit, Uranus refused to behave as expected. Until 1822 it moved faster than predicted, but after 1822 it seemed to slow down. By the mid-1830s, a number of scientists were wondering whether the unexpected errors in its position were caused by another, more remote, planet pulling on Uranus.

Would it be possible to calculate the position of such an object from its perturbations of Uranus, and then recognize it among the plethora of background stars? A number of mathematicians decided to attempt the incredibly difficult calculation. By 1845, John Couch Adams, a student at Cambridge University, England, informed James Challis, Professor of Astronomy at Cambridge, that he had come up with a solution to the problem, but little action was taken.[1]

By then, another brilliant mathematician was causing a stir on the other side of the English Channel. Unaware of Adams' work, Urbain Jean Joseph Le Verrier produced his own independent solution, which predicted that the new planet was located about 5° east of the star delta Capricorni – a very similar answer to that of his English rival.

Le Verrier wrote to Johann Galle at the Berlin Observatory, requesting that a search be carried out. With the backing of the observatory's director, Johann Encke, Galle and graduate student Heinrich d'Arrest immediately started to scan Le Verrier's predicted location with a 23 cm refractor. As Galle stared through the telescope and called out each star in the field of view, d'Arrest compared them with a new, detailed map of the heavens. Within one hour they found an object that was not on the chart.

The following night (September 24), they were able to confirm that the object had moved and that it displayed a tiny disk. The discovery had been made within one degree of Le Verrier's prediction, and was a great triumph for Newton's gravitational theory.

The solution to the Uranus problem was now clear (see Fig. 11.1). Before 1822, Uranus was lagging behind Neptune in its orbit, but slowly catching up with the outer planet. As a result, Neptune's gravity was accelerating Uranus' motion. In 1822 the two planets were at opposition. Thereafter, although Uranus began to pull ahead of Neptune, the outer planet's gravitational pull slowed it down, causing it to lag behind its predicted position.

The Eighth Planet

After the usual debate over what to call the newly discovered world, it was eventually decided to name it Neptune, after the

[1] Adams' solution was actually within two degrees of arc of Neptune's actual position. If a search had begun straight away, he would probably have gone down in history as the unchallenged discoverer of the eighth planet from the Sun. Instead, as a result of bad luck and incompetence by his seniors, the search by Challis did not begin until late July 1846.

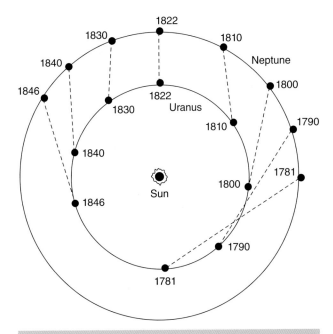

Figure 11.1 Unexpected perturbations in the motion of Uranus caused astronomers to recognize that the gravitational pull of an unknown giant planet could be influencing its orbit. Until 1822, Neptune's gravity caused Uranus to move faster than expected. After their opposition in 1822, Uranus overtook Neptune, but the pull of the unseen planet caused it to slow down. As the distance between the pair increased, Neptune's gravitational influence gradually diminished.

Table 11.1
Neptune Summary

Mass	102.43×10^{24} kg
	(17.1 Earth masses)
Equatorial Diameter	49 528 km
Polar Diameter	48 682 km
Axial Inclination	28.32 degrees
Semimajor Axis	4495.06 million km (30 AU)
Rotation Period (internal):	16.11 hours
Sidereal Orbit Period	164.79 years
Mean Density	1.64 g/cm³
Number of known satellites	13

Roman god of the seas and brother of Jupiter. Neptune is too faint to be seen with the naked eye, although it is often brighter than eighth magnitude. With an apparent diameter of little more than three arc seconds, it is 600 times smaller than a Full Moon. However, with a telescopic magnification of more than 100, the planet clearly shows a bluish disk.

The eighth planet turned out to be another giant, similar in size and outward appearance to Uranus. For more than a century, astronomers were unsure which of the pair was the larger, and it is only in recent years that Uranus has been given the nod. Neptune's equatorial diameter is about 49,528 km (almost four times the diameter of Earth). 57 Earths would fit inside Neptune.

On the other hand, it has a greater mass than Uranus and its bulk density is actually the highest of the Jovian planets. Neptune has a mass more than 17 times that of Earth, although the gravity at Neptune's cloud tops is only 1.19 times Earth's surface gravity. This is because it is such a large planet (the gravitational force a planet exerts upon an object at the planet's surface is proportional to its mass and to the inverse of its radius squared). Hence a 10 kg object would weigh 11.9 kg at Neptune's visible surface.

These physical properties suggest slightly different compositions and internal structures for the two blue giants. This seems to be confirmed by measurements indicating that – unlike Uranus, but similar to Jupiter and Saturn – Neptune radiates more than twice as much energy as it receives from the Sun.

The difficulty in observing long-lived cloud features meant that the planet's rotation period was uncertain for many years. (As with all gas giants, clouds swept along by strong zonal winds are unreliable markers for the rotation of the planet as a whole.) It was not until the Voyager 2 flyby of Neptune that the internal rotation period was established to be 16 hours 7 minutes – considerably faster than that of Uranus.[2]

Neptune has an axial inclination of 28.3°, only a little more than Earth and Mars. This means that it experiences seasons that last approximately 41 years during its 164 year circuit of the Sun. At the time of the Voyager 2 flyby the south pole was illuminated by early spring sunshine, with an absence of solar heating at the wintry north pole. Midsummer in the south (and midwinter in the north) will arrive in 2020.

Neptune's orbit is almost circular, with a semimajor axis of about 4.5 billion km (30 AU). At such a vast distance, the planet crawls around the Sun at 5.5 km/s, taking almost 165 years to complete one orbit. Neptune will complete its first circuit of the Sun since its discovery on July 11, 2011.

Atmosphere

Hydrogen and helium, the main constituents of Jupiter, Saturn and Uranus, were also expected to be the major components of Neptune's atmosphere, but spectroscopic evidence was hard to obtain. However, by the late 1940s, techniques had improved sufficiently to recognize the spectral signature of molecular hydrogen (H_2) and methane (CH_4). The abundance of hydrogen inferred that helium, the second most common gas in the Sun, was also present in large quantities.

This was confirmed by data from Voyager 2, which indicated that the upper atmosphere (above the methane ice clouds) is 84% molecular hydrogen, 14% helium and 2% methane. As is the case with Uranus, the scattering of blue light and absorption of red light by methane gives the planet its characteristic blue color.

---

[2]**Neptune's rapid rotation causes an equatorial bulge, so the polar diameter is about 850 km smaller than the equatorial diameter.**

Box 11.1 Voyager 2's Grand Finale

After the flyby of Uranus in January 1986, NASA's Voyager 2 spacecraft set course for Neptune on the final leg of its Grand Tour. In order to prepare for the high speed flyby of Neptune, where light levels are 900 times lower than on Earth, the spacecraft's computers were reprogrammed to allow longer imaging exposures and greater spacecraft stability, and to compensate for the relative motion of the spacecraft and its targets.

Typical camera exposures at Neptune were to last for 15 seconds or longer, compared with 5 seconds at Uranus. However, exposures of up to 61 seconds were now possible, with additional extensions in multiples of 48 seconds.

In order to ensure that Voyager's weak signals were safely received, all three of NASA's 64 meter Deep Space Network antennas were enlarged to 70 meters and made as efficient as possible. Other antennas and ground stations were enlisted, including the Parkes radio telescope in Australia and another dish in Usuda, Japan.

Figure 11.2 Voyager 2 was targeted for closest approach only 4500 km above Neptune's northern hemisphere. This would bend its trajectory south of the ecliptic, enabling it to pass only 38 500 km from the centre of Triton less than six hours after the Neptune encounter. This flyby sent Voyager 2 heading south, out of the Solar System, with no possibility of redirecting it towards Pluto. It also meant that Voyager was unable to make a close approach to the other known moon, Nereid. (NASA-JPL)

The main objective was to find the closest approach to Neptune that would result in a grandstand view of Triton. On the date of the flyby (August 25, 1989) Triton would be south of Neptune's orbital plane, so Voyager had to sweep low over the north pole, allowing the planet's gravity to bend its path southward. In the event, Voyager passed a mere 4500 km above Neptune's cloud tops.

By May 1988, images from Voyager were already showing more detail than any obtained by ground-based telescopes. Then, in January 1989, a huge dark spot was discovered in the planet's southern hemisphere. In early July a 400 km diameter satellite was found inside the orbit of Triton. Five more small moons were found close to Neptune's rings in the weeks that followed.

The spacecraft crossed the bow shock – the boundary where the solar wind plowed into Neptune's magnetosphere – about 9½ hours before closest approach. It eventually exited the magnetosphere 38 hours later, but multiple bow shock crossings were subsequently experienced almost four million km from the planet. The magnetic dipole was found to be highly inclined and offset from the planet's center.

One of the most intriguing puzzles solved by Voyager was the apparent presence of ring arcs – partial rings – around Neptune. Images and occultation observations showed that the arcs are actually elongated clumps within one of the five dark, narrow, ringlets. Voyager crossed the ring plane twice, about 60 minutes before and 85 minutes after its closest approach to Neptune.

Voyager's trajectory was bent about 48° south of the ecliptic, and, 110 minutes after leaving Neptune behind, its instruments began to observe Triton. The spacecraft eventually passed 38 500 km from the satellite's center, less than six hours after the Neptune encounter. Images of the sunlit southern hemisphere revealed an extremely cold, icy world, with a surface of pink ice, cantaloupe patterns and dark streaks.

Voyager's farewell view of Neptune and Triton was taken on August 28, when the narrow angle camera captured two pale crescents hanging in a black sky.

Figure 11.3 This false color image from Voyager 2 shows a high level haze (red), particularly over the south pole and near the limb, where the instruments were looking through a greater thickness of atmosphere. At the center of the disk, sunlight penetrates the haze and is absorbed by the methane, producing a blue color. Highly reflective ice clouds appear as patches of white. (NASA-JPL)

Figure 11.4 Voyager 2 imaged bright, linear clouds of methane ice casting shadows on a lower cloud deck of uncertain composition. These clouds were observed at latitude 27°N and extended parallel to the equator for thousands of kilometers. The width of the cloud streaks ranged from 50–200 km and the shadows were 30–50 km across. The clouds were above most of the methane, hence their whitish hue. Their height above the main cloud deck was estimated to be 50–100 km. The Sun is toward lower left. (NASA-JPL)

There are also trace amounts of carbon monoxide (CO), hydrogen cyanide (HCN), acetylene (C_2H_2) and ethane (C_2H_6) – the last two being products of methane photochemistry. As is the case with Uranus, ultraviolet light from the Sun destroys the methane high in Neptune's atmosphere, converting it to acetylene and ethane. These gases diffuse down to a lower, colder part of the stratosphere, where they condense into tiny ice particles. On reaching the warmer region of the upper troposphere, the hydrocarbon ice particles evaporate and are converted back to methane gas.

Voyager's instruments were only able to probe the atmosphere down to a level where pressure reached a few bars. Beneath one or more layers of ethane-acetylene haze is a physically thick but optically transparent layer of methane ice haze. Discrete, denser, clouds of methane ice created by localized upwelling sometimes cast shadows on the main cloud deck 50–75 km below. This lower, blue-tinted, opaque layer has an unknown composition, but may be made of hydrogen sulfide (H_2S) ice or ammonia (NH_3) ice.

Below this, conditions may be suitable for the existence of ammonium hydrosulfide (NH_4SH) clouds – a mixture of water, ammonia and hydrogen sulfide. The deepest hypothesized cloud layer is thought to exist between 40 and 400 bars, at a depth of several hundred kilometers.

Voyager also measured the temperature distribution in the atmosphere. The infrared data showed that the upper atmosphere is warmer near the equator, cooler in mid-latitudes and warmer again above the south pole (which was enjoying its long summer). This pattern resembled that of Uranus, even though their axial inclinations are very different. In the troposphere, there was hardly any temperature difference between the south pole and the equator – a characteristic common to all four giant planets.

The temperature of the stratosphere fell from about −155°C at an altitude of 100 km above the 100 millibar level to a minimum of −220°C at the 100 millibar level. It then increased steadily through the troposphere, reaching a temperature of −146°C at the 5 bar level. The planet was notably warmer than expected, based on its distance from the Sun, with an average surface temperature of −220°C, compared with −214°C for Uranus.

The relative warmth of Neptune is accounted for by the fact that, unlike Uranus, it has a substantial internal heat source. Neptune emits about 2.5 times as much heat as it receives from the Sun. This internal source of warmth may explain why Neptune's winds are much stronger than those on Uranus.

By studying the motions of long-lived cloud features moving parallel to the equator, it was possible to determine that – as on Uranus – zonal winds in equatorial and temperate latitudes are retrograde – opposite to the eastward rotation of the planet. (On Jupiter and Saturn the equatorial jets are prograde, while retrograde winds are weak and scarce.)

At the equator, winds are blowing towards the west at speeds of up to 450 m/s (1600 km/h), comparable to the near-equatorial winds of Saturn. Clouds caught up in this wind take more than

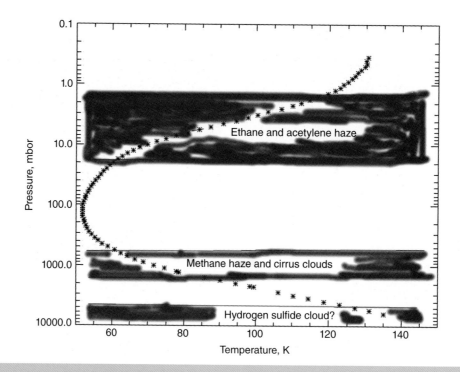

Figure 11.5 A profile of Neptune's upper atmosphere showing changes in temperature, as well as probable haze and cloud layers. The upper region (stratosphere) is dominated by methane photochemistry. Models suggest that other cloud layers exist beneath the methane ice. The uppermost of these may be composed of hydrogen sulfide.

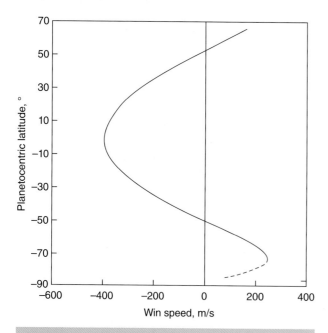

Figure 11.6 Neptune is the windiest planet in the Solar System. Most of the cloud features are blown westward – the opposite direction to the planet's rotation – with peak wind speeds of more than 400 m/s close to the equator. Only at high latitudes is the predominant wind towards the east in relation to the planet's rotation.

20 hours to travel around the planet, compared with Neptune's rotation period of a little over 16 hours.

In contrast, winds at latitudes of 70° peak at about 200 m/s in a prograde direction, sweeping clouds around the planet in a mere 11 hours. This means that the variation of wind speeds on Neptune far exceeds that on any other planet.

The most noticeable cloud features imaged by Voyager were in the southern hemisphere. The largest of these was the **Great Dark Spot**, located at about 22°S. Blown westward at about 30 m/s by the zonal wind, it completed one circuit in just under 18 hours.

In many ways, the oval storm, which measured some 15,000 km by 7000 km, resembled Jupiter's Great Red Spot. The counter-clockwise motion of clouds around its rim indicated an anticyclonic weather system, an area of rising gas that was creating a high pressure system.

Bright, cirrus-like clouds of methane ice crystals were seen above and around the Great Dark Spot. They appeared to be similar to orographic clouds on Earth, which form when moist air is forced upwards over a mountain. In this case, the Spot was apparently forcing methane-rich gas to higher levels, resulting in cloud formation.

Two significant differences between the huge anticyclones of Jupiter and Neptune were their stability and longevity. Not only was the Great Dark Spot "floppy" – changing shape and orientation over a period of about eight days – but it was drifting toward the equator at a rate of 15° per year. Scientists were not too surprised when Hubble images taken in 1994 indicated that the Spot had disappeared.

Figure 11.7 Images of the Great Dark Spot show bright cirrus-like clouds above and around the anticyclone that formed and dissipated quite rapidly. The sequence spans about 36 hours, a little over two rotations of Neptune. (NASA-JPL)

Figure 11.8 A Voyager 2 image showing features in Neptune's southern hemisphere. Most prominent is the Great Dark Spot and its associated cirrus ice clouds. Also visible are a bright cloud feature nicknamed Scooter (lower left) and a small dark spot (D2) with a bright core. Driven by zonal winds, all the features were moving around the planet at different speeds. The maximum diameter of the Great Dark Spot was about 15 000 km. (NASA-JPL)

Two more striking cloud features were seen in the Voyager images. One, a triangular white cloud at latitude 42°S, was nicknamed Scooter because every few revolutions it caught up with and overtook the Great Dark Spot, completing one circuit in about 16 hours.

At latitude 54°S was another dark spot, known as D2, which displayed a bright core of methane ice above a dark oval. Unlike its giant cousin, D2 appeared to rotate in a clockwise direction, indicating that it was a cyclone or region of low atmospheric pressure. Gases would, therefore, be expected to rise in the cloudy centre and descend in the dark oval around it. D2 almost matched the planet's rotation period, reappearing roughly every 16 hours.

Although the detailed observations made by Voyager are unique, an observation campaign has been undertaken since the early 1990s with modern ground-based instruments and the Hubble Space Telescope. These found evidence that the Great Dark Spot and Scooter had disappeared by 1993. However, images taken in late 1994 did reveal a new dark spot in the northern hemisphere that appeared to resemble its southern counterpart in terms of size and appearance. Once again, high altitude clouds around the edge of the spot were thought to be cooled clouds of methane ice crystals.

Storm activity also seems to have been increasing as the seasons slowly change. The planet has been growing brighter since the early 1980s, as it moves from the onset of southern spring toward southern midsummer. There is also some evidence that the planet's overall brightness varies over a period of about 10 years, possibly related to the cycle of solar activity.

In the early 1990s, cloud activity shifted from the southern to northern mid-latitudes. Since the mid-1990s, there has been a distinct increase in the area and brightness of banded cloud features, particularly in the planet's southern hemisphere.

Bright cloud features observed in 1996 were found consistently at latitudes of 45°S and somewhat less frequently at 30°S and 30°N. By 1998, the clouds at both of these southern latitudes were more numerous and a new cloud feature, similar to one seen in Voyager 2 images, was seen near 67°S. A much more dramatic change was apparent by 2002, when individual clouds appeared to be much brighter at 30°S, 45°S, and 67°S, and the spatial extent of the bright clouds was much greater.

The changes are especially prominent at near-infrared wavelengths, which are much more sensitive to high altitude clouds. They are thought to be a response to seasonal variations in sunlight, particularly at higher latitudes, where the seasons tend to be more pronounced. The result is stronger upwelling of convectional currents, leading to more condensation and increased cloud cover. Since the response of the massive atmosphere to the small change in insolation is thought to be very slow, the planet is likely to continue brightening for another 15–20 years.

Figure 11.9 A large dark spot at 30°N was imaged with the Hubble Space Telescope on November 2, 1994. Unfortunately, because the planet's northern hemisphere was tilted away from Earth, the new feature appeared near the limb of the planet and was difficult to observe. Like the Great Dark Spot, it was outlined by methane ice clouds that formed at high altitude. (H. Hammel/Massachusetts Institute of Technology and NASA)

2002

1998

1996

| 0 ROT | 1/4 ROT | 1/2 ROT | 3/4 ROT | AVG |

Figure 11.10 Changes in Neptune's appearance 1996–2002, as seen by the Hubble Space Telescope. The individual images were processed to reduce limb darkening and enhance contrast. The first four columns display views at 0, 1/4, 1/2, and 3/4 rotations to show the distribution of discrete features. The last column displays the longitudinal average, showing little change at the equator but dramatic increases in brightness near 30°S, 45°S, and 67°S. (L. Sromovsky, P. Fry / Univ. of Wisconsin, NASA, ESA)

Interior

The internal structure of Uranus and Neptune is difficult to determine (see Chapter 10) because their bulk composition seems to be more complex than for Jupiter and Saturn, with a smaller fraction of hydrogen and helium, and their gravitational moments are known with a lower accuracy. Both are thought to be "ice giants," but there are some differences.

Neptune's mass and density are higher than those of Uranus, despite it being slightly smaller. This is partly due to greater compression, but could also be the result of small differences in composition, perhaps with a slightly higher proportion of rock to ice. (Assuming that the ice/rock ratio of both planets is similar to the protosolar nebula, their overall compositions, by mass, should be about 25% rock, 60–70% ice, and 5–15% hydrogen and helium.)

Beneath the atmosphere of hydrogen, helium and methane, there is probably a layer of liquid hydrogen that also includes some helium and methane. Most of Neptune's interior is thought to be made up of an "ocean" of molten "ices," probably a mixture of water, methane and ammonia. At the center may be a rocky core, perhaps slightly larger than the core of Uranus.

There has been considerable debate over the possible chemical reactions that take place at high temperatures and pressures inside Neptune and Uranus. One theory suggests that the outcome would be molecular hydrogen and pure carbon in the form of diamond crystals. The latter, if large enough, would irreversibly sink as sediment toward the center of the planet, whereas the hydrogen would rise into the atmosphere. Another theory favors the dissociation of methane into free hydrogen and ethane.

Why is Neptune releasing more than twice as much energy as it receives from the Sun, so that its temperature is comparable to that of Uranus? As with Jupiter and Saturn, much of Neptune's internal heat is thought to be produced by continuing, slow contraction because of its strong gravitational field. Differentiation in the icy mantle also allows heavier molecules to sink to lower levels, liberating heat energy.[3]

This heat seems to escape to the surface more easily in Neptune than in Uranus, possibly because convection is less inhibited by differential layering in the latter's interior. If this layering exists at deeper levels in Neptune, it would allow more heat to be transported outward.

Magnetic Field

Like the other Jovian planets, Neptune has a powerful magnetic field and a large magnetosphere. Voyager found that Neptune's magnetic field was 25 times stronger than Earth's, though the planet is so large that the field is actually weaker at its cloud tops than on the surface of our planet.

One of the biggest surprises of the Voyager encounter was the lopsided nature of the magnetic field, which bore many similarities to that of Uranus. The axis of the magnetic dipole is tilted 47°

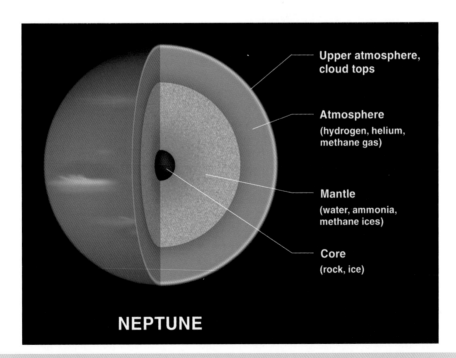

Figure 11.11 Like Uranus, Neptune consists mainly of a vast internal ocean of water, methane and ammonia "ices" which become liquid at the high temperatures (several thousand degrees Celsius) and pressures deep in the interior. The presence of a rock-ice core about the size of Earth is inferred. (LPI)

[3] Energy is released when mass sinks toward the center of an object, effectively converting gravitational potential energy into heat.

to the rotation axis and offset from the planet's center by 55% of Neptune's radius.

This means that the field originates in a region of the interior that is actually closer to the cloud tops than to the center, and the field is much stronger in one hemisphere than the other. In fact, the field strength at the surface varies from more than 1 gauss in the southern hemisphere to less than 0.1 gauss in the northern hemisphere. (Earth's equatorial magnetic field at the surface is 0.32 gauss.)

The large tilt of the magnetic dipole axis and the planet's rotation result in a dynamic magnetosphere which changes from an Earthlike configuration (when the angle between the dipole axis and solar wind is about 90°) to a pole-on configuration and back every 16 hours or so. As a result, the cusp regions above the poles change location and size as the planet rotates while the magnetic equator and field lines appear to gyrate wildly in the solar wind as the planet rotates. When Voyager 2 passed the planet, one pole of the highly inclined magnetic field was pointing toward the Sun.

The field seems to be generated by slow moving convectional currents of ionized material in the planet's icy mantle. Recent studies suggest that the magnetic fields of both ice giants are generated in a thin outer layer of the mantle, rather than in the lower regions where convection does not take place.

Compared with other planetary environments, the magnetosphere is actually quite empty, with fewer protons and electrons per unit volume than any of the other gas giant planets. The main source of nitrogen ions is Triton's sparse atmosphere or ionosphere, while the hydrogen ions come from a large, diffuse cloud that extends inward from the satellite's orbit for more than 150,000 km.

The high inclination of Neptune's magnetic field means that charged particles trapped in the magnetosphere are repeatedly swept past the satellites and rings. Many of these charged particles are absorbed by the satellites and rings, effectively removing a large fraction of its charged particle population.

Weak auroral emissions were identified on the night side of Neptune, and they are thought to occur close to both magnetic poles.

Triton

Prior to the Voyager 2 flyby, only two satellites of Neptune were known. The largest of these, Triton, was found by William Lassell in October 1846, only 17 days after the planet's discovery.

Triton was quite bright for its distance, suggesting that the satellite rivaled Titan and Ganymede in size. However, its precise diameter remained uncertain until the Voyager encounter, when it became apparent that ice-covered Triton acts like a mirror, reflecting 72% of sunlight at visible wavelengths. This unexpectedly high albedo explained why astronomers had seriously overestimated Triton's size. With a diameter of 2710 km, Triton is smaller than the Galilean satellites, Titan and Earth's Moon, but larger than Pluto.

Based on its size and density (derived from its effect on Voyager's trajectory), it is possible to model the moon's interior. With

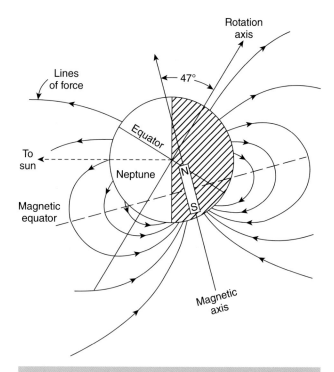

Figure 11.12 Like the other giant planets, Neptune's magnetic polarity is the opposite to that of Earth. The field is tilted 47° with respect to the rotation axis and offset 0.55 Neptune radii from the center. (NASA-JPL)

a density about twice that of water, Triton has approximately the same density as Ganymede and Callisto, suggesting that it is rocky rather than icy. Models indicate that there is a thick layer of water ice above a rocky central region, which occupies about two thirds of its radius. There may also be a small iron-rich core.

Triton is unusual in other ways. It orbits about 355,000 km from Neptune, completing one circuit every 5.9 days in a retrograde direction – opposite to the planet's rotation.[4] A season on Triton lasts about 41 years, and Triton passed its southern summer solstice in 2000. At the time of Voyager's flyby Triton's southern hemisphere was experiencing high summer (and still is), while the northern hemisphere was in the middle of a prolonged winter.

However, the satellite also experiences a much more complex, 688-year-long cycle of seasons, resulting from Neptune's 164-year passage around the Sun, an orbital inclination of 23° to the planet's equator, and the precession of Triton's orbit about Neptune's axis of rotation. This means that the seasons vary considerably in length and intensity in the long term.

Despite the seasonal variations in sunlight, solar heating appears to have very little impact on Triton, with only minor variations in temperature from one hemisphere to the next. The average temperature of −235°C means that Triton is so cold that most of the nitrogen in its tenuous atmosphere freezes out onto the surface, making it the only satellite in the Solar System known to have a surface made mainly of nitrogen ice. (Infrared spectra

[4]Triton was the first satellite found to have a retrograde orbit. It is also the only large satellite in the Solar System to orbit a planet in a retrograde direction.

TRITON LATITUDE OF SUBSOLAR POINT

Figure 11.13 Triton experiences a remarkably complex succession of seasons as the position of the overhead Sun (the subsolar point) shifts from the southern to the northern hemisphere. At the time of the Voyager encounter in 1989, it was early summer in the southern hemisphere, whereas the northern hemisphere was in prolonged darkness. The subsolar point reached latitude 50°S in late 2000, the first time in over 350 years that it had been so far from the equator. (NASA-JPL)

Figure 11.14 A photomosaic of Triton centered near 20°N, 0°W. The equator lies approximately along the pale green region (centre). The surface is mainly nitrogen ice. Pinkish deposits in the southern hemisphere contain methane ice which has been modified by ultraviolet sunlight to form organic compounds. The dark streaks are probably carbonaceous dust deposited from geyser-like plumes. The pale green band that extends all the way around Triton is relatively fresh nitrogen frost. The greenish areas include the "cantaloupe" terrain and cryovolcanic plains produced by icy material erupted from Triton's interior. (USGS/NASA-JPL)

have also revealed the presence of frozen methane, carbon monoxide and carbon dioxide.)

Triton's atmosphere is composed mostly of molecular nitrogen, with a trace of methane near the surface, but it is extremely thin. The surface pressure determined by Voyager was about 16 microbars – about the same as that 70 km high in the Earth's atmosphere. Nevertheless, Voyager imaged discrete clouds up to 8 km above the surface, and haze was detected up to altitudes of 20–30 km.

Triton is one of only four objects in the Solar System known to have a nitrogen-dominated atmosphere. The others are Earth, Pluto and Saturn's giant moon Titan. Nitrogen, carbon dioxide and carbon monoxide are known to exist in interstellar clouds, so it is likely that the very cold objects in the outer Solar System have probably retained a lot of the original molecular mix of the nebula from which the planets formed.

When Triton's southern hemisphere receives more direct sunlight as summer approaches, the thin layer of frozen gases on the surface sublimates into gas, thickening the atmosphere. Confirmation of seasonal variations came in November 1997 when Triton passed in front of a star. Astronomers took advantage of the rare occultation to study its atmospheric density by measuring the gradual decrease in the star's brightness as its light traveled through the satellite's sparse atmosphere.

The data indicated that Triton's atmosphere had doubled in density. Based on the unusually strong link between the temperature of its surface ice and atmospheric pressure, the moon had apparently warmed by almost 2°C since the Voyager encounter nine years earlier, resulting in a "balmy" temperature of −234°C. By 2010, it was estimated that Triton's atmospheric pressure had risen by a factor of four since 1989, so that the atmospheric pressure had increased to between 40–65 microbars (still 20,000 times less than on Earth).

Voyager's cameras unveiled a strange landscape shaped by the seasonal variations in ice cover and by internal processes. The visible area was dominated by two dramatically different types of

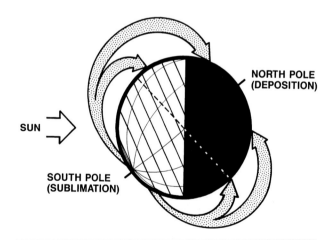

Figure 11.15 Sublimation of nitrogen frost in Triton's summer hemisphere is counterbalanced by condensation and deposition of ices in the winter hemisphere. (NASA-JPL)

terrain: the so-called "cantaloupe" terrain close to the equator and a receding ice cap.

The bright southern polar cap appeared to be composed largely of nitrogen frost. There was some evidence that this ice

Figure 11.16 Two images of the dark Mahilani plume on Triton, imaged by Voyager 2. The plume (arrowed) rises rapidly to a height of about 8 km – presumably the top of the troposphere – and is then blown westward for about 100 km by winds of 10–20 m/s. (NASA-JPL)

was sublimating – turning directly from ice to vapor – in the equatorial region, presumably to recondense in the dark northern polar region.

More than 100 dark streaks were visible on the ice cap, mainly between 15–45°S. Ranging from 4 km to over 100 km in length, they were often fan-shaped and aligned toward the north east. Their appearance suggested that they were wind-blown deposits – despite the sparse atmosphere on Triton. This interpretation was confirmed when stereo images of some streaks revealed that they were associated with plumes that soared to an altitude of 8 km before being diverted horizontally by high level wind.

The Mahilani Plume at 48°S 2°E appeared as a very narrow, linear cloud at least 90 km in length, whereas the Hili Plume at 57°S 28°E comprised several much broader streamers up to 100 km long. In both cases the clouds were moving toward the west. It seems that the wind direction shifts dramatically at different levels on Triton, with a north eastward flow close to the surface, an eastward flow above 1 km and a westward flow at the height of the plumes.

The origin of the plumes remains uncertain. Most models assume that they are somehow driven by the heat of the overhead Sun in summer. The weak solar radiation could penetrate the transparent nitrogen ice to a depth of several kilometers. Since the nitrogen ice is an excellent insulator, subsurface ices may melt or sublimate. Pressure building up over a period of time could suddenly be released when liquid nitrogen (and perhaps other materials) finds its way to the surface and erupts as a boiling geyser. Traveling rapidly upward, the plume could pick up dust from the surface, while nitrogen and other gases making up the plume would freeze and condense. Such an eruption could last for at least a year before it subsided.

It is also possible that they are geysers powered by cryovolcanic activity involving localized internal heat sources, or that the thickness of the nitrogen ice cap plays a role. Another possibility is that the plumes are an atmospheric phenomenon, perhaps resembling some form of dust devil.

There is a certainly a strong case for arguing that many of the surface features seen beyond the edge of the polar cap are associated with cryovolcanism and internal activity. Voyager found that the western half of Triton's visible equatorial region comprises a unique surface of large, elliptical or kidney-shaped "dimples" criss-crossed by linear ridges. The depressions (known as cavi) that make up this "cantaloupe" terrain are typically 25–35 km across, with no overlapping.

The favored explanation is that these dimples are created by diapirs – blobs of warm, buoyant, less dense material rising to the surface through denser overlying material. This theory implies that Triton has distinct crustal layers which allow ammonia-water ices to respond to heating from below and move toward the surface.

The network of ridges bears some resemblance to features on the surface of Jupiter's moon Europa. Some parallel ridges 6–8 km wide occur in pairs either side of a central trough; others are triple or in the form of single, narrow valleys 2–3 km in width. All of them appear to be tectonic features created as the result of extension of the crust. The medial ridges may be the result of subsequent intrusions of material, similar to volcanic dikes on Earth.

Most of Triton is fairly flat, with a difference in relief of no more than 1 km. This lack of topography is particularly noticeable to the east of the cantaloupe terrain. The smooth plains close to the eastern limb in Voyager's close-up images are thought to be the oldest, based on their higher crater count. They include Mozamba, the largest known impact crater on Triton, which has a diameter of 27 km and a prominent central peak.

Nearby are some smooth, relatively dark patches surrounded by brighter aureoles. Unlike any features seen on other objects, these "maculae" (spots) are of unknown origin. It may be that the central patches are made of carbonaceous material (presumably including some methane ice) bounded by brighter, nitrogen-rich ice.

Equally intriguing are four walled plains whose edges rise in steps (scarps) to the surrounding, terraced terrain. Their floors are the flattest places on Triton, implying that they were flooded by very fluid material. Clusters of small pits in their central regions may be associated with cryovolcanic vents.

Is this activity still ongoing? The absence of large impact craters excavated by comets from the Kuiper Belt, together with the relatively low crater density (only a few hundred craters more than 3 km across have been identified), indicate that the surface is relatively young. Estimates range from 100 million years to only 10 million years, but if the latter figure is accepted, surface activity is almost certainly continuing at the present day.[5]

Although there have been various theories concerning the origin of Triton, its unusual retrograde orbit suggests that it was captured by Neptune, rather than formed in the planet's vicinity. Remarkable similarities with Pluto – including size, density, surface composition and atmosphere – suggest a common birthplace in the Kuiper Belt.

According to this view, Triton was a planetesimal in the inner Kuiper Belt that approached too close to Neptune. The proto-Triton could have been slowed by a collision with an existing satellite or interaction with a disk of material orbiting the young planet. As time went by, the moon's highly elliptical orbit became smaller and more circular due to tidal forces generated during each close passage. A periodic rise and fall of the bulge in Triton raised by Neptune's gravity at each approach would dissipate energy and modify the orbit.

Critics argue that the likelihood of a collision which is large enough to slow down Triton, but small enough not to destroy it, is extremely small. Similarly, aerodynamic drag from a disk of gas around Neptune would have to occur at a particular stage in the planet's early evolution. It would have to be surrounded by a disk that was dense enough to slow Triton down and bring it into a circular orbit, but the gas would have to disperse before Triton's orbital motion was slowed so much that it was sent crashing into Neptune.

An alternative scenario suggests that Triton may once have belonged to a binary pairing, similar to Pluto and its large moon Charon. In a close encounter with Neptune, the pair were pulled apart by the planet's gravitational forces and Triton was captured by the giant planet.

Tidal forces are still influencing Triton today, causing it to spiral slowly towards the planet. In 10–100 million years, it will be so close that Neptune's gravity will tear it apart, forming a new, spectacular ring.

Small Satellites

Before the Voyager flyby, the Neptune system was known to have one small moon, Nereid, which was found by Gerard Kuiper in

Figure 11.17 Ruach Planitia is a 175 km wide walled plain bounded on all sides by rougher plains. The extremely flat floor may have been flooded by fluid, icy material. Terraces and benches around the edges may be evidence of previous flooding followed by collapse. In this computer-generated view, topography has been exaggerated roughly 30 times; the actual range of relief is about 1 km. (NASA-JPL)

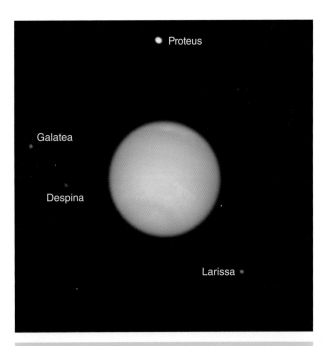

Figure 11.18 A Hubble Space Telescope image taken in April 2005 revealed high altitude ice clouds on Neptune. Also visible are four small moons close to the planet. (NASA, ESA, E. Karkoschka/University of Arizona, and H. B. Hammel/Space Science Institute)

[5] All of the craters are on Triton's leading hemisphere, implying that it is sweeping up debris – possibly from a shattered satellite – as orbits Neptune.

1949. Nereid follows one of the most elliptical orbits of any satellite, traveling between 1.4 million km and 9.7 million km from Neptune over a period of 360 days. The orbit is inclined about 7° to Neptune's equator and is prograde.

The fuzzy Voyager images suggest that Nereid is roughly spherical, with a diameter of about 340 km. It is brighter than the other small satellites and gray in color, suggesting an icy surface. This seems to be confirmed by the discovery of water ice in its spectrum. Unusually, the rotation period, based solely on changes in its light curve, is almost certainly not synchronous. One estimate suggests a period of 13.6 hours, but others have failed to find any regular pattern, possibly pointing to a chaotic motion resembling that of Saturn's satellite, Hyperion.

Voyager revealed six more satellites inside the orbit of Triton. They are arranged in order of size, with the smallest (Naiad) closest to Neptune and the largest (Proteus) furthest away. It is assumed that all of them are in synchronous rotation.

Proteus, orbiting at a distance of 117,650 km, is the second largest satellite of Neptune. With a diameter of more than 400 km, it has an odd, box-like shape, with a large impact crater, called Pharos. With a slightly higher mass, its gravity would have been sufficient to pull the body into a sphere.

Like Proteus, all of the other small, inner moons are dark and follow circular orbits with low inclinations. Apart from Proteus, they all scoot around the planet in less than one Neptunian day.

Perhaps the most important characteristic of the five inner satellites is their shepherding interaction with the planet's narrow rings (see Chapters 9 and 10). Naiad, Thalassa and Despina orbit between the innermost ring (Galle) and the Le Verrier ring. Galatea patrols between the Arago and Adams rings, while Larissa shapes the outer edge of the Adams ring.

With the advent of CCD cameras and ground-based telescopes fitted with adaptive optics, five more, irregular moons were discovered far beyond Triton and Nereid in 2002 and 2003, boosting the number of known satellites to 13. This was something of a surprise, since it was thought that cataclysmic events related to the capture of Triton would have dispersed any outer satellites.

Little is known about these moons, but they are all thought to be 60 km or less in diameter. Their average orbital distances are between 15 million km and 49 million km. Indeed, Neso is the furthest known satellite of any planet, taking about 25.6 years to complete one orbit.

The satellites follow highly elliptical orbits steeply inclined to Neptune's equator. One of them (Halimede) even crosses the orbit of Nereid. Two of the satellites follow prograde paths, while the other three have retrograde orbits. The irregular nature of their orbits indicates that they have been captured by Neptune, rather than forming in a disk around the young planet.

The outer satellites are regarded as remnants of a system that has been modified over billions of years by collisions and gravitational interactions with Triton and Nereid, together with solar tides. Dynamical studies show that the eccentricities and inclinations of their orbits undergo large variations over time. Collisions between Nereid and the innermost of the outer satellites, particularly Halimede, are quite probable over the lifetime of the Solar System.

Figure 11.19 Proteus imaged on August 25, 1989, from a range of 146,000 km. Spatial resolution is about 2.7 km. The half-illuminated satellite has a diameter of more than 400 km. It is dark (albedo 6%) and spectrally gray. Apart from a large crater, hints of craters and groove-like features can be seen. (NASA-JPL)

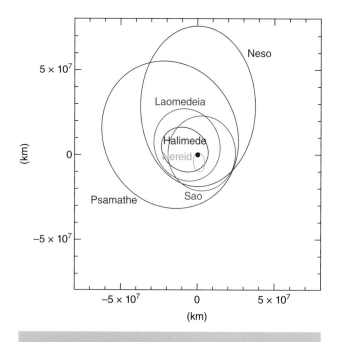

Figure 11.20 The orbits of Neptune's five outer moons are all highly elliptical and inclined to its equator. Two of them (Sao and Laomedeia) follow prograde orbits. The others are retrograde. (After S. Sheppard)

Rings

With the discovery of rings around Uranus in 1977, it seemed possible that similar, dark rings might exist around Neptune. The best chance of finding them was to examine light curve data obtained during a stellar occultation.

During the 1980s, many occultations were meticulously studied, but definitive proof that Neptune had a ring system was hard to find. Although more than 90% of the observations yielded no detection, at least five occultations seemed to provide evidence of material orbiting Neptune.

One observation indicated the presence of either a small satellite or an optically thick, incomplete ring with a radial width of about 80 km.[6] Other confirmed sightings also detected dimming on only one side of the planet. The puzzling data led to the theory that Neptune was orbited by several discontinuous ring arcs. These were thought to be narrow (15–25 km), at least 100 km in length, and located between 41,000 and 67,000 km from Neptune.

Voyager 2's cameras soon solved the mystery by discovering five charcoal black, but continuous, rings between 41,900 km and 62,933 km from the center of Neptune. Images confirmed the existence of elongated arcs of material as distinct sections of the outer, Adams ring. There was also some evidence for a faint, partial ring sharing the orbit of Galatea – about 1000 km inside the Adams ring.

The ring nomenclature is a little easier to memorize than for Uranus, since they were named after some of the leading characters involved in the planet's discovery. In order of distance from the planet, they are: Galle, Le Verrier, Lassell, Arago and Adams.

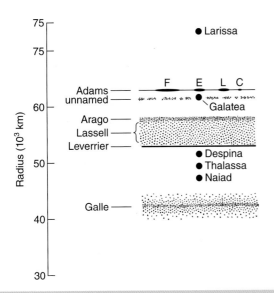

Figure 11.21 The five inner satellites shepherd the rings of Neptune. Galatea may shape the arcs of the Adams ring. There may also be a partial ring which shares the orbit of Galatea. (From Neptune and Triton, Univ. of Arizona press)

Three of the rings are extremely narrow, with a radial width of less than 100 km. Only Galle and Le Verrier measure several thousand kilometers across. Radio and optical occultation data indicate that the Adams and Le Verrier rings are composed of material no more than a few centimeters across. There is a much higher proportion of micron-sized dust than in the rings of Uranus or Saturn. Their overall mass is about 10,000 times lower than the rings of Uranus. All the material gathered together would form a ball only a few kilometers across.

One similarity with the Uranian system is the low albedo (less than 5%), possibly due to the alteration of methane ice by ultraviolet light and high energy particles, leading to a coating of elemental carbon similar to graphite.

The Adams ring is particularly interesting, not only because it is the densest and slimmest member of the ring system, but also because it contains five arcs that are noticeably brighter than the remainder of the 50-km-wide ribbon. All are contained within a 40° section of the ring.

At the time of the Voyager flyby, the arcs made up about one tenth of the ring's circumference, sufficient to explain the puzzling occultation observations. Leading the orbital procession of "beads" was an arc called Courage, followed by Liberté, Egalité 1 and 2 (sometimes regarded as subdivisions of the same arc) and Fraternité.

The reasons for the formation and continuing existence of the rings and the arcs are not clear. The presence of a large amount of dust, together with the presumably ephemeral ring arcs, favors a system that formed fairly recently. The most likely explanation is the destruction of a small satellite.

The evolution of the rings might be expected to be affected by the nearby presence of five small, icy moons. Naiad, Thalassa and Despina travel between the innermost, Galle ring and Le Verrier, while Galatea patrols between the Arago and Adams rings. A fifth satellite (Larissa) moves outside the Adams ring. (The four inner satellites and the rings all reside within the Roche limit for objects composed of water ice – see Chapter 9)

However, most of these moons do not behave like the shepherds that confine the rings of Uranus and some of Saturn's rings. Apart from Galatea, which shares its orbit with a partial ring, there is no obvious relationship between the inner moons and the narrow rings. The only orbital resonance that has been confirmed involves Galatea and the Adams ring, with the particles in the ring moving around the planet 42 times for every 43 orbits of Galatea. This repetitive resonance enables Galatea's gravitational influence to limit the longitudinal spread of the arcs and confines the radial width of the Adams ring.

However, this does not explain why the Adams ring does not migrate outward as energy is transferred from Galatea to the particles in the Adams ring, or the absence of arcs in the other rings. Neither does it explain why some of the other rings are so narrow and sharply defined. This has led to suggestions that one or more unseen moons, less than 10 km across, may be embedded either in the arcs or at a Lagrangian point in the ring, 60° ahead or behind the arcs.

Neptune's inner moons and ring arcs are difficult to observe because they are faint (below magnitude 20) and close to the

---

[6]**Data obtained during the Voyager 2 flyby confirmed that this was a small satellite, Larissa, which orbits just outside the ring system.**

Figure 11.22 Two long exposures of Neptune's rings taken by Voyager 2 from a distance of 280,000 km. The period between exposures was one hour and 27 minutes. The arcs in the Adams ring were not visible, since they were on the opposite side of the planet during each exposure. The image shows the faint inner ring (Galle), about 42,000 km from the center of Neptune, and Lassell, a broad, faint band between the Le Verrier and Arago rings. The rings were back-lit by the Sun, allowing fainter, dusty regions to be seen. The two gaps in the outer ring (left image) are due to computer processing. (NASA-JPL)

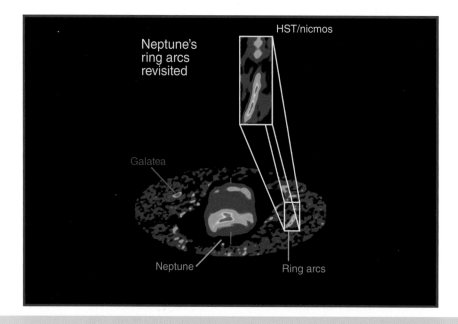

Figure 11.23 Near-infrared observations made with the Hubble Space Telescope have provided information about the orbital periods of the inner moons and subtle changes in the rings. The data indicate that the Liberté ring arc has edged forward, towards Courage. (Christophe Dumas, NASA/JPL, University of Arizona Steward Observatory/HST-NICMOS)

planet. However, monitoring with the HST and ground-based instruments has shown that the rings are gradually evolving. The narrow Le Verrier ring has brightened by a factor of four since 1989. By 2003 Galatea had moved 8° ahead of its predicted position, relative to the arcs. This led to the hypothesis that the arcs are massive enough to affect the satellite's orbit, possibly because of the presence of sizeable objects hidden in the ring, which act as the source of the visible dust.

Major changes in the arcs are also taking place, both in their location and intensity relative to the trailing arc, Fraternité, which seems to be the most stable. In contrast, the leading arcs, Liberté and Courage, have shifted forward and decayed.

The most noticeable change was in Liberté, which displayed a forward shift of about 2° and was considered to be on the verge of disappearing completely. At the same time, there was a slight lengthening of Egalité, possibly as the result of material being transferred from Liberté. The sub-arcs Egalité 1 and 2 appeared to have reversed in relative intensity, perhaps as the result of material migrating between resonance sites, A forward shift in location of 8° by Courage could also be associated with a movement over one full resonance site.

No missions to Neptune are planned for the foreseeable future.

Questions

- The discovery of Neptune was said to be a triumph for gravitational theory. Do you consider this to be true? Explain your answer.
- Why is scientists' knowledge about the Uranus and Neptune systems much less than for the other planets?
- Summarize the main similarities and differences between Uranus and Neptune. Suggest reasons for these similarities and differences.
- Compare and contrast the magnetospheres of Uranus and Neptune.
- Four planets possess ring systems. What are the main similarities and differences between the rings of Neptune and the other ring systems?
- Why is Triton unique (as far as we know) among the major satellites of the Solar System?
- Why is a future mission to Neptune unlikely to take place for several decades?

TWELVE

Pluto and the Kuiper Belt

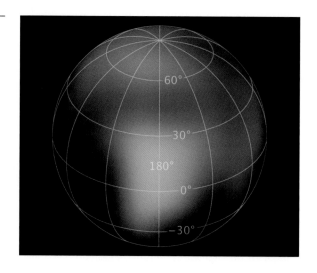

For more than 80 years, Neptune was considered to be the outermost planet in the Solar System. Then, in 1930, a young research assistant at Lowell Observatory discovered Pluto, a small planet in an eccentric orbit that crossed the orbit of Neptune. 48 years passed before a large satellite was found. 14 years on and the first of many objects that belong to the so-called Kuiper Belt was discovered beyond Neptune. Today, the number of known Kuiper Belt objects exceeds 1300, though the total population is probably above 200 million. Some of these objects rival Pluto in size, and at least one is even larger. Today, Pluto and the other large Kuiper Belt objects are classified as dwarf planets.

The Search for Planet X

With the discovery of Neptune, the catalog of major planets seemed once again to be complete. However, almost from the time of Galle's triumph (see Chapter 11), some individuals began to speculate that a ninth planet, presumably another giant, was lingering unseen in the outer Solar System. This suspicion was reinforced in the late 1800s, when observers reported that Uranus was once again straying, very slightly, from its predicted position, after all known gravitational influences were taken into account.[1]

The errors, or "residuals", were much smaller than those that prompted a search for Neptune, and might be accounted for by minute inaccuracies in the mapped positions of the background stars. However, there remained the tantalizing possibility that they might be caused by the pull of another planet.

The quest for "Planet X" was taken up by Percival Lowell, a wealthy amateur astronomer who had built his own observatory in Flagstaff, Arizona, and become a celebrity with his theories and observations concerning the existence of artificial canals on Mars. Lowell instigated a photographic survey along the ecliptic plane in 1905, the first of many attempts to find Planet X, but none of them were successful.

After a long hiatus, the Flagstaff observatory resumed the quest in 1929. The equipment included a 33 cm wide field photographic refractor and a blink comparator that would enable rapid, efficient comparisons of photographic plates of the same region of the sky taken a few days apart.

The search was entrusted to Clyde Tombaugh, a 23-year-old farmer's son and keen amateur astronomer. Tombaugh began his search in the constellation of Gemini, where Lowell had predicted Planet X would be. He soon decided to increase the chances of a discovery by concentrating on the sky around the "opposition point," the part of the sky directly opposite the Sun, where any hidden planet would appear brighter and move a greater angular distance against the background stars over a period of a few days.

Tombaugh's routine involved taking exposures of the night sky that lasted one hour or longer. The process would be repeated a few days later, if possible, for the same area of sky, so that the plates could be exposed and developed under similar conditions. (Tombaugh photographed around the time of New Moon.)

There was then the small matter of manually comparing the plates, each one containing between 50,000 and 900,000 stars. Although the blink comparator helped considerably, there was always the problem of microscopic flaws in the plates and false

---

[1] The orbit of Neptune was less well known – it had completed only about one third of an orbit since its discovery – so calculations of Planet X's location were based on the measured position of Uranus.

Exploring the Solar System, First Edition. Peter Bond.
© 2012 by Peter Bond. Published 2012 by Blackwell Publishing Ltd.

alarms due to wandering asteroids. Fortunately, a typical asteroid trail would be about 10 times longer than the shift of a few millimeters which Tombaugh's much more remote Planet X would be expected to display.[2]

Then, on February 18, 1930, the young man's dogged persistence paid off. Having worked his way around the ecliptic, Tombaugh had returned to the region near Delta Geminorum, a

Figure 12.1 Clyde Tombaugh (1906–1997) using the blink comparator to search for Planet X among the thousands of stars on each photographic plate. The instrument enabled him to detect small, faint objects moving against the background of "fixed" stars by comparing plates taken a few days apart. (Lowell Observatory)

bright star in the constellation of Gemini. He noticed a star-like object that shifted position against the stars. Careful study of its motion confirmed that this was no ordinary asteroid, and on March 13, 1930, the discovery of the ninth planet was officially announced. (The newcomer had actually been photographed in early 1929, but it had not been recognized.)

What to call Planet X? Suggestions poured in from all over the world, but it was eventually decided to name it Pluto – a proposal sent in by Venetia Burney, a young schoolgirl from Oxford, England. The name was considered particularly appropriate for such a distant planet, since it was associated with the Roman god of the underworld. The symbol for Pluto comprises the letters P and L – the initials of Percival Lowell.

Puzzling Pluto

At first, it seemed as though the discovery was another triumph for gravitational theory, since the planet had been found within 6° of the locations predicted by both Lowell and Pickering, his main rival. However, Lowell had predicted that Planet X would have more than six times the mass of Earth.

Unfortunately, in even the largest telescopes, Pluto showed no disk or evidence of an atmosphere. Neither did the tiny point of light appear to have any satellites that would make it possible to calculate its mass.

The first fairly reliable measurements of its size were made by Gerard Kuiper in 1950. Using the new 5 meter Palomar reflecting telescope, Kuiper estimated that Pluto's diameter was a modest 5860 km, which placed it midway between Mercury and Mars.

With the 1976 discovery of methane frost on its surface, Pluto was revealed to be a highly reflective object. Calculations based on its high albedo and apparent brightness showed that it must

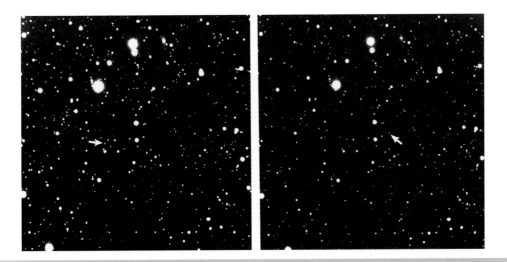

Figure 12.2 These photos, taken on January 23 and 29, 1930, with the 33 cm refractor at Lowell Observatory, are small sections of the photographic plates on which Pluto was discovered. Pluto (arrowed) is a star-like point of light. By rapid "blinking" of the two images, Clyde Tombaugh was able to detect its movement against the "fixed" background stars. (Lowell Observatory)

[2]In the course of his search, Tombaugh discovered numerous clusters of stars and galaxies, hundreds of asteroids, two comets and one nova.

Figure 12.3 Pluto follows an eccentric orbit, with perihelion at 29.6 AU and aphelion at 50.3 AU. For 20 years of its 248 year orbital period it is closer to the Sun than Neptune, but their different orbital inclinations and 3:2 orbital resonance mean that they never come close to each other. Pluto's orbital inclination (17°) carries it well above and below the orbits of the major planets. Also shown is the orbit of Eris, another dwarf planet and the largest known member of the Kuiper Belt. (NASA-JPL)

be smaller than the Moon, with a mass only a few thousandths that of Earth. This was confirmed in 1978, when a large satellite, named Charon, was discovered in close orbit around Pluto.

Between 1985 and 1990, there was a rare opportunity to observe mutual occultation events as Charon and Pluto took it in turns to eclipse each other.[3] These events made it possible to constrain their sizes. Subsequent HST imaging and stellar occultation measurements have made it possible to tie down Pluto's equatorial diameter at approximately 2320 km.[4] With a mass 500 times smaller than Earth's, and an orbit that ensures it never comes close to Neptune (see below), Pluto's gravitational influence on Neptune is negligible. Indeed, Earth has a greater gravitational pull on Neptune than Pluto ever can.

Clearly, Pluto is not Lowell's Planet X. Instead, its discovery close to the predicted positions must rank as one of the greatest coincidences in the history of science.

Orbit

Piecing together the nature of this dim, distant world has been a major challenge, although pre-discovery observations going back to 1914 soon enabled a reasonably accurate orbit to be calculated.

Although its mean distance from the Sun is 5.9 billion km (39.7 AU), well beyond Neptune, Pluto has a far more eccentric orbit than any other planet in the Solar System. At perihelion, it approaches to within 4.5 billion km, while at aphelion it recedes to 7.4 billion km. Perihelion last took place in 1989 and aphelion

will be reached in 2114. One entire orbit – the Plutonian year – lasts for 248 Earth years, so Pluto has completed less than one third of an orbit since its discovery. Its extreme distance from the Sun means that its orbital velocity is very low, averaging about 4.7 km/s.

The high eccentricity means that for 20 years of its 248 year orbit, Pluto is closer to the Sun than Neptune, an event that last occurred between 1979 and 1999. However, there is no possibility of a collision because Pluto's orbit is inclined 17° to the ecliptic and because the two objects are locked in a 3:2 orbital resonance. This means that Neptune completes three orbits in the time it takes Pluto to complete two. Conjunctions always occur near Pluto's aphelion, so that they never come closer than 2.5 billion km (17 AU), whereas Pluto can approach to within 1.6 billion km (11 AU) of Uranus.

Pluto's peculiar orbit is best explained if it formed in a low-eccentricity, low-inclination orbit within a primordial disk of planetesimals beyond Neptune. According to this model, outward migration of Neptune during the late phases of giant planet formation caused mean motion resonances with many of the planetesimals. The outcome was the eventual capture of Pluto (and other trans-Neptunian objects) in the 3:2 resonance, and an increase in Pluto's orbital eccentricity. The detection of many Kuiper Belt objects (Plutinos) locked in the 3:2 resonance with Neptune supports this scenario (see below).

Long-term orbit integrations have uncovered other subtle resonances and near-resonances which indicate that, although Pluto's orbit is regarded as chaotic, it remains stable over billions of years (i.e. most of the lifetime of the Solar System).

---

[3]The previous series of eclipses had taken place about 120 years earlier.
[4]Pluto is 2/3 the diameter of Earth's Moon but 15 000 times farther away. Viewing surface detail is as difficult as trying to read the print on a golf ball located more than 50 km away.

Seasons

For many years, Pluto's axial inclination and rotation period were unknown. The first clue came from measurements of its albedo, which showed fluctuations of more than 10% – a bigger variation than shown by all of the planets except Mars. Moreover, the fluctuations always repeated over the same period – 6 days 9 hours 17 minutes, plus or minus four minutes. It seemed clear that the observations were caused by variations in the light reflected from a solid surface rather than clouds in an atmosphere. If this was the case, then Pluto had a surprisingly long "day."

This rotation period was supported by the discovery of Charon, which orbits Pluto every 6.3872 days (see below). This is identical to Pluto's day – an example of **synchronous rotation**.

The next problem was to identify Pluto's axial inclination. Light curve studies over several decades had suggested that the axis was tilted at least 50°, but more precise calculations were impossible.

Once again, the discovery of Charon proved to be the key. Most major satellites orbit above or very close to the equatorial plane of their primaries, and the synchronous rotation of Charon indicated that it was no exception to the rule. Since Charon's orbit is highly inclined to the plane of Pluto's orbit, it follows that – like Uranus – Pluto rotates with its poles almost in its orbital plane.

In fact, Pluto's rotational axis is currently tipped 122°, although this obliquity varies considerably over a period of about 3.7 million years. Like Uranus and Venus, its rotation is retrograde, so that, if you were to view Pluto from above the ecliptic, it would spin from east to west. This means that each polar region experiences prolonged "summers" and "winters," with long periods in between when the Sun is nearly overhead at the equator.

This is thought to be the explanation for the variations in brightness that have been observed since the 1950s. When Pluto was first discovered, its relatively bright north polar region – possibly coated in fresh, reflective frost – was visible from Earth. Pluto appeared to dim as the visible surface gradually shifted from nearly pole-on in 1954 to equator-on in 1987. Since the autumnal equinox, Pluto's southern hemisphere has started to receive more sunlight than the northern hemisphere.[5]

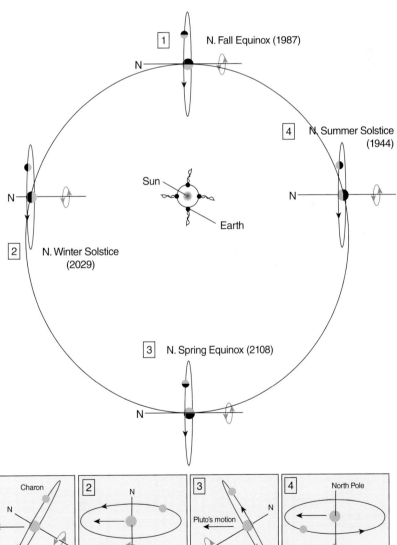

Figure 12.4 The changing alignment of the orbit of Pluto's moon Charon. The satellite's orbit is nearly circular, but tilted to our line of sight, so that it appears elongated. (JHU-APL)

However, the seasons on Pluto are not all the same length. The eccentric orbit causes it to receive almost three times more sunlight at perihelion than at aphelion (the variation on Earth is a mere 5%). As it happens, perihelion almost coincides with the northern autumnal equinox, while aphelion is close to the northern spring equinox. Since planets move faster when they are closer to the Sun, Pluto's northern spring lasts much longer than the relatively short northern fall.

Surface Conditions

Pluto is so small and far away that it is extremely difficult to see much detail on its surface. The apparent diameter of about 0.1 arcsecond, (18,000 times smaller than the width of a full Moon) makes it impossible to resolve any features on the surface with ground-based telescopes.

As a result, little was known about Pluto's surface for many years. The main conclusion, based on studies of the planet's light curves over many years, was that Pluto is highly reflective, with an average albedo of about 55%, similar to that of freshly fallen snow. (The Moon's average albedo is 12%.)

However, there were major differences. A dimming in its albedo as the aspect changed from near-polar to equator-on indicated that the poles were brighter than the equator. Separate dips in the light curve during each rotation suggested there were two dark regions on opposite hemispheres.

During the occultations of 1985–1989, when Charon moved directly in front of and behind Pluto (as seen from Earth), it was possible to obtain precise information on the surface markings of both Pluto and Charon. Several maps of Pluto's surface reflectivity were produced as a result of these efforts (see Box).

Although there were some significant differences poleward of 45°N, the maps showed general agreement on the broad distribution of light and dark regions. In particular, areas near the equator were generally dark, reflecting only 15% of incoming sunlight, while the north pole had an albedo of more than 70%.

Then, in 1994, the Hubble Space Telescope was able to image almost the entire surface at visible and ultraviolet wavelengths. Although each pixel (picture element) covered an area more than 160 km across, the images revealed at least 12 major regions defined by dark and light surface features. These included a permanent bright spot, a cluster of dark spots, and a bright linear marking. The darkest area of the equatorial zone appeared to lie very close to the sub-Charon point and to be ringed by a bright fringe. Both poles were bright and displayed a "ragged" border, but they were not identical. The northern polar region appeared to be both larger and brighter.

Subsequent observing campaigns with Hubble have shown more details of the surface markings and indicated how the frosts are changing with the seasons. Pluto has become significantly redder, probably the result of solar ultraviolet radiation breaking up methane frost on Pluto's surface and leaving behind a dark, reddish carbon-rich residue.

Figure 12.5 The first image-based map of Pluto was produced by computer processing of images taken with the Hubble Space Telescope in 1994. The map, which covers 85% of its surface, shows a dark sub-equatorial belt and bright polar regions. Some brightness variations may be due to topographic features, such as impact craters. (Alan Stern / SWRI, Marc Buie / Lowell Observatory, NASA and ESA)

---

[5] Traditionally, the north pole of a planet was defined as the one at which the planet rotated to the left and the stars to the right, as on Earth. For a world with retrograde rotation, such as Pluto, this placed the north pole on the south side of the Solar System. To avoid confusion, this definition has been changed so that the north poles of retrograde rotating objects are on the north side of the Solar System, as defined by Earth's rotation.

Figure 12.6 This map of Pluto, released in 2005, is the result of computer processing of HST images taken between July 2002 and June 2003. The red areas indicate methane ice and possibly other organics (carbon-based molecules). The dark areas may be dirty water ice. Lighter areas indicate nitrogen frost. The bright spot near the center of the map could be frozen carbon monoxide. (Marc Buie / Lowell Observatory and STScI)

The southern hemisphere has been getting brighter and the northern hemisphere has darkened. These are most likely seasonal changes, associated with sublimation and refreezing of ices. However, the brightening of the southern hemisphere is rather surprising, since it is experiencing a substantial increase in solar insolation, which might be expected to darken the surface. It has been suggested that may be due to a change in texture and reflectivity of the sublimating nitrogen ice.

Particularly noticeable in the Hubble images is a bright spot that is unusually rich in carbon monoxide frost, which contrasts with a nearby region covered in pitch-black surface material.

The reasons for many of the surface markings are still unclear. At least some may be associated with topographic features, such as impact basins or craters. However, if Neptune's moon Triton is any guide, most of the surface features, including the prominent polar caps, could be produced by the complex distribution of frosts that migrate across Pluto's surface during its orbital and seasonal cycles (see Chapter 11).

Support for this theory comes from studies of the ices on Pluto and their behavior as the seasons come and go. Clearly, its surface temperature is always extremely cold. Even on the inward leg of its orbit it receives 900 times less solar radiation than our Earth. At aphelion, the incoming insolation is more than 2400 times weaker than Earth receives. As a result, even the mid-summer temperature registers only −233°C, 40 degrees above absolute zero.

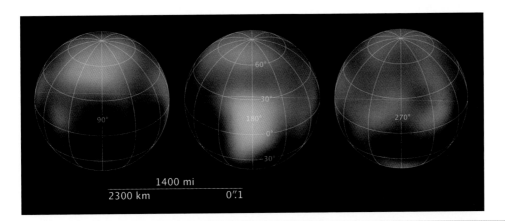

Figure 12.7 HST images of Pluto showing how the surface appeared 2002–2003. The north polar region (bottom) is not visible. The complex pattern reveals white, dark-orange, and charcoal-black terrain. Note the bright south polar region, and the mainly darker equator and northern hemisphere. (NASA, ESA, and M. Buie / Southwest Research Institute)

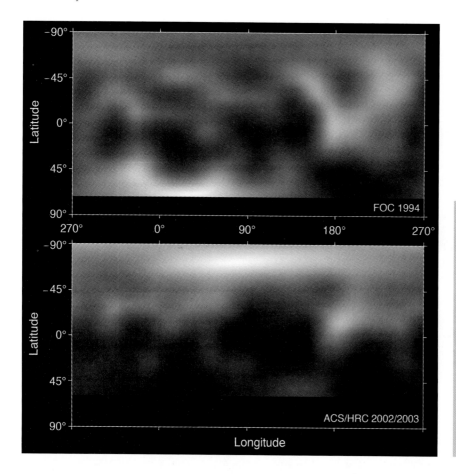

Figure 12.8 A comparison of HST photo maps of Pluto, obtained in 1994 (above) and 2002–2003 (below). The white areas are frosts, and the dark areas are a carbon-rich residue caused by sunlight breaking up methane. Summer is approaching in the south (top), which should cause ices to vaporize, then freeze out in the colder, shaded region. Surprisingly, the south pole has become brighter and the northern hemisphere is darker. More of the northern hemisphere was no longer in sunlight by 2003. (NASA, ESA, and M. Buie / SwRI)

However, the reality is more complex. Infrared measurements show that the darker regions on Pluto are up to 20°C warmer than the bright regions. This could be explained if the brighter ices were not only reflecting more sunlight, but also rapidly sublimating (changing directly from ice to gas).

Meanwhile, spectroscopic observations at infrared wavelengths have shown that the surface is covered with a veneer of different ices, particularly frozen nitrogen, carbon monoxide and methane. By far the most abundant of these is nitrogen ice. Studies of changes in Pluto's reflectance spectrum as it rotates make it possible to distinguish the compositions of different regions.

One bright area on the side of Pluto facing away from Charon (longitude 180°) is especially rich in carbon monoxide ice, whereas the Charon-facing hemisphere is dominated by methane ice. Another bright region shows evidence for nitrogen ice while a darker region is relatively ice-free and dominated by dark, organic residues resulting from processing of hydrocarbons and nitrogen ice by ultraviolet light and cosmic rays.

These ices are volatile, even at Pluto's frigid temperature, so they sublimate to create a thin atmosphere. The gases eventually find their way to the planet's colder night side, where they condense.

Pluto's Atmosphere

Like most of its other physical characteristics, the question of whether Pluto possessed an atmosphere remained uncertain for many years. Its low mass and density indicated that any atmosphere must be very sparse, even though the escape velocity is very low. The apparent absence of any transient features in the light curve also suggested an absence of clouds or thick hazes.

The first evidence of a possible atmosphere came in 1976, when the spectral signature of methane ice was recorded on Pluto's surface. Since its temperature is below the freezing point of methane, there was some doubt over whether any significant sublimation of the methane would occur, thereby creating a gaseous blanket. However, much of Pluto has a high reflectivity, something that would not be expected if methane ice had been reacting with ultraviolet sunlight over millions of years. The resultant layer of hydrocarbons, such as ethane and acetylene, would result in a dark surface.

The first unambiguous detection of an atmosphere came during a stellar occultation in June 1988. Instead of suddenly disappearing behind Pluto, there was a gradual dimming of the star's light, followed by a more rapid change closer to the surface – evidence of some form of atmospheric layering. This could be associated either with layers of haze at an altitude of 30–40 km or an increase in temperature with altitude due to absorption of infrared radiation by methane gas.

The surface pressure was obviously very low, probably between 3–100 microbars. (One microbar is a millionth of the surface air pressure on Earth.) However, there was evidence for a sparse, gaseous halo of escaping molecules extending into space for thousands of kilometers.

The occultation data caused a problem for theorists, who were still uncertain about Pluto's surface temperature. If the

Box 12.1 Eclipse Season

The Pluto-Charon eclipse season took place between 1985 and 1990, when Pluto's equator was aligned with the Earth. (Fortuitously, it was also around the time of Pluto's closest approach to the Sun, so that the largest possible disk was visible.) Since Charon orbits above the equator, astronomers were able to observe it moving in front of and behind Pluto every 6.4 days.

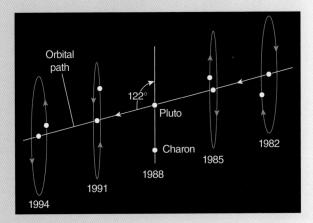

Figure 12.9   Pluto is tipped on its side with its spin axis close to the plane of its orbit. Like most satellites, Charon orbits above Pluto's equator. From 1985 to 1990, Pluto's equator and Charon's orbital plane were aligned with Earth, so that Charon passed in front of, or behind, Pluto every 3.2 days. (JHU)

The location of Charon – and its shadow – varied slowly over the eclipse season. The first eclipses (also known as occultations or "mutual events") began with Charon covering more and more of Pluto's north polar region. Later eclipses blocked the equatorial region, and the final occultations took place over the south polar region.

At the beginning and end of the eclipse season, Charon blocked very little of the light reflected from Pluto, so there was little change in the light curve. Around the peak of the eclipses, in 1987, the reflected sunlight dimmed by more than 30%. The eclipses lasted up to four hours and careful timing of their beginning and end made it possible to obtain the first accurate measurements of the diameters of the two objects.

Changes in their combined brightness made it possible to produce maps of the surface reflectivity on the Charon-facing hemisphere. It was found that Pluto had a highly reflective north polar cap, a dimmer south polar cap, and a mixture of bright and dark features in the equatorial region. Pluto's geometric albedo is 0.49–0.66, so it is much brighter than Charon, whose albedo ranges from 0.36–0.39.

temperature was in the range −218 to −213°C (55–60 K) there should be a methane-rich atmosphere with a pressure of perhaps 500 microbars. The explanation came in 1993, when the signatures of frozen nitrogen and carbon monoxide were found in infrared spectra of Pluto. Nitrogen accounted for 99.5% of the surface ices.

The following year saw the first spectral detection of methane gas in Pluto's atmosphere – but only in trace amounts. Clearly, as on Neptune's moon Triton, gaseous nitrogen must be the major atmospheric constituent, with a small amount of methane and carbon monoxide.

More recent observations have shown that dramatic seasonal changes can take place. Two stellar occultations in 2002 provided evidence that, instead of cooling as it moved away from the Sun, Pluto's lower atmosphere (and presumably the surface) had actually warmed by about 2°C since perihelion in 1989.

At the same time, atmospheric pressure had doubled over a period of 14 years, presumably through the sublimation of surface ices (though some cryovolcanic activity cannot be ruled out.)

There was also a reduction in the high level hazes inferred in the 1988 observations, possibly caused by dispersion or condensation on the surface. At the highest levels the temperature was measured as −163°C, 10–30°C colder than in the past.

This surface warming was probably the result of "thermal lag." Just as the surplus of incoming solar heat over heat radiated to space results in the warmest time of the day on Earth occurring in mid-afternoon rather than at noon, so Pluto's highest temperature occurs after its closest approach to the Sun.

This trend cannot continue for too long. Pluto will inevitably cool as its orbital distance increases. Whether this will be a gradual cooling or a dramatic collapse remains uncertain. Some estimates suggest that the atmosphere could freeze out completely by 2020.

Meanwhile, with a denser atmosphere fed by sublimation of ices and significant variations in surface temperature, Pluto may be imagined as a world swept by strong winds and with a complex atmospheric circulation.

The current picture is of a world whose sparse atmosphere is fed by sublimation of volatile ices for a period of about 50 years

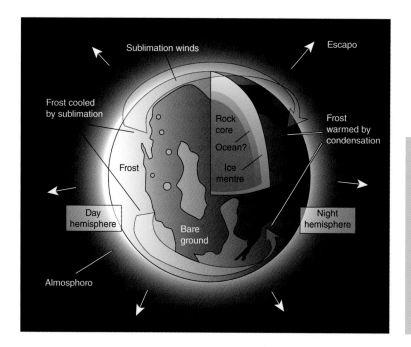

Figure 12.10 For perhaps 20% of its orbital period, Pluto is a dynamic world with strong winds and active recycling of its atmosphere. Ices warmed by the Sun sublimate and are transported to the colder night side where they condense back onto the surface as fresh white frost. This reflects more of the Sun's heat, making the surface still colder. With less fresh ice on its Sun-facing side, Pluto becomes slightly darker. Older surfaces, darkened by radiation and space dust absorb more heat, accelerating the warming effect. (JHU-APL)

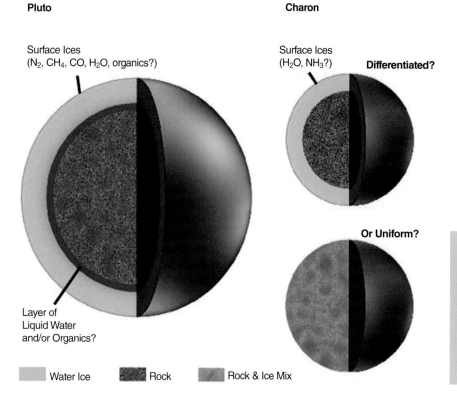

Figure 12.11 One model of Pluto's interior includes a large core of hydrated rock, overlain by a thick mantle of water ice and a surface layer of exotic ices. There may also be a layer of liquid water, possibly mixed with, or overlying, a layer of organic (carbon-based) compounds. (JHU-APL)

around the time of perihelion. As these old, dirty surface ices vaporize, the gases are transported toward the colder, night hemisphere where they are redeposited as bright frost. The residual ices, particularly those containing frozen methane, are exposed as dark areas of hydrocarbons that have been modified by cosmic rays and ultraviolet radiation. In this way, the surface is constantly changing and most of the planet is resurfaced over thousands of years.

Interior

Although there are still some uncertainties about the mass of Pluto, it is clear that its bulk density is between 1.75 and 2.1 g/cm³, or about twice the density of water. Models of the interior indicate a mixture of 55–70% rock (including metal) and about 30–35% ices. Most of this is thought to be water ice, though more exotic ices similar to those found on the surface may also be present.

Whether Pluto is large enough for internal differentiation to have taken place is uncertain. Assuming that heat from its formation and from the natural radioactivity of the rock have warmed the interior sufficiently to melt the ice and allow it to separate from the rock, Pluto probably has a large core of partially hydrated rock, surrounded by a thick mantle of ice. Close to the surface should be an icy crust, coated with volatile ices (nitrogen, methane and carbon monoxide).

Does Pluto possess an ocean of liquid water beneath its icy shell? Observations of other icy objects, such as the Galilean moons of Jupiter, indicate that such oceans are not as rare as once believed. It is possible that such an ocean could be mixed with, or overlying, a layer rich in organic (carbon-based) compounds.

Box 12.2 The Discovery of Charon

The existence of a large satellite close to Pluto came as a major surprise. The discovery was made by James Christy, an astronomer at the US Naval Observatory, on June 22, 1978. During a routine study of photographic plates taken to refine Pluto's orbit, Christy noted that many of them had been labeled "defective." A closer look showed that Pluto seemed distorted, although the nearby stars looked perfectly normal.

Since the "blob" on Pluto's limb seemed to shift position, the problem was clearly related to the planet rather than flawed instruments. Christy eventually realized that Pluto had a large moon that completed one orbit in about six days, the same as Pluto's period of rotation. Checks of previous photos confirmed the discovery.

Christy suggested the name Charon, in honor of his wife Charlene. However, aware that this was unlikely to sway the nomenclature committee of the International Astronomical Union, he also searched for a suitable mythological name. To his delight, he found that, in Roman mythology, Charon was the boatman who ferried the souls of the dead across the River Styx into Hades, the realm of Pluto, god of the underworld. The name could hardly have been more appropriate.

Charon

Pluto's largest satellite was discovered by chance in 1978 (see Box). Measurements of its size, based on stellar occultations and direct imaging with the Hubble Space Telescope, suggest its diameter is about 1207 km, approximately half the diameter of Pluto. With a density of 1.6–1.75 g/cm^3, Charon's mass is about one eighth that of Pluto.

Of all the major satellites in the Solar System, Charon is the largest in comparison with its primary, so Pluto-Charon are sometimes considered as a double planet, rather than a planet with a large moon (such as the Earth-Moon system).

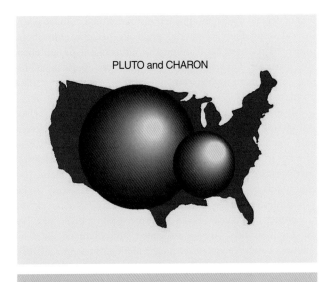

Figure 12.12 The relative sizes of Pluto (left) and Charon compared with the United States. (JHU-APL)

Furthermore, Charon orbits at a distance of only 19,570 km or 17 Pluto radii. Its relatively large mass and close proximity mean that their common center of gravity – or barycenter – actually lies outside Pluto. This situation does not occur in any planetary system.[6] Charon follows a near-circular, synchronous orbit above Pluto's equator, so it is always above the horizon on one hemisphere of Pluto, but never visible from the opposite hemisphere.

Little is known about Charon's surface, although some vague light and dark markings have been mapped by light curve studies during the eclipses on 1985–1990 and by direct imaging with the Hubble Space Telescope. Its overall appearance is gray and bland, with a significantly lower albedo than Pluto (35% compared with 55%).

The first measurements of Charon's spectrum were made during the eclipse season of 1987 and 1988 (see above). A spectrum of Pluto alone, obtained when Charon was out of sight, was subtracted from their combined spectrum. The difference yielded a Charon spectrum which showed a surface dominated by water ice, with little, if any methane.

This came as a surprise, since Pluto's spectrum is dominated by methane, carbon monoxide and nitrogen ices. One explanation is that Charon's lower escape velocity has caused it to lose its more volatile surface ices, leaving behind the much less volatile water ice. The form of the water ice, revealed by observations with the HST and the Keck telescope in Hawaii, was also a surprise. It is in a crystalline (as opposed to amorphous) state.[7]

Water ice found on satellites of Jupiter, Saturn, and Uranus is always crystalline, because amorphous ice rapidly crystallizes if the temperature is above −153°C. However, on Charon, where the surface temperature should never exceed −193°C and is usually

---

[6]Typically, a planet has a much bigger mass than any of its satellites. If the objects are far apart, their center of mass is close to the center of the planet, so the planet hardly moves. In the case of the Earth and Moon, the barycenter is about 4640 km from Earth's center (i.e. inside the planet.)

[7]Crystalline water ice has since been found on other large Kuiper Belt objects, notably Quaoar.

tens of degrees colder, this phase transition should not take place. Instead, crystalline ice should gradually become amorphous under the constant bombardment of ultraviolet light and energetic particles. This radiation breaks hydrogen bonds in the ice that subsequently reform, but not in their original crystalline positions.

The presence of crystalline ice may mean that Charon's surface is fresh or that its temperature has exceeded −193°C in the recent past. One suggestion is that a comet impact vaporized a great deal of water ice, which then spread across Charon before condensing over its entire surface. Alternatively, the ice may be maintained by continuous bombardment from tiny meteorites, resulting in the vaporization and recrystallization of surface ice.

Another possibility is an internal heat source, possibly the result of tidal stresses associated with Pluto or the slow sinking of rocky material through a homogeneous mix of rock and ice.

One other observation of interest was that Charon's spectrum is slightly darker than that of pure ice. This can be explained by the presence of ammonia (NH_3) or ammonia hydrates (which form if the ammonia molecules lie within a water matrix). These contaminants can lower the melting temperature of water ice by around 100°C, making water ice more ductile.

Ammonia is quite easily destroyed by radiation, so how could it survive on Charon? And why does it exist on Charon, but not on nearby Pluto? Its presence implies that comets have delivered interstellar ammonia ice to Charon.

No evidence of an atmosphere has been found during stellar occultations, with an upper limit on the surface pressure of 0.1 microbar. However, its average surface temperature is −220°C, about 10°C warmer than Pluto.

Since Charon is smaller and less dense than Pluto, its internal structure should be easier to model. If internal heating has allowed Charon to differentiate, it should be composed of about 60% rock, with the remainder largely in the form of a water ice mantle. However, there is also the possibility that Charon is undifferentiated, with a roughly uniform mixture of rock and ice throughout.

Small Satellites

On May 15, 2005, during a search with the Hubble Space Telescope, two more satellites (later named Hydra and Nix) were found orbiting Pluto. Subsequent observations, combined with images taken in 2002, which also showed the newcomers, confirmed that they orbit in the same plane as Charon (i.e. above Pluto's equator).

Hydra orbits approximately 64,800 km from Pluto, with an orbital period of about 38.2 days. Nix orbits Pluto at a distance of about 48,700 km, completing one circuit every 25.5 days. Both of them are in a resonant dance with Charon. Nix completes one orbit for every four Charon orbits, while Hydra orbits Pluto in the time it takes Charon to complete six orbits.

Calculations of their sizes depend on their apparent brightness and surface albedos. The brighter satellite, Hydra, has a visual

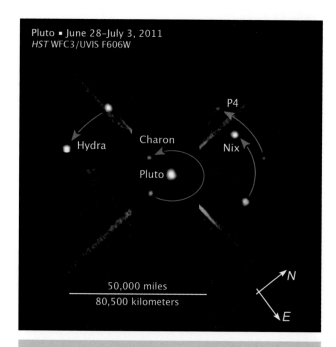

Figure 12.13 This composite HST image of the Pluto system, taken June 28 – July 3, 2011, confirms the existence of a fourth moon (labeled P4) in orbit around Pluto. P4 orbits between Nix and Hydra. (NASA, ESA, and M. Showalter / SETI Institute)

magnitude of about 23.0.[8] Based on a surface reflectivity ranging from 4% (like the darkest Kuiper Belt objects), 15% (typical KBOs) to 35% (Charon), the diameter of Hydra could be 50–160 km.

Nix is about 25% fainter, so, assuming they have the same surface albedo, it could be 10–15% smaller than Hydra. Clearly, both satellites are much smaller than either Pluto or Charon, with masses no larger than 0.0003 that of Pluto or 0.003 that of Charon.

Like Charon, they are neutrally colored. Since Charon's surface is known to consist primarily of water ice, the similar color of Hydra and Nix indicates that their surfaces are composed of water ice.

A fourth satellite, provisionally named S/2011 P1, or P4, was discovered with the HST in June 2011. It completes an orbit around Pluto in about 31 days and is estimated to be 13–34 km across.

The Origin of the Pluto System

Ever since its discovery, astronomers have been debating the origin of Pluto. A popular theory that attempted to explain its eccentric, inclined orbit, suggested that Pluto was an escaped satellite of Neptune. However, we now know that their orbital resonance means that the two worlds can never come anywhere

---

[8]The satellites are about 600 times fainter than Charon and 5000 times fainter than Pluto.

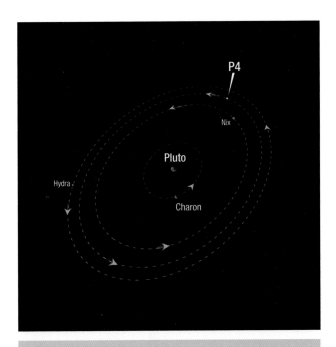

Figure 12.14 The orbits of Pluto's moons: Charon, Nix, P4, and Hydra. The orbits are nearly circular and tilted to our line of sight, so they appear elongated. (NASA, ESA and A. Feild / STScI)

near each other. It is impossible for Pluto to have originated in orbit around a planet which it can never approach.

One current theory suggests that Pluto (and the other Kuiper Belt objects – see below) moved into its current orbit more than four billion years ago, during the late stages of planetary accretion. At that time, the Jovian planets were undergoing significant orbital migration as a result of encounters with large, residual planetesimals. As Neptune moved outwards, a small planetesimal such as Pluto, which was initially following a roughly circular orbit, could have been captured into the 3:2 resonance, causing its orbital eccentricity to rise rapidly until it reached its current Neptune-crossing orbit.

This theory has been questioned, however, because some 30% of the objects in the so-called "Cold Classical Kuiper Belt" occur in widely separated binaries which would have been disrupted during their relocation by Neptune. This suggests that they formed near their present locations and have remained largely undisturbed over the age of the Solar System (see Chapter 1).

The stable, circular orbit of Charon suggests that the two objects have been linked for a very long time. Two leading theories have been put forward. One favors the idea that, like Pluto, Charon was a member of the icy debris that populated the Kuiper Belt beyond Neptune's orbit. A close encounter resulted in Pluto capturing its smaller neighbor. However, computer simulations show that gravitational capture is extremely unlikely.

Today, it is widely believed that the Pluto-Charon system formed in a manner similar to the Earth and Moon – through a giant impact. The recent discovery of other satellites in normal orbits that are in resonance with Charon further supports this theory.

According to this scenario, the system was created during a collision between two Pluto-sized objects over four billion years ago. Simulations show that some of the debris from the collision would go into an orbit around the protoplanet and coalesce to form one or more satellites. In this scenario, the circular orbits of Hydra and Nix imply that they formed much closer to Pluto, subsequently moving out to their present-day positions as tidal forces caused Charon to drift to its current orbit.

However, recent simulations of the orbital evolution of Pluto's system suggest that Nix and Hydra may have formed from a disk of captured planetesimals that surrounded Pluto early in its history. Collisions between the small objects in the disk would have led to the growth of the two satellites. The damping effect of these collisions with the other members of the disk would have circularized their orbits above Pluto's equator. If this scenario is true, small moons could be common around large Kuiper Belt objects.

The Edgeworth-Kuiper Belt

For decades, Pluto was regarded as a unique maverick, since it could not be classified either as a small, rocky planet or as a gas giant. In some ways it most closely resembled a periodic comet, since it follows an inclined, elliptical orbit, periodically warming and losing its atmosphere into space. However, Pluto is far larger and denser than any known comet nucleus. Until the discovery of Charon, the only object that seemed comparable in size and density was Neptune's satellite, Triton.

Was Pluto really a planet that accreted from the solar nebula about 4.5 billion years ago, or was it some kind of protoplanetary debris that happened to have been shifted into a strange orbit? Some astronomers found it hard to believe that Pluto was the outermost member of the planetary system, and there were various attempts to predict the locations of unseen planets beyond Pluto. Clyde Tombaugh continued to search for such a planet for 14 years after he found Pluto, eventually concluding that no large worlds exist in the outer Solar System.

Nevertheless, the apparent emptiness of the region beyond Neptune was a puzzle. Apart from the occasional comet, there seemed to be few left-overs from the planetesimals that thronged the early Solar System.

The first suggestion that a reservoir of comets should exist beyond Neptune came from British astronomer Kenneth Edgeworth in 1943. He noted that collisions and coalescence between planetesimals would be so infrequent in the outer Solar System that only small bodies would be likely to form. The Dutch-American astronomer, Gerard Kuiper, came to a similar conclusion in 1951, when he proposed that enormous numbers of icy bodies populate the region beyond Neptune, possibly extending as far out as 120 AU (three times Pluto's mean distance from the Sun).

During the 1980s, various researchers suggested that a flattened, donut-shaped zone of icy planetesimals could be the source of the short-period comets that occasionally enter the inner Solar System. One key piece of evidence was that these comets travel around the Sun in the same direction as the

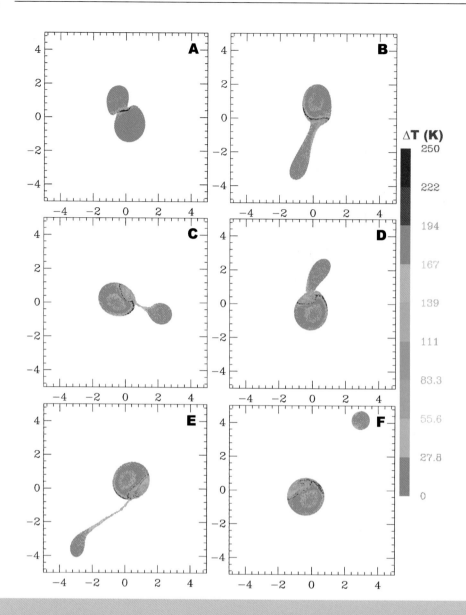

Figure 12.15   A computer simulation showing how Pluto and Charon may have formed. After a very oblique initial impact (A), the two objects separate (B and C). During this period the smaller impactor receives a net torque from the distorted target object. After another, even more grazing encounter (D), more of the impactor is accreted onto Pluto. The remainder accretes into a moon containing 12% of the central object's mass. This is torqued once more by the ellipsoidal target (D and E), eventually forming a moon with a stable orbit (F). (Robin Canup / SWRI and AAS)

planets, following orbits that are only slightly inclined to the ecliptic plane. Unfortunately, the instruments of the time were not sufficiently advanced to find any proof that these dormant comets actually exist.

The breakthrough came on August 30, 1992. After five years of searching the ecliptic with a new-generation CCD camera linked to the 2.2 meter University of Hawaii telescope, David Jewitt and

Jane Luu discovered 1992 QB1 drifting slowly among the background stars. The newcomer measured about 250 km across and took 296 years to complete one orbit. It was the first object in the Solar System known to spend its entire life beyond Neptune, and the first recognized member of the Kuiper Belt.[9]

Today, over 1300 occupants of the belt have been found, although the total population probably exceeds 200 million. Of

[9]Today, the name of the Edgeworth-Kuiper Belt is usually shortened to simply Kuiper Belt. Its members are called Kuiper Belt objects (KBOs), Edgeworth/Kuiper Belt objects, or trans-Neptunian objects (TNOs).

Box 12.3   New Horizons

All of the major classes of objects in the Solar System have been explored by spacecraft, with the exception of the millions of Kuiper Belt objects. This is about to change, following the launch of NASA's New Horizons spacecraft on January 19, 2006.

The nuclear-powered spacecraft has a mass of 478 kg. Most of its subsystems, including the computers and propulsion control system, are based on designs used in other spacecraft. A 2.1 meter high gain antenna is available to communicate with Earth from a distance of 7.5 billion km.

New Horizons carries four scientific instruments. A miniature, high resolution camera will image selected regions, revealing objects down to 50 meters in diameter. An infrared imaging spectrometer will map the surface at a resolution of one kilometer and study the composition and temperature of Pluto's ices and atmosphere. A radio science instrument will probe the structure of Pluto's tenuous atmosphere and measure the surface temperatures on both day and night hemispheres. A suite of charged particle detectors will sample gases escaping from the rarefied atmosphere and determine their escape rate.

A major factor in the success of the mission was the ability to meet the primary launch window. If New Horizons did not lift off in January 2006, it would miss the opportunity for a gravity assist from Jupiter, delaying a Pluto flyby until 2019–2020, when it might be too late to study its transient summer atmosphere. The viewing conditions would also be less favorable at that time – because of Pluto's extreme axial tilt and inclined orbit, much of the northern hemisphere would be hidden in darkness.

The Atlas V – Centaur rocket sent the spacecraft speeding away from Earth at approximately 58 000 km/h – the fastest human-made object ever built. New Horizons conducted an intensive, four month reconnaissance of the Jupiter system in early 2007, putting it on course for a flyby of Pluto and Charon in the summer of 2015. Observations will begin six months before closest approach. Once the distance is down to 100 million km – about 75 days before close encounter – New Horizons will begin to send back images superior to anything yet seen.

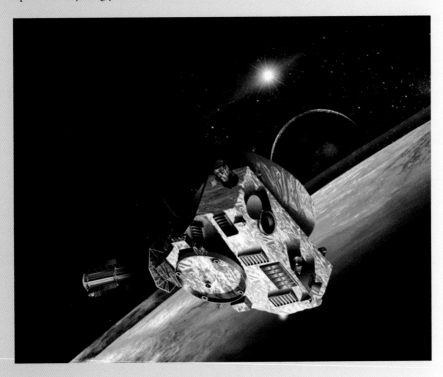

Figure 12.16   The New Horizons spacecraft during its encounter with Pluto (foreground) and Charon. (JHUAPL / SwRI)

As the two worlds loom larger, their surfaces will be mapped in increasing detail, along with any hazes, clouds or other variable phenomena, such as cryovolcanism. On the day of closest approach, New Horizons will sweep within a few thousand kilometers of Pluto. It will turn to image the planet's night side, illuminated by dim moonshine from Charon. Measurements of the behavior of radio signals transmitted by the spacecraft will reveal the vertical structure of the sparse atmosphere. Similar observations will then be made of Charon.

It is hoped that the probe will go on to complete at least one close encounter with a Kuiper Belt object. The number of flybys will depend on the available targets close to the spacecraft's trajectory, the supply of hydrazine propellant and the amount of power still being generated by the single radioisotope thermoelectric generator.

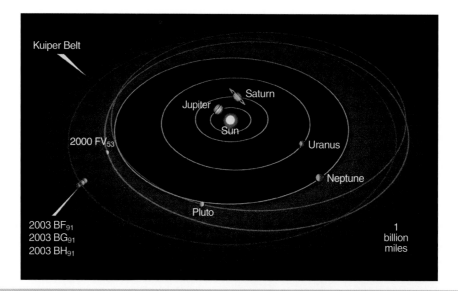

Figure 12.17   The Edgeworth-Kuiper Belt is named after Kenneth Edgeworth and Gerard Kuiper. They were the first to propose that enormous numbers of icy bodies exist beyond the orbit of Neptune. After many years of searching, the first object was found in 1992. Since then, more than 1300 have been discovered. The orbits of four KBOs are shown on this diagram. (NASA and A. Field / STScI)

these, perhaps 35,000 measure 100 km or more in diameter, compared with 200 objects that size in the main asteroid belt. The overall mass of the Kuiper Belt objects (KBOs) is probably only a few percent of Earth's mass.

Only half of the known KBOs have well defined orbits. However, observations show that they are mostly confined within a thick, torus-shaped region that is centered on the ecliptic. The vast majority of them exist between the orbit of Neptune at 30 AU and an apparent outer edge about 50 AU from the Sun. Beyond this outer limit there appears to be a sparse population of scattered KBOs with large orbital eccentricities.

The Kuiper Belt population is thought to comprise the last surviving icy planetesimals that existed 4.5 billion years ago, when the planets were forming from the solar nebula. Whereas the inner, dense parts of the pre-planetary disk condensed into the major planets, probably within a few millions to tens of millions of years, the outer reaches were less dense, and collisional accretion progressed more slowly. Instead of coming together to form large planets, a multitude of small objects formed.

Some of these ice dwarfs were probably ejected to their present remote location billions of years ago by gravitational interactions with the giant planets, particularly Uranus and Neptune. Models suggest that this could have happened during the planets' outward migration. Today, when they are occasionally disturbed by the outer planets, they enter Sun-approaching orbits and become short-period comets.

Alternatively, they may have formed *in situ* during the era of planet formation some 4.5 billion years ago. How the larger members of the Belt could have grown to their present size in the vast, near-empty region beyond Neptune has yet to be fully explained, although it has been suggested that turbulence and vortexes in the protoplanetary nebula might have enabled many tiny particles to coalesce and grow quite rapidly.

The Classification of Kuiper Belt objects

By studying the orbits of KBOs, it has been possible to recognize four main categories:

1.   Cold Classical KBOs. The most numerous group so far detected, these occupy a narrow region which is about 7 AU thick and stretches between 42 and 48 AU from the Sun. They generally follow orbits of modest inclination, which suggests that they were born outside Neptune's orbit and have been largely unperturbed by the giant planet. They also tend to be smaller and browner than other KBOs. Examples include 1992 QB1, the first KBO ever discovered, which has a semimajor axis of 43.7 AU and an eccentricity of 0.07, and Quaoar, one of the largest KBOs yet discovered.

2.   Hot Classical KBOs. These have similar average distances from the Sun to the Cold Classical KBOs, but their orbits are more eccentric and/or inclined. This suggests that their orbits have been disturbed and altered by Neptune's gravity. Their sizes and colors vary, but they include larger and grayer objects than Cold Classicals. Examples include 1996 RQ20 and 1997 RX9, which have inclinations greater than 30°.

3.   Resonant KBOs (Plutinos and Twotinos). Like Pluto, Plutinos follow orbits that have been modified by Neptune's gravity, so they have 3:2 orbital resonances with the ice giant (i.e. they make two orbits for every three of Neptune). They include Orcus and Ixion. Another, more remote group ("Twotinos") has 2:1 resonances with Neptune. With semimajor axes at about 47.7 AU, they are close to the outer edge of the main Kuiper Belt. Less significant resonances also exist at 3:4, 3:5, 4:7 and 2:5.

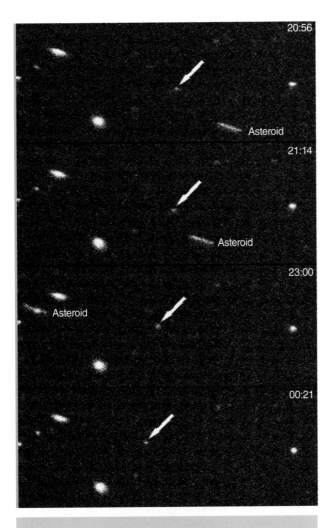

Figure 12.18 The first Kuiper Belt object, known as 1992 QB1, was found by David Jewitt and Jane Luu on August 30, 1992. These images taken with the University of Hawaii's 2.2 meter telescope at Mauna Kea shows its motion over a period of about 3½ hours. With a magnitude of about 23, 1992 QB1 was six million times fainter than the faintest star visible to the naked eye. Also visible is a streak caused by a faster-moving asteroid. (David Jewitt)

4. Scattered KBOs. Although their perihelion distances of around 35 AU are not too unusual, they have inclined and eccentric orbits that may take them hundreds of AU from the Sun. Along the inner part of their orbits, Neptune may be close enough to exert weak dynamical control over their motion. Over billions of years, perturbations by Neptune around perihelion change their orbits. They account for only 3–4% of the known Kuiper Belt objects, but this apparent depletion may be explained by the fact that they spend most of their time at large distances and are difficult to detect. Eris and Sedna are examples of this group.

Surfaces

Little is known about the surface properties of KBOs. Most of them are too small and distant for spectral data to be obtained, and broad spectra are available for a mere handful. Albedos and sizes have been estimated for a small sample, but thermal emissions are difficult to measure. Visible colors have been reported for about 200, with near-infrared colors for about 50.

Visible photometry shows that their colors cover a wide range, from neutral to very red, indicating that KBOs have very diverse surfaces. It seems unlikely that this diversity is related to widely differing chemical compositions, since few chemical reactions are likely in a region where the temperature is only 30–50°C above absolute zero. Besides, there is no apparent correlation between color and current distance, as might be expected if location and temperature were important. However, the color range might be explained if some of the objects were born in different regions of the outer Solar System and subsequently relocated to their present positions.

One possibility is that the colors represent different evolutionary stages – as the surface ages, it becomes darker and more reddish. Many KBOs may have similar compositions, a mix of solid organic compounds, silicates and volatile ices derived from the primordial solar nebula. Over time, irradiation by high energy cosmic rays, the solar wind and solar ultraviolet light, combined with sublimation of the volatiles and non-disruptive collisions with other objects, would change their surface compositions.

In this way, bright surfaces coated with fresh ices would be transformed by the loss of hydrogen atoms and carbonization. The end result would be a low albedo surface "crust" of red, organic-rich sludge. Grazing or non-disruptive collisions could remove some of this outer layer, revealing more reflective, neutrally colored, ices beneath. There is also the possibility that resurfacing through deposition of fresh icy material, resulting from cryovolcanism, might also play a role.

The Largest KBOs

Despite their remoteness, it has been possible to determine that some KBOs are comparable in size to Pluto. The largest yet discovered is Eris. First imaged with the 1.2 meter Samuel Oschin Telescope at the Palomar Observatory, California, on October 31, 2003, it was rediscovered by Mike Brown (California Institute of Technology) and colleagues on January 8, 2005.

Eris is an example of a scattered KBO, and it follows a remarkably eccentric orbit. It was discovered near aphelion, 97 AU from the Sun, making it the most distant object ever observed in the Solar System. When it reaches perihelion in 2257, it will close to about 38 AU – so Eris is sometimes much closer to the Sun than Pluto, although never closer than Neptune.

Observations with the Keck Telescope in September 2005 led to the discovery of a satellite, named Dysnomia. Its diameter is estimated to be 100–150 km. It orbits about 37,000 km from Eris over a period of about 15.8 days. Like the Earth-Moon system, Eris and Dysnomia may have formed after a massive collision about 4.5 billion years ago.

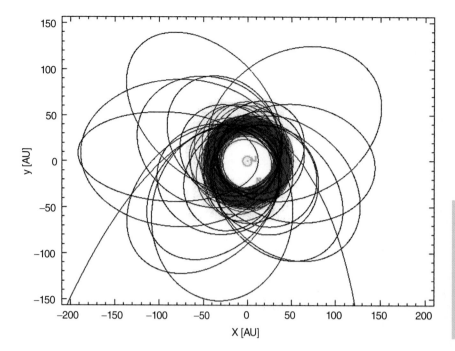

Figure 12.19 Typical orbits of Kuiper Belt objects projected onto the ecliptic, together with the orbits of Jupiter (J) and Neptune (N). Red orbits are Plutinos. Blue orbits are Classical Kuiper Belt objects. Black orbits are Scattered KBOs. (David Jewitt / IFA)

Calculations based on its orbital period show that Eris is 27% more massive than Pluto. With a density of about 2.5 g/cm³, Eris likely contains more rock than Pluto, with a comparatively thin outer layer of ice.

Despite its vast distance, Eris is the third brightest of the Kuiper Belt objects. Studies of its apparent brightness and the amount of heat it appears to radiate suggested that its diameter is about 3000 km, some 700 km larger than Pluto. However, subsequent stellar occultation data obtained in 2010 indicate that its diameter is 2326 km, almost identical to that of Pluto. Eris is, therefore, much brighter than expected. (If so, the only object with a higher albedo is Saturn's icy satellite, Enceladus.) The high albedo is probably due to the presence of methane and nitrogen ices on its surface.

Since Eris takes about 560 years – twice as long as Pluto – to orbit the Sun and is now very close to aphelion, its atmosphere probably froze onto the surface hundreds of years ago. This could result in a layer of solid nitrogen several centimeters thick, or a layer of fresh methane ice tens of microns thick. Beneath this seasonal ice cover may be a darker, methane-rich layer which is exposed around perihelion and darkened when it is broken down by solar ultraviolet light.

Variations in surface ices and albedo are also apparent on Makemake, another dwarf planet discovered by Brown's team, whose diameter seems to be about 1500 km. Currently at a distance of 52 AU, this classical KBO should have a surface temperature approximately midway between Pluto and Eris.

Another large member of the Belt is Haumea, which is about one-third as massive as Pluto. Studies of its light curve show great variations in reflected sunlight that repeat every three to four hours. This has led to suggestions that it is spinning so quickly that it has assumed an elongated, cigar shape, with diameter of the long axis twice that of the short axis. Haumea has two known satellites, one of which has water ice on its surface. They follow very different, inclined orbits that argue in favor of a collisional origin.

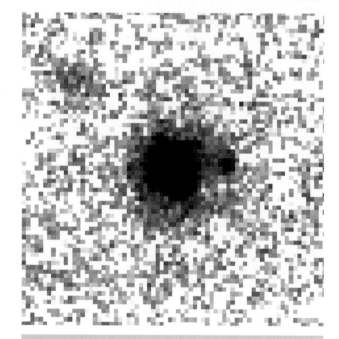

Figure 12.20   Eris is about the same size as Pluto. It was discovered near aphelion – 97 AU from the Sun – making it the most distant known object in the Solar System. A satellite (right) was discovered in 2005. (W. M. Keck Observatory)

Another unusual object is Quaoar, which has an estimated diameter of about 890 km but an albedo of only 12%. It also has a small satellite, named Weywot, which follows an eccentric, 12.4 day orbit. Calculations based on the moon's orbital period indicate that

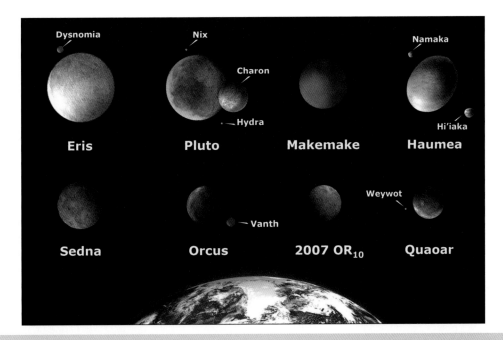

Figure 12.21 The largest known KBOs compared with Earth. Eris has a probable diameter of 2326 km, so it is comparable in size to Pluto. Haumea is thought to be elliptical in shape. Known satellites are also shown. Apart from Eris, Pluto and Charon, the sizes are uncertain. (NASA, ESA and A. Field / STScI)

Quaoar has a surprisingly high density of about 4.2 g/cm³, making it the most dense Kuiper Belt object currently known. This implies that Quaoar is mainly rocky, with only the thin veneer of water, methane, and ethane ices (which appear in its spectrum).

One suggestion is that it originated far away in the asteroid belt. It may then have been scattered outward into the Kuiper Belt region by one or more gravitational interactions with giant planets, for example, during the migration of Jupiter during the early history of the Solar System. Quaoar might once have been an asteroid which lost most of its ice content due to rapid sublimation from solar insolation, or it may always have been a rocky object.

It is also possible that Quaoar's high density resulted from a collision that stripped away almost the entire surface layer of the parent body, leaving Quaoar as the virtually intact rocky core. This may explain the origin of Weywot.

Box 12.4 Dwarf Planets

For 76 years, Pluto was regarded as the ninth major planet in the Solar System. However, the definition of a planet was not very precise, referring simply to a large, non-stellar object in orbit around a star. Then, in 2006, the International Astronomical Union (IAU) devised and introduced new classifications for Solar System objects, effectively demoting Pluto and reducing the number of planets to eight (see Chapter 1).

The problem arose partly because Pluto is not only smaller than the other planets, but also smaller than seven of their satellites, including the Moon. Furthermore, it is only one member of a vast population of Kuiper Belt objects. The 2005 discovery of Eris, a KBO even larger than Pluto, brought the issue to a head. If Pluto is a planet, then Eris must also be one, along with the numerous Pluto-sized objects yet to be discovered.

New IAU resolutions reclassified Pluto as a "dwarf planet", as well as one of the largest members of the Kuiper Belt. Ceres, the largest main belt asteroid, was also classified as a dwarf planet, along with the large KBOs Eris, Haumea and Makemake.

The IAU subsequently decided that newly discovered trans-Neptunian objects with an absolute magnitude brighter than +1 (and hence with a minimum diameter of 838 km) will be classified (and named) as dwarf planets.[10]

The term "*plutoid*" was also introduced for trans-Neptunian dwarf planets. The IAU definition states: "Plutoids are celestial bodies in orbit around the Sun at a semimajor axis greater than that of Neptune that have sufficient mass for their self-gravity to overcome rigid body forces so that they assume a hydrostatic equilibrium (near-spherical) shape, and that have not cleared the neighborhood around their orbit." The currently recognized plutoids are Pluto, Eris, Haumea and Makemake. Other candidates are 2007 OR10, Sedna, Quaoar and Orcus.

[10] Absolute magnitude is defined as the apparent magnitude (brightness) that an object would have if it were one astronomical unit (AU) from both the Sun and the observer, and at a phase angle of zero degrees (i.e. directly overhead).

Peculiar KBOs

A handful of objects which do not fit the major groupings above have been found in recent years. One of the most unusual is Sedna, which remains far beyond the outer edge of the Kuiper Belt, never coming closer than 76 AU and reaching 975 AU at the most distant point of its 12,050 year orbit.[11] Despite its eccentric orbit, Sedna stays fairly close to the ecliptic, with an orbital inclination of 12°.

Estimated to be at least half the diameter of Pluto, Sedna has a strong reddish color. (It is the second reddest object in the Solar System, after Mars.) Infrared spectra show its surface contains little methane ice or water ice. Instead, it seems to be covered with a coating of hydrocarbon "sludge" produced when the Sun's ultraviolet radiation and charged particles alter the chemical bonds between atoms in the ice. The absence of major collisions may explain the apparent absence of bright, icy patches.

There is some question over whether Sedna should be considered part of the Kuiper Belt, since it spends its entire life well beyond the main belt's outer edge, though it also lies well inside the postulated Oort Cloud of comets. It may be the first known member of a huge population of undiscovered objects in an "inner Oort Cloud" (see Chapter 13).

Neptune shapes the dynamics of most KBOs, but Sedna always orbits far beyond the planet's gravitational influence. Since an object as large as Sedna is unlikely to have formed by accretion in the sparsely populated outer regions of the protoplanetary disk, it must have formed closer to the Sun or in the Kuiper Belt, and then been ejected outwards.

Since there are no known planets beyond the Kuiper Belt large enough to have reshaped Sedna's orbit, it has been suggested that passing stars could have created the Belt's outer edge, at the same time scattering some KBOs, such as Sedna, into large, eccentric orbits.

According to this hypothesis, a star passing at a distance of 150–200 AU could have captured objects from the outer Kuiper Belt, while at the same time diverting Sedna into its present orbit, without affecting Neptune or the inner planets. (Computer models also suggest that some KBOs – though not Sedna – could be captured from the passing star's system in a two-way swap of material.)

At least one other object, 2000 CR105, has an orbit similar to Sedna's, though a little less eccentric, with perihelion at 45 AU, aphelion at 415 AU, and an orbital period of 3420 years. Its orbit may have resulted from the same processes that influenced Sedna.

Another object which has been puzzling theorists is 2004 XR190, which is quite bright, with a mass perhaps half that of

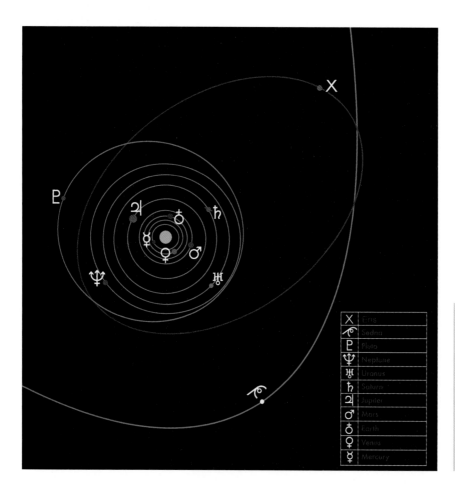

X	Eris
⯲	Sedna
P	Pluto
♆	Neptune
⛢	Uranus
♄	Saturn
♃	Jupiter
♂	Mars
⊕	Earth
♀	Venus
☿	Mercury

Figure 12.22 The orbit of Sedna ranges between 76 AU and 975 AU, so that it is always far outside the main Kuiper Belt, but well inside the postulated Oort Cloud. It takes about 12 050 years to complete one orbit of the Sun. Here the orbit of Sedna is compared with those of the planets, Pluto and Eris. (Gemini Observatory)

---

[11] Many comets have far more eccentric orbits. Kuiper Belt object 2000 OO67 also has an extremely elongated orbit, with aphelion at 1005 AU and perihelion at 20 AU. However, Sedna is always extremely remote.

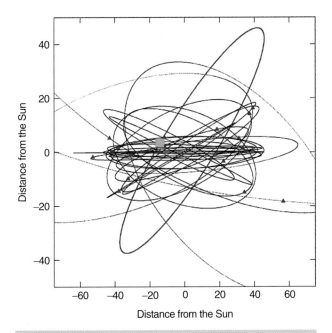

Figure 12.23 2004 XR190 has a unique orbit (red) which is highly inclined but relatively circular compared with other KBOs. (CFEPS)

Pluto. The KBO travels between 52 and 62 AU from the Sun, making it the first known object with a fairly circular orbit at the outer edge of the Kuiper Belt. Furthermore, its extreme orbital inclination takes it 47° above and below the ecliptic. It is currently too far from the Sun to have been affected by Neptune's gravity, and its orbit remains a mystery.

Binaries

Pluto and its satellites make up the only known five-body system in the Kuiper Belt, but many more KBOs are known to have at least one satellite. Indeed, some 30% of the known Cold Classical KBOs are binaries.

The first of these, 1998 WW31, was found on April 16, 2001. The two objects are thought to be approximately 150 and 130 km in diameter. By studying their orbital motions, it was calculated that the pair's combined mass is about 5000 times smaller than Pluto–Charon. The mutual orbit of 1998 WW31 is the most eccentric ever measured for any binary Solar System object or planetary satellite, with a variation in distance between 4000 and 40,000 km. (The eccentricity is approximately 0.82).

Some of the largest KBOs either have companions or resemble contact binaries (see Eris and Haumea above). Their orbital periods are highly variable, ranging from 6 days (for Charon) to 876 days (2001 QT297). In cases where the pair is widely separated, the systems may be unstable over a timescale of billions of years. However, some binaries orbit only a few thousand kilometers apart and must be very stable. How did the binaries originate?

Gravitational capture of one KBO by another is considered to be impossible without some process that can dissipate some of the kinetic energy present in the bodies. On the other hand, the fact that three of the four largest objects in the Kuiper Belt have moons suggests that they may be produced by accretion of material from a disk of debris created by a cataclysmic collision. Unfortunately, this process is unlikely with the smaller binaries, since they would seem to have insufficient mass and gravity for the creation of a sizeable accretion disk.

Another explanation is that they might have formed as the direct result of low velocity (100 m/s or less) collisions between KBOs. When the relative velocities are smaller than, or comparable to, the gravitational escape speeds from the colliding objects,

Box 12.5 Measuring the size of a KBO

The small size and distance of KBOs mean that it is very difficult to measure their size. This calculation is further complicated by uncertainties over their surface brightness or reflectivity (albedo).

Direct optical measurements of large KBOs can be made by using the Hubble Space Telescope. Since it orbits above the turbulence of the atmosphere, the HST's angular resolution is much higher than that of most ground-based telescopes. By very carefully measuring the size of an object about 10 times over the course of an hour and comparing it to a nearby star, it is possible to calculate the diameter.

Thermal infrared measurements made by ground-based or orbiting telescopes measure the heat coming from a large KBO. This makes it possible to distinguish between a smaller, reflective object and a larger, darker object which has the same apparent brightness. For example, a dark object absorbs much more light than a white object, so it will be warmer. The diameter can be estimated by measuring the heat coming from a KBO and comparing it with the reflected optical light.

However, such estimates are difficult to make if the only information available is an object's approximate distance from the Sun and apparent brightness. Astronomers then have to estimate the object's most probable albedo, based on the likely nature of its surface, for example, icy or covered in dark organic compounds.

It is often assumed that KBOs are like comets, with albedos of around 4%, and this figure has been used to calculate their diameters. However, observations have shown that true albedos are often much higher, so that many estimated KBO diameters have been revised downwards.

One example is 2002 AW197, which has been found to reflect 18% of its incident light, leading to an estimated diameter of about 700 km. By assuming that it reflects only 4% of its incident light, this diameter would balloon to 1500 km.

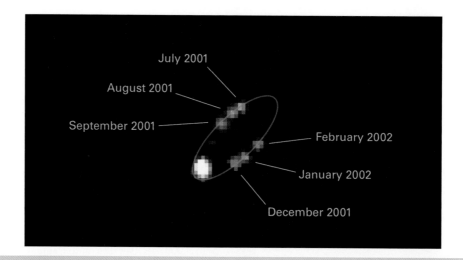

July 2001

August 2001

September 2001

February 2002

January 2002

December 2001

Figure 12.24 This HST composite shows the apparent orbit of one member of the binary KBO known as 1998 WW31. In fact, the objects revolve around a common center of gravity. The image is a composite of six exposures taken between July 2001 and February 2002. Their mutual orbital period is about 570 days. (NASA and C. Veillet / Canada-France-Hawaii Telescope)

peanut-shaped contact binaries may result. Alternatively, the objects may rebound with low enough energy for a binary to form (see also Chapter 13).

Even these hypotheses are not without their difficulties. Calculations indicate that collisions between 100-km-sized KBOs are currently too rare to account for the assumed number of binaries. Another problem is that the known Kuiper Belt binary objects often have comparable masses, something which would not be expected from a collisional origin.

The solution may lie in the existence of a much denser Kuiper Belt in the past – perhaps 100–1000 times more massive than it is today. In those conditions, collisions would be more frequent. This might also make it possible for dynamical friction – the net effect of many small KBOs tugging on nearby objects – to cause a net loss of energy that would result in objects slowly spiraling together.

What happened to this massive Kuiper Belt? Some of the objects would have been removed by gravitational interactions with Neptune – either toward the Sun, or into the Oort Cloud of comets and beyond. Many others may have been ground down or shattered by collisions. The fine dust that resulted would have been blown into interstellar space by the solar wind.

Questions

- Why has been Pluto been "demoted" to a dwarf planet?
- Why did Kenneth Edgeworth and Gerard Kuiper believe that a large population of objects existed beyond the orbit of Neptune?
- Why did it take so long to find the first members of the Kuiper Belt?
- Describe and attempt to explain the differences between Pluto and Charon.
- What new information was provided by the mutual eclipses of Pluto and Charon 1985–1990?
- What is the most likely explanation for the origin of the Pluto-Charon system?
- How may binary KBOs have come about?
- What are the major similarities and differences between the main asteroid belt and the Kuiper Belt?
- Suggest reasons for the differences in size, density, color and orbit between KBOs.

THIRTEEN

Comets, Asteroids and Meteorites

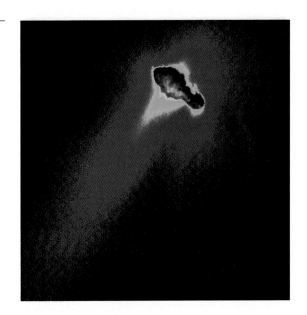

Once the Solar System settled down after the traumas of its birth and the final late asteroid/comet bombardment, the formation of the Sun, planets and planetary satellites was largely complete. Yet, even today, the Sun's domain contains untold billions of pieces of planetary debris, much of it left over from the processes of accretion and collision that occurred more than four billion years ago. This debris has traditionally been classified as rocky material, which dominates the main asteroid belt and the inner Solar System, and icy material which pervades the cold, outer regions. We are not just disinterested observers of these ancient fragments. Earth is bombarded each day by interplanetary dust and meteorites. On a longer time scale, our planet periodically suffers large scale destruction and even mass extinctions as the result of impacts by comets and asteroids.

By the late 18th century, the known population of the Solar System had risen to seven planets and 14 satellites. However, astronomers had become convinced that there must be another planet between the orbits of Mars and Jupiter. Their belief was based on a numerical relationship, known as Bode's Law, which seemed able to predict the distances of planets from the Sun.

The so-called Law was an arithmetic progression that seemed to predict the relative distances of the planets from the Sun. In the numerical series 0, 3, 6, 12, 24 . . . each successive number after 3 was double the previous number. After adding 4 to each number and dividing by 10, this mathematical sequence provided a good approximation to the actual distances of the planets in astronomical units (AU).

Its apparent predictive power was reinforced in 1781, when William Herschel discovered Uranus at 19.2 AU from the Sun, very close to the Law's prediction of 19.6. The only major flaw was a missing planet in the numerical gap between Mars and Jupiter.

Among the believers in Bode's Law was a group of six astronomers, headed by Johann Schröter, who called themselves the "celestial police." In 1800, they prepared a systematic search for the missing planet along the ecliptic plane. However, before the survey could get under way, some unexpected news arrived from Sicily. On January 1, 1801, Giuseppe Piazzi, director of the Palermo Observatory, had discovered a strange object in the constellation of Taurus.

At first, Piazzi thought he had found a comet, but the absence of a gaseous coma caused him to have some doubts. Unfortunately, while Piazzi was suffering from a serious illness, the mysterious object disappeared in the Sun's glare. With limited orbital information to go on, frustrated astronomers all over Europe began to search for the newcomer.

The breakthrough came when a brilliant young mathematician, Carl Gauss, found a way of predicting planetary positions from a limited set of observations. Using his calculations, Franz Xavier von Zach, one of Schröter's group, rediscovered the faint object on 31 December, only half a degree from where Gauss had predicted.

At Piazzi's suggestion, the name Ceres – after the Roman goddess who was the patron of Sicily – was soon adopted. The object was found to follow a nearly circular path between Mars and Jupiter at a distance of 2.77 AU – an almost exact fit with the empty 2.8 AU slot in Bode's Law. The only concern was that the 8th magnitude object was obviously very small – less than 1000 km across.

Then, on March 28, 1802, Wilhelm Olbers (another member of the "celestial police") found a second minor planet at roughly the same distance. Pallas, as it was named, was smaller and fainter

Exploring the Solar System, First Edition. Peter Bond.
© 2012 by Peter Bond. Published 2012 by Blackwell Publishing Ltd.

than Ceres, and traveled in a fairly eccentric, inclined orbit, though it did approach Ceres from time to time.

It was already apparent that these objects were very different from their larger neighbors. William Herschel's suggestion that they be called "asteroids," because of their star-like appearance and absence of visible disks in even the largest telescopes, was quickly accepted.

Two more minor planets, Juno and Vesta, were discovered in 1804 and 1807. Juno was notably smaller than Pallas, while Vesta did not even approach the other known asteroids, and was sometimes bright enough to be visible with the unaided eye.

The scattered family seemed complete for nearly 40 years until Astraea was found in 1845, followed by another three objects in 1847. Since then, at least one asteroid has been discovered every year, with a marked leap in the rate of discovery after the introduction of photographic techniques in the late 19th century. The first asteroid to be discovered photographically was 323 Brucia, found by Max Wolf on December 20, 1891.

Unfortunately, unwanted asteroid tracks showed up on exposures at the most inopportune times, causing one astronomer to dub them "the vermin of the skies." Most scientists lost interest in the thousands of small, seemingly unimportant, chunks of rock that occupied the broad zone between the four terrestrial planets and their giant, gaseous companions. Only the discovery of

sizeable near-Earth objects, such as 433 Eros, caused a stir, when it was realized that asteroids sometimes paid a visit to the inner Solar System, even threatening catastrophic impacts with our planet.

The desire to track down and characterize the near-Earth marauders has led to the introduction of new instruments and observing techniques. Although amateur astronomers continue to play a role, by far the majority of asteroids have been discovered since the mid-1990s using automated search programs. Most of these are located in the United States and funding is provided by NASA, the US Air Force and various astronomical observatories. The preponderance of observing sites in the northern hemisphere means that coverage of the night sky is far from ideal.

The most successful ground-based effort has been the Lincoln Near-Earth Asteroid Research (LINEAR) program, operated by MIT's Lincoln Laboratory in co-operation with the US Air Force. Located in Socorro, New Mexico, it discovered almost 100,000 asteroids 1997–2008.

The most recent addition to the search network is the 1.8 meter Pan-STARRS 1 telescope, located on Mount Haleakala in Maui, which began full operation in May 2010. It has an enormous field of view – six times the width of the full Moon – and a huge 1.4 gigapixel camera that makes it the world's most powerful survey telescope.

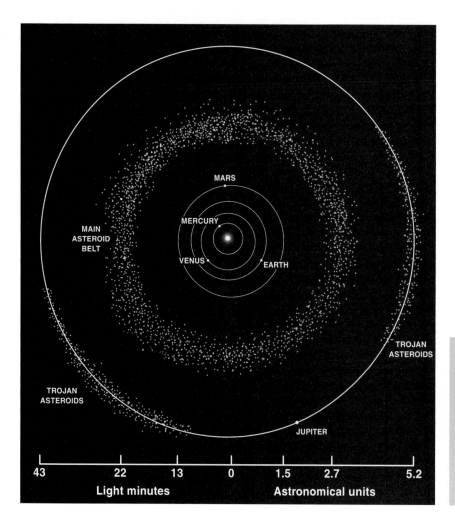

Figure 13.1 The vast majority of asteroids travel around the Sun in the "main belt,, between the orbits of Mars and Jupiter. Two groups of Trojan asteroids lie along Jupiter's orbit, centered 60° ahead and behind the giant planet. Not shown here are the numerous near-Earth asteroids which approach or cross Earth's orbit. (NASA)

Box 13.1 Naming Asteroids

After a particular asteroid has been observed on at least two nights, and it cannot immediately be matched to any known object, it is given a provisional designation by the Minor Planet Center (MPC) at the Smithsonian Astrophysical Observatory. (Note that it may take many weeks or months to determine its orbit with reasonable accuracy.) This designation is based on the date of its discovery and its place in the order of discoveries for that month and year.

The first four numbers give its year of discovery. The fifth character – a letter of the alphabet – represents the two-week period of the discovery month. A equals the first two weeks in January (1–15), B marks the third and fourth weeks in January (16–31), C the first two weeks of February, and so on. The letter I is not used and Z is not needed.

The sixth character (also a letter) tells the order of discovery during each two-week period. Hence, asteroid 1999 DA is the first asteroid discovered during the second half of February 1999. If there are more than 25 discoveries in any half-month period, the second letter is recycled and the number "1" (usually written in subscript) is added. When the number of discoveries in that fortnight exceeds 50, the number "2" is used, so 1999 DA_1 is the 26th asteroid found in that two-week period and 1999 DA_2 is the 51st. Today, so many asteroids are being discovered that frequent recycling of the alphabet is the norm, for example, 2008 TN_{166}. A lot of asteroids are reported several times in different epochs, so they may receive multiple provisional designations.

Objects with well-determined orbits are given a catalog number, and may eventually receive a name. The MPC assigns numbers in the order that observations are received and catalogued. Since the number generally gives an indication of how long ago the object was discovered, a lower number often indicates a discovery many decades ago. There are a few exceptions, such as 69230 Hermes, which was originally discovered in 1937 when it passed close to the Earth, but then lost. Only after its rediscovery in 2003 could its orbit be determined and a number assigned. Before this, it was known by its provisional designation, 1937 UB.

In most cases, the name is chosen by the astronomer who first provides enough observations to calculate the orbit with precision, NOT the first person to see the asteroid. Following the established tradition, the early discoveries were named after characters from classical mythology. Hence, the first asteroid to be discovered was designated 1 Ceres, and the second was 2 Pallas.

Today, the International Astronomical Union's Committee on Small Body Nomenclature is responsible for assigning names to asteroids. In most cases, the name proposed by the discoverer is accepted. Although controversial political, religious or military figures such as Hitler and Stalin are not allowed, there have been some interesting – and unusual – choices. For example, asteroid 6042 is named CheshireCat and 1625 is named NORC, after the Naval Ordnance Research Calculator. Among those honored are popular singers, science fiction characters, social clubs, shipping lines and foods.

Sometimes the IAU committee takes the initiative for special numbers. Asteroid 1000 was named Piazzia, in honor of the discoverer of Ceres. All Trans-Neptunian Objects are named after deities associated with creation myths. Thus, 50,000 Quaoar, one of the largest known members of the Kuiper Belt, is named after a Native American creation deity.

By February 2011[1] there were over 230,000 numbered minor planets, but only about 15,600 had been given names.

Searches by spacecraft are a more expensive, but highly effective alternative. During a full survey of the sky in 2010, NASA's Wide-field Infrared Survey Explorer (WISE) spacecraft detected more than 150,000 main belt asteroids, including 33,000 which had never before been detected.

Asteroid Origins

Where did the asteroids come from? The realization that most of them are confined to a broad belt between Mars and Jupiter led to speculation that a small planet had once occupied that zone, then somehow been destroyed, perhaps by an explosion or a catastrophic collision with another object.

As the population of the main belt was cataloged in greater detail it became clear that the mass of the entire community – perhaps a million objects with diameters of 1 km or more – was only about 5% of the Moon's mass. Any single ancestor must have been far too small to be classified as a planet.

Furthermore, most of the main belt's mass is actually accounted for by the three largest asteroids, Ceres, Pallas and Vesta. Spectral studies show that these objects are very different in composition and bulk density, so they are unlikely to have been derived from a single parent body.

Today, the consensus is that the vast majority of the bodies orbiting in the main belt are primordial debris left over from the formation of the planets, some 4.5 billion years ago.[2] They were formed through collisional accretion in the solar nebula, probably in the vicinity of the present main belt (see Chapter 1).

Despite following the first steps towards planetary formation, the rocky planetesimals were prevented from accumulating into a full-grown planet because of the gravitational influence of massive Jupiter. Over the eons, the nearby giant has kept the main belt objects dynamically agitated so that collisions between them often result in their erosion and destruction, rather than the gradual agglomeration of a single large body. (Nevertheless, a number of asteroids have been shown to have very low densities, indicating that they are rubble piles with large empty spaces in

[1] http://www.minorplanetcenter.net/iau/lists/NumberedMPs.html
[2] Some asteroids may be the nuclei of dead comets, for example, 3200 Phaethon.

Figure 13.2   Most asteroids are so small and far away that they are visible as mere points of light. However, during a long-exposure photograph, an asteroid's motion results in a linear streak against the background of "fixed" stars. This Hubble Space Telescope image shows a fairly bright asteroid, with a visual magnitude of 18.7, in the constellation Centaurus. The blue trail was caused by the motion of a 2 km diameter object in the main asteroid belt, 250 million km from the Sun. (R. Evans and K. Stapelfeldt, NASA-JPL)

Figure 13.3   Asteroids are the remnants of rocky planetesimals that were formed through collisional accretion in the solar nebula. Over millions of years, embryonic planets collided and shattered into pieces which repeatedly impacted other fragments to create ever finer debris. In the main asteroid belt, planetesimals were prevented from accreting into a full-grown planet because of the gravitational influence of massive Jupiter. (NASA/ JPL-Caltech)

their interiors. Others, such as 4769 Castalia and 4179 Toutatis, appear to be touching binaries that have come together through gentle collisions.)

Spectral studies show that the main belt asteroids are broadly separated into zones that may be associated with different conditions in the nebula: most of the stony S-type asteroids lie in the inner regions, while the carbonacous C types are more numerous further from the Sun. This suggests that the temperature gradient played a key role, with volatile-rich objects forming in the cooler reaches of the main belt (see Box 13.2).

Not all of the differences in composition and physical characteristics are accounted for by their formation in different parts of the solar nebula. Some appear to be the result of melting in the cores of newly forming asteroids. This melting was due to high energy collisions and the decay of radioactive isotopes, such as aluminium-26 (^{26}Al), that heated the interiors of the larger rocky objects. Confirmation comes from the presence of ^{26}Al in ancient meteorites known as carbonaceous chondrites.

Many asteroids may have developed a molten core with a mantle and a crust – rather like miniature Earths. Others, such as Vesta, were probably totally molten, with short-lived surface oceans of molten rock, or magma. The variety of meteorites found on Earth shows that they originated from different regions of these parent bodies after they were partially or completely shattered by subsequent collisions.

Why do other asteroids, such as Ceres, have very low densities – which suggest that they are 25–30% water ice – and not show any evidence of melting or internal differentiation? Since ^{26}Al has a half-life of only 730,000 years, it may be that too little remained to warm any asteroids that formed later in the aggregation process.[3] Another possibility is that Ceres, along with many smaller icy objects, was displaced from the outer Solar System when Uranus and Neptune were rampaging along chaotic orbits around 3.9 billion years ago (see Chapter 1). As a result, they ended up joining the rocky asteroids that were born in the main belt.

Impacts and Rubble

The most obvious characteristics of nearly all asteroids are their rocky appearance, irregular shape and numerous craters – evidence of numerous impacts since they formed 4.5 billion years ago. The largest asteroids possess sufficient gravity to have retained roughly spherical shapes.

---

[3] The half life is the period it takes for half of the atoms in a radioactive isotope to decay. Aluminium-26 decays into magnesium-26.

Only a handful of asteroids have been seen at close quarters. Some of these, such as Mathilde, display large impact features that dominate their topography. These impacts must have been so energetic that they almost destroyed the asteroid. However, most of the asteroids imaged by spacecraft display numerous small and medium-sized craters rather than huge impact features. The presence of many sizeable craters indicates ancient surfaces, the exception being Gaspra, whose craters are relatively sparse and small. Interestingly, although the average asteroid's gravity is very weak, many of these seem to be secondary craters caused by impact ejecta falling back onto the surface.

Close-up views of Eros and Itokawa also show numerous boulders, further evidence of impact debris being scattered over the surface. Occasionally, these have been disturbed and roll down the slopes, leaving trails in the dusty regolith. Fine, loose material also seems widespread on many asteroids (including the Martian moon Deimos, which may be a captured main belt asteroid). Itokawa, for example, was found to be covered with unconsolidated, millimeter-sized particles and larger "gravels". This material may be deep enough to fill and cover small depressions, giving them a generally smooth, rounded appearance (see Box 13.3).

Most surprising of all is the presence of localized plains that cover 20% of Itokawa's surface and are formed from the finest granules. This was a surprise because impact ejecta on a small asteroid were expected to spread globally over its surface, resulting in continuous regolith.

One possibility is that the unconsolidated gravels have globally migrated and segregated due to fluidization caused by vibrations induced by numerous impacts of small meteoroids. When gravel is vibrated, it can be fluidized and behave as granular fluid. The most popular phenomenon related to this is the "Brazil nut effect," which causes the biggest particles to end up on the surface when granular material is shaken.

This theory is supported by close-up images which show that Itokawa's gravels are generally aligned in the same direction – exactly coinciding with the directions of local gravitational slopes.

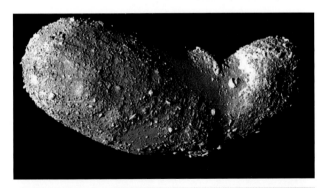

Figure 13.4 A view of near-Earth asteroid Itokawa taken by Japan's Hayabusa orbiter. Itokawa measures only 0.5 × 0.2 × 0.3 km and has an unusual shape, with a narrow waist and variable terrain that includes smooth plains, boulder-covered regions and no obvious impact craters. (JAXA)

When the asteroid is shaken by an impact with a small meteoroid, the fine regolith shifts position, slowly migrating down slope and collecting in the nearest hollow. Over time, the larger boulders mixed in with the finer material are left stranded on the higher slopes, forming an area of rough terrain. Such granular sorting processes may be a major resurfacing mechanism for all small asteroids possessing regolith.

Some asteroids, including Itokawa and Mathilde, have very low bulk densities. For example, Mathilde seems to be composed of material similar to that found in carbonaceous chondrite meteorites, but calculations of its volume and mass (derived from spacecraft tracking data) indicate a bulk density of only 1.3 g/cm^3, about half that of the meteorites. This suggests that Mathilde may have a rubble pile interior characterized by large pores or empty spaces.

It seems that such objects were once shattered but slowly reformed through their mutual gravitational attraction and low velocity collisions. Today, the loosely packed rubble is barely held together by the asteroid's gravity. If an object collided with Itokawa, it would probably resemble a rock landing in a bucket of sand. The signs of impacts with small meteorites are erased as the rubble shifts after the impact, so there are very few visible craters on Itokawa.

Eros is a much larger representative of the near-Earth asteroid population, but it, too, has a surprising lack of small craters and some remarkably smooth patches. Across nearly 40% of its surface, all craters up to about 0.5 km wide have been erased. Instead of the expected 400 craters per km^2 with diameters of only 20 meters, it averages only about 40.

On the other hand, one of its most striking features is a fairly recent, 5.3 km diameter impact crater. Much of the surface is covered with piles of rubble, blocks of ejecta between 30–100 meters across. However, they are not uniformly distributed and the NEAR spacecraft revealed some smooth, flat areas, dubbed "ponds" by mission scientists. As with Itokawa, these are attributed to the downslope movement of fine material due to seismic activity generated by meteoroid impacts.

There were few visible albedo variations, except on the inner walls of some craters where mass movement of loose material was taking place. Spectra indicate that it is probably similar in composition to ordinary chondrite meteorites. A mean density of 2.67 g/cm^3 implies an internal porosity of 10–30%, and analysis suggests that Eros has been internally fractured by large impacts. Supporting evidence is seen in a series of grooves and ridges that run across the asteroid's surface both globally and regionally.

Two-part asteroids that resemble a peanut or hourglass are also quite common. In the case of Itokawa it has been surmised that the asteroid's head-and-body structure might have arisen as the rubble pile shifted in response to impacts. Alternatively, the two rounded segments may once have been separate rubble piles that slowly collided and gradually merged.

Asteroid Families and Kirkwood Gaps

The orbits of more than 230,000 asteroids have been cataloged since the discovery of Ceres, and the number is rising inexorably, with thousands more reported every month. Analysis of the

Box 13.2   Asteroid Classes

Asteroids are classified according to their spectral and physical characteristics. Attempts to link these with the various groups of meteorite (see below) have not always been successful. The overall composition of the main belt asteroids also varies. Stony, S-type asteroids are more common near the inner edge, while darker, carbon-rich C-types are more numerous further out. The most common classes are C, S and M.

- C (carbonaceous). The most numerous group, accounting for 75% of the main belt population. The proportion increases from about 10% at 2.2 AU to about 80% at 3 AU. They are dark gray and reflect very little sunlight (albedos 3–5 %). Their flat, generally featureless, spectra resemble those of carbonaceous meteorites, a link that was confirmed when the NEAR Shoemaker spacecraft landed on Eros.
- S (stony or "silicaceous"). About 15% of the total population, they are most numerous in the inner main belt, accounting for some 60% of the population at 2.2 AU, decreasing to 15% at 3 AU. They are reddish and moderately reflective (albedos 15–25%). Their spectra resemble those of metal-bearing meteorites known as chondrites. Notable examples are 3 Juno and 5 Astraea. S-type 25143 Itokawa was the target for the first asteroid sample return mission.
- M (metallic). The third most populous. They have moderate albedos and are linked to iron-nickel meteorites. They are thought to represent the remnants of the metal-rich cores of ancient asteroids that were large and hot enough to differentiate internally. 16 Psyche is thought to be almost pure iron-nickel alloy.
- E ("enstatite"). These are fairly rare. They have high albedos (30% or more) and may contain silicates with a fairly high amount of the mineral enstatite. They are thought to be the source of aubrite meteorites. E-type asteroids include 64 Angelina, 214 Aschera and 620 Drakonia. The first example to be observed from close range was 2867 Steins, imaged by Rosetta in October 2008.
- V. Very rare, they are highly reflective and were once molten. They are associated with basaltic meteorites known as eucrites. The main example is 4 Vesta, which is probably the parent body of most of the eucrites. However, several V-type asteroids do not belong to the Vesta family, including 1459 Magnya, the first basaltic object detected in the outer asteroid belt.

Many other classes and subdivisions are recognized, though they are often given different designations since a number of classification schemes based on different criteria are in use. For example, Ceres belongs to class G, a subcategory of C-class asteroids which is distinguished by a very strong ultraviolet absorption feature at wavelengths shorter than 0.4 microns due to water ice on its surface.

computed orbits shows that the vast majority lie in the main belt between 2.2 and 3.3 AU, so they are generally closer to Mars than to Jupiter.

About half of all main belt asteroids belong to families that have similar orbital characteristics. Similarities in their reflection spectra indicate that each family is made up of fragments of a particular parent body that was shattered by a catastrophic collision billions of years ago. Certain types of meteorites have also been matched to individual large asteroids and their families (see Box 13.2).

The existence of such groups was first recognized by Japanese astronomer Kiyotsugo Hirayama (1874–1943), so they are now known as Hirayama families. Approximately 100 such families have now been identified, each being named after its largest member. The most populous examples, such as the Flora, Eos, Themis and Koronis groups, have hundreds of members.

Jupiter is one of the major drivers behind the distribution of main belt asteroids and families. Some regions of the asteroid belt – called Kirkwood gaps, after US astronomer and mathematician Daniel Kirkwood (1814–1895) – are relatively empty, due to the influence of the planet's enormous gravity. The gaps occur where an asteroid's orbital period would be an exact multiple of Jupiter's

Figure 13.5   The distribution of asteroids in the main belt, located 2.2–3.3 AU from the Sun. Repeated gravitational interactions with Jupiter have caused asteroids to be ejected out of certain regions, creating the Kirkwood gaps (e.g. where their orbital periods are 3:1, 5:2, 7:3 and 2:1 of Jupiter's orbital period). There are also several peaks corresponding to groups of asteroids with nearly the same orbital distance. The Hilda group and Jupiter's Trojan asteroids are not shown. (Open University)

orbital period. The most important ratios are 2:1, 3:1, and 5:2. This means that any asteroid which orbits the Sun exactly twice in the time it takes Jupiter to go around once will be removed by the repeated influence of the giant planet on the same sector of the asteroid's orbit. Jupiter's gravity simply keeps the gaps clear by changing the orbits of any asteroids within them.

One interesting exception is the region where 700 Hilda asteroids are trapped in a 3:2 mean resonance with Jupiter, at a distance of about 4 AU – well outside the main belt. Almost all Hildas have orbital eccentricities between 0.1 and 0.25. Models show that, during the early evolution of the Solar System, Jupiter probably migrated inward by perhaps 0.35–0.45 AU over a period of at least 100,000 years. During this inward movement, the planet would have shepherded many asteroids into 3:2 resonant orbits. Any family members with low eccentricity would have been ejected – hence their absence in the Hilda group today.

Not all asteroid families are ancient. In 2006, a team of US scientists found a group of six objects orbiting close to the asteroid Datura, suggesting that fragments were broken off from the main member of the group during a collision some 450,000 years ago. Their continuing proximity supports the youthful age of the family, since they would be expected to drift apart and eventually mix with the general asteroid population. At least two more co-orbiting family groups with ages a little over one million years old have been identified.

On a larger scale, the collisional destruction of a 150 km diameter main belt asteroid is thought to have created the Veritas family 8.2 million years ago. Analyses of sea floor sediments show that the collision may have been responsible for a marked, prolonged, increase in the amount of interplanetary dust falling to Earth.

Asteroid Moons

When the Galileo spacecraft flew past the 31 km diameter main belt asteroid Ida in 1993, astronomers were surprised to see that it had a small companion. This discovery of 1 km wide Dactyl came as a surprise because, until then, it was considered that the weak gravity of an asteroid might be insufficient to hold a satellite in orbit.

Today, the number of asteroids encountered by spacecraft remains very small, so they provide a very limited and selective source of data. Meanwhile, advanced ground-based instruments have enabled dozens of asteroid satellites to be detected. In 1998, the second of these objects, 13 km wide Petit-Prince, was found in orbit around a main belt asteroid, 215 km diameter Eugenia.[4]

This discovery was made possible through the use of a revolutionary technique called adaptive optics. When stars or asteroids are viewed through the Earth's atmosphere, turbulence makes them appear blurred. Adaptive optics remove the blurring by use of an adjustable mirror, resulting in images that are comparable with those taken from by the Hubble Space Telescope.

Although this method works well for large, remote asteroids, ground-based radar is superior for observing small near-Earth asteroids. With this technique, radar waves transmitted by a large dish antenna are bounced off the asteroid and the echoes are detected. The size, surface characteristics and orbital speed of an asteroid can be detected in this manner.

Another technique that has recently been introduced is optical interferometry, which combines the light from two or more telescopes. This method enabled two of the European Southern Observatory's 8.2 meters VLT telescopes to discover that main belt asteroid 234 Barbara is composed of two bodies with diameters of 37 and 21 km. It was either shaped like a gigantic peanut or made up of two separate bodies orbiting each other.

Through these various efforts, satellites have been found orbiting asteroids of many different sizes – between 15 and 50% of all asteroids with diameters less than 10 km are estimated to have companions.

One of the strangest "binaries" is near-Earth asteroid 4179 Toutatis. Radar images indicate that it comprises two asteroids that are either loosely attached, rolling around against each other, or in close orbit. "Yam-shaped" Toutatis measures $1.92 \times 2.29 \times 4.6$ km and has a very unusual rotation. Whereas most Solar System objects spin around a single axis, Toutatis tumbles as the result of two different types of motion with periods of 5.4 and 7.3 Earth days. For anyone standing on its surface, the Sun and the stars would appear to cross the sky following different paths and moving at different rates during each new Toutatis "day." The asteroid follows a very eccentric orbit close to the ecliptic. It barely crosses Earth's orbit at perihelion, and retreats to the main belt at aphelion, taking 3.98 years to orbit the Sun.

Another interesting binary is 90 Antiope, which was discovered in 1866. In 2000, a team led by William Merline (Southwest Research Institute) found that it actually comprises two objects that measure about 88 km and 85 km across. Their densities are not much higher than that of water, suggesting that they are very porous. The distance between them is 171 km, with one orbit taking 16.5 hours. Their rotations are tidally locked, so they always present the same side to each other. Moreover, the asteroids rotate in the same plane as they orbit each other.

Such discoveries provide valuable insights into the nature of asteroids. Astronomers can use the orbital period of a companion to determine the mass, and then the density, of the asteroids involved (see Chapters 1 and 8). The density gives a clue to the asteroid's composition and structure. For example, thanks to the discovery of its moon, we know that Eugenia is rather lightweight, with a lower density than its dark, rock-like appearance might suggest.[5]

Multiple asteroids are not confined to the main belt. In September 2000, NASA's Goldstone radar facility in California detected 2000 DP107, the first near-Earth binary asteroid. Subsequent observations made at Arecibo in Puerto Rico showed that its components orbit about 3 km apart, and the 300 meters wide

---

[4] Petit-Prince was named after the son of the French Empress, Eugenie.

[5] Spacecraft flybys have also made it possible to calculate asteroid masses from the slight orbital changes caused by the object's gravitational tug. Previously, although the approximate size of a large, individual asteroid could be obtained through direct measurement or studies of stellar occultations, there was no accurate way to determine its mass.

Box 13.3   Hayabusa Samples Itokawa

Although many asteroids have been studied by ground-based remote sensing or by spacecraft, the only way to determine an asteroid's composition is to retrieve a sample of pristine material and return it to a laboratory on Earth for analysis. The first attempt to achieve such a sample return was made by Japan's Hayabusa (Falcon) spacecraft, which was launched toward near-Earth asteroid 25143 Itokawa on May 9, 2003.

Despite the failure of several reaction wheels and damage from a large solar flare, which degraded the solar panels and caused a severe loss of power to its ion engines, the spacecraft successfully rendezvoused with Itokawa on September 12, 2005. Remaining in a station-keeping position 20 km from the asteroid, it was able to map the surface of the tiny object at different wavelengths.

In addition to a multi-spectral camera, Hayabusa carried a laser altimeter (LIDAR) which determined Itokawa's size and shape by constantly measuring the distance between the spacecraft and the asteroid. A near-infrared spectrometer measured the mineral composition and properties of the surface. The link between asteroids and meteorites was investigated by an X-ray fluorescence spectrometer designed to detect different elements and measure their abundance.

Itokawa was found to measure only $540 \times 310 \times 250$ meters, with a surface gravity 750,000 times lower than on Earth. Other obstacles to a successful sample return included a 12-hour rotation period and a rugged, boulder-strewn surface.

An attempt to deploy a small, hopping robot, named Minerva, on the surface failed on November 12 after the lander was jettisoned at too high an altitude. The efforts to conduct one dress rehearsal and two sampling runs also proved problematic. These involved releasing a softball-sized target marker made of highly reflective material. A flashlight-like device was to illuminate the surface marker and assist in the landing.

During a fleeting touchdown on the surface, Hayabusa would fire a "bullet" made of tantalum into the surface at a speed of 300 m/s. Tiny rock fragments ejected by the impact would enter a funnel-shaped horn and pass into a sample container which would then be sealed, ready for return to Earth.

The first dry run had to be abandoned. During a second rehearsal, the spacecraft descended to a height of 55 meters but lost Minerva. The first sampling attempt on November 20–21 saw Hayabusa remain on the surface of the flat "MUSES Sea" for half an hour – much longer than planned – but the sampling gun failed to fire.

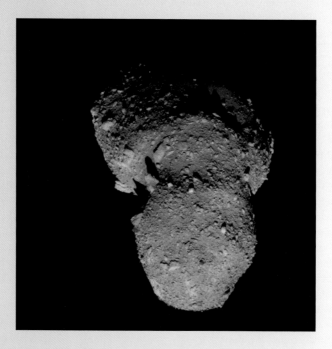

Figure 13.6 Itokawa is the smallest asteroid ever seen from close range. Despite its almost imperceptible gravity, this NEO has a very diverse surface, with numerous large boulders and several smooth areas. Note the double asteroid appearance, suggesting that it may comprise two objects which gently collided. (ISAS / JAXA)

Hayabusa made a second, brief landing on November 26. At first, it appeared that two bullets had fired and sample retrieval had been successful. However, subsequent analysis indicated that the sampling gun may not have fired, with the possibility that no samples were sealed inside the return capsule. The spacecraft then suffered leaks in its chemical thrusters and began to spin out of control, losing contact with the ground. The Japanese space agency delayed the return to Earth for two years in order to give engineers more time to regain control.

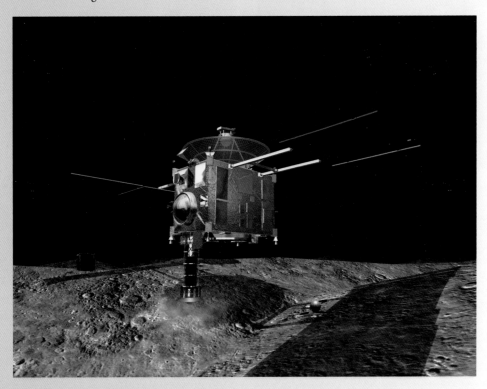

Figure 13.7 Hayabusa was the first mission designed to land on an asteroid and return a rock sample to Earth. The sampling horn is shown beneath the spacecraft. The spacecraft touched down twice and successfully obtained fragments of surface material. The sample return capsule landed in Australia on June 14, 2010. (JAXA)

Hayabusa finally began its return journey on April 25, 2007, but it faced an uncertain journey with only one of four ion engines working properly. Overcoming all the odds, it safely returned to Earth on June 14, 2010, three years later than originally planned. The small re-entry capsule was ejected for a parachute descent near Woomera in Australia. The capsule was opened in Japan under controlled conditions at a new national facility built for the purpose. Preliminary analysis revealed some 1500 tiny particles of asteroid material inside one compartment of the capsule. Their mineral content matched that of ordinary chondrite meteorites – an important clue to what happened during the earliest epoch of the Solar System's history. The data confirmed a link between S-type asteroids and ordinary chondrite meteorites (see below).

satellite always presents the same face to its 800 meters diameter companion.

Even more complex systems are possible. In August 2005, US and French astronomers observed the first triple asteroid system. The largest member, known as 87 Sylvia, is 280 km in diameter and orbits in the outer part of the main belt, about 3.5 AU from the Sun. It spins quite rapidly, with a 5 hour 11 minute period.

In 2001, Sylvia was found to have a much smaller satellite in a four-day orbit. Four years later, a second moon was discovered. Since Sylvia was named after Rhea Sylvia, the mythical mother of the founders of Rome, the companions were named after her sons, Romulus and Remus. The small moons follow nearly circular, prograde orbits in the same plane.

How these gravitationally bound systems originate is not clear. At first it was thought that asteroid moons formed solely through collisions and/or close encounters with planets. Theoretical models indicate that an asteroid which resembles a pile of loose rubble can be disturbed during close encounters with the inner planets of the Solar System. By passing within 16,000 km of a planet's surface, a piece of the asteroid may break away, forming a satellite.

However, it was found that such encounters cannot account for the large number of binary asteroids that have been discovered. Some other process must have been responsible. One of the favored candidates is the YORP effect, in which sunlight spins up a small asteroid so much that it splits apart or ejects material from

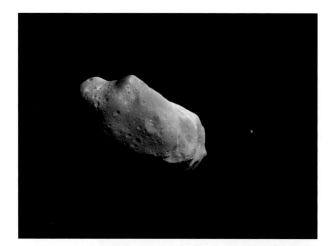

Figure 13.8 Main belt asteroid 243 Ida is a member of the Koronis family. This false color mosaic was made from images taken by the Galileo spacecraft at a distance of about 10,500 km, on August 28, 1993. Ida measures 56 × 24 × 21 km. It is probably composed of silicates and covered in impact craters, so its surface is very ancient. Brighter, bluish areas around craters (upper left, centre and upper right) suggest a difference in the abundance or composition of iron-bearing minerals. This was the first image of an asteroid moon, now named Dactyl (far right). Both are probably fragments of a much larger asteroid. (NASA-JPL)

its equator that eventually coalesces into a satellite (see Box 13.4). Another possibility is that a rubble pile asteroid is caused to spin faster by an oblique impact. The spinning asteroid would become elongated and eventually shed one or more small moons.

Ceres

Ceres is by far the dominant member of the main asteroid belt, accounting for about 25% of the belt's total mass. It is also the only asteroid that is deemed worthy of being classed as a "dwarf planet" (see Chapter 1). Yet, in many ways, Ceres remains a mystery. Until recently, even its size was a matter of debate. Stellar occultations, together with studies by the Hubble Space Telescope and Keck telescopes, have shown that Ceres is a slightly flattened sphere with a diameter of about 910 × 970 km. Its day lasts a little over nine hours.

More intriguing is the nature of its surface and interior. Hubble and Keck images have shown some low albedo surface features. One hemisphere appears slightly brighter than the other, while a dark region running north-south is located at the boundary between them. Several circular features are visible, including one dark region with a bright central spot (named "A") and another, some 250 km in diameter, named Piazzi. These may be large impact craters. Two other dark features are visible in the southern hemisphere. Whether these are craters or manifestations of surface processes will be revealed by NASA's Dawn spacecraft, which is due to enter orbit around the dwarf planet in 2015 (see

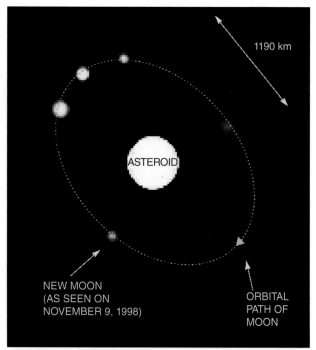

Figure 13.9 This infrared composite image, obtained with the Canada-France-Hawaii telescope in Hawaii, shows Petit-Prince, a small satellite orbiting main belt asteroid 45 Eugenia at a distance of about 1200 km. One orbit of the moon (green dashed line) takes 4.7 days. The near-circular orbit is tilted about 45° with respect to the line-of-sight. Eugenia has a diameter of about 215 km, while the satellite's diameter is about 13 km. The moon is 300 times fainter than the asteroid. (Laird Close/ESO and Bill Merline/SwRI)

Box 13.5). If an incoming object did create Piazzi, it must have been about 25 km across.

Studies of its gross properties indicate that Ceres has a density of about 2.1 g/cm^3, similar to Ganymede and Callisto, suggesting that its composition is at least 25% water ice. The spectral characteristics of the brightest regions are consistent with minerals that have been altered by water (e.g. phyllosilicates) and carbonate compounds. Darker, isolated regions could be ice or frost mixed with other minerals.

Ceres may well be one of the most important left-over relics of the early Solar System, large enough to have experienced many of the processes normally associated with planetary evolution and to have survived the early accretion phase in which most of its companions were incorporated into the planets.

Models suggest that Ceres grew from a mix of silicates and water ice, possibly with considerable amounts of carbonaceous material. Early in its history, internal heating from short- and long-lived radionuclides would have caused the water ice to quickly melt, forming a mantle beneath the icy crust. The circulating water would alter the silicates. Internal thermal processing and mixing with the water could also create methane and CO_2 clathrates, in which gases are trapped within crystalline ice. As heat was lost by conduction through the frozen surface, water began

Figure 13.10 These radar images of Toutatis were obtained by ground-based radar in December 1992, when it passed about 2.5 million km from Earth. They reveal two irregular, cratered objects about 4 km and 2.5 km in diameter that are probably in contact with each other. The largest crater (top right image) is about 700 meters across. The asteroid's tumbling rotation and cratered surface suggest a history of frequent collisions. A 4 kW coded radio transmission was beamed at Toutatis from the 70 meter antenna at Goldstone, California. The echoes took about 12 seconds to return to Goldstone, where they were decoded and processed. (Steve Ostro / NASA-JPL)

to freeze out at the base of the crust. As the crust grew in thickness, solid-state convection would become continuous, transporting more heat as well as altered materials toward the surface. After perhaps two billion years, Ceres' water layer eventually froze, forming a layered structure within the asteroid.

In addition, melting and freezing plus mineralization and frequent impacts would, over time, create surface topographic features. If clathrates were brought near the surface, they would suddenly vaporize, possibly producing explosive release of gases, bringing considerable quantities of altered materials to the surface, and creating surface landforms.

Ceres would also have shrunk as internal ice melted. Then, as the water froze again, the asteroid would have expanded, perhaps creating topographic features associated with extensional forces, such as cracks, faults and dropped blocks of crust.

With so much water present and the energy to distribute it in liquid form, Ceres probably experienced complex chemistry – at least in its interior – that may have included organic materials. It seems that Ceres may be a perfect place to study the evolution of objects at the interface between the terrestrial planets and the colder, icy objects of the outer Solar System.

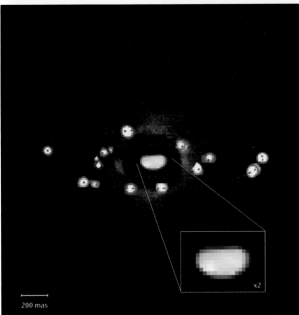

Figure 13.11 Main belt asteroid 87 Sylvia was the first triple asteroid system to be discovered. This composite image shows the positions of its two moons, Remus and Romulus over nine nights. The inset shows the potato shape of Sylvia. The discovery was made with one of the 8.2 meter telescopes of ESO's Very Large Telescope Array in Chile, using adaptive optics. (ESO)

Vesta

In contrast with ice-rich Ceres, 4 Vesta, the other object on Dawn's itinerary, is much denser and drier, and seems to have experienced a very different early history. The third largest of the asteroids,

Box 13.4   Solar-Powered Asteroids

The rotation rates and orbits of asteroids are influenced by external gravitational forces and collisions. However, sunlight has also been shown to put asteroids in a spin.

Incoming solar radiation is absorbed on the day side and radiated back into space as the asteroid rotates, generating a minuscule recoil effect. Since small asteroids are irregular in shape, the recoil affects one part of the asteroid more than another. The object's thermal inertia also means that the maximum surface temperature occurs on its afternoon side, rather than the morning side or sub-solar point (midday), because it has been in the Sun for longer. This is similar to Earth, where 2–3 pm, rather than noon, is the warmest time of day. As a result, the asteroid tends to emit more heat from its afternoon side. This unbalanced thermal radiation produces a tiny acceleration or thrust, known as the Yarkovsky effect. It also results in a twisting force or torque known as the YORP (Yarkovsky-O'Keefe-Radzievskii-Paddack) effect.[6]

The momentum generated by the Yarkovsky effect over each orbit of the Sun causes gradual drifting of the orbits of kilometer-sized asteroids, causing objects to spiral inward or outward at different rates as a function of their spin, orbit and material properties. Since every action has an equal and opposite reaction, according to Isaac Newton's laws of motion, heat radiating from one side of the asteroid will push it out of its existing orbit in the opposite direction – towards the Sun if the asteroid spins retrograde, and away from the Sun if prograde.

Not only does it influence the dynamical spreading of asteroid families, but it also plays an important role in changing the orbits of main belt objects so that they are redirected into planet-crossing orbits (e.g. near-Earth asteroids).

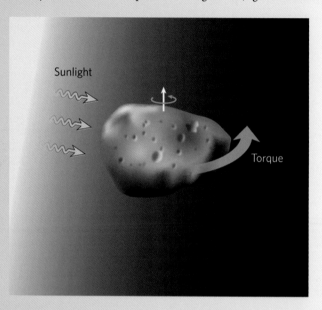

Figure 13.12   An unevenly shaped asteroid heated by solar radiation re-radiates the thermal energy at right angles to its surface. The resultant net torque can provide a small thrust (the Yarkovsky effect), causing the asteroid to drift towards or away from the Sun. It can also change the asteroid's spin rate – the YORP effect. (Nature)

The YORP effect has a subtle influence on the rotation rate of asteroids. In 2003, William F. Bottke (Southwest Research Institute) and colleagues discovered that members of the Koronis family have their spins clustered into several groups, rather than the expected random distribution. They concluded that the slow, but steady, recoil force of thermal re-radiation can be more effective than collisions in controlling the direction of asteroid spins.

The force generated by sunlight and its effect on asteroid spin rate was measured directly for the first time by a team led by Stephen Lowry (Queen's University, Belfast, UK). They monitored the spin rate of near-Earth asteroid 2000 PH5 from 2001 to 2005. The 120 meter long rock rotated once every 12 minutes or so, but its spin period was being shortened by one millisecond every year, a rate that matched the predicted YORP effect. (It is also possible, of course, that the effect may cause asteroids to slow down.)

Similar studies have been made of a much bigger Earth-crossing asteroid, 1862 Apollo. Between 1980 and 2005 its rotation period shortened by about four milliseconds per year, also in line with predictions. Although it would take about 2.6 million years for YORP to double Apollo's rotation rate, the change is highly significant over the age of the Solar System.

The YORP effect is also important for understanding binary asteroids. Many binaries form from asteroid collisions or gravitational capture. However, a single asteroid may split in two if the YORP effect causes it to spin too fast. Lowry predicts that in about 14 million years, 2000 PH5 will spin once every 20 seconds or so, faster than any asteroid yet observed. If the object is a rubble pile, rather than solid rock, it could fly apart.

This process may sometimes create an asteroid moon. When solar energy "spins up" a rubble pile asteroid to a sufficiently fast rate, material may be thrown off from its equator. If the ejected material is able to shed a lot of its momentum through mutual collisions, then it may coalesce and form a satellite in orbit around its parent.

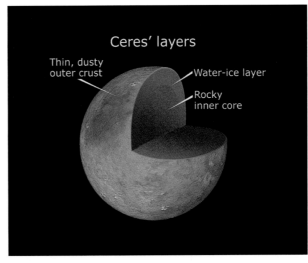

Figure 13.13   An enhanced color view of Ceres obtained with the Advanced Camera for Surveys on the Hubble Space Telescope. The colors show variations in surface material. The brighter and darker regions may be impact features. The observations were made in visible and ultraviolet light between December 2003 and January 2004. (NASA-ESA, J. Parker / Southwest Research Institute, P. Thomas / Cornell University, L. McFadden / University of Maryland, M. Mutchler and Z. Levay / STScI)

Figure 13.14   Computer models based on HST observations suggest Ceres has a differentiated interior, with a dense, rocky inner core and a thin, "dusty" crust. In between there may be a mantle rich in water ice. If this mantle is composed of at least 25% water, it would contain more fresh water than Earth. Ceres is nearly spherical, which suggests that gravity controls its shape. However, small surface irregularities indicate that material is not evenly distributed throughout the interior. (NASA, ESA, and A. Field / STScI)

Vesta is roughly spherical, with dimensions of $578 \times 560 \times 458$ km, and an estimated density of about 3.9 g/cm^3, corresponding to a rocky composition.[7]

Vesta has a spectrum that is unique among all the larger asteroids and shows evidence of past geologic activity. It is apparently covered with basalt, leading researchers to conclude that Vesta's interior was once molten – at least for a brief period. The cause of the heating may have been the presence of short-lived radioisotopes in the predominantly rocky body.

Observations with the Hubble Space Telescope show that it has one dark hemisphere and one bright hemisphere, representing two different types of solidified igneous rock. The lighter hemisphere is thought to be composed of basalt which is rich in the mineral pyroxene, formed when magma cooled and solidified beneath the surface. This region may have been heavily excavated by impacts, resulting in the exposure of ancient, subsurface material.

The darker hemisphere is interpreted to be composed of a type of basalt which is a mixture of pyroxene and feldspar. It may be the remains of Vesta's ancient crust, formed from lava which cooled and solidified on the surface.

The HST also revealed several likely impact features, including a giant crater near the south pole that dominates the entire hemisphere. About 500 km wide and more than 15 km deep, it has a raised rim that is up to 17 km above the surrounding terrain. Since Vesta is large enough to be internally differentiated, the impact may have penetrated almost all the way through the volcanic crust to expose the mantle. The asteroid's small diameter and low gravity enabled a major crustal rebound after the impact, creating a towering, 22 km high central peak (see Chapter 4). It is almost as tall as the highest mountain in the Solar System, Olympus Mons on Mars.

The vast amount of debris that was blasted into space from the crater is also the likely source of many basaltic meteorites. Approximately 6% of the meteorites that fall to Earth have a similar mineralogical signature, as indicated by their spectral characteristics. Vesta is the only major world, other than Earth, Moon and Mars, for which there are samples of known origin.

---

[6]These physical processes are named after Ivan Osipovich Yarkovsky, a 19th century Russian civil engineer and part-time scientist. Other contributors to understanding of the YORP effect were American planetary scientist J. A. O'Keefe, Russian astronomer V. V. Radzievskii and NASA aerospace engineer S. J. Paddack. They proposed that diurnal solar heating of a rotating object in space would cause it to experience a tiny force that could lead to large secular effects in the orbits of small bodies, especially meteoroids and small asteroids.

[7]The mass, density and composition of Vesta will be refined by the Dawn spacecraft.

Box 13.5   Dawn

Ceres and Vesta are two of the largest and most intriguing members of the main asteroid belt. In order to learn more about these contrasting objects, NASA developed the Dawn spacecraft, which was launched in September 2007. Dawn is the first spacecraft to orbit a main belt object and is planned to become the first to orbit two bodies. Its mission is to study two protoplanets that were never given the opportunity to grow fully. This will provide vital clues about their divergent evolutionary paths, together with the conditions and processes associated with planetary formation 4.5 billion years ago.

Figure 13.15   Dawn was launched in September 2007. After a four-year journey using ion propulsion, it entered orbit around Vesta on July 15, 2011. At the end of a year-long study of the basaltic asteroid, it will depart for water-rich Ceres, with arrival due in February 2015. (NASA/William Hartmann)

Dawn uses an innovative ion propulsion system that comprises three xenon-fuelled engines powered by electricity generated by two large solar arrays. After a journey of 2.8 billion km, which included a Mars gravity assist in February 2009, it entered orbit around Vesta in July 2011. Over the next year, it will conduct an in-depth investigation of the basaltic asteroid. Dawn will then depart for Ceres, with arrival due in February 2015. The primary mission is expected to end in July 2015.

Dawn carries three science instruments:

1.   a visible/near-infrared camera;
2.   a visible and infrared mapping spectrometer to map mineral distribution, and;
3.   a gamma ray/neutron spectrometer to determine the elemental composition of the surface.

Radiometric and optical navigation data will provide information about the asteroids' gravity fields, bulk properties and internal structure.

© NASA/JPL-Caltech/UCLA/MPS/DLR/IDA   -22 —————— 17   Elevation [km]   200 km

Figure 13.16 This false-color map of Vesta was created from stereo images obtained by the framing camera on NASA's Dawn spacecraft. The image shows the surface elevation with a horizontal resolution of about 750 meters per pixel. The terrain model of Vesta's southern hemisphere shows a circular impact structure with a diameter of about 500 km. Its rim rises above the interior of the basin for more than 15 km. The impact excavated about 1% of Vesta's entire volume. The Vesta asteroid family and V-type asteroids are probable products of this collision. (NASA/JPL-Caltech/UCLA/MPS/DLR/IDA)

The Dawn spacecraft has also revealed a landscape of grooves lying parallel to the equator, cliffs, and smooth areas, as well as numerous impact craters of different sizes.

Other Large Asteroids

Information about the other major asteroids is even more limited. Estimated to account for 7% of the total mass of the asteroid belt, 2 Pallas is one of three objects in the main belt that can be classified as a protoplanet, rather than mere debris from the early Solar System.

Pallas is the largest of the B-type asteroids. Until, recently, the only way to determine its size with any accuracy was through stellar occultations and speckle interferometry (a technique for overcoming atmospheric turbulence by combining short photographic exposures of an object). More recent imaging with the largest ground-based telescopes and the Hubble Space Telescope indicate that Pallas is a spheroid that measures approximately $582 \times 556 \times 500$ km. This makes it slightly larger than Vesta, though it is some 20% less massive.

Based on its size, the mean density of Pallas is in the range $2.4–2.8$ g/cm^3 – consistent with a body that formed from a rock-ice mixture. This density is significantly lower than Vesta's, but greater than Ceres, suggesting that it is largely rocky and contains less water than its larger neighbor. The spectrum and estimated density are comparable to those of carbonaceous chondrite meteorites. However, multispectral analysis suggests that its surface is covered with hydrated minerals, consistent with both Ceres and Pallas forming from a common reservoir of water-rich material. Both objects also appear to have differentiated interiors. The hydrated material that covers Pallas suggests that its surface and interior were thermally altered by impacts or internal heat while it contained a large amount of liquid water.

Few details can be seen on the fairly dark surface of Pallas (albedo is 12%). HST images show little variability in visible and infrared light, but significant variations at ultraviolet wavelengths suggest large surface or compositional variations. There may be a 240 km wide impact crater within Pallas's ultraviolet-dark terrain at about 30°S, 75°E, as well as smaller impact features.

Rotation appears to be prograde, with a very large axial tilt, indicating that it is spinning on its side. This means that it experiences long summers and winters, with each hemisphere in constant sunlight or darkness for about one Earth year. The rotation period is about 7.8 hours, midway between the periods of Vesta (5.3 hours) and Ceres (9 hours).

Figure 13.17 Images taken with the Hubble Space Telescope in 2007 show Pallas's largest crater-like feature, appearing face-on (upper right) and edge-on along the limb (lower right). The flat region indicated by the arrow in the digital model (left) and in two images is visible as a 240 km wide depression in the center of the model. (B. E. Schmidt, et al., in *Science*, Oct. 9, 2009)

The orbit of Pallas is inclined at an angle of 34.8° to the plane of the main asteroid belt, and the orbital eccentricity is nearly as large as that of Pluto, making Pallas a difficult target for visiting spacecraft.

Like Vesta, Pallas is linked to an asteroid family that shares its orbital and spectral parameters. The largest of these is 5222 Ioffe, with a diameter of 22 km. They may all have originated from the large crater seen in HST images.

The third asteroid to be discovered, 3 Juno, is considerably smaller than Vesta or Pallas, but it is also more reflective. Juno has a diameter of about 240 km with an area that appears dark at near-infrared wavelengths. It seems that the asteroid has collided with another object fairly recently (in astronomical terms), resulting in a 95 km wide crater, or possibly a smaller crater that is surrounded by a 95 km blanket of ejecta.

Juno follows an even more eccentric orbit than Pallas, inclined about 13° to the ecliptic. It, too, is thought to be linked with a family of small asteroids. The members of the "Juno Clump" are probably derived from impact debris ejected from the parent object.

Trojan Asteroids

Although most asteroids are found between the orbits of Mars and Jupiter, a significant population can be found elsewhere in the Solar System. Many of these cross the orbits of the inner planets (see Box 13.6). Others have been shepherded into groups that are co-orbital with the planets.

Known as Trojan asteroids, these latter objects are in gravitationally stable orbits around the L4 and L5 Lagrange points, 60° behind and 60° ahead of the planet in its orbit. They typically follow a path that takes them up to 30° either side of the equilibrium points over a period of 150–200 years. Many of them have

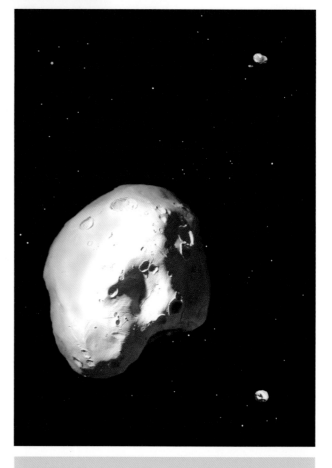

Figure 13.18 Images of asteroid 3 Juno taken with an adaptive optics system on the 2.6 m Hooker telescope at Mt. Wilson Observatory show what appears to be a 95 km wide crater. This artist's concept shows the large basin excavated by the impact. (D. A. Aguilar, Harvard-Smithsonian Center for Astrophysics)

large orbital inclinations, up to 40° relative to the orbital plane of Jupiter.

The first Trojan, was discovered in 1906. By February 2011, 4822 such asteroids had been discovered, nearly all of them associated with Jupiter. Four Trojans have also been found along the orbit of Mars (three at the L5 point and one at L4), one orbits ahead of Earth, and seven are associated with Neptune (all but one at L4). However, some estimates suggest there may be in excess of one million Trojans. None are known to be linked with Saturn, presumably because its Lagrangian points are destabilized by Jupiter's gravity.

Since most of the Trojans are more remote and considerably fainter than the main belt asteroids, they are more difficult to study. However, they are generally very dark (albedos of 2–5%) and classified as primitive, carbonaceous, D type asteroids which are redder than the C type asteroids of the main belt. They have been trapped around the Lagrange points over the lifetime of the Solar System, although those in marginally stable orbits may be ejected.

Jupiter's Trojans are subdivided into the Achilles group (60° ahead) and the Patroclus group (60° behind), though their individual positions can vary considerably.[8] About two thirds of them belong to the Achilles group. 624 Hektor is the largest known Trojan, as well as the most elongated known asteroid of its size. Measuring roughly 150×300 km, it seems to have a dumbbell shape, probably as a result of the partial coalescence of two spheroidal planetesimals during a relatively low-speed collision.

Until recently, the Trojans were generally regarded as planetesimals that formed near Jupiter during the early accretion phase of planet formation. They would have been captured into their current orbits while Jupiter was growing, possibly with the help of gas drag and/or mutual collisions which slowed their motion and modified their trajectories. This theory, however, cannot explain some properties of the Trojans, particularly the range of about 40° in orbital inclinations.

An alternative explanation envisages their formation in more distant regions of the Solar System. They were subsequently scattered and captured into co-orbital motion with Jupiter when the giant planets migrated outward to their current orbits. Such captures would have been possible during a brief period of time, just after Jupiter and Saturn crossed their mutual 1:2 resonance, when the dynamics of the Trojan region were completely chaotic.

One of the most interesting Trojans is 617 Patroclus, which is now known to comprise two objects that are nearly identical in size. The larger object has a maximum diameter of 122 km while its companion is 112 km across. They orbit their center of mass every four days, separated by a distance of about 680 km. Both of the binary components are less dense than water, suggesting that they are porous and made out of water ice.

It may be that Patroclus originated far from the Sun, about 4.5 billion years ago. Like millions of other planetesimals, it may have been redirected inward and then pulled apart when it passed close to Jupiter. Such tidal splitting has also been suggested for the formation of binary near-Earth asteroids, which often comprise objects of similar size.

The first Mars Trojan to be discovered was 5261 Eureka, whose infrared spectrum is typical of an A-class asteroid. It trails Mars at the L5 point at a distance varying by 0.3 AU during each revolution and has probably been in this orbit for much of the Solar System's lifetime. Three other Martian Trojans have been confirmed – 1999 UJ$_7$ at the L4 point, and 1998 VF$_{31}$ and 2007 NS$_2$ at the L5 point.

Neptune's first Trojan asteroid was first imaged on August 21, 2001 and confirmed during observations over the next 16 months. Named 2001 QR$_{322}$, it is located at the L4 Lagrange point and is estimated to be about 230 km in diameter. Five more have since been found at the L4 point. 2007 VL$_{305}$ has an inclination of 28 degrees, the highest of the known Neptune Trojans. With an absolute magnitude of 7.9, its diameter is 65–150 km. There may be many thousands of these objects yet to be found.

Near-Earth Asteroids

Most asteroids spend their entire lives within the main belt. However, there are many millions of small rocky objects – most of them originating in main asteroid belt – which follow orbits that take them into the inner Solar System. Many of these approach or cross Earth's orbit.

It is estimated that there are about 1000 near-Earth asteroids (NEAs) larger than 1 km. All of them are capable of causing large-scale devastation on our planet through the heat and shock wave generated by their immediate surface impact and the subsequent ignition of huge fires which create clouds of smoke and other pollutants that may trigger rapid climate change.[9] Fortunately, "extinction-class" objects measuring at least 10 km across are rarer still. An impact of this scale in Mexico's Yucatan peninsula is widely accepted as the cause of the dinosaurs' demise 65 million years ago (see Chapter 3).

As their size and explosive potential decreases, the number of NEAs increases dramatically. There are thought to be at least 100,000 objects larger than 100 meters, each capable of causing regional destruction or generating a large tsunami after splashing into one of Earth's oceans. Furthermore, it is estimated that a 100-meter-class asteroid passes closer than the Moon 50 times a year, while 100 fist-sized meteorites, fragments of NEAs, fall to Earth each day – although most of these break up in the atmosphere, producing bright meteors or fireballs.

NEAs typically have a survival time in deep space of about 10 million years. Since their population is continually being depleted through collisions with terrestrial planets or gravitational interactions that send them into different orbits, there must be some means of replenishing the population. Some of the newcomers may be extinct comet nuclei (e.g. 3200 Phaethon), but many are probably removed from the Kirkwood gaps in the main belt by Jupiter's gravity and slung into highly eccentric orbits that send them into the inner Solar System. Others may be derived from

---

[8] The names of these objects are associated with the heroes of Homer's Iliad.

[9] A 100 meter rocky object has a mass of the order of one million tons, whereas a 1 km object has a mass of about one billion tons.

Box 13.6 NEAR-Shoemaker at Eros

NASA's Near Earth Asteroid Rendezvous (NEAR) mission was the first to orbit and land on an asteroid. Launched on February 17, 1996, it was later renamed NEAR-Shoemaker in tribute to planetary geologist Eugene Shoemaker. The chosen target was the second largest near-Earth asteroid, an S-type object known as 433 Eros.

Simplicity and low cost were the main driving factors behind NEAR's design. The science instruments, the solar panels and the high-gain antenna were fixed and body mounted. The six science instruments were a mix of tried-and-tested hardware and innovative designs. The multispectral imager and near-infrared spectrometer were adapted from military remote sensing systems. Together, they would map the surface and determine the mineralogical composition of Eros. The X-ray and gamma ray spectrometers measured and mapped elemental abundances. A laser rangefinder would measure surface topography, while a magnetometer searched for a magnetic field.

Restricted by the capability of the Delta 2 launch vehicle, the spacecraft followed a low energy trajectory designed to last three years. En route, it flew past main belt asteroid 253 Mathilde on June 27, 1997, at a distance of 1212 km, followed by an Earth flyby in January 1998. The Mathilde flyby was the closest spacecraft encounter carried out up to that time, and the first close-range reconnaissance of a C-class asteroid. NEAR discovered that Mathilde has at least five sizeable craters and a low density, indicative of a rubble pile. Its 17.4 day rotation is the third longest period of any known asteroid.

On February 18, 1997 – NEAR reached its most distant point from the Sun (2.18 AU), setting a new distance record for a spacecraft powered by solar cells. Unfortunately, a computer malfunction meant that the planned orbit insertion in January 1999 had to be postponed. When communications were re-established, NEAR was able to observe Eros at a distance of 3827 km. The flyby images acquired on December 23 showed about two-thirds of its surface. The asteroid, which measured about 33 × 13 × 8 km, was shown to rotate once in just over 5 hours.

Figure 13.19 NEAR-Shoemaker investigated two asteroids, shown here at the same scale. En route to Eros (right), it took 534 images of main belt asteroid 253 Mathilde (left). Mathilde was found to be a very dark, heavily cratered object that measured 66 × 48 × 46 km. The carbon-rich asteroid reflected just 3–5% of incoming sunlight, making it twice as dark as charcoal. There were at least five craters more than 20 km across on the sunlit side alone. Mathilde displayed no color or albedo variations over the 60% of its surface that was visible to NEAR. (NASA/JHU-APL)

After another circuit of the Sun, NEAR was successfully inserted into a 321 × 366 km orbit around Eros on February 14, 2000 – the first time that a spacecraft had been captured by the gravitational pull of such a small body. During its initial high-orbit phase, thousands of images of Eros' illuminated northern hemisphere were obtained. Over the next 2½ months, a series of small maneuvers lowered the orbit to an altitude of 50 km, close enough to resolve objects 5–10 meters across and map surface composition with the X-ray/gamma ray spectrometer. Subsequent low-altitude passes resulted in images of 0.5–1 meter resolution.

NEAR showed that much of Eros is nearly saturated with craters smaller than 1 km, although its mid-section is marked by a large, saddle-shaped depression, named Himeros, which resembles a degraded crater. The largest crater, Psyche, is 5.3 km across and located on the opposite side to Himeros. Surface grooves and ridges are common, and may follow lines of weakness produced by ancient collisions. Much of the surface is covered with piles of rubble, blocks of ejecta between 30–100 meters across. However, they are not uniformly distributed. NEAR also revealed some unusual, smooth areas, dubbed "ponds" by mission scientists.

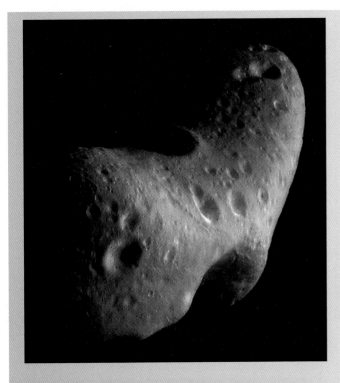

Figure 13.20   A mosaic of six images taken by NEAR-Shoemaker on February 29, 2000, from a distance of about 200 km. The northern hemisphere of Eros is dominated by Psyche, a 5.3 km wide crater, with Himeros, the 10 km wide "saddle" at the bottom. A major ridge links the two features and spans at least one-third of the asteroid's circumference. (NASA-JPL, JHU-APL)

Although NEAR was never designed to land, it was decided to end its mission by attempting a controlled descent to Eros' surface. On February 12, the spacecraft began a five-hour descent from a height of 36 km, eventually touching down close to the Himeros depression. 69 images were returned during the descent, with the last of them taken 125 meters above the surface.

Since the solar panels were still pointing toward the Sun, it was possible to establish a 10 bit/s telemetry link that sent back useful data from the gamma ray spectrometer and the magnetometer. NEAR was finally shut down on February 28.

high velocity impacts or drift toward the Sun as a result of the Yarkovsky effect (see Box 13.4)

The first asteroid found to cross the orbit of Mars was 433 Eros, discovered by Gustav Witt in 1898. With a maximum diameter of 33 km, it is now recognized as the second largest NEA.[10] A great deal is now known about Eros, since it was explored by the NEAR-Shoemaker spacecraft (see Box 13.6).

Eros is not likely to collide with Earth any time soon, but there are many objects which offer a potential threat. The first object to be discovered on a path that crossed the Earth's orbit was 1862 Apollo, found in 1932 (see below).

NEAs are classified according to their orbital distances from the Sun at perihelion and aphelion. The three main groups are named after a representative asteroid.

The Amors always remain further than Earth's aphelion distance, with perihelia of 1.017–1.3 AU. Many of them cross the orbit of Mars and they may approach Earth. Close encounters with Mars or Earth can perturb them into Earth-crossing, Apollo-type orbits. They are named after 1221 Amor. Eros is the most famous example, but more than 3200 Amors have so far been found.

The Apollos cross Earth's orbit. They have semimajor axes greater than 1 AU and perihelia less than 1.017 AU (Earth's aphelion distance). Since they spend most of their time beyond Earth's orbit, their orbital periods are longer than one year. The first to be found was 1862 Apollo, which is now known to have a small satellite. Other notable members include 1566 Icarus, 1866 Sisyphus (the largest, with a diameter of about 8 km), 3200 Phaethon, 1685 Toro and 4179 Toutatis. Some Apollos have very eccentric orbits. Icarus, for example, crosses the orbit of Mercury. 3870 Apollos have so far been found.

The Atens also cross Earth's orbit, but, with orbital periods of less than one year, and semimajor axes smaller than 1 AU, they spend most of their time closer to the Sun than Earth. They are difficult to find, because they are usually hidden in the Sun's glare and rarely appear in a dark sky. They are named after 2062 Aten. The current record holder for closest approach to the Sun is 1995 CR, with a perihelion distance of 0.12 AU. By February 2011, 641 Atens had been found.

The first representative of a fourth group, dubbed Apoheles, was found in 2003. These asteroids follow orbits that are entirely within Earth's orbit. Indeed, 2004 JG6 crosses the orbits of

---

[10] 1036 Ganymed is thought to be slightly larger than Eros.

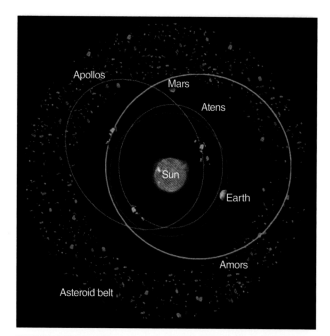

Figure 13.21 There are three main groups of near-Earth asteroids: the Apollos, Atens and Amors. The Apollos and Atens currently cross Earth's orbit. The Amors remain outside Earth's orbit, but may be diverted into Earth-crossing orbits in the future. All of these are a potential impact threat to our planet. (ESA)

Mercury and Venus, and has the shortest known orbital period of any asteroid – just six months. Apoheles are thought to account for perhaps 2% of the total NEA population, so they are rare, as well as difficult to discover because they stay in the daytime sky almost all of the time.

Although the Apoheles are unlikely to threaten Earth, long-term planetary perturbations can alter the orbits of the Atens and Apollos, as well as about 50% of the Amors, eventually sending them toward Earth. One consequence of this potential threat is that NEAs are being considered as targets for manned space exploration, not only for scientific research but for the potential exploitation of minerals and other resources.

The Impact Threat to Earth

Fortunately, collisions of asteroids and comets (together they are known as near-Earth objects or NEOs) with our planet are rare. Estimates based on lunar crater counts and studies of the orbital dynamics of the NEO population suggest that Earth currently suffers globally catastrophic impacts with objects larger than 1 km once every 500,000 to 700,000 years on average. Regionally devastating impacts (impact energy at least 4×10^{18} J) occur on average every 41,000–53,000 years. Events with energies similar to that of the object that exploded over Tunguska, Siberia, in 1908 are thought to occur once every 200–300 years on average (see below).

Close approaches by Potentially Hazardous Asteroids (PHAs) occur every few weeks or months. In order to establish the actual

Box 13.7   The Torino and Palermo Scales

The Torino Scale is a color-coded advisory system that enables NEO researchers to place objects within a potential threat range from zero (virtually no chance of collision) to 8 and 9 (a collision is certain) and 10 (global catastrophe is certain). It was first adopted in 1999 by a working group of the International Astronomical Union (IAU) at a meeting in Torino (Turin), Italy. It is also of value in describing the impact threat to the media and the general public.

The highest Torino score yet given was a level 4, assigned briefly to asteroid 2004 MN4 (99942 Apophis) in December 2004, with a forecast 2% chance of hitting Earth in 2029. After extended tracking of the asteroid's orbit, it was reclassified to level 1, with an extremely low chance of collision.

The logarithmic Palermo Technical Impact Hazard Scale has also been developed to enable NEO specialists to categorize and prioritize potential impact risks spanning a wide range of impact dates, energies and probabilities. There is no direct correlation between the Palermo Scale and the Torino Scale.

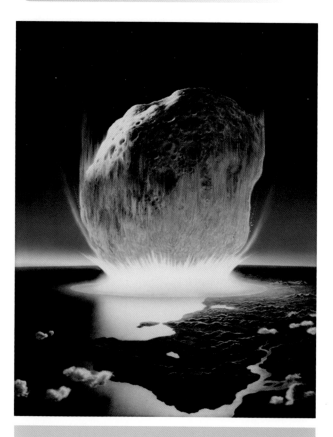

Figure 13.22 The extinction of the dinosaurs 65 million years ago is widely attributed to the impact of a 10 km diameter asteroid which created the 180 km wide Chicxulub crater in the Yucatan Peninsula, Mexico. Such impacts with Earth are now very rare, taking place on time scales of many millions of years. (NASA)

impact threat for the next century or so, NASA has been given the task of drawing up an inventory of the current NEO population. The first goal established in 1998 was to find 90% of all NEOs larger than 1 km within 10 years. In 2005, Congress mandated NASA to extend the search by finding 90% of the much more numerous objects larger than 140 meters across (see above).

Currently, no NEOs are known to be on a collision course with our planet, although some 1250 potentially hazardous asteroids are known.[11] The most notable close encounter so far predicted involves 99942 Apophis. It is now expected to skim past Earth at a distance of about 22,000 km on April 13, 2029, passing inside the orbit of geostationary satellites. There is also a 1 in 45,000 probability that this 270 meter asteroid might return and collide with Earth in April 2036. The energy release from an impact by Apophis would be in the 10,000 megaton range (see Chapter 3).

Dense, metallic asteroids pose the greatest threat, since they generally survive all the way to the surface. Fortunately, the damage from stony asteroids falls off very rapidly for sizes smaller than Tunguska because the energy release is smaller and the explosion takes place at such a high altitude (15 km+) that there is no noticeable damage on the ground. Similarly, an ice-rich object tends to explode at very high altitude.

Despite their differences in size and density, the consequences of comet and asteroid impacts on Earth are roughly comparable. The key lies in the amount of energy they release. Since the kinetic energy of an object is determined by its mass and velocity (K.E. $= \frac{1}{2} mv^2$), a large icy object traveling very fast may pack at least the same punch as a much smaller rocky object traveling at a slower speed.

The average speed of asteroids colliding with Earth is about 20 km/s, whereas long period comets arrive at much greater speeds, often around 50 km/s. Even though a comet is much less dense than an asteroid, its faster speed means that impacts involving comets and asteroids of similar size release comparable amounts of energy. The energy released by the impact of a 100 meter asteroid is comparable to a 50 megaton bomb, whereas a 1 km diameter asteroid releases 250,000 megatons of explosive energy – enough to wipe out most life on Earth. [12]

Tunguska

The most famous impact-related event in modern times occurred in 1908, when a stony asteroid (a comet is now generally discounted) about 60 meters across exploded above Tunguska, Siberia, causing devastation over an area about 50 km in diameter.[13]

At around 7:17 local time on the morning of June 30, 1908, a huge explosion occurred above the vast boreal forest near the Podkamennaya Tunguska River, north of Lake Baikal. Subsequent investigation showed that the dramatic event was caused by the destruction of an incoming cosmic object – the only entry of a large meteoroid in the modern era to be witnessed at first hand.

The explosion flattened some 80 million trees over an area of more than 2000 km². (If it had occurred over a city such as London, the entire urban area and its population would have been wiped out.) Moreover, they were aligned in a radial pattern, pointing directly away from the blast's epicenter.

The scene changed at ground zero: the trees were standing upright, but their limbs and bark had been stripped away, resembling a forest of telephone poles. Such debranching requires fast moving shock waves that cause the branches to snap off before they can transfer the impact momentum to the tree trunk. There was no sign of an impact crater and no fragments of an incoming object could be found.

The after-effects of the explosion were experienced around the world. A seismic shockwave was registered by instruments as far away as England. Massive, silvery clouds formed at high altitudes above the Tunguska region, reflecting light from the Sun after sunset. Some people saw brilliant, colored sunsets, others who lived as far away as England and Asia witnessed luminescent skies that enabled them to read newsprint at midnight. Locally, hundreds of reindeer were killed, but there seems to have been no loss of human life.

Figure 13.23 This photo taken during the 1927 expedition shows parallel trunks of trees that were flattened by the blast from the Tunguska event. Note how the branches have been stripped off the trees. (ESA)

---

[11] Only 1% of the impact threat to Earth is thought to come from comets. The frequency of such impacts involving long period comets is about one every 32 million years. The downside is that they usually arrive at much higher speeds from the outer Solar System, providing only a few months of advance warning.

[12] One megaton is equivalent to the explosive power of one million tons of TNT – comparable to roughly 77 Hiroshima atomic bombs.

[13] A less well-known example is the February 1947 Sikhote-Alin strike in Siberia, about 430 km north east of Vladivostok, which created a cluster of small craters.

Sandia National Laboratories supercomputer simulations completed in 2007 suggest that the explosion – previously estimated to be between 10–20 megatons – was more likely to be three to five megatons (though still equivalent to a modern nuclear bomb).

Many explanations have been proposed over the years, but it is now generally agreed that an asteroid entered the atmosphere above Siberia and then exploded at an altitude of approximately 8 km. The rocky object may have measured 30–60 meters across, depending on its speed, porosity, water content, composition and other material characteristics.

Traveling at a speed of about 53 600 km/h, the asteroid would heat the surrounding air to 25,000°C before the combination of pressure and heat caused it to blow itself apart in a huge fireball. A high-temperature jet of expanding gas struck the ground beneath, causing blast waves and thermal pulses that flattened the forest beneath.

Box 13.8   Saving Earth

Relatively small-scale explosive events such as Tunguska may occur on average once every 200–300 years. Since larger objects are much less common, the expected frequency of Earth impactors decreases as the size of the incoming object increases (see above). However, this is counterbalanced by the much larger number of deaths and more widespread devastation associated with a mega-impact. Calculations show that the probability of someone dying in any one year due to a cosmic impact is of the order of one in one million. Surprisingly, this is comparable with the likelihood of dying in a plane crash or more likely than death due to a tornado.

Although no one in living memory has died from an asteroid impact, the subject is now being taken quite seriously, with NASA and other agencies beginning to consider how best to deal with the potential threat. Given sufficient warning, modern technology should be able to prevent humanity from suffering a grisly extinction like the dinosaurs.

An overall Earth protection system must have three components. The first step is to identify and catalog any potential NEO impactors above a certain size, although there will always be an additional threat from long period comets that may appear with little advance warning. This cataloging is based primarily on ground-based observations, though spacecraft, such as NASA's WISE (which operated January 2010 to February 2011) and ESA's Gaia (scheduled for launch in 2012) can also be used to search for small asteroids in the inner Solar System. By the end of its mission in early 2011, WISE had discovered 134 NEOs.

Second, a series of space missions should be launched to understand the structure, composition, rotational state, and other physical properties of potential impactors. Although a few asteroids have already been studied from close range by spacecraft, much more needs to be learned about the nature of potentially hazardous objects. One important mission already under way is ESA's Rosetta, which will attempt to place a lander on the surface of comet 67P/Churyumov-Gerasimenko in 2014.

Finally, technologies must be available to prevent the intruder from impacting Earth. Several Hollywood blockbusters have shown astronauts being dispatched to save Earth from imminent destruction by a NEO, but, in reality, no practical systems to deflect or destroy such an object exist at the present time.

The most obvious solution would seem to be launching a missile with a nuclear warhead to break up a large asteroid. Unfortunately, this is likely to create a swarm of smaller objects that would continue on a similar course, resulting in multiple impacts with similar total energy release around the world – rather like the collision of comet Shoemaker-Levy 9 with Jupiter (see below).

The less drastic alternative of deflecting an object from its collision course has been widely studied. A change in orbital speed of the order of a few cm/s will usually be sufficient. Deflection techniques generally involve the sudden introduction of a fairly large lateral force or a slow, relatively gentle, but prolonged thrust.

Once again, nuclear devices have been suggested, since they supply around a million times more energy per unit of mass than conventional explosives. A thermonuclear explosion would release highly energetic particles, X-rays and gamma rays. The resultant vaporization of the surface would cause the object to recoil, similar to when a rifle is fired (see Newton's laws of motion, Chapter 1).

The effect of such an explosion would be substantially increased if it took place beneath the surface. In this case, most of the explosive energy would be transferred to the target, resulting in the high-speed expulsion of a great deal of material. This would result in a much larger change in orbital velocity than a similar explosion close to the asteroid. Unfortunately, it would not be easy to place a thermonuclear charge deep inside a NEO, and the level of unpredictability and risk is very high, with the possibility that the fragile object could fragment.

A less drastic deflection method would involve the transfer of kinetic energy through an impact by a non-explosive object. The amount of energy exchanged would depend on the mass and relative impact speed of the projectile. For example, in order to divert a 2 km diameter asteroid following a prograde circular orbit at 1 AU, it is necessary to deliver a 60 ton projectile at a velocity of 10 km/s. The mass of the projectile could be greatly reduced by launching it into a retrograde orbit, so that a head-on impact speed of 60 km/s could be achieved.

ESA has suggested a mission to test such a technique. Don Quijote would involve the launch of two spacecraft in separate interplanetary trajectories. One spacecraft (Hidalgo) would impact an asteroid at high speed while the other (Sancho) would detect any subtle changes in its orbit.

Figure 13.24 Artist's impression of a gravity tractor, involving a spacecraft using an ion engine to hover near an asteroid. The asteroid feels the gravitational pull of the spacecraft and accelerates slightly towards it. (Dan Durda – FIAAA / B612 Foundation)

A more subtle approach involves a gravity tractor, in which the tiny gravitational pull provided by a nearby thrusting spacecraft would very slowly accelerate the asteroid in the spacecraft's direction. Although the acceleration imparted to the asteroid would be very small, the technique would be able to slightly alter an asteroid's orbit. Calculations show that a 20 ton gravity tractor could deflect a 200 meter asteroid after a year of such "towing."

One of the earliest deflection proposals involved placing an electromagnetic "mass driver" on the surface of a NEA. Material excavated on the asteroid's surface would be fired into space, creating an equal and opposite reaction that gradually modifies the NEA's orbit. A similar idea involves anchoring ion engines or a reflective solar sail on an asteroid in order to apply a weak propulsive force over a long period of time.

The difficulty in anchoring such devices can be overcome if a solar sail or mirror is placed in orbit. By focusing incoming solar radiation onto the NEA, some of the surface would be strongly heated and vaporized. The resultant jet of material would modify the object's trajectory.

Whichever method is chosen, it should be borne in mind that a deflection mission may take up to 5 years to reach its target, and the propulsive procedure itself may require many years to have the desired effect – so the threat must be recognized well in advance.

Meteorites

Meteorites are small, rocky objects that enter Earth's atmosphere at high speed and survive their fall to reach the surface. They are classified according to their chemical and physical characteristics. Some are metallic, consisting mostly of iron alloyed with nickel, but most are stony. These characteristics provide clues to their place of origin and the conditions which prevailed in their parent body. Most meteorites originated in the main asteroid belt.

There are estimated to be about 20,000 meteorite falls involving objects heavier than 100 g every year. They occur largely at random over Earth's surface, although the number is slightly reduced near the poles, since most meteorites (and asteroids) travel close to the ecliptic plane.

The majority of meteorites splash down in the oceans or arrive on land without being noticed. However, many thousands have been found on every continent. The best-known hunting ground for meteorite hunters is Antarctica, where the slowly moving ice sheet preserves the cosmic pebbles and slowly concentrates them near mountains which protrude above the ice, much as waves wash pebbles to the shore. Each summer, expeditions scoop up the dark rocks which stand out so clearly in the white wilderness.

Flat desert regions are similarly favored, although it is not always so easy to recognize a meteorite if numerous local rocks are also scattered across the surface. Deserts offer ideal prospecting places, since the dry climate prevents chemical weathering. Parts of the Sahara have proved particularly productive, notably

Figure 13.25 The Hoba meteorite (also known as Hoba West) is the largest single meteorite on Earth. The 60 ton iron-nickel slab was found by a farmer in 1920, near Grootfontein, in Namibia. It is thought to have arrived over 80,000 years ago. It landed at such a slow speed that it survived the impact and no crater was found. Flat on both major surfaces, the meteorite's unusual shape may have caused it to skip across the upper atmosphere like a flat stone on water. (Patrick Giraud)

the Dar al Gani (Libya), Hammada al Hamra plateau (Libya) and Acfer (Algeria) areas.

Most meteorites are small, partly because the weaker stony objects fragment during their high speed fall to Earth, creating a meteorite shower. Iron meteorites are stronger and generally survive intact. The largest single meteorite on Earth is the 60 ton Hoba iron-nickel meteorite which remains where it fell, near Grootfontein in Namibia, and is now a tourist attraction. Meteorites typically enter the upper atmosphere at about 30 km/s, but their descent is slowed by friction. This causes the outer surface to heat up and melt, although the interior remains cool and unchanged. Its passage usually results in a bright fireball as molten droplets are carried away in its wake.

By the time it hits the ground, the temperature of the outer skin has dropped sufficiently for it to feel cold. The only outward signs of its fiery entry are a dark, glassy, fusion crust, possibly allied to a curved or conical shape which formed as the projectile traveled through the air.

The Origin of Meteorites

Studies of the chemical compounds that make up meteorites show that they are the oldest available samples of Solar System material, providing invaluable information about the conditions that prevailed 4560 million years ago.

Almost all meteorites are fragments of asteroids that broke free or were ejected during collisions. These rocks were then hurled toward Earth, primarily by the disruptive effects of Jupiter's gravity or by the influence of sunlight (see Box 13.4)

Confirmation of the asteroid belt as the main source region has come from the calculation of some meteorite orbits, based on photographic images or video records (e.g. Příbram, Lost City, Neuschwanstein and Peekskill). Dedicated fireball detection networks, such as one established on the Nullarbor Plain in Australia, have also had some success. However, establishing a direct link

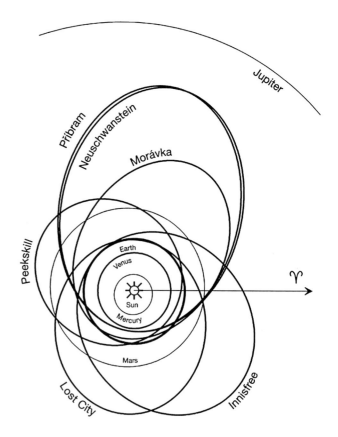

Figure 13.26 The calculated orbits of five meteorites, inferred from their trajectory before hitting the ground. All of them originated in the asteroid belt, indicating that these meteorites are chips off asteroids. (Nature)

between particular parent asteroids and the majority of meteorites that fall on Earth is almost impossible.[14]

The first task is to determine (or at least estimate) an asteroid's size / volume and composition, in order to calculate its mass. (This is easier for asteroids than planets or spherical moons since they are too small to compress gravitationally.) Spectroscopic data are then used in an attempt to verify differences in composition.

The next step is to compare these data with different meteorite samples in an effort to find links between them and their parent asteroid types. This is often problematic, since prolonged space weathering alters the nature of asteroid surfaces. However, there are good matches between C class asteroids and some stony meteorites. Similarly, many basaltic meteorites are thought to have originated from 4 Vesta.

Other exceptions are the small number of meteorites known to have originated on Mars and the Moon. Ejected by high energy collisions, they were accelerated to beyond escape velocity, and traveled on a circuitous route through space until they encountered Earth. Their origins have been confirmed by chemical analysis (see below).

Figure 13.27 An ordinary chondrite found in Plainview, Texas. Numerous small, round chondrules are visible in this cut and polished specimen, which measures about 10 cm top to bottom. Flecks of metal are also present, but difficult to see in this photo. (D. Ball, ASU)

Types of Meteorites

There are three main types of meteorites: stones, irons, and stony-irons. Primitive stony meteorites are known as chondrites, so-called because they contain spherical grains known as chondrules. Made of silicates, these chondrules are typically about 1 mm in diameter.[15] Largely made of olivine and pyroxene (minerals rich in magnesium, iron and silicon) they appear to have been small balls of dusty material in the primordial solar nebula that were suddenly heated to over 1400°C, possibly due to shock waves in the solar nebula. They became molten before rapidly cooling and recrystallizing. During this process, water and other volatiles were vaporized, leaving behind the grains of silicates and iron.

There are three main classes of chondrites, based on their chemistry. Ordinary chondrites are the most common meteorites found on Earth, accounting for about 87% of all falls. Most of them experienced high temperatures before they recrystallized. They can be distinguished from terrestrial rocks because their iron content makes them slightly attractive to a magnet.

Carbonaceous chondrites are made of undifferentiated silicates (including chondrules) and their chemical compositions match the chemistry of the Sun more closely than any other class of chondrites. Hence, they represent the primordial material from which planets accreted. Less than 5% of all meteorites belong to this class.

These primitive and undifferentiated meteorites formed in oxygen-rich regions of the early Solar System, so that most of the material is found as silicates, oxides, or sulfides. Most of them contain water, or minerals that have been altered in the presence of water, and some contain large amounts of carbon. Specimens such as the Murchison meteorite contain amino acids and

Figure 13.28 A rare carbonaceous chondrite, part of the large fall at Allende in Mexico. It contains rounded chondrules (note the cavity left by the removal of a large chondrule) and calcium-aluminum-rich inclusions (CAIs) which are white or light gray. CAIs are believed to pre-date chondrules by at least two million years. This specimen is about 11 cm wide. (D. Ball, ASU)

complex organic compounds that are important building blocks for life on Earth. They may also contain numerous white "inclusions" rich in aluminum and calcium. Dating of some inclusions has shown that they formed 4560 million years ago – the oldest known solids in the Solar System.

---

[14] The first definitive asteroid / meteorite link was made after a small asteroid known as 2008 TC3 was detected before it exploded over the Nubian Desert of northern Sudan on October 7, 2008. 208 fragments were subsequently recovered. 2008 TC3 had the spectral signature of an F-class asteroid, while the meteorite was a very rare and unusually fragile, rock known as a polymic ureilite.

[15] The term chondrule comes from the Greek word "chondros," meaning seed or "little grain."

Figure 13.29 An iron meteorite found at Henbury, Australia, displays numerous rounded indentations known as regmaglypts. These features are most likely due to the selective melting and ablation of areas with different compositions during its descent though the atmosphere. This specimen is about 26 cm long. (D. Ball, ASU)

Figure 13.30 A pallasite stony-iron meteorite found at Albin, Wyoming. Pallasites are composed of about half metal and half olivine, a greenish silicate mineral. This slice, which is about 18 cm long, is illuminated from behind to distinguish the olivine from the surrounding metal. (D. Ball, ASU)

Based on their distinctive compositions, there are at least seven groups that formed on different parent bodies in different regions of the solar nebula. Some groups have retained more of their volatiles in the form of carbon compounds and water because they have not undergone significant heating since their formation. It seems that their low temperature enabled water to produce clays, sulfates and carbonates. Perhaps the most famous specimens are from the Allende meteorite shower that fell in Mexico on February 8, 1969.

Enstatite chondrites differ by being poor in volatiles and less oxidized. They contain up to 15% iron-nickel, very small chondrules and sulfides.

All other meteorite types (achondrites, stony-irons and irons) are differentiated. Since they were once completely molten, they contain no water or other volatiles. Achondrites, which account for about 9% of falls, are mainly igneous silicates that probably originated in the mantle or crust of differentiated asteroidal parent bodies. They rarely contain chondrules and have a low metal content.

Irons are made of nickel-iron, with some sulfur, so they are much denser than other meteorites. They probably originated in the differentiated cores of large asteroids. About 42% of all meteorite finds are irons, although they only account for about 3% of all observed falls. This is because metallic meteorites are stronger and more resistant to frictional heating, so they are more likely to survive to reach the surface.

As their name implies, stony-irons are composed of silicates mixed with iron and nickel. Their compositions range from droplets of olivine imbedded in pure nickel-iron to droplets of nickel-iron in pure olivine. The stony-irons are rare, accounting for only about 1% of all falls. Some (called pallasites) probably came from a core-mantle transition region inside a large asteroid, where metallic core components had not entirely separated from the silicate mantle. Others (called mesosiderites) have a jumbled texture comprising angular grains of rocky minerals and metal, suggesting they are the product of collisions between asteroids of different composition.

Meteorites from the Moon and Mars

More than 30,000 meteorites have been discovered on Earth, but only a few dozen have been identified as originating from the Moon or Mars. All of these are pieces of debris that were ejected by a sizeable impact which excavated a large crater, then accelerated them beyond escape velocity. After a lengthy journey through space, they entered Earth's atmosphere and survived the heat of descent to reach the surface. (It is also possible that many terrestrial meteorites have made the reverse journey from Earth to the Moon or Mars during the history of the Solar System.)

Small meteoroids impact the Moon every day and the flashes of these impacts have been observed and imaged on a number of occasions. Far less frequent, particularly in the present epoch, are impacts that are sufficiently energetic to accelerate ejecta to the lunar escape velocity of 2.38 km/s – a few times the velocity of a rifle bullet – so that they can escape the Moon's gravitational influence. The lunar rocks may eventually be captured by Earth's gravity after a journey lasting a few hundred thousand years (or less).

It is a similar story with Martian impacts. Although the escape velocity is higher (5 km/s), it is still possible for ejecta to be blasted into an interplanetary trajectory that may eventually take it close to Earth.

The history of these meteorites can often be determined from their composition and exposure to cosmic rays. By studying their textures and mineral content, it is clear that they come from different impact events. Indeed, it has been argued that most lunar meteorites are derived from relatively small, recent, impacts that created craters only a few kilometers in diameter. Studies of Martian meteorites indicate that they originated in at least three fairly recent craters whose diameters are greater than 12 km.

How is it possible to know whether a meteorite came from the Moon? First, its abundances of major minerals and elements match those established through remote sensing and analysis of Apollo samples. Their oxygen isotopic compositions are also identical. Finally, nitrogen and noble gases implanted in them by the solar wind whilst they were on the lunar surface show that they came from a world with no atmosphere.

In the case of the SNC meteorites from Mars, all are igneous and extremely young compared to other achondrites.[16] Dating of their crystals shows that most of them solidified between 1.3 and 0.16 billion years ago, long after most stony meteorites. They obviously originated on a parent body that supported volcanic activity until recent times, which favors either Venus or Mars. However, analysis of gases trapped in the rocks shows that they match the composition of the Martian atmosphere, as measured by the Viking landers (see Chapter 7).

Since impacts may occur at random over the entire Moon (or Mars), meteorites could come from anywhere on their surfaces. Although it is almost impossible to directly link an individual rock with a particular crater, some assumptions can be made about the general source region. For example, meteorites that are rich in the mineral anorthite and have high concentrations of aluminum and calcium probably originated in the lunar highlands, which are also composed predominantly of the same material. Such rocks are not widely found on other Solar System bodies.

It is possible to determine how long ago a rock left the Moon or Mars by studying its isotope or nuclide content. Rocks lying on the lunar surface or traveling through space are exposed to cosmic ray hits. These are so energetic that they cause nuclear reactions in the elements, changing one nuclide into another.

Some of the isotopes produced are stable, and become more abundant over time. Hence, the duration of the meteorite's exposure to cosmic rays in space can be determined by measuring the abundance of these rare stable isotopes and elements, such as ^{3}He (helium 3), neon, argon, krypton and xenon. This is used to calculate their time of flight to Earth. Using this technique, it has been calculated that the Dhofar 025 meteorite left the Moon at least 13 million years ago, whereas Kalahari 008/009 took only a few hundred years to reach Earth.

Some of the nuclides produced are radioactive. However, the production of these nuclides ceases as soon as the meteorites fall to Earth, because the atmosphere absorbs nearly all cosmic rays. By measuring how much they have decayed, it is possible to determine the length of time that has passed since they arrived on Earth. Radionuclides used for dating of meteorites' arrival on Earth include ^{14}C (carbon 14), ^{10}Be (beryllium-10), ^{26}Al (aluminum 26), ^{36}Cl (chlorine 36), and ^{41}Ca (calcium 41).

As might be expected from the greater distance and higher escape velocity of Mars, the number of known meteorites from the red planet (55) is even smaller than the lunar collection (65). Some of them show evidence of having cooled rapidly from molten lava, while others are more coarse-grained, suggesting that they cooled more slowly at depth before being excavated. Some also contain minerals, such as carbonates, that were produced during exposure to water at fairly low temperatures (see Chapter 6).

Meteors

Stare at the sky on a clear, Moonless night and you are likely to see a meteor, commonly known as a "shooting star." These brief light shows are produced by grains of cosmic debris, perhaps 2–5 mm across, most of which have been jettisoned into space by passing comets. The meteors leave glowing trails as they are incinerated by frictional heating during their high speed (11–76 km/s) entry into Earth's upper atmosphere. The result is a short-lived streak of light caused by ionized and excited atmospheric atoms. Occasionally, larger objects may produce fireballs, or even brighter bolides, whose passage is marked by a sonic boom.

A few sporadic meteors per hour may be seen at any time during the year. However, the number of visible meteors rises considerably at certain times of year, during meteor showers. There are some 20 major showers each year, when shooting stars seem to radiate like the spokes of a wheel from one point in the sky, although the particles are actually traveling along parallel paths. The showers are named after the constellation from which they appear to originate (e.g. the Leonids, which radiate from Leo). However, they are (with one exception) all caused by Earth plowing through streamers of ice and dust that periodic comets have left trailing in their wake.[17]

Table 13.1
Major Meteor Showers

Shower	Dates	ZHR[1]	Parent Comet
Quadrantids	Jan 1–6	100	96P Macholz 1?
Lyrids	Apr 19–25	10–15	C/1861 G1 Thatcher
Eta Aquarids	Apr 24–May 20	50	1P Halley
Delta Aquarids	Jul 15–Aug 20	20–25	96P Macholz 1?
Perseids	Jul 25–Aug 20	80	109P Swift-Tuttle
Orionids	Oct 15–Nov 2	30	1P Halley
Leonids	Nov 15–20	100	55P Tempel-Tuttle
Geminids	Dec 7–15	100	Asteroid 3200 Phaethon

[1] Approximate zenithal hourly rate.

[16] The SNCs are named after the first three subgroups of Martian meteorites that were identified (Shergotty, Nakhla and Chassigny).

[17] The exception is Apollo asteroid 3200 Phaethon, which follows a similar orbit to the debris that causes the Geminid meteor shower. Phaethon's association with a meteor stream suggests that it may actually be an old comet nucleus.

The showers occur when Earth encounters a stream of comet debris around the same dates each year. However, they may vary in timing, brightness and hourly rate of meteors. This is particularly true if the associated comet has made numerous journeys through the inner Solar System, leaving behind many different streams. Meteor showers are most prominent if the comet has passed close to Earth's orbit in the last few years. The duration of each shower ranges from a few hours to several days.

One of the most famous annual meteor showers is known as the Leonids, associated with dust particles ejected from comet P/55 Tempel-Tuttle, which visits the inner Solar System once every 33.25 years. They appear November 15–20, when Earth passes very close to the comet's orbit. However, the numbers of meteors on view vary tremendously.

In most years, observers may see a peak of perhaps 5–10 per hour around November 17. But when Tempel-Tuttle approaches the Sun, it can generate a magnificent storm, with thousands of shooting stars illuminating the night sky. The most spectacular apparitions came in 1833 and 1966, with dazzling displays of over 100,000 meteors an hour.

The Leonids are also renowned for producing bright fireballs which outshine every star and planet. Their long trails are often tinged with blue and green, while their vapor trains may linger in the sky like enormous smoke rings for five minutes or more.

Although the incoming particles are small, ranging from specks of dust to the size of small pebbles, the Leonids glow brightly because they are the fastest of all the meteors. A typical Leonid, arriving at a speed of 71 km/s (more than 200 times faster than a rifle bullet), will start to glow at an altitude of about 155 km and leave a long trail before it is extinguished.

Figure 13.31 This drawing shows the famous Leonid meteor storm that took place on November 12, 1833, when the skies over the United States were ablaze with shooting stars. It was, one eyewitness said, as if "a tempest of falling stars broke over the Earth. . . . The sky was scored in every direction with shining trails and illuminated with majestic fireballs." (ESA)

Long-Haired Stars

Comets are defined as small Solar System bodies consisting of frozen volatiles and dust. They are thought to be ice-rich planetesimals left over from the formation of the planets, about 4.5 billion years ago.

At the heart of all comets is a solid nucleus, which is usually invisible through even the largest telescopes. This is partly because comets spend most of their lives far from the Sun (and Earth) and partly because their nuclei are typically only a few kilometers across. Furthermore, any comets that enter the inner Solar System are masked by a surrounding cloud of gas and dust known as a coma. The roughly spherical coma is fed by jets of material that erupt into space as the surface of the nucleus is warmed by solar radiation.

The coma is mainly composed of water vapor and carbon dioxide. Some comas display the greenish glow of cyanogen (CN) and carbon when illuminated by sunlight. This is called "resonant fluorescence." Other compounds of carbon, hydrogen and nitrogen have been found. Ultraviolet images from spacecraft have also shown that the visible coma is surrounded by a huge, sparse cloud of hydrogen gas.

Some comets become splendid naked eye objects, so it is not surprising that their existence has been known since time immemorial. Many ancient civilizations saw them as portents of death and disaster, or omens of great social and political upheavals. Shrouded in thin, luminous veils with tails streaming behind them, they were termed "kometes" by the ancient Greeks, meaning "long-haired stars." In the pre-telescopic era this name was particularly appropriate, since naked eye comets often display one or more diaphanous tails that extend for millions of kilometers from the nucleus and surrounding coma.

Despite their size, the tails are clearly insubstantial – stars can be seen shining through them. Indeed, in 1910, news stories warning about the imminent "end of the world" proved to be unfounded when Earth passed unscathed through the tail of comet Halley. It was rather like a bowling ball speeding through a cloud of smoke.

Two tails are generally recognized. One of these is the yellowish dust tail, usually broad, stubby and curved, which is formed when tiny dust particles in the coma are pushed away by solar radiation pressure, as photons of light impact the grains. The particles

Box 13.9 Meteors, Meteoroids and Meteorites

Even experts are sometimes confused by the subtle differences in the meaning of these three terms. A meteoroid is a small rocky object (typically less than 100 meters in diameter) traveling through space, particularly one with the potential to encounter Earth. If it enters Earth's atmosphere, its passage usually results in a meteor – a luminous trail across the sky (commonly called a "falling star" or "shooting star"). Tiny meteoroids usually vaporize completely in the atmosphere due to friction with air molecules, but larger specimens may survive to become stony or metallic remnants on Earth's surface. These rocks are known as meteorites.

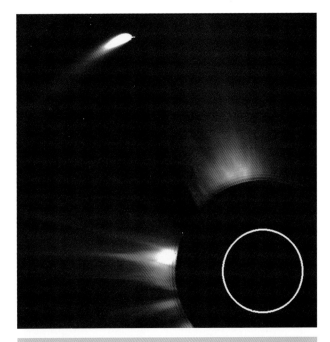

Figure 13.33 Comets travel around the Sun in highly elliptical orbits. As they near perihelion, the warmth of the Sun causes the icy nucleus to vaporize, producing a bright coma and tails of gas and dust. This image of comet 96P/Machholz (top) was taken on January 8, 2002, by the LASCO C2 instrument on the SOHO spacecraft. The mask in the LASCO coronagraph blocks light from the Sun, shown by the white circle. Comet Machholz has a 63-month orbit, passing within 18.5 million km of the Sun. (SOHO/LASCO, ESA, NASA)

Figure 13.32 A 10 micron interplanetary dust particle collected in the stratosphere with a U2 aircraft. This particle has a similar elemental composition to primitive meteorites, but with higher abundances of carbon and volatile elements. The particle is composed of glass, carbon and many types of silicate mineral grains. Its porosity and unusual composition suggest that it may be of cometary origin. (NASA-JPL)

It is estimated that more than 30,000 tonnes of meteoroid material enter Earth's atmosphere every year. Of this, perhaps 10% reaches the surface in the form of meteorites and micrometeorites. The latter group comprises particles of dust which are so tiny – typically less than 120 microns across – that their energy is dissipated before they burn up in the atmosphere. These particles have been collected by spacecraft and by high flying aircraft, as well as on the ground, where they are often recognized through their magnetic properties. Some 4000 tonnes of micrometeorite material may fall to Earth each year.

ejected during a comet's passage through the inner Solar System eventually spread along its orbit. When Earth plows through the cloud of debris, the particles enter the upper atmosphere and burn up as meteors (see above).

Meanwhile, gases released by the vaporizing nucleus are ionized by solar ultraviolet light. These electrically charged particles are influenced by the magnetic field carried by the solar wind and swept out of the coma into a long, distinctive ion tail (also called a gas or plasma tail). Since the most common ion, CO+, scatters blue light better than red, the ion tail often appears blue to the human eye. Gusts in the solar wind can cause the ion tail to swing back and forth, often developing temporary ropes, knots and streamers that sometimes break away and then reform. These features are not seen in the dust tail. The ion tail is usually narrow and straight, sometimes streaming away from the nucleus for many millions of kilometers.[18]

Until recently, the Great March Comet of 1843 held the record for the longest ion tail, stretching 330 million km – more than two astronomical units – across space. Then, in 1998, analysis of data from the Ulysses spacecraft indicated that it had passed through the ion tail of comet Hyakutake at the remarkable distance of 570 million km from the nucleus.

[18] A third tail, made of neutral sodium atoms, was discovered in comet Hale-Bopp in 1997.

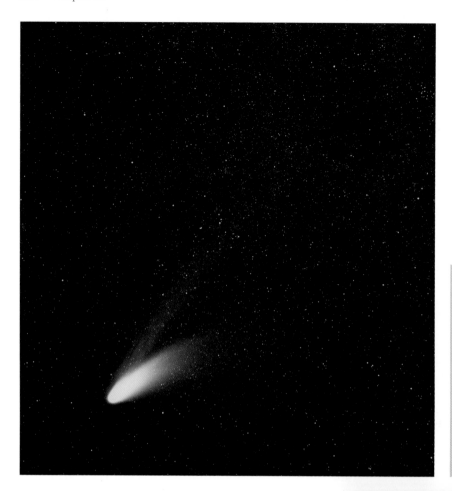

Figure 13.34 Comet Hale-Bopp, discovered by Alan Hale and Thomas Bopp on July 23, 1995, was one of the great comets of the 20th century. As it approached the Sun, it became extremely bright and active, developing an 8° long ion tail (blue) and a yellowish, 2° long dust tail. The nucleus was estimated to be 40 km in diameter, huge compared with most comets that reach the inner Solar System. (Peter Stättmayer)

One of the most interesting characteristics of the tails is a dramatic shift in their alignment as the comet moves along its orbit. As the solar wind sweeps past the comet at about 500 km/s, it shapes the tails and causes them (particularly the ion tail) to always point away from the Sun. This means that, during the outward leg of the orbit, the outward flow of the solar wind causes the tails to point ahead of the comet, rather than trail behind it.

In extreme cases, comets have been seen to lose their tails temporarily, due to strong gusts in the solar wind. In 2007, a similar disconnection was observed by NASA's STEREO spacecraft when a coronal mass ejection (CME) – a huge cloud of magnetized gas ejected into space by the Sun – collided with the tail of comet Encke and literally cut it in two. The disconnection was triggered by a process known as magnetic reconnection, when the magnetic fields around the comet and the CME were spliced together.

Icy Dirtballs or Dirty Snowballs?

For hundreds of years, scientists have tried to unravel the true nature of comets. The introduction of two powerful new techniques – spectroscopy and photography – in the second half of the 19th century showed that comets are highly complex, not only in the variability of their comas, jets and tails, but also in their composition. Bright bands in the spectrum of reflected light indicated the presence of numerous compounds made of hydrogen,

Figure 13.35 The ROSAT satellite discovered X-ray and extreme ultraviolet (EUV) emission from comet Hyakutake. Note the bright emission on the sunlit side of the nucleus. Since then, X-ray emissions have been detected from several other comets. Most of the X-rays are thought to be produced by a charge exchange caused when heavy, ionized particles carried by the solar wind collide with neutral atoms and molecules in the comet's coma. (NASA-GSFC/MPE)

Box 13.10   Giotto

On the night of March 13–14, 1986, the Giotto spacecraft encountered comet Halley.[19] ESA's first deep space mission was part of an international effort to investigate this famous object. Navigation to the nucleus was assisted by the Soviet Union's two Vega spacecraft, which also returned images and other data.

Giotto was launched on July 2, 1985. On March 12, 1986, after a flight of almost 150 million km, the spacecraft's instruments detected hydrogen ions at a distance of 7.8 million km from the comet. About one day later, Giotto crossed the bow shock, where a shock wave is created as the particles in the supersonic solar wind slow to subsonic speed. When Giotto entered the densest part of the coma, the camera began tracking the brightest object (the jets of the nucleus) in its field of view.

Figure 13.36   This composite image, the first detailed view of a comet's nucleus, was taken by ESA's Giotto spacecraft on March 13, 1986, from a distance of about 18,000 km. The dark nucleus measured about 8 × 8 × 16 km. The sunlit side (left) was illuminated by bright jets of gas and dust. The rugged surface revealed sizeable valleys and depressions. A "mountain" on the dark side was illuminated by sunlight. The rotation axis is approximately horizontal in this view. (ESA/MPI)

The first of 12,000 dust impacts was recorded 122 minutes before closest approach as the spacecraft closed on the nucleus at a speed of 68 km/s relative to the comet. The rate of dust impacts rose sharply when Giotto passed through a jet of material that streamed away from the nucleus.

Only 7.6 seconds before closest approach, the spacecraft was sent spinning by an impact from a "large" (one gram) particle. Contact with Earth was temporarily lost, but, over the next 32 minutes, the spacecraft's thrusters stabilized its motion and contact was fully restored. By then, Giotto had passed within 596 km of the nucleus and was heading back into interplanetary space.

The resilient spacecraft continued to return scientific data for another 24 hours. The last dust impact was detected 49 minutes after closest approach. The historic encounter ended on 15 March, when the experiments were turned off. However, Giotto went on to encounter a second comet, Grigg-Skjellerup, on July 10, 1992.

carbon, oxygen and nitrogen, as well as the silicate dust particles that produce the meteor swarms.

Then, in 1950, American astronomer Fred Whipple (1906–2004) turned cometary science on its head. Whipple knew that some periodic comets have orbited the Sun 1000 times or more. He realized that they would have broken apart if they comprised only a large pile of sand mixed with hydrocarbons.

Whipple argued that comet cores must be "icy conglomerates" made of water ice and dust, mixed with ammonia, methane and

carbon dioxide – popularly termed "dirty snowballs". As they approach the Sun and become warmer, their outer ices begin to vaporize, creating jets that eject large amounts of dust and gas. The result is the formation of an all-enveloping coma, together with two characteristic tails if dust and gas production is sufficiently high.

Whipple's model also made it possible to understand the variations in comet orbits and predict their orbital motion by taking into account the thrust from the gaseous jets that erupt from the nucleus.

---

[19] Giotto was named after the Italian Renaissance painter, Giotto di Bondone, who included Halley's comet in one of his paintings.

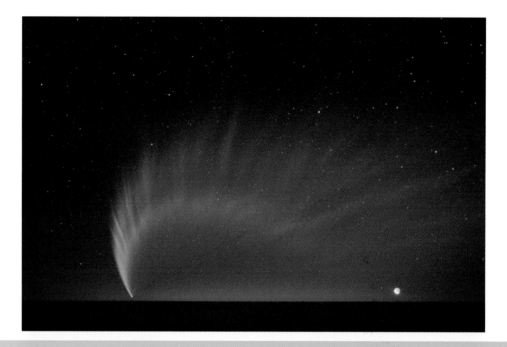

Figure 13.37 Comet C/2006 P1 (McNaught) provided a spectacular sight close to the horizon in the southern hemisphere during January and February 2007. At least three jets of gas and small dust particles were seen to spiral away from the nucleus as it rotated, stretching over 13,000 km into space. The larger dust particles, ejected on the sunlit side of the nucleus, followed a different pattern. They produced a bright fan, which was then blown back by the pressure of sunlight. (ESO/Sebastian Deiries)

Whipple's theories have since been tested by observing comets with orbiting observatories such as Hubble, Spitzer and the Ultraviolet Explorer, as well as dispatching spacecraft to study them at close range.

The first close encounters came during an international campaign to study comet Halley during its return to perihelion in 1986 (see Box 13.10). Since then, spacecraft have flown past the periodic comets Borrelly, Wild 2 and Hartley 2, collected samples from the coma of Wild 2, and delivered an impactor into the nucleus of comet Tempel 1. The results have been surprising, and sometimes perplexing.

Comets are now known to display a wide range of physical characteristics, although their nuclei are generally quite small. At the top end is Hale-Bopp, with a diameter of perhaps 35 km, followed by the 16 × 8 km, potato-shaped Halley. More typical are Borrelly, which measures 8 × 4 km, the roughly spherical Tempel 1, which measures 5 × 7 km, and Wild 2 which resembles a thick hamburger with a diameter of about 5 km. The smallest nucleus yet observed belongs to Hartley 2, which is about 2 km long, narrowing to 0.4 km in its mid-section.

Of course, the size of the nucleus shrinks each time it approaches the Sun and becomes active. During peak activity near the Sun, Halley's comet was releasing about 20 tonnes of gas and 10 tonnes of dust per second, though the amount of dust ejected varied considerably. Calculations suggested that the surface of the nucleus is lowered by about 10 meters during each apparition. However, since the mass of the nucleus was estimated at 10^{14} kg, it is likely to remain active for many millennia. On the other hand, smaller comets trapped in shorter periodic orbits will shrink much more rapidly.

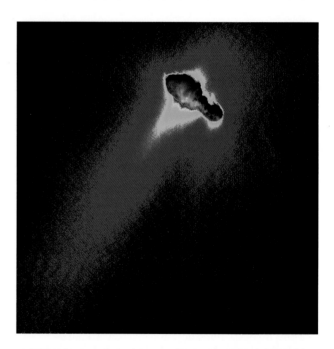

Figure 13.38 A false color, composite view of comet Borrelly taken by NASA's Deep Space 1 spacecraft on September 22, 2001. Apart from the dark nucleus, it shows details in the dust jets (purple and blue) and the coma of dust and gas. The nucleus was the blackest object in the Solar System, in places reflecting less than 1% of the sunlight that it receives. It is about 8 km long and 4 km wide. (NASA-JPL)

Figure 13.39   A composite image of comet Wild 2 taken during Stardust's flyby on January 2, 2004. The unusual surface displayed several large crater-like features, depressed regions, cliffs and pinnacles. Shaped like a thick hamburger, Wild 2 is about 5 km in diameter. Some of the main features are named in the sketch at right. (NASA-JPL)

Measuring the density of a nucleus is not easy, but estimates for various comets indicate that they are typically only 0.3–0.5 g/cm³, well below the density of water. This may be due to a porous, fluffy texture, or to a rubble pile structure with large empty spaces, in combination with a largely icy composition.

Their surfaces are blacker than coal with albedos of 2–4%. Some spots on comet Borrelly were so unreflective that their albedos were only 0.8%, making it the darkest object known in the Solar System. Despite their considerable ice content, comets all appear to be coated with a "crust" of carbon-rich organic material.

The prevalence of fine dust in comet nuclei was demonstrated when the impactor probe from NASA's Deep Impact spacecraft hit comet Tempel 1. Observations showed only weak emissions from water vapor and all the other gases that were expected to erupt from the impact site. The most conspicuous feature of the blast was brightening due to sunlight scattered by the extremely fine, ejected dust, which was described as more like talcum powder than beach sand. It seems that comets should be regarded more as "icy dirtballs" than "dirty snowballs."

The main Deep Impact spacecraft went on to complete a second comet flyby on November 4, 2010, as part of its extended EPOXI mission. Although 103P/Hartley 2 had the smallest nucleus yet observed from close range, it was emitting numerous jets of gas and ice particles.

As anticipated, water and carbon dioxide dominated the outgassing, with organics, including methanol, present at lower levels. However, there was a lot more carbon dioxide escaping than expected. This was associated with numerous bright jets that were erupting from the rough surfaces at either end of the nucleus. Most of the water vapor was derived from beneath the smooth central "waist," as water ice vaporized before escaping through the overlying porous material. The smooth surface in the middle was lower than the rest of the comet and may be coated by fine dust.

Previously, it was thought that water vapor from sublimating water ice was the main propulsive force behind jets of material

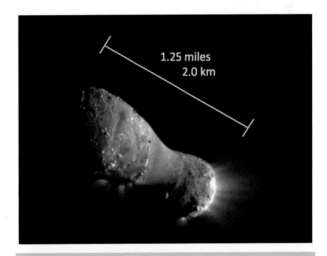

Figure 13.40   Comet Hartley 2 was imaged by the Deep Impact orbiter from a distance of about 700 km. The nucleus is approximately 2 km long, but only 0.4 km wide at its smooth "waist." Jets of carbon dioxide gas and ice particles can be seen streaming from the rough parts of the nucleus. The Sun is to the right. (NASA/JPL-Caltech/UMD)

coming from comet nuclei, but the data from the Deep Impact / EPOXI flyby provided unambiguous evidence that solar heating causes sublimation of subsurface frozen carbon dioxide. It is this primordial gas, rather than water vapor, which powers the numerous jets of material that erupt from comet nuclei.

Further, dramatic insights into the nature of comet nuclei and the processes that transform them as they near perihelion are expected from ESA's Rosetta mission. Launched in 2004, Rosetta is on a 10-year trek to comet 67P/Churyumov-Gerasimenko, which it will encounter more than 5.2 AU from the Sun. In 2014, it will become the first spacecraft to enter orbit around a comet. After gradually closing to within 25 km of the nucleus, it will release a small lander, named Philae, which will attempt to make the first soft touchdown on a comet.

Philae is designed to send back pictures and data about the temperature, composition and other physical properties of the surface. Meanwhile, the Rosetta orbiter will accompany the comet on the inward leg of its orbit, until it passes perihelion. Over an entire year, its 11 experiments will investigate every aspect of the comet as it is transformed from an inert chunk of ice to a warm, active mini-world surrounded by a coma that is fed by jets of gas and dust.

Breaking Up Is Easy To Do

How many passages through the inner Solar System can comets complete before they cease to illuminate the sky? What happens to them when their activity ceases?

There is no doubt that some comets break apart and disintegrate as they approach the Sun. Indeed, catastrophic breakups may be the ultimate fate of most comets. One of the first objects to reveal the fragility of nuclei was 3P/Biela, which orbited the Sun roughly once every six years. During its apparition of 1846, observers were surprised to discover two comets on almost identical paths, each with its own tail. For a while, the two nuclei were enveloped in a common coma.

Six years later, the pair was still flying in tandem, but some two million kilometers apart. They were never seen again. However, starting in 1872, thousands of meteors illuminated the November sky at six-yearly intervals. They became known as the Andromedid shower. When the orbits of the meteors were traced

back, they matched the path followed by comet Biela. Clearly, the comet(s) had disintegrated, leaving only a swarm of debris. After the turn of the 20th century, the meteor shower largely disappeared.

Several comets have been seen to break into multiple fragments in modern times. Comet 73P/Schwassman-Wachmann 3 began to splinter in 1995 during one of its many passages through the inner Solar System. Shortly after experiencing a huge outburst of activity, four separate nuclei were identified. Since then, the comet has disintegrated into dozens of fragments.

Images taken by the Hubble Space Telescope showed that a hierarchical destruction process was taking place as large fragments continued to break into smaller chunks. Several dozen "mini-fragments" were found trailing behind each main object. The images suggested that the pieces were propelled down the tail by outgassing from their sunward-facing surfaces. Since the smaller chunks had the lowest mass, they were accelerated away from the parent nucleus faster than the larger fragments. Some of the smaller pieces seemed to dissipate completely over the course of several days.

Occasionally, close encounters with planets can have a similar effect. Nuclei are probably porous, fragile mixes of dust and ices and they can be broken apart by gravitational tidal forces when they pass near large bodies. This is what happened to comet Shoemaker-Levy 9 when it passed near Jupiter in 1992, prior to plunging into the planet's atmosphere two years later (see chapter 7).

An even more catastrophic recent example was comet LINEAR (C/1999 S4), which suddenly disintegrated on July 25, 2000, during its first passage through the inner Solar System. Having previously appeared completely normal, the comet was seen to undergo a rapid change as it evolved into a fuzzy, extended and much fainter object. By early August, all that could be seen was a

Figure 13.41 This Spitzer Space Telescope infrared image, taken May 4–6, 2006, shows at least 36 sizeable fragments of comet 73P/Schwassman-Wachmann 3 amidst a stream of dust and smaller debris. The flame-like tails were warmed by sunlight so that they glowed at infrared wavelengths. The debris ranged in size from pebbles to large boulders. (NASA/JPL-Caltech)

Figure 13.42 The Hubble Space Telescope imaged 22 pieces of comet Shoemaker-Levy 9, each with its own coma, after the nucleus broke apart during a close approach to Jupiter in 1992. The remaining "string of pearls" was made up of one remnant around 3 km across and other fragments measuring 0.5–2 km. (NASA/STScI)

cloud of debris, with no sign of the nucleus or any active sub-nuclei larger than a few meters across.

Analysis of the sequence of events led to the conclusion that rapid vaporization near perihelion caused the small comet to run out of ice. When all of its ice was exhausted, nothing remained to bond the solid material together. The nucleus began to fall apart, leaving behind a loose conglomerate of particles that dispersed into space.

Comet nuclei can also be broken up by rapid rotation, thermal stresses as they pass near the Sun, or explosive disruption – like corks from champagne bottles – due to a sudden outburst of trapped volatile gases.

One of the most surprising examples is comet 17P/Holmes, which has exploded on two separate occasions (November 1892 and October 2007) as it approached the asteroid belt. During its 2007 return, Holmes unexpectedly changed from magnitude 17 to 2.8, becoming almost a million times brighter in only 42 hours so that it was visible to the naked eye. Its coma gradually expanded outward, eventually growing larger than the Sun before it finally dissipated.

Observations made with Spitzer in November 2007 revealed a lot of fine silicate dust, or crystallized grains smaller than sand. The dust was apparently created by the recent violent explosion that destroyed larger particles in the interior of the nucleus. This was confirmed when observations made in March 2008 found only large dust particles, about one millimeter in size, in the coma. Calculations indicated that the energy of the blast was about 10^{14} joules – equivalent to 24 kilotons of TNT – and the total mass of ejected material was some 10 million tonnes.

The presence of jets and a spherical cloud, not to mention two similar events more than a century apart, indicated that a collision with a meteoroid was an unlikely explanation. Thermal stresses are also unlikely to be the cause since, in both cases, the comet was well past perihelion and heading away from the Sun. Rather, it seems more likely that the nucleus behaved like a pressure cooker, in which trapped gases suddenly erupted through weaknesses in the overlying surface.

Other comets have undergone violent eruptions far from the warmth of the Sun. Asteroid 2060 Chiron is currently in an eccentric orbit between the orbits of Jupiter and Uranus, so observers were surprised in 1988 when it suddenly doubled in brightness during the inward leg of its journey. A few months later, it developed a coma – Chiron had been transformed into a comet (see below).

Even more impressive was the discovery on February 15, 1991, that comet Halley was surrounded by a huge coma, at least 300,000 km in diameter – evidence of a tremendous outburst. At the time of the observation, the comet was at a heliocentric distance of 14.3 AU.

This renewed activity was attributed to sublimation of a more volatile compound than water, for example CO or CO_2, which had built up sufficient pressure beneath the dust mantle to initiate a major outburst.

Centaurs

The term Centaurs is applied to objects that have been found orbiting among the giant planets of the outer Solar System. The first Centaur, Chiron, was discovered in 1977 by Charles Kowal. It was found that Chiron follows a highly elliptical (e = 0.383) orbit with a perihelion of 8.46 AU and an aphelion of about 19 AU. Its orbital period is 50.7 years and the orbit is inclined about 7° to the ecliptic.

Although it followed such an unusual orbit, Chiron was originally classified as an asteroid, with an estimated diameter of 148–208 km. Light curve studies gave a rotational period of 5.9 hrs. This classification came into question in 1988 when Chiron unexpectedly brightened. The following year, a faint, coma-like, shroud of dust and gas was discovered, suggesting that it was more likely to be a very large comet nucleus. Today, it has a classification both as an asteroid (2060 Chiron) and as a periodic comet (95 P/Chiron).

The diameter of the coma has occasionally expanded to almost two million km and spectroscopic analysis has revealed the presence of carbon monoxide (CO) and cyanogen (CN) radicals. The existence of surface activity and coma production at the low temperatures beyond Saturn suggests that sublimation of supervolatile substances, such as methane, carbon monoxide, and molecular nitrogen, is the source of the coma. Curiously, observations show that Chiron's activity was lower near perihelion (February 1996) than near aphelion.

One estimate puts the number of Centaurs with diameters larger than 1 km at about 44,000, assuming an inward flux of one new short-period comet every 200 years. Although their orbital inclinations vary considerably, they are all thought to originate in the Kuiper Belt. Their chaotic orbits are influenced by the giant planets, so that objects such as Chiron will ultimately either collide with a planet or be ejected from the Solar System.

The Centaurs are now generally regarded as icy objects with diameters typically greater than 100 km – transitional in size between typical short-period comets and the even larger Kuiper Belt objects that orbit beyond Neptune.

They have reddish surfaces – though not as dark or red as the Kuiper Belt objects. Near-infrared spectroscopy has revealed water ice on the surfaces of a number of Centaurs, while methanol (CH$_3$OH) ice (or another light hydrocarbon or oxidized derivative) has been seen in the near-infrared spectrum of Pholus. (Pholus' orbit takes it from inside the orbit of Saturn to beyond the orbit of Neptune. Like many Centaurs, no comet-like activity has been recorded for Pholus.)

Comet or Asteroid?

Calculations related to the rate of capture of long-period comets from the Oort Cloud imply that there should be about 1000 times more Halley-type objects than are actually observed. If, on the other hand, they disintegrate to dust, the debris would create a bright, near-spherical zodiacal cloud[20] and some 15–30 strong annual meteor showers, also contrary to observation.

One possibility is that many active comets rapidly become dormant when they lose most of their volatiles and/or develop a thick, dark surface deposit of organic material that insulates the inner nucleus and prevents all further solar heating of the interior. Such inactive comets may resemble cometary meteoroids, with extremely low reflectivities in visible light. This implies a large population of fast-moving, multi-kilometer-sized bodies that are too dark to be seen with current near-Earth object surveys.

A number of extremely dark objects that may belong to this postulated population of dead comet nuclei have been identified, including some transitional objects that were originally classified as asteroids and then observed to undergo comet-like outbursts of activity.

The first indications that the classical distinction between rocky asteroids and icy comets was inadequate came in the 1970s with the discovery that many comets (including Halley) have black surfaces, often as dark as carbonaceous asteroids. Since a comet loses its coma and tails as it retreats from the Sun, it becomes very difficult to distinguish between an asteroid and a bare comet nucleus – particularly if the latter is in an asteroid-like orbit.

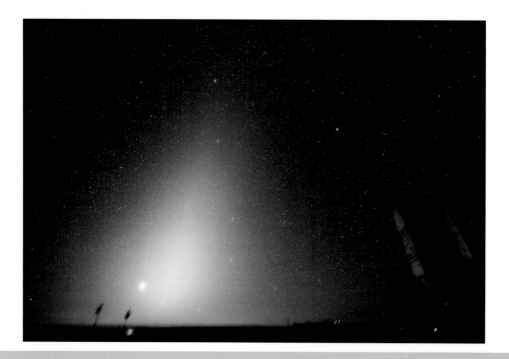

Figure 13.43 The zodiacal light is a cone of faint light which is visible above the horizon in the tropics. It lies along the ecliptic and is caused by scattering of sunlight due to dusty debris from asteroid collisions and comets. (ESO/Yuri Beletsky)

[20]The zodiacal cloud is a disk of dust in the inner Solar System. It is believed to be derived from comet dust and the debris of asteroid collisions, and causes a band of light (the zodiacal light) that extends along the ecliptic and is visible in the west after sunset or in the east before sunrise.

One famous example is asteroid 4015, which was discovered in 1979. Subsequent checks revealed that it was actually the periodic comet 107P/Wilson-Harrington, which had displayed a diffuse coma and tail when it was first found in 1949. Since then, all activity had subsided, enabling the object to take on the external mask of an asteroid.

More indirect evidence has pointed to the likelihood that Earth-crossing asteroid 3200 Phaethon is also a dead (or dormant) comet. Phaethon follows an unusual orbit that takes it to within 0.14 AU of the Sun and it is linked with the Geminid meteor stream that follows the same orbit around the Sun. Although it is currently a faint, dark object, Phaethon must have been active in the not-too-distant past in order to have generated such a major trail of debris.

Other comets that may be masquerading as Apollo asteroids include 2101 Adonis, which may be associated with the fairly unspectacular Capricornid-Sagittariid meteor shower, and 2201 Oljato, which has been linked with minor outgassing as well as a minor meteor shower. Modeling of Oljato's chaotic orbital evolution also suggests a cometary origin.

Another group of objects that fits this category is the Damocloids, named after 5335 Damocles, a peculiar asteroid that follows a highly inclined (61.95°) elliptical orbit that takes it from just outside Earth's orbit almost to the orbit of Uranus.

As of February 2011, 41 Damocloids were known. Although they are often classified as asteroids, they follow eccentric orbits typical of longer period comets such as Halley. About 25% of their orbits are retrograde – unlike any other asteroids. They also seem to be similar in size to Halley's nucleus, with an average radius of 8 km. The albedos of some Damocloids have been measured, and they are among the darkest objects known in the Solar System. Damocloids are reddish in color, though not as red as many Kuiper Belt objects or Centaurs.

Damocloids are believed to be the nuclei of Halley-type comets that have lost most of their volatile materials during many approaches to the Sun over thousands or millions of years. This hypothesis is strengthened by the fact that some objects thought to be Damocloids subsequently developed a coma and were confirmed to be comets.

Such cases have led to the suggestion that perhaps half of the near-Earth asteroid population consists of rocky objects, while the remainder are dead comets.

Main Belt Comets

The population of active comets that we see today consists almost entirely of objects from the Kuiper Belt and Oort Cloud that have been scattered into Jupiter-crossing orbits by gravitational interactions with the giant planets. Even the dynamically peculiar comet 2P/Encke which has an orbital period of only 3.3 years, is believed to have originated in the Kuiper Belt, though its orbital evolution has been strongly influenced by non-gravitational forces induced by cometary outgassing.

However, while the dominant cometary reservoirs are located beyond the orbit of Neptune, their main volatile constituent, water ice, is stable at much smaller heliocentric distances and it has long been suspected that dormant comets might exist closer to the Sun (e.g. the Hilda asteroids at 4 AU and the Jovian Trojans at 5 AU).

There is clear evidence that the definitions of a comet and an asteroid are not as clear-cut as once thought. As mentioned above, a number of asteroid-like objects are almost certainly dead or dormant comets. It has also been discovered that comets can inhabit the main asteroid belt between Mars and Jupiter.

Seven objects are currently known to occupy a typical orbit in the main belt while sometimes displaying comet-like behavior by ejecting clouds of dust and sprouting tails. The first to be recognized, in 1996, was asteroid 7968 Elst-Pizarro (also known as comet 133P/Elst-Pizarro). Since then, five more objects have exhibited similar activity while following stable orbits within the asteroid belt: P/2005 U1 (Read), asteroid 118401 (176P/LINEAR), P/2008 R1 (Garradd), P/2010 A2 (LINEAR), P/2010 R2 (La Sagra), and 596 Scheila. Together, these objects form a new class known as main belt comets (MBCs). In some cases, their activity persists for several weeks or months, consistent with ejection caused by sublimation of volatiles. Others are more likely to be activated by collisions that release carbon-rich dust and ice.

The discovery of comets which are in stable orbits in the main asteroid belt sheds some doubt on the idea that these primitive members of the Solar System only come from the Kuiper Belt and Oort Cloud. Instead, they appear to have formed within the main belt and presumably have remained there all their lives. This would suggest that they formed in a warmer environment than other comets, and that they may differ in composition from the Kuiper Belt and Oort Cloud comets. They may even have been a major source of water for the early Earth. The existence of these objects for billions of years can be explained if a thin, insulating layer of dusty material has, until relatively recently, prevented solar heating from triggering the vaporization of their ices and the creation of gaseous comas and tails.

Sungrazing Comets

Although the nuclei of most comets that enter the inner Solar System measure only a few kilometers in diameter, there is clear evidence that much larger objects exist. Direct measurements of Kuiper Belt objects such as Eris and Pluto show that icy planetesimals can have diameters exceeding several thousand kilometers. Furthermore, analysis of comet orbits shows that many thousands of icy fragments may be derived from one large parent body.

The most notable example of this is the Kreutz comet family, first recognized by German astronomer Heinrich Kreutz (1854–1907). They were all found to have similar retrograde orbits with orbital inclinations of about 142°, longitudes of perihelion at 280–282° and orbital periods of more than 500 years. They are often dubbed "sungrazers" because (like several other comet families) their perihelion distances lie within 0.02 AU.

The remarkable similarities in their orbits have led to the belief that they are all derived from a single, large parent comet which fragmented in the remote past, probably more than 1000 years ago. Detailed analysis indicates that there are two principal subgroups, possibly the result of separate fragmentation events at different perihelia. Since then, there has been progressive

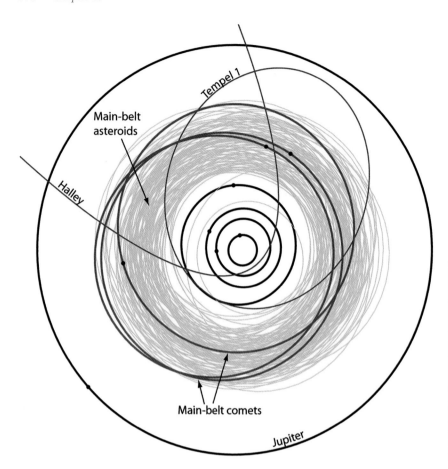

Figure 13.44 Orbits of the first three known main belt comets (red), the five innermost planets (black), a sample of 100 main-belt asteroid orbits (orange), and two "typical" comets – Halley and Tempel 1 (blue). Positions of the main belt comets and planets on March 1, 2006, are plotted with black dots. (Pedro Lacerda / Univ. Hawaii and Univ. Coimbra, Portugal)

fragmentation of the remnants, creating smaller and smaller pieces of debris. Most of these tiny objects, only a few meters across, plunge into the Sun or vaporize completely during their passage through the inner Solar System.

Kreutz-group comets have been observed for many hundreds of years. The number of known Kreutz family comets has soared since 1995, following the launch of ESA's SOHO spacecraft, which takes images of the Sun's neighborhood, using a coronagraph to blank out the brilliant solar disk. By the end of December 2010, the number of SOHO comets had exceeded 2000, with more being discovered (mainly by amateur web watchers) on an almost daily basis. This meant that SOHO had been indirectly responsible for the discovery of more comets than all other observers throughout history. Kreutz comets have also been discovered in images taken by NASA's twin STEREO spacecraft, launched in October 2006.

Not all of them are invisible to the unaided eye. Larger fragments have produced some of the brightest comets in history, notably comet C/1965 S1 Ikeya-Seki. Some unusual features have been observed as Kreutz objects sweep towards the Sun. For example, on May 24, 2003, two comets swept to within 0.1 solar radii (70,000 km) of the Sun's visible surface, deep within its searing multimillion-degree atmosphere, the corona. As usual, one of the intruders was vaporized by the intense heat and radiation pressure. However, SOHO-614 left a faint, tail-like feature that moved away from the Sun, apparently emanating from a point in the orbit beyond the comet's closest approach. Although

the intense solar heat and radiation vaporized its nucleus and coma, the dust tail survived the encounter and was pushed outwards by the radiation pressure. Such headless comets are very rare.

The Riddle of Stardust

One of the most ambitious space missions to explore a comet was NASA's Stardust. Launched in 1999, the spacecraft encountered comet P81/Wild 2 in January 2004. Passing through the coma, it captured an estimated 1000 particles, ranging in size range from 5–300 microns, in an incredibly low density material known as aerogel. The comet dust and samples of interstellar dust were successfully brought back to Earth in January 2006, when the return canister parachuted onto the Utah desert.

Early analysis of the comet samples led mission scientists to the surprising conclusion that Wild 2's coma contained minerals that formed at very high temperatures – despite a likely birthplace in the frigid outer regions beyond the orbit of Neptune.

Another surprise was the discovery of mineral particles that had formed in the presence of liquid water, presumably in pockets inside the nucleus. This indicated that the comet's internal temperature had – at least locally – risen to temperatures of 50–200°C.

Until then, the consensus opinion had been that comets are made of pristine, pre-solar materials that have never been altered

Box 13.11 Naming Comets

Comets have traditionally been named after their discoverers. If a comet suddenly brightened so that its discovery was made by several people, independently or jointly, the comet's name could include up to three of them. If the same team made more than one discovery, a number was added after their names (e.g. comet Shoemaker-Levy 9). Sometimes, as in the case of comets Halley and Encke, they were named after the astronomer who carried out special studies of them, since the actual discovery could not be attributed to any particular individual.

In the case of a periodic comet (with a revolution period of less than 200 years), the letter P is placed before the name (e.g. 160P/LINEAR). Hence, 1P/Halley was the first periodic comet to be discovered and confirmed. Other prefixes are C/ for a comet that is not periodic, X/ for a comet for which a meaningful orbit can not be computed, D/ for a periodic comet that no longer exists or is believed to have disappeared, and A/ for an object first thought to be a comet but later reclassified as an asteroid.

Some comets have never been named, including most comets from the 18th century or earlier, comets discovered years after they were first visible (e.g. from photographs) and others whose orbits cannot be determined due to a lack of observations.

In March 2003, the procedure for naming comets was made more formal by the International Astronomical Union, and now resembles the system applied to asteroids. A comet is immediately given a provisional designation, comprising the year of its discovery, followed by a capital letter and a number. The letter (in alphabetical order) identifies successive periods of 15 consecutive days or half months, while the number indicates its order of discovery during that time. Hence, comet 1996 B1 was the first to be discovered in the second half of January 1996.

After its perihelion passage, each comet receives a definitive designation. This includes the year of perihelion passage, followed by a Roman numeral specifying the order of such passage in that year. Obviously, the year of discovery does not always coincide with the year of perihelion passage. The full designation of Halley's comet is 1P/1682 Q1 (Halley), since Edmond Halley predicted its return to perihelion on September 15, 1682 – the first time such a prediction had been successfully fulfilled.

With the arrival of the Space Age and dedicated sky surveys, the number of discoveries made by a single spacecraft or ground-based telescope has often led to large numbers of comets with the same name, for example, SOHO or LINEAR. In addition to the normal designation, numbers are often added to clarify their order of discovery, for example, LINEAR 43.

Figure 13.45 The Stardust spacecraft brought back samples of comet (and interstellar) dust after passing through the coma of P81/Wild 2 in January 2004. During closest approach, the spacecraft was protected from incoming particles by bumper shields. (NASA)

Figure 13.46 In order to prevent damage and preserve them during and after the collection process, the dust particles from comet Wild 2 were captured in a low density material known as aerogel. Despite their high velocities, the particles were gently slowed, leaving trails in the aerogel from which they could be extracted back on Earth. (NASA-JPL)

by thermal processes. In addition to the icy material that forms comets at a temperature of around 30K (−243°C), their dust component was thought to be primarily interstellar material, tiny (submicron) grains that formed around other stars. However, Stardust has shown that, although it does contain some interstellar dust, Wild 2's dirt content is composed mainly of relatively large silicate particles. Their composition indicates that some of these formed at temperatures of 1500K (1227°C) in the innermost part of the Solar System, very close to the Sun.

The analysis of comet dust also identified a few grains that contain calcium/aluminum-rich inclusions (CAIs) similar to those found in chondrite meteorites. Thought to be the oldest objects in the Solar System, chondrites have always been a puzzle because they contain minerals that formed at low temperatures, as well as minerals, like CAIs, that must have formed at high temperatures (see above).

Some high-temperature, crystalline, minerals – rich in calcium, aluminum and titanium – apparently formed near the Sun during the early history of the Solar System. They were subsequently transported as far out as the Kuiper Belt and then incorporated into the object which eventually became comet Wild 2. The majority of the dust particles appear to have formed over a very broad range of solar distances and perhaps over an extended time range.

The process which enabled this transportation is not clear. Large-scale turbulence in the solar nebula has been suggested, although other evidence, such as the way in which planetary and asteroidal composition changes with solar distance, suggests that the cloud was not well mixed.

Another theory favors ballistic transport above the main disk of the nebula by a strong solar wind outflow, called the X-wind. This theory suggests that some of the dust falling into the young Sun was melted and ejected outward in a fiery spray. This condensed into small chondrules, or beads of melted rock, which were pushed out of the nebular disk by the X-wind. When they eventually reached the distant reaches of the Solar System, they combined with colder material to form chondrites, the building blocks for asteroids and planets. Some particles were also incorporated into comets. Alternatively, there may even have been ejection of material by bipolar jets shooting out of the young Sun.

Whichever theory is closest to the truth, the unexpected origin of the dust in comet Wild 2 adds up to more evidence for the "icy dirtball" theory of comet composition. In recent years, astronomers have come to think of comets not as snowballs coated in dust, but as dirtballs crusted with ice.

Deep Impact

Comet nuclei have been observed by several spacecraft, but there is still considerable uncertainty about the nature of their interiors. In an attempt to address this knowledge gap, NASA decided to send a spacecraft that would send a high speed projectile into periodic comet 9P/Tempel 1 as it neared perihelion.

The Deep Impact spacecraft, which would observe the impact from a safe distance, carried two imagers. The High-Resolution Instrument comprised a 30 cm diameter telescope attached to a multispectral camera and an infrared spectrometer. This system would take high resolution pictures of the comet and the impact crater. The Medium-Resolution Instrument took visible light images with a wide field-of-view, enabling it to observe the cloud of ejected material as well as the crater created by the impact event. A duplicate of the Medium Resolution Instrument on the impactor recorded the vehicle's final moments before it collided with Tempel 1.

On July 4, 2005, the Deep Impact spacecraft released the 1 meter wide, copper-fortified impactor. Over the next 22 hours, the mother craft and 364 kg impactor headed toward the comet using an auto-navigation system. During most of the encounter phase before closest approach, the instruments on the flyby spacecraft were aimed at the comet in order to capture the impact. Protected by dust shields, the spacecraft was oriented so that its cameras could continue operating until it came to within about 700 km of the nucleus. At this point, the spacecraft stopped taking pictures and turned so that its dust shields could provide as much protection as possible during the passage through the inner coma.

Meanwhile, the impactor continued to send back images until three seconds before its head-on collision. The final image was taken from a distance of about 30 km above the comet's surface, resolving features on the surface less than four meters across. It then smashed into the nucleus – with kinetic energy comparable to detonating 4.5 tons of TNT – and vaporized.

Figure 13.47   Deep Impact comprised a flyby spacecraft and a 1 meter diameter copper probe (left) that would collide with the nucleus of comet Tempel 1. The impactor vaporized deep below the comet's surface when the two collided on July 4, 2005, at a speed of 10 km/s. The cloud of ice and dust ejected from the impact site completely obscured the crater that was excavated. (NASA)

The Deep Impact cameras and other telescopes on the ground and in Earth orbit observed a cone-shaped cloud of ejecta rising from the nucleus at an average speed of about 110 m/s. The cone was seen to rise from the impact site for more than an hour, evidence that the comet's surface must consist of weak, easily pulverized material, and continued to expand outward over the following days.

The impact blew out so much fine, powdery material that the crater remained totally hidden. However, a subsequent visit by the Stardust-NExT mission revealed a crater about 150 meters in diameter, from which perhaps 1000 tonnes of material had been excavated. The amount of dust compared with the water vapor content indicated that Tempel 1 is best described as an "icy dirtball" rather than a "dirty snowball". Based on the rate at which the ejecta cone expanded near its base, the science team calculated that the comet has an extremely low bulk density, roughly 0.6 g/cc. 50–70% of Tempel 1 must, therefore, consist of empty space.

Spitzer Space Telescope observations after the impact returned detailed spectra of the ejecta, showing that the nucleus was composed of primordial material formed in the solar nebula. Emission signatures due to amorphous silicates and carbon, and crystalline silicates, carbonates, phyllosilicates, polyaromatic hydrocarbons (PAHs), water gas/ice and sulfides were found. The presence of crystalline silicates indicates that they were heated to temperatures of more than 700°C before somehow being transported beyond the orbit of Neptune, the formation region of the comet.

Three small patches of water ice were observed on the surface of the nucleus, surrounded by dark, organic material. This was the first definitive observation of water ice on the surface of any comet (although it has been spectrally identified in near-infrared observations of Kuiper Belt objects and Centaurs). The fact that the nucleus did not exhibit any enhanced activity in the days after the impact suggested that, in general, meteoroid impacts are not the cause of comets'outbursts.

Periodic Comets

By the beginning of the 18th century, it was understood that comets were celestial objects that occasionally appeared without warning, illuminated the skies for a few weeks or months as they moved closer to the Sun and then disappeared whence they came – presumably never to be seen again. However, their composition and origin remained a mystery.

The breakthrough came in 1705, when British astronomer Edmond Halley (1656–1742) began to calculate the orbits of 24 comets. He noticed that the path followed by a bright comet observed in 1682 was very similar to the orbits of other comets recorded in 1607 and 1531. Halley concluded that the only reasonable explanation was that the same comet had reappeared over

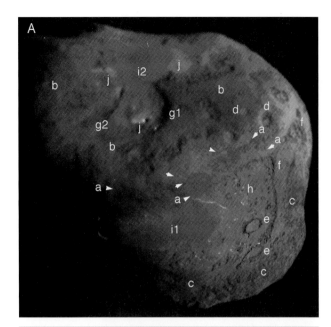

Figure 13.48 A composite image of comet Tempel 1 taken by the high resolution camera on Deep Impact's flyby craft, before the arrival of the impactor. The nucleus has ridges, scalloped edges and possible impact craters that formed long ago. Examples of different features are marked: (a) linear outcrops; (b) intermediate roughness pitted surface; (c) pitted and rough surface; (d) isolated, rimless depressions; (e) rim remnants; (f) close-packed depressions; (g) large arcuate depressions; (h) stripped surface; (i) smooth units; (j) exposures of ice. (NASA/UMD/Cornell/Peter Thomas)

a period of 75–76 years. The slight variations in the timing of each return were accounted for by small tugs on the comet by the giant planets.

Working forward in time, Halley predicted that the comet should return in December 1758. Although he did not live to see the event, his theory was proved to be correct when the comet duly reappeared on schedule. The first periodic comet to be recognized was named 1P/Halley in his honor.

Trawls through ancient records have shown that the famous comet was recorded by the Chinese as long ago as 240 BC. Halley is currently heading outward, and will reach its furthest point from the Sun (aphelion) in December 2023. It will then begin its long return journey, arriving in the inner Solar System in 2062.

Since Edmond Halley successfully predicted the return of a comet in 1682, more than 230 periodic comets have been discovered. As time went by and more orbits were calculated, it became clear that these comets were hard to classify. At one extreme are the Jupiter family comets, which typically take no more than 20 years to circle the Sun. Most of these have orbits that have been drastically altered by the gravitational pull of the largest planet. The shortest period belongs to comet Encke, which scoots around the Sun every 3.3 years.

About 20 known comets, including Halley, follow a more leisurely route, traveling beyond the orbit of Neptune before returning to the inner Solar System. Although these short-period comets have also been influenced by encounters with the giant planets, their orbits are more random, often tilted steeply to the ecliptic. Many of these – including Halley's comet – travel in a retrograde (backward) direction.

New comets arrive from the Oort Cloud, far beyond the orbit of Pluto, and, after making a brief appearance in our skies, disappear from view for thousands of years. The orbits of others are

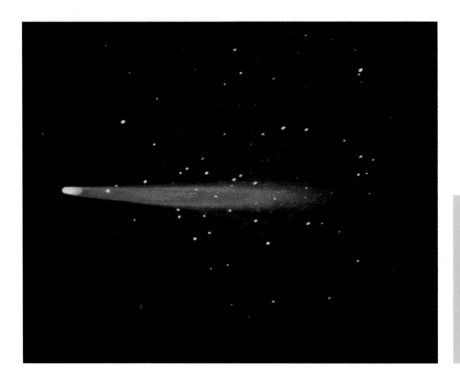

Figure 13.49 This image of comet 1P/Halley was processed at the Kitt Peak National Observatory from an original black-and-white plate taken at Lowell Observatory in 1910. The false colors indicate varying levels of brightness. A broad dust tail is visible above a narrow ion (gas) tail. (Lowell Observatory/NOAO/AURA/NSF)

modified by close encounters with the planets, particularly Jupiter. These periodic comets may reappear every five or six years.

Comets with orbits of less than 200 years are thought to originate in the Kuiper Belt, a doughnut-shaped region that begins at the orbit of Neptune and extends to at least 50 Sun-Earth distances. They were probably ejected to their present location billions of years ago by gravitational interactions with Uranus and Neptune. The first Kuiper Belt object was discovered in 1992 and many hundreds are now known (see Chapter 12).

The Oort Cloud

Approximately a dozen "new" comets are discovered every year. In contrast to the relatively predictable, well-behaved periodic comets, these intruders arrive without warning. They can appear in any part of the sky, heading sunwards from all directions. Known as long period comets, they follow highly elliptical or near-parabolic orbits which carry them many billions of kilometers from the Sun, and may take many thousands or millions of years to return. Some may never return.

By analyzing the paths of 19 such comets, Dutch astronomer Jan Oort (1900–1992) calculated that they must originate in a huge spherical reservoir located 100,000 to 300,000 AU. Based on the frequency of their appearance in our skies, he estimated that this cloud may contain around 190 billion unseen, dormant comets. Oort suggested that every 100,000 to 200,000 years, the gravity from a star passing within 200,000 AU of the Sun would send some of the comets toward the inner Solar System.

In the last half century, the availability of more detailed information and new analysis of comet orbits has led to a modified version of Oort's cloud. It is now thought that the average aphelion distance of a new long period comet lies at about 44,000 AU, markedly closer than Oort suggested. The outer edge of the cloud is now thought to reside at about 200,000 AU (3 light years) from the Sun, since comets beyond this distance are so loosely bound by the Sun's gravity that they are easily nudged into interstellar space.

Meanwhile, analysis of stellar orbits shows that it is quite common for a star to pass right through the Oort Cloud. On average, a star will approach within 10,000 AU of the Sun once every 35 million years, with an even closer passage (about 3000 AU) once every 400 million years.

Tidal forces related to the motion of the Solar System through the galactic plane and encounters with molecular clouds also influence the objects within the Oort Cloud. A giant molecular cloud made of cold hydrogen gas is likely to be encountered about every 300–500 million years, but its massive gravity is capable of altering the orbits of the Oort population, sending comets toward the Sun or outward into interstellar space.

The most widely-accepted theory of the Oort Cloud's formation is that the objects initially formed much closer to the Sun – probably in the Uranus-Neptune region – as part of the same accretion process that formed the planets and asteroids, but that gravitational interaction with the young gas giants ejected them into extremely elliptical or parabolic orbits. This process also scattered the objects out of the ecliptic plane, explaining the cloud's spherical distribution. The ejection of comets to the Oort Cloud may have been a violent process involving numerous

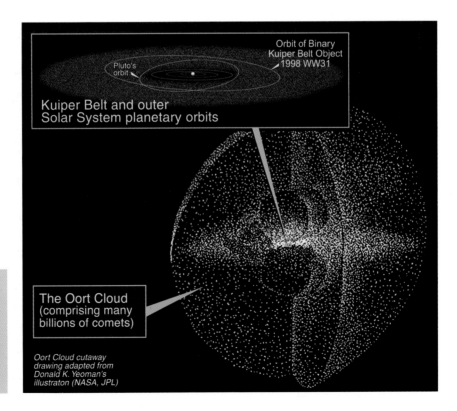

Figure 13.50 The Oort Cloud is a hypothetical, spherical swarm of icy bodies located 10,000 to 200,000 AU from the Sun. This diagram shows its size and shape in relation to the Kuiper Belt and the region inside Pluto's orbit. (NASA/JPL/Yeomans)

collisions, so that many of the objects which survived to reach the Oort Cloud were quite small. As time went by, gravitational interaction with nearby stars further modified their orbits to make them more circular.

Although there are probably billions of objects in the cloud, the area they occupy is so vast that they rarely approach to within tens of millions of kilometers of each other. Collisions must be extremely rare. Since the majority of the objects are expected to be small (most cometary nuclei have diameters of ~20 km or less), the amount of material contained in the Oort Cloud is thought to be fairly modest – several times Earth's mass. However, even that amount of material is more than would be expected from current theories about the Solar System's formation.

No members of the main Oort Cloud have yet been observed or identified, but an unusual object known as Sedna has been found in an extremely elliptical orbit which resembles that of objects thought to lie in the Oort Cloud (see Chapter 12). However, Sedna's 10,500 year orbit takes it out to only around 990 AU at aphelion, some 10 times closer than the predicted distance of the cloud. This has led to the suggestion that it may belong to an inner cloud, made up of bodies trapped between the Kuiper Belt and the classical Oort Cloud.

Questions

- What are the links between (a) asteroids and meteorites, (b) comets and meteors?
- Why are Ceres, Vesta and Pallas of particular interest to planetary scientists?
- Describe and explain the distribution of asteroids and comets in the Solar System.
- Describe the main characteristics of a comet. Is the term "dirty snowball" an adequate description of comets?
- What evidence is there for the existence of the Oort Cloud? What are the main differences between the Oort Cloud and the Kuiper Belt?
- Asteroids have traditionally been described as small, rocky objects in the inner Solar System, while comets are small icy objects in the outer Solar System. Explain why these descriptions have been challenged in recent years.
- If a comet or asteroid was predicted to impact Earth in 10 years' time, what scientific advice would you give your government and the United Nations?

FOURTEEN
Exoplanets

Since ancient times, astronomers and philosophers have debated and speculated about the existence of exoplanets (also known as extrasolar planets) – planets in orbit around distant stars. Is our Solar System unique, or is it just one of many such systems scattered throughout the Galaxy? Are there worlds out there that support alien life forms, and are there other advanced civilizations in the Galaxy?

The distances between the stars are so immense that, for many centuries, there seemed little likelihood of answering any of these questions. Then, as measurements of stellar positions became more accurate, suspicions arose that planets might be the cause of the irregular proper motions of some stars. Barnard's Star, one of our closest stellar neighbors, was a favorite candidate, but its postulated planetary companions proved to be elusive and have yet to be found.

Then came the breakthrough. In 1991, Aleksander Wolszcan (Pennsylvania State University) and Dale Andrew Frail (National Radio Astronomy Observatory) were using the giant radio telescope at Arecibo Observatory in Puerto Rico to study a rapidly rotating neutron star known as PSR B1257 + 12. Tiny variations in the arrival time of its radio signals led them to the surprising conclusion that the dense, dead star was orbited by at least three fairly small planets.

Four years later, the first undisputed planet orbiting another main sequence star (51 Pegasi) was found by Michel Mayor and Didier Queloz of Geneva Observatory, Switzerland. Detected indirectly by observing tiny changes in the star's radial velocity, their planet was at least half the mass of Jupiter. The rapidly orbiting world was very close to 51 Pegasi and blisteringly hot.

Three months later, a team led by Geoffrey W. Marcy and Paul Butler of San Francisco State University and the University of California at Berkeley confirmed the Swiss discovery and turned up two more planets.

These announcements marked the beginning of a flood of discoveries. Since then, well over 600 exoplanets have been confirmed – mostly by indirect observations – and the total number is rising rapidly. The Kepler spacecraft alone has identified 1235 candidate exoplanets, leading to speculation that there may be up to fifty billion alien worlds in the Milky Way galaxy.

Many of the confirmed objects are like 51 Pegasi b, giant planets with short periods (up to about 10 days) and orbits that bring them extremely close to their star. However, improved detection techniques are now discovering planets that more closely resemble those in our Solar System, with circular orbits and longer orbital periods.

Beta Pictoris

Until recently, exoplanets had always been too small and distant to be observed directly. However, since planets were thought to be born inside huge, dusty nebulas surrounding young stars, the obvious starting place was to look for circumstellar disks that might contain fledgling planetary nurseries.

Prospects for finding new worlds around other stars brightened in 1984 when infrared observations by Bradford A. Smith (University of Arizona in Tucson) and Richard J. Terrile, (Jet Propulsion Laboratory) showed a disk of dust surrounding the young star Beta Pictoris, located 63 light years from Earth.

Exploring the Solar System, First Edition. Peter Bond.
© 2012 by Peter Bond. Published 2012 by Blackwell Publishing Ltd.

Box 14.1 Pulsar Planets

The first exoplanets to be discovered were detected in 1991, during a search for pulsars conducted with the 305 meter Arecibo radio telescope in Puerto Rico. The find was most unexpected, since pulsars are rapidly spinning neutron stars that emit regular, pulsing radio emissions, rather like a rotating lighthouse beam. These extremely small, dense objects are the remnants of supernova explosions that mark the deaths of massive, normal stars. The collapsed core measures only about 15 km across, but a teaspoonful of their matter would weigh a billion tonnes.

The discovery was made by a team led by Polish astronomer Aleksander Wolszczan. They were studying an old neutron star, known as PSR B1257 + 12, which is spinning once every 6.22 milliseconds. He recognized tiny variations in the time of arrival of the pulses from the star, instead of their usual precise regularity. Wolszczan analyzed these variations and came to the remarkable conclusion that the small, but unexpected, departures from the regular beat of the radio signals were caused by three small planets that were tugging on the pulsar and causing slight wobbles as it rotated.

Figure 14.1   An artist's impression of a planetary system around pulsar PSR B1257 + 12. The planet nearest the pulsar is very lightweight, while the other two (foreground) are more massive than Earth. Radiation from charged pulsar particles probably rains down on the planets, causing their night skies to light up with auroras similar to our Northern Lights. One such aurora is illustrated on the planet at the bottom of the picture. (NASA/JPL-Caltech/R. Hurt -SSC)

Skepticism about the reality of these unexpected worlds was overcome when predictions of how they should gravitationally interact were subsequently confirmed. The perturbations between the two larger planets were used to measure their masses and orbital inclinations.

The calculations indicate that the three pulsar planets have masses of 0.02, 4.3 and 3.9 Earth masses, and their orbital periods are 25, 66 and 98 days. Compared to our Solar System, they would all fit within the orbit of Mercury, and the spacing of their orbits is similar in proportion to those of Mercury, Venus and the Earth. Their orbits are inclined about 50° with respect to the plane of the sky. There is also some evidence that the pulsar may have an asteroid belt similar to the one that occurs beyond the orbit of Mars.

The fact that the planets have almost coplanar orbits indicates that they all evolved from a protoplanetary disk, very much like the planets around our Sun. However, it seems unlikely that could have existed before the supernova occurred and then survived the stellar holocaust. Indeed, they may well represent a second generation of planets, replacing an earlier family that was destroyed by the cataclysmic explosion.

One possibility is that metal-rich debris from the supernova formed an orbiting disk around the neutron star. Small, but massive planets with large iron cores could have coalesced from the disk. Powerful dynamos in the cores would generate very strong magnetic fields, so, although the dead star would be a poor source of illumination, dense streams of energetic particles emitted by the pulsar would fill the sky with magnificent auroras. However, the highly energetic particles would also gradually erode each planet's surface, generating a low-level haze.

Their discovery provided the first unambiguous proof that flattened disks of matter exist around stars other than the Sun. The Beta Pictoris disk appeared to be a young planetary system in the making, and thus supported the standard model of Solar System birth, which theorizes that planets accrete from a disk of dust and gas surrounding a young star.

Since then, Beta Pictoris has been the subject of intense study. In 1995, Hubble Space Telescope observations revealed an apparent warp in the disk, which was later confirmed. Observers suggested that the warp might be associated with a secondary disk tilted about 4° from the main disk, possible due to a planet in an inclined orbit.

Further Hubble observations made in 2003 confirmed the existence of two dust disks encircling the star. To see the faint disks, astronomers used the Advanced Camera for Surveys' coronagraph, which blocked the light from Beta Pictoris. The secondary disk was visible out to roughly 250 AU from the star, and probably extended even farther.

Since dust grains are only expected to survive for a few hundred thousand years, the continuing existence of these disks around a star no older than 20 million years means that the dust is probably being replenished by collisions between larger planetesimals.

The observers theorized that Beta Pictoris may have two planet-forming regions, one associated with each disk. An unseen planet, up to 20 times the mass of Jupiter and orbiting within the secondary disk, was sweeping up planetesimals from the primary disk. When these collide, they create the secondary dust disk. The pale red color of the disk may be due to the presence of tiny grains of graphite and silicates, a theory supported by data from NASA's Far Ultraviolet Spectroscopic Explorer (FUSE) spacecraft, which discovered abundant amounts of carbon gas in the dust disk.

To complicate matters still further, infrared images taken in 2002 by the Keck II telescope in Hawaii showed that another, smaller, inner disk may exist around the star in a region the size of our Solar System. The possible inner disk was tilted in the opposite direction from the secondary disk seen in the Hubble images.

In November 2008, a team of French astronomers using ESO's Very Large Telescope reported the discovery of a postulated giant planet, about eight times as massive as Jupiter (see Fig. 14.3). With a projected distance from the star of 8-15 AU, comparable to Saturn's distance from the Sun, it orbits inside the main dust disk of Beta Pictoris. Further observations have confirmed the existence of the planet, known as Beta Pictoris b.

Dusty Disks

Since the discovery of the complex disk surrounding Beta Pictoris, similar structures have been found around numerous young stars. Many of these embryonic planetary systems – also known as protoplanetary disks (proplyds) – are buried deep within dense nebulas where large scale star formation is taking place. Although they appear as fairly indistinct clouds surrounding newborn stars, the proplyds are helping to clarify the mechanisms behind planet formation.

The most famous examples of proplyds are found in the Orion Nebula, a huge star-forming region about 1500 light years away, where dozens of examples have been found with the Hubble Space Telescope. Estimates indicate that at least half of the stars in the nebula are surrounded by such disks (see Fig. 14.4).

Two different types of disk have been identified. Those close to the brightest star in the Trapezium cluster, Theta 1 Orionis C, shine brightly as a result of the energetic radiation they receive. These hot, glowing disks – some seen edge-on and some face-on – are sculpted into different shapes by the stellar wind and radiation, a process known as photoevaporation. In some cases

Figure 14.2 Images of Beta Pictoris, taken by the Advanced Camera for Surveys' coronagraph on the Hubble Space Telescope, confirmed the existence of two dust disks encircling the star and suggested the presence of at least one giant planet. The visible-light image shows a distinct secondary disk that is tilted by about 4° to the main disk. The secondary disk may be created by a giant planet that is removing material from the primary disk. (NASA, ESA, D. Golimowski/Johns Hopkins University, D. Ardila/IPAC, J. Krist/JPL, M. Clampin/GSFC, H. Ford/JHU, G. Illingworth/UCO/Lick and the ACS Science Team)

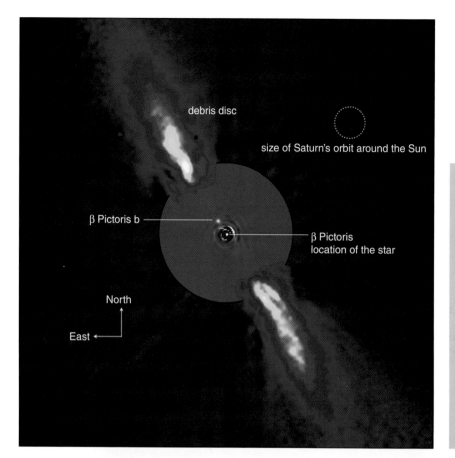

Figure 14.3 This composite infrared image shows a giant exoplanet inside the dust disk of Beta Pictoris. This very faint planet was revealed after subtraction of the much brighter stellar halo. The outer part of the image shows light reflected from the dust disk, as observed in 1996 by ESO's 3.6 meter telescope. The inner part of the system was imaged at 3.6 microns with the NACO instrument on the Very Large Telescope. The newly detected source is more than 1000 times fainter than Beta Pictoris, and aligned with the disk, at a distance of about 8 AU. (ESO/A.-M. Lagrange et al.)

Figure 14.4 The 30 protoplanetary disks in the Orion Nebula, imaged by the Hubble Space Telescope's Advanced Camera for Surveys. The images show two different types of disk around young and newly forming stars: those that lie close to the brightest star in the cluster (Theta 1 Orionis C) and those further away from it. The star heats up the gas in the nearby disks, causing them to shine brightly. The cooler, more remote, disks can only be detected as dark silhouettes against the background of the bright nebula. Some are viewed edge-on, some face-on. (NASA, ESA and L. Ricci/ESO)

emerging jets and curved shock waves are apparent. The shock waves occur when the stellar wind collides with the gaseous disks. Proplyds further away from Theta Orionis C are not illuminated and appear as dark smudges silhouetted against the bright nebula in the background.

The dark proplyds are composed of very cold gas and dust with a temperature of only 20K. Their masses vary considerably, but those that are in the range 0.01 – 0.025 solar masses are thought to have sufficient disk material bound to their central stars to form planetary systems on the scale of our Solar System. However, those that are being blasted by intense radiation from Theta 1 Orionis C are likely to have much of their material dissipated and vaporized, possibly destroying their potential for planets to form.

It is interesting to note that dusty disks have also been found around brown dwarfs, often known as "failed stars." Although the disks are often less massive than those around ordinary stars, the material is less likely to be blown away by the weak radiation from the small, cool brown dwarfs. Some crystallization and clumping of dust particles has been detected, showing that the disks are undergoing at least the same initial steps of the planet-building process as circumstellar clouds.

Disk Evolution and Exoplanet Formation

If the disks survive, significant changes take place over time, as the central stars and their disks evolve, and accretion processes begin (see Chapter 1). Structure within the disks appears, often in the form of belts of asteroidal or cometary debris. "Empty" zones which have been largely cleared of material, probably by one or more large protoplanets, become apparent. Although they are extremely difficult to detect, exoplanets orbiting within the disks have occasionally been found.

The structure and composition of protoplanetary disks can be studied with infrared telescopes, such as the Spitzer Space Observatory. Infrared observations detect dust that is created during collisions between planetesimals left over from planetary formation. In the majority of cases, the warm dust in the inner regions of the disk is made of common terrestrial minerals such as olivine, pyroxene and silica. Further out, the colder material is dominated by ices.

The formation of dust disks around stars seems to be almost universal. However, studies of intermediate mass stars similar to the Sun, or slightly larger, show that dust within the terrestrial planet zone soon disappears. Only stars with ages in the range 10–30 million years display dust in their inner disks, created during the epoch of final catastrophic mass accretion for rocky, terrestrial planets.

Observations of star forming regions have shown that about 10% of type A and B stars, about two to fifteen times as massive as the Sun, appear to possess dusty disks. However, there is a fairly small chance that the disks will survive long enough for planets to evolve. Planet formation around massive stars is a race against time. On one hand, the circumstellar disks tend to be larger and contain more of the raw materials to build planets. On the other

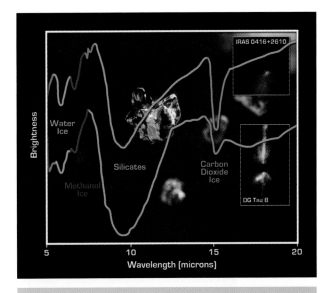

Figure 14.5 Spectra obtained by the Spitzer infrared space observatory reveal the chemical composition of protoplanetary disks around two young stars. The broad depression in the center of each spectrum signifies the presence of crystalline silicates. Tiny olivine crystals are thought to make up at least some of the dust grains, becoming coated with ice deep within the disk. The depth of the silicate absorption feature indicates that the dusty cocoon around the embedded protostar is extremely thick. Other dips due to absorption are produced by water ice (blue), methanol ice (red), and carbon dioxide ice (green). Their existence in solid form suggests that the material close to the protostar is cold. (NASA/JPL-Caltech/D. Watson – University of Rochester)

hand, fierce UV radiation and stellar winds tend to destroy the disks rapidly. Within two to five million years of their birth, most of the massive stars have already lost the raw materials needed to form planets. This indicates that, at least for type A and B stars, planets must form quickly or not at all.

Although direct imaging of exoplanets around dwarf stars is very difficult with current instrumentation, the coalescence of planets from the less massive nebulas surrounding smaller, cooler stars seems to be a much easier proposition. Observations indicate that at least 20%, and possibly as many as 60%, of stars similar to the Sun are candidates for forming rocky planets. One possibility is that the most massive disks form their planets quite rapidly, whereas the low mass disks take from 10 to 100 times longer.

The composition of the disk is also an important factor. When a star forms by collapsing from a dense cloud into a luminous sphere, the new arrival and its circumstellar disk reflect the composition of the original molecular cloud. Some nebulas are much richer in "metals" than others.[1] Elements such as iron and oxygen are the raw materials for rocks and ice, so it is not surprising that the stars with more heavy elements tend to have more planets. In

---

[1] **Astronomers refer to all the elements heavier than helium as "metals."**

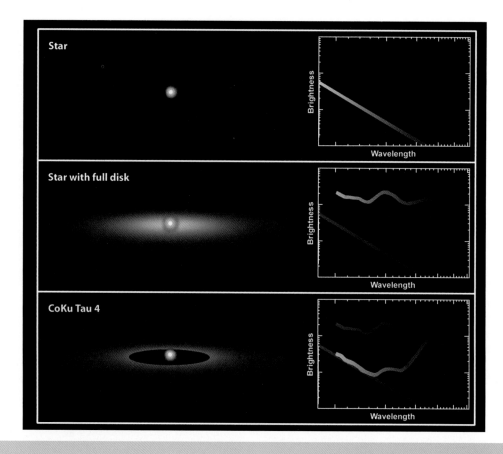

Figure 14.6 Spectroscopy enables scientists to deduce the temperature and chemical composition of material around a star, even if the disk itself cannot be seen. In the case of a star with no circumstellar disk (top), most of the light is produced at shorter wavelengths (left side of graph), due to the high temperature of the star's surface. If a star has a disk around it (center), the warm dust and gas produce their own infrared light, which changes the shape of the spectrum. The circumstellar material is cooler than the star, so it emits most at longer, infrared wavelengths. If the inner part of the disk has been cleared (bottom), perhaps by the formation of a planet, there is a normal spectrum near the star, with an excess of infrared light further away. (NASA/JPL-Caltech, D. Watson/University of Rochester)

contrast, there are few opportunities for planets to coalesce and grow from gaseous disks that contain very little solid material.

The dust particles, which are the building blocks of planets, accumulate into comets and small asteroid-size bodies, and then clump together to form planetary embryos, and finally full-fledged planets. In the process, some rocky objects collide and grow into planets, while others shatter into dust or larger debris.

Observations of fledgling systems show bands of material separated by regions that have been cleared by planets. In the case of Epsilon Eridani, a younger, slightly cooler and fainter version of the Sun, infrared observations indicate the presence of two asteroid belts and a broad, outer ring of icy comets similar to our own Kuiper Belt, in addition to at least two possible planets.

By the time a Sun-like star reaches an age of around 300 million years, the dusty disk has almost completely dissipated, either because the dust particles are blown away or dragged onto the star, or the particles clump together to form much larger objects.

It is interesting to note that systems containing asteroids, comets, and, possibly, planets are at least as abundant in binary star systems as they are in those with only one star. Since more than half of all stars are binaries, this suggests that planets with two Suns may be quite common across the Universe.

Disks have been found not only around widely spaced pairs, but also around very tight binaries separated by only a few astronomical units. Far fewer disks have been found in intermediately spaced binary systems, three to 50 AU apart. In other words, if two stars are as far from each other as the Sun is from Jupiter (5 AU) or Pluto (40 AU), they would be less likely to host a family of planetary bodies.

Spitzer data have also revealed that debris disks circle all the way around both members of a close-knit binary – these are called circumbinary disks – but only a single member of a wide duo. This could explain why the intermediately spaced systems are inhospitable to planetary disks: they are too far apart to support one large disk around both stars, and they are too close together to have enough room for a disk around a single star.

Another curiosity is that at least some systems may contain planets that orbit the central star in opposing directions. The first evidence for such a system was provided by radio observations of

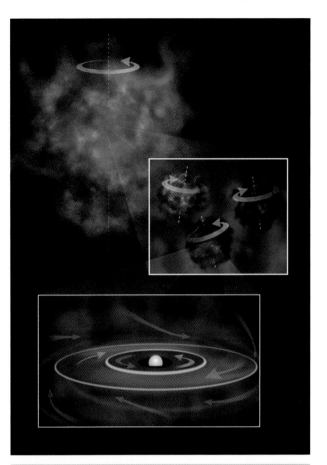

Figure 14.7 Infrared observations with NASA's Spitzer Space Telescope show that debris disks are common around binary stars. In particular, they are more abundant around pairs that are 0–3 AU apart (top panel) and 50–500 AU apart (bottom) than binary stars whose orbits lie 3–50 AU apart. (NASA/JPL-Caltech/T. Pyle-SSC)

a disk around a young protostar some 500 light years from Earth in the constellation Ophiuchus. Doppler shifts in the frequencies emitted by molecules within the cloud show the direction in which the gas is moving relative to Earth. The shifts show that the inner and outer parts of the disk are rotating in opposite directions. This is likely to result in the formation of a system where the inner planets orbit in one direction and the outer planets go around their star in the opposite direction, unlike our own Solar System, where all the planets travel in the same direction.[2]

The Sun and its planets are thought to have formed during the collapse of a single molecular cloud, so they all rotate in the same direction. However, the system in Ophiuchus lies in a large, star-forming region where chaotic motions and eddies in the gas and dust result in small clouds with independent rotations. The protoplanetary disk may, therefore, contain substantial amounts of material from two such clouds, each rotating in a different direction.

Figure 14.8 The stages in the formation of a counter-rotating protostellar disk. (Top) A huge star-forming nebula is rotating as it gives birth to multiple stellar systems. (Middle) A detailed view inside the nebula shows three protostars forming as the cloud collapses. The chaotic collapse process creates eddies, allowing newly-forming stars to rotate in different directions and at different speeds. (Bottom) One protostellar cloud collapses further into a disk-like structure that rotates counter-clockwise about the newly-formed star. The protostar also draws material from a second, passing, cloud which is rotating in the opposite direction. As a result, the outer part of the disk rotates clockwise, so any planets forming in this region will orbit the star in the opposite direction from the inner planets. (Bill Saxton, NRAO/AUI/NSF)

Detecting Exoplanets: Radial Velocity

Until recently, it was thought that current technology was not sufficiently advanced to achieve direct imaging of an exoplanet. The only alternative was to search for indirect evidence for the existence of such objects. The first exoplanets were detected by measuring the tiny wobbles that they induced in the motion of their parent stars. Since then, the vast majority of planetary

detections has been achieved with ground-based telescopes, using the so-called radial velocity technique.

The method requires the light from a star to be split into a spectrum, similar to the way water droplets in the atmosphere separate sunlight into a colorful rainbow. When the spectrum is magnified, straight black lines can be seen superimposed on the colors. These spectral lines correspond to the wavelengths of light that have been absorbed by chemicals in the atmosphere of the star from which the light originated.

[2]It is not too surprising to find counter-rotation in a protostellar disk, since the phenomenon has been previously reported in the disks of galaxies.

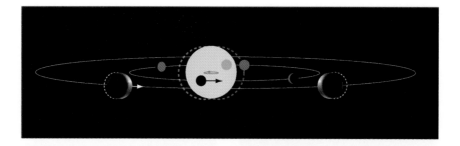

Figure 14.9   An illustration of how an Earth-sized planet orbiting close to a star may be discovered and studied. The planet lies inside the orbit of a giant planet and both are shown at different positions in their orbits. The presence of the unseen planets can be inferred from the periodic Doppler shift of starlight (blue and red dashed circles) as the star is pulled toward and away from the observer. The smaller planet's diameter can be estimated when it transits the star and blocks a tiny fraction of the starlight. At the opposite phase, the planet is eclipsed by the star, and its infrared flux can be measured by comparing the total flux before and during eclipse. The inner planet perturbs the outer giant planet on its orbit (dashed lines), causing times of transit of the latter to deviate from simple periodicity by a measurable amount. (Eric Gaidos et al / Science / AAAS)

By studying a sequence of tiny Doppler shifts (frequency shifts) in these lines, it is possible to show that a star is orbited by one or more planets. When a star is moving toward Earth, the wavelengths of incoming light "bunch up" and shorten. Astronomers say they are "blue-shifted." When the star is moving away from Earth, the wavelengths stretch out, so they are "red-shifted." A sequence of abnormal, but predictable, shifts in the spectrum can reveal minor changes in a star's motion in response to the gravitational pull of a nearby planet. The star appears to wobble, or pirouette, around a point in space.

This technique should eventually be able to detect Earth-sized worlds if they have very short periods and orbit low-mass stars. At present, the best spectroscopes can confidently detect motions of about 0.5 m/s. Although Earth's modest gravity only causes the Sun to wobble at 0.1 m/s, new instruments under development will soon be able to detect such tiny motions. Techniques are also being introduced to extract the effects of the bubbling activity of the star's gaseous surface, which tends to hide the Doppler effect of such a planet.

Since the discovery of 51 Pegasi in 1995, hundreds of planets have been discovered using this radial velocity (RV) technique. Many of them occur in multiplanet systems that occasionally lie in resonant orbits with one another. Given the period of the orbit and the maximum radial velocity observed for a system, it is possible to deduce the mass of the exoplanet (assuming the star's mass is known) as well as its orbital distance.

Unfortunately, the inclination of the planet's orbit is often unknown, since the only data available relate to the observed radial velocity changes of the star. As already mentioned, the other disadvantage of the RV technique is that it is most sensitive to large, Jupiter-like, exoplanets that are orbiting close to their star. Small, rocky worlds located 1 AU or more from the star cause such small RV shifts that they are very difficult to detect. As a result, the present census of exoplanets is dominated by gas giants, with a marked absence of Earthlike worlds.

The most successful exoplanet finder to date has been the High Accuracy Radial Velocity Planet Searcher (HARPS) spectrograph on ESO's 3.6 meter telescope at La Silla, Chile. An exact copy of

the HARPS instrument is to be installed on the Telescopio Nazionale Galileo in the Canary Islands. The duplicate HARPS will allow observers to survey stars in the northern sky. HARPS is capable of detecting back-and-forward motions in a star's radial velocity as small as 3.5 km/h – a slow walking pace.

Its extremely precise measurements also make it possible to search for less massive planets. By September 2011, HARPS had discovered over 150 exoplanets, many of them comparable in mass to Neptune, or smaller. Some of these orbit low mass stars known as M dwarfs, or red dwarfs.

Transits

Another method for detecting small worlds is to look for a slight dip in a star's brightness when a planet passes in front of its star. Such a celestial alignment is known as a transit. From Earth, both Mercury and Venus occasionally pass across the front of the Sun, appearing as tiny black dots passing across the brilliant photosphere.

Unfortunately, this possibility only opens up for star–planet systems that happen to be viewed nearly edge-on from Earth. Furthermore, detection is not easy, since transits only block a tiny fraction of the star's light. Even if a distant star is transited by a giant planet like Jupiter, only 1% or less of the starlight will be lost from view. Detection is usually limited to planets that orbit close to their primary. However, if one or more exoplanets can be observed in transit across the star's disk, then a much clearer understanding of the system can be obtained.

In some cases, a discovery made using the radial velocity method may be confirmed by subsequently detecting a transit. One example was the star HD 209458, which was discovered to possess a hot Jupiter in a tight circular orbit. Its existence was confirmed in 1999, when the planet passed in front of the star, exactly as predicted.

The key is to measure the time each transit takes and the relative diameters of the planet and the star. The diameter of a planet can be derived from the star's light curve during the transit. Since

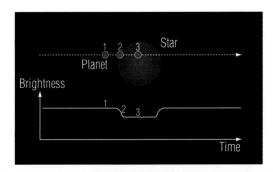

Figure 14.10 An exoplanet in transit across a star's disk causes a tiny drop in the apparent brightness of the star. Such transits are only possible if the planet and star are in our line of sight. However, they enable the diameter of the planet to be calculated, from which the mean density of the planet can be derived. It is also possible to study the exoplanet's atmosphere by comparing the combined spectrum of the star and planet with a spectrum taken when the planet passes behind the star. (NASA)

radial velocity data enable the minimum mass to be calculated, by combining the minimum mass and the orbital inclination, it is possible to obtain the true mass. Then, by combining the latter with the diameter, the mean density can be calculated.

By studying transits it is also possible to learn about the exoplanet's atmospheric composition and weather, by comparing the combined spectrum of the star and planet with a spectrum taken when the planet passes behind the star. By subtracting the second spectrum from the first, it is possible to derive the spectrum of the planet itself.[3]

By September 2011, 179 transiting exoplanets had been detected. Nearly all of the transiting planets so far observed are quite close to their central star: among the furthest examples are the furthest examples are CoRoT-9b, a gas giant with a 95 Earth day orbit, and HD 80606b, a gas giant that orbits a star in a wide binary system. HD 80606b has the most eccentric orbit of any exoplanet yet detected and a 111 day period.

Many more transiting exoplanets are now being discovered by the Kepler spacecraft, which is designed to detect tiny dips in luminosity among 150,000 stars in the Milky Way. The most remote example yet found is Kepler 16b, which orbits 0.7 AU from its star, and takes 228 Earth days to complete one circuit.

Direct Imaging

Predictions based on radial velocity measurements are, of course, very useful precursors to follow-up observations by the most powerful telescopes. If images of exoplanets can be obtained, it is then possible to measure their orbits and extrapolate various physical characteristics. Infrared imaging will also give an idea of an object's temperature, as well as its chemical composition and physical state.

Although direct imaging would seem to be an obvious method to use, it is extremely difficult to carry out, mainly due to the overwhelming brightness of the central star. It is rather like looking for the light from a glow worm flying next to a searchlight. (Of course, the problem diminishes if the central star is a small, faint object, such as a red dwarf or a white dwarf.)

At visible wavelengths, a star like the Sun will outshine a planet like the Earth by a thousand million times. This is because planets simply reflect some of the star's light. If, however, observations are made at longer wavelengths, such as the mid-infrared, the contrast in luminosity between the star and the planet drops to a million times, because the amount of infrared light given out by the star goes down while the planet itself begins to glow.

With the advent of adaptive optics, high resolution observations in near-infrared light can be undertaken from large ground-based telescopes, whereas mid-infrared radiation is most easily viewed from space. However, in the former case, it is essential to use a coronagraph to block out the light from the star. In recent years, 25 exoplanets have been imaged directly, either through the detection of their infrared emission or from their reflected visible light.

The first exoplanet candidate to be imaged in infrared light was 2M1207b, a reddish object which orbits a young "failed star" known as a brown dwarf. Located about 41 AU from its companion, the object has about five times Jupiter's mass but it is more than 100 times fainter than the brown dwarf. This planetary mass object is now thought to be more likely to be a sub-brown dwarf than a *bona fide* planet (see below).

Since then, infrared images taken with the Keck II telescope have revealed a multiple planet system around a young, massive star called HR 8799. Images taken at different times show four exoplanets moving around their primary (see below).

In 2008, the Hubble Space Telescope was used to image a planet orbiting within the dust disk around the naked eye star Fomalhaut, located 25 light years away in the southern constellation Piscis Australis. This was the first extrasolar planet to be imaged directly by an optical telescope, demonstrating the advantages of observing from above the turbulent atmosphere. The detection was also made easier by the relatively large angular separation (more than 12°) between the planet and its star.

Fomalhaut b is estimated to be three times the mass of Jupiter. Observations taken 21 months apart by the Advanced Camera for Surveys coronagraph on Hubble showed that the planet is 17.1 billion km (119 AU) from the star, or about three times the distance of Pluto from our Sun. It completes one orbit every 872 years.

Such discoveries can cause scientists to question the current theories of planet formation. The extreme faintness of Fomalhaut b and the fact that it could not be seen in infrared light seem to remove any doubts over whether it is a brown dwarf. Since the planet must have cooled to its current temperature since the

---

[3] The first direct spectrum of a giant exoplanet was announced in January 2010. The planet orbits the bright, very young star HR 8799. Unfortunately, the spectrum was difficult to interpret, possibly because of large amounts of dust in the upper atmosphere.

Figure 14.11   The first visible-light images of an exoplanet were taken by the Hubble Space Telescope. A white dot (center) marks the location of the star Fomalhaut. The inset (bottom right) is a composite image showing the planet's position in 2004 and 2006. Fomalhaut b appears as a tiny point of light which has cleared a path along the inner edge of a vast, dusty ring. The planet lies 2.8 billion km inside the ring's inner edge and orbits 17.1 billion km from the star. The black region around Fomalhaut shows where a coronagraph blocked the star's glare. The radial streaks are scattered starlight. The red dot at lower left is a background star. (NASA, ESA, P. Kalas, J. Graham, E. Chiang, E. Kite / University of California, Berkeley, M. Clampin / Goddard Space Flight Center, M. Fitzgerald / Lawrence Livermore National Laboratory, K. Stapelfeldt and J. Krist / JPL)

system formed, 100–300 million years ago, it is likely to contain up to three Jupiter masses. Nevertheless, it is hard to explain how such a gas giant could form so far from the star.

The ultimate goal of obtaining images and spectra of terrestrial planets around nearby stars will require new instruments and techniques developed for a future generation of extremely large ground-based facilities, such as the European Extremely Large Telescope, as well as sophisticated space telescopes.

One example is a technique known as nulling interferometry. This method relies on the wave nature of light. A wave has peaks and troughs. Usually when combining light in an interferometer, the peaks are lined up with one another, boosting the signal. In nulling interferometry, however, the peaks are lined up with the troughs so that they cancel each other out and the star disappears. Planets in orbit around the star show up, however, because they are offset from the central star and their light takes different paths through the telescope system. Several ground-based nulling interferometers are being developed around the world to perfect the technique.

Astrometry

Astrometry involves precise measurement of a star's position in the sky and detection of any changes in its position over time. The technique is related to the radial-velocity detection method, and is used to detect any wobbling of a star due to the gravitational influence of one or more planets. Such observations are restricted by Earth's atmosphere, so they have to be made from space.

Various mistaken claims over the years show how difficult it is to use this technique to detect exoplanets. For instance, in 2009, it was announced that a world named VB 10b was the first distant planet to be discovered through astrometry, but this was later retracted after radial velocity measurements. However, astrometric measurements with the HST have been used to characterize a previously discovered planet around the star Gliese 876.

ESA's forthcoming Gaia spacecraft is designed to be the most precise astrometric satellite ever launched. During its detailed survey of a billion stars, Gaia is expected to find between 10,000 and 50,000 giant exoplanets, although even it will not be sufficiently sensitive to detect the tiny wobbling motion caused by an Earth-sized planet.

Gravitational Microlensing

Gravitational microlensing also has potential to find exoplanets, though the number of objects found is always likely to be modest. The effect comes into play when two stars are almost exactly aligned to an observer on Earth. When one of the stars moves in front of the other, its gravitational field acts like a lens, magnifying the light of the more distant star.

If the foreground lensing star is orbited by a planet, then that world's gravity can make a detectable contribution to the lensing effect, but only if it happens to be in the right place at the right time. Since such alignments are highly improbable, it is necessary to monitor continuously millions of stars in order to detect a handful of planetary microlensing events. The chances of detection are improved considerably by looking towards the galactic centre, where there is a huge population of background stars.

Such lensing events may last a matter of days or weeks, and more than one thousand such events have so far been observed. However, the duration of an event decreases when the mass of the lens decreases. In the case of low-mass planets, the duration may be only a few hours, so they are extremely difficult to detect unless continuous observation is possible through networks of telescopes.

The first detection of an exoplanet by this method was announced in 2004 by two international research teams – Microlensing Observations in Astrophysics (MOA) of New Zealand / Japan, and the Optical Gravitational Lensing Experiment (OGLE). The Jupiter-sized object (OGLE-2003-BLG-235 / MOA-2003-BLG-53Lb) and its red dwarf star are three times farther apart than Earth and the Sun. Together, they magnify a distant, background star some 24,000 light years away, near the Milky Way's center.

OGLE alone monitors the brightness of four hundred million stars, and by March 2011, twelve exoplanets had been found by studying OGLE and MOA data, including eight that transit their star. They include a probable brown dwarf, MOA-2007-BLG-192L, which is the lowest mass primary known to have a companion with a planetary mass, and its companion, MOA-2007-BLG-192Lb, which is one of the lowest mass exoplanets found to date.

Another unusual discovery was OGLE-2005-BLG-390Lb, a super-Earth which orbits a red dwarf once every 10 years at a distance of about 2.6 AU. The planet's large orbit and its dim parent star make its likely surface temperature a frigid −220°C.

Exoplanet Missions

Exoplanet science is advancing rapidly with the launch of space-based missions designed to search for such objects. The first of these was the European CoRoT (Convection, Rotation and planetary Transits) spacecraft, which was launched on December 27, 2006. CoRoT carries a 27 cm-diameter telescope designed to detect tiny changes in the brightness of nearby stars. The light is

Figure 14.13 The CoRoT space observatory is designed to study the internal structure of stars and seek out planets orbiting nearby stars. Launched into a 900 km polar orbit around Earth, its mission was to study slight variations in the light output of about 120,000 stars. From this orbit it can observe continuously the same area of the sky for 150 days without its view being obstructed by Earth or the Sun. The variations caused by internal stellar vibrations provide data on their internal structure, age and composition – stellar seismology. CoRoT also detects a slight dimming caused by transiting planets. (CNES)

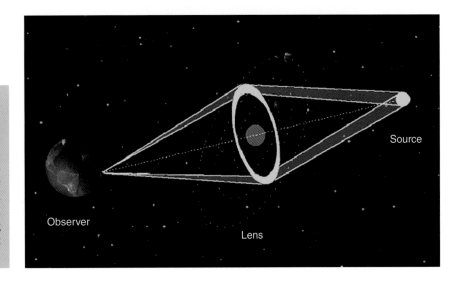

Figure 14.12 Gravitational microlensing occurs if the light rays from a star pass sufficiently close to a massive foreground object (e.g. another star, so that their path is bent, or lensed to create multiple images). If the foreground star has a planetary companion, additional lensing may occur, producing another small change in the light curve. (STScI)

Figure 14.14 Kepler is designed to survey a single star field in the Cygnus-Lyra region of the Milky Way. The single instrument, a photometer, has a 0.95 meter aperture and a 1.4 meter primary mirror. It has an array of 42 CCDs and a very large field of view (105 deg²). It stares at the same star field, continuously monitoring the brightness of more than 100,000 stars over more than three years, in a search for transiting planets. (NASA)

fed into a special two-part camera. One half of the camera is designed to look for planets; the other half is optimized to detect the subtle variations in a star's light caused by sound waves rippling across the surface. By September 2011, 21 exoplanets had been discovered from the CoRoT data and confirmed by ground-based follow-up campaigns.

These included CoRoT-7b, one of the smallest exoplanets ever found, with about 5 Earth masses. Its density has also been calculated to be high (about 5.6 g/cm³), showing that CoRoT-7b is a rocky planet, just like Earth. This made it the first rocky exoplanet to be confirmed. A second planet has also been found in the CoRoT-7 system. Known as CoRoT-7c, it is another super-Earth of about 13 Earth masses.

Many more discoveries are coming from NASA's Kepler mission, launched in March 2009 into an Earth-trailing heliocentric orbit. In this orbit, it slowly drifts away from our planet, reaching a distance of up to 0.5 AU after four years. Unlike CoRoT, this is a dedicated, space-based transit survey designed to hunt for planets. It uses a special 1.4 meter telescope, called a photometer, to measure small changes in brightness caused by exoplanet transits.

Equipped with a 95-megapixel camera, the largest ever launched into space, Kepler can detect changes in a star's brightness of only twenty parts per million. The spacecraft will spend at least 3.5 years simultaneously measuring the variations in brightness of more than 100,000 stars every thirty minutes, and should be capable of detecting dozens, or even hundreds, of Earth-sized planets. The large sample of stars should also help to determine

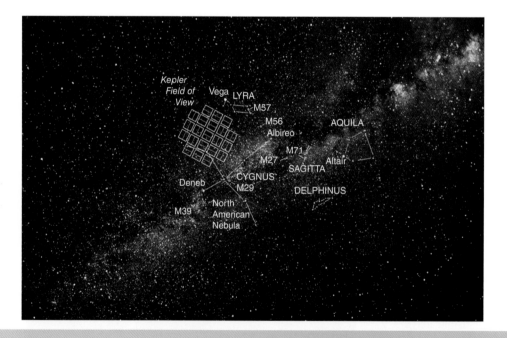

Figure 14.15 Since exoplanet transits are very brief, Kepler must monitor the brightness of all the stars in its field of view at least once every few hours. This continuous monitoring of the targeted stars means that the field of view (FOV) must be clearly visible throughout the year. To avoid the Sun, the telescope points well away from the ecliptic plane. In order to include the largest possible number of stars in the FOV, a region along the Cygnus arm of our Galaxy was selected. The squares show the FOV of each of Kepler's 21 CCD camera modules. Each covers 5 deg². (NASA – Carter Roberts)

the percentage of exoplanets in or near the habitable zones around many different types of stars, and make it possible to compare the sizes and shapes of the orbits of these planets. Within its first four months of operation, Kepler detected 1235 candidate planets, of which 68 were approximately Earth-sized. Of the 54 candidates found in habitable zones, one was smaller than Earth, and four others were slightly larger.

While Kepler is capable of determining the frequency of terrestrial planets within 1 AU of Sun-like stars, more advanced space missions could pave the way for future discoveries. One mission studied by NASA was SIM Lite (formerly called the Space Interferometry Mission), which would determine the positions and distances of stars several hundred times more accurately than any previous program. As a result, SIM would have been able to determine the distances to stars throughout the Galaxy and detect Earth-sized planets around nearby stars..

This breakthrough in capabilities was to have been made possible by the use of optical interferometry. By combining light from two or more telescopes up to 6 meters apart, as if they were pieces of a single, much larger mirror, this technique would eventually lead to the development of telescopes powerful enough to image Earth-like exoplanets and determine whether these planets sustain life as we know it. However, development of the mission ceased in 2010, when funding was withdrawn.

One recent proposal is the Pupil-mapping Exoplanet Coronagraphic Observer (PECO) mission. This concept uses a coronagraphic, 1.4 meter space-based telescope to image the habitable zones of about twenty known F, G, K stars with sufficient sensitivity to detect and characterize Earth-like planets at optical wavelengths. For exoplanet characterization, PECO would acquire narrow field images simultaneously in sixteen spectral bands over wavelengths from 0.4 to 0.9 micron, utilizing all available photons for maximum wavefront sensing and sensitivity for imaging and spectroscopy.

Another mission studied by NASA was the Terrestrial Planet Finder (TPF). TPF was proposed as two separate and complementary observatories: a coronagraph operating at visible wavelengths and a large-baseline interferometer operating in the infrared.

The Terrestrial Planet Finder Coronagraph (TPF-C) would be a moderate-sized, visible light telescope. A central disk and other specialized techniques would block the glare of a star, allowing detection and characterization of dimmer planets around it. The Terrestrial Planet Finder Interferometer (TPF-I) would consist of multiple spacecraft carrying 3–4 meter infrared telescopes flying in precise formation. Using interferometry to combine the infrared radiation gathered by flotilla of telescopes, it will be possible to simulate a much larger telescope and thereby detect and study individual planets orbiting a parent star observed by TPF-C as well as others that are beyond the reach of TPF-C.

They would study all aspects of exoplanets, including their formation and development in nebulas around newly forming stars, detection and characterization of the features of planets orbiting the nearest stars, and their suitability to support life.

By combining the high sensitivity of space telescopes with revolutionary imaging technologies, the TPF observatories would measure the size, temperature, and location of planets as small as the Earth in the habitable zones of distant stars. In addition, TPF's spectroscopy would detect the relative amounts of gases like carbon dioxide, water vapor, ozone and methane to find whether a planet has life, or could support it in the future.

Figure 14.16 SIM Lite was envisaged as the first space-based, long baseline Michelson interferometer designed for precision astrometry. Equipped with two visible-wavelength interferometer sensors located up to 6 m apart and an extremely stable spacecraft bus, this pathfinder mission would make the most accurate measurements yet of stellar positions and detect minor variations over many years. Development studies ceased in 2010 when funding was withdrawn. (NASA-JPL)

Figure 14.17 Terrestrial Planet Finder was envisaged as a suite of two complementary space observatories: a visible-light coronagraph and a mid-infrared interferometer comprising multiple spacecraft carrying 3–4 meter infrared telescopes and flying in formation. They would be capable of detecting and characterizing Earth-like planets around as many as 150 stars up to 45 light years away. (NASA-JPL)

Weird Worlds

Although the study of exoplanets is still in its infant stage, it is already clear that they are far from uniform. More than half of the exoplanets so far discovered are at least as massive as Jupiter and almost all of these are located within 5 AU of their star. The proportion rises considerably if the search is extended to include planets the size of Saturn or Neptune within a radius of 10 AU.

Of those so far discovered 38% are less massive than Jupiter and located within 5 AU of their star. The difficulty in detecting small, low mass planets means that the vast majority of those so far detected lie within 1 AU. The proportion rises considerably include planets the size of Saturn or Neptune, within a radius of 10 AU.

Although lower mass planets are currently in the minority, it is almost certain that they far outnumber the gas giants. However, an extrapolation of the number of low mass planets found by radial velocity shows that at least 30% of stars have at least one super-Earth companion. Furthermore, 80% of circumstellar disks contain enough material to form Earth-like rocky planets.

What can we say so far about the physical characteristics of exoplanets? All giant planets seem to be made of a gaseous envelope of hydrogen and helium – the same elements that make up our Sun – surrounding a central dense core that is probably made of compressed water (ices) and rock. This core is at the heart of the formation of giant exoplanets.

Standard planet formation theories predict that the core equals about 10 Earth masses. This works well for Uranus and Neptune, but not so well for Jupiter and Saturn. Jupiter appears to have a rather small core of a few Earth masses, while Saturn seems to have a larger core, around ten to twenty-five times the mass of the Earth.

What about exoplanets? Although they are more difficult to model, astronomers have confirmed that they are indeed made mostly of hydrogen and helium. However, according to recent studies, some of them possess surprisingly large cores, up to 100 times the mass of the Earth. Furthermore, there seems to be a correlation between how rich the star is in "heavy" elements, such as iron, silicate and oxygen, and the mass of the planetary cores. This supports the theory that such cores grow by the sticking together of grain particles and subsequent accretion, but the process may be more effective than previously thought.

The range of densities observed to date is also startling, varying by more than a factor of twenty. This suggests that planetary systems may form, as well as evolve, in different ways. Our experience is already broadening far beyond the three broad categories of planet that occur in the Solar System: those made mostly of water ice; those composed largely of iron, and those made up mainly of silicates. The menagerie of exoplanets already contains everything from gigantic balls of gas that would float in water, like TrES 4, to dense rocks that would sink like stones, for example, CoRoT-7b (see below).

Modeling of all the possible exoplanets results in some very weird worlds. The models take into account how the size of a planet with a given mass varies according to its composition. This relationship is not as straightforward as might be expected because some substances can be compressed more easily than others.

One exotic possibility is a planet composed mainly of carbon monoxide. This could form in a disk around a star that is stealing material from a nearby white dwarf, the remnant of a Sun-like star that is made mostly of carbon and oxygen.

Another extreme environment associated with a super-dense neutron star could result in a helium planet, which started life as

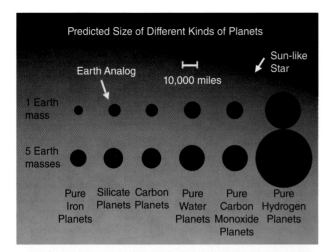

Figure 14.18 If a planet transits the face of its star, it is possible to calculate its diameter, mass and density. However, the composition of planets with similar masses may vary considerably. Hot gas giants may also become bloated if they orbit close to their star. This diagram shows the predicted sizes of transiting planets according to their composition. (Marc Kuchner/NASA-GSFC)

Figure 14.19 HD 80606b has the most eccentric orbit yet found for an exoplanet. During its 111 day orbit, it travels along a comet-like path that takes it from 0.85 AU to 0.03 AU from its star. (NASA-JPL)

a binary companion but lost all of its outer layers to the overwhelming gravitational attraction of the nearby stellar cinder.

Carbon planets, with more than 50% of their mass in the form of carbon compounds such as graphite, could also potentially form under certain circumstances, for example a circumstellar disk that is rich in carbon grains and methane, like the disk around Beta Pictoris. This has even led to speculation that planets formed in such environments might be covered by tar, with mountains made of diamonds.

Exoplanets can differ drastically from the worlds of the Solar System in other unexpected ways. Although many comets, plus Triton and many smaller, captured satellites, follow retrograde orbits, all of the major planets orbit in the same direction as the rotation of the Sun. However, at least nine exoplanets have now been found to orbit the "wrong way" around their star.

To account for these retrograde hot Jupiters orbiting close to their stars, it has been suggested that their orbits moved slowly inward, over hundreds of millions of years, due to a gravitational tug-of-war with more distant planetary or stellar companions. These disturbances would move a planet into a tilted, elongated orbit that would cause it to experience tidal friction, losing energy every time it swung close to the star. It would eventually enter a near circular, but randomly tilted, orbit close to the star. This theory is supported by the fact that more than half of all known hot Jupiters have orbits that are misaligned with the rotation axis of their stars.

Similarly, although comets, together with some asteroids and small satellites, have highly eccentric orbits, the planets follow near-circular orbits around the Sun, and all of them orbit close to the plane of the ecliptic. Numerous exceptions to this are now being found in the exoplanet community. They include XO-3, WASP-14b and, CoRot-1b. The most unusual example is HD

80606b, which follows a highly eccentric orbit that is inclined to the star's rotational axis by about 89° (see below).

Hot Jupiters

Partly due to the limits of current technology, most of the exoplanets so far discovered lie close to their central star and are sufficiently massive to cause a noticeable wobble in that star's motion. This combination of factors has inevitably led to a bias in the population of known exoplanets: the vast majority of them are gas giants, many of them orbiting closer to their star than Mercury's orbit around the Sun.

Apart from the fact that they are very large and massive, sometimes dwarfing Jupiter, and many of them zip around their orbits in a matter of days, little is known about the majority of hot Jupiters. However, it is clear that those orbiting very close to their star are tidally locked, so they always keep the same face toward the star. Their proximity to their star also means that they are blasted by thousands of times more energy per second than Jupiter, so their surface temperatures soar to hundreds, or even thousands of degrees.

Some hot Jupiters follow surprisingly eccentric orbits, very different from the well behaved, almost circular orbits of the Sun's retinue. As already mentioned, the most extreme example is HD 80606b, a planet four times the mass of Jupiter which follows a path that is almost as elongated as that of Halley's comet and markedly inclined to the star's equator.

During its 111 day orbit, HD 80606b travels as far out as 0.85 AU (almost equal to Earth's distance from the Sun) before swinging back to a mere 0.03 AU. An observer sitting on its cloud tops would see the star's apparent size increase almost 1000 times in only a few weeks. At the same time, its temperature rockets from about 100°C to 1700°C as the planet rushes towards its star. The extreme temperature swing indicates that its outer atmosphere must absorb and lose heat very quickly. It may even sprout a giant, comet-like tail (see below).

The driving force behind this unusual arrangement may be a binary companion star, known as HD 80607. It seems likely that the nearby star caused long-term gravitational perturbations to the planet, not only stretching its orbit but also causing it to be tilted.

With such a massive bully sweeping through the star's inner system, any smaller planets would be thrown into chaotic orbits, and possibly even ejected from the system altogether. Under such conditions, it seems unlikely that any habitable worlds will exist around HD 80606.

Another consequence of the elliptical orbit is that, unlike hot Jupiters which remain in tight orbits around their stars, HD 80606b does not always keep the same face towards its star. Instead, it rotates around its axis roughly every 34 hours. This means that it experiences the most extreme temperatures swings on any known planet. Observations with the Spitzer Space Observatory, made when HD 80606b disappeared behind its star around the time of periastron (closest approach), show that the planet's temperature rose from about 230°C to 1230°C in just six hours.

Although simple weather patterns have been detected on HD 80606b, little is known about the atmospheres and appearance of the so-called "hot Jupiters." However, some interesting – and unusual – discoveries have been made by space observatories. For example, Spitzer has observed HD 189733b, a transiting planet, which is located 60 light years away in the constellation Vulpecula. The gas giant crosses in front of and behind its star every 2.2 days.

Observations of longitudinal surface temperature variations on HD 189733b have shown that there is a fairly modest difference between the day and night hemispheres: The temperatures range from 650°C on the dark side to 930°C on the sunlit side. Since the planet's overall temperature variation is much less than predicted by numerical models, it is thought that winds reaching speeds of 9600 km/h or more must be spreading heat from its permanently sunlit side around to its night side.

HD 189733b has a warm spot 30° east of the substellar point, the point directly below the star. Assuming the planet is tidally locked to its parent star, this implies that fierce winds are blowing eastward. Direct heating from the star combined with the rapid delivery of hot gas by the hurricane force wind results in a particularly warm region.

The redistribution of heat observed on HD 189733b does not apply to all giant exoplanets. One example is another transiting planet known as HD 149026b, located 279 light years away in the constellation Hercules. Its diameter is thought to be similar to Saturn's, but with a large, dense core suspected to be seventy to ninety times the mass of Earth. It speeds around its star once every 2.9 days.

HD 149026b is one of the hottest exoplanets yet discovered, with a surface temperature of 2038°C – even hotter than some

Figure 14.20 These computer-generated images chart the development of thermally driven storm patterns on the night side of the highly eccentric exoplanet HD 80606b. The simulations were based on measurements of heat radiating from the planet obtained by the Spitzer Space Telescope. The six frames are evenly spaced in time, between 4.4–8.9 days after closest approach. The blue glow of the crescent is starlight that has been scattered and reflected by the planet. The blue color is mainly caused by absorption of red light by sodium and potassium in its atmosphere. The night side appears reddish orange as it glows with its own internal heat. (NASA/JPL-Caltech, J. Langton/UC Santa Cruz)

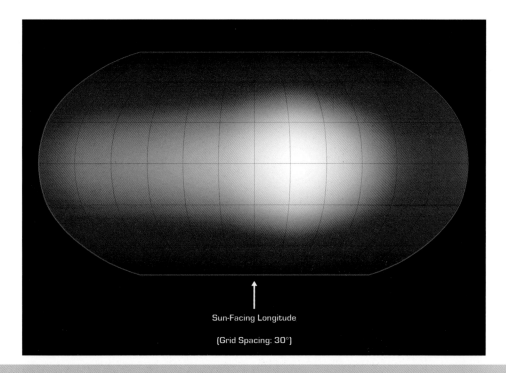

Figure 14.21 Jupiter-sized HD 189733b orbits close to a star 63 light years from Earth. Like other hot Jupiters, its rotation is tidally locked, so one side always faces the star during each 2.2 day orbit. Infrared data from the Spitzer Space Telescope enabled its surface temperature variations to be mapped – the first map ever made of an exoplanet. The brighter colors show higher temperatures. The hottest region is about 30° to the east (right) of the substellar point, evidence that fierce, planet-circling winds transfer heat toward the east. The temperatures vary from about 930–650°C. (Heather Knutson/Harvard-Smithsonian CfA et al., NASA/ JPL-Caltech)

low mass stars. The temperature was calculated by observing the drop in infrared light that occurs as it moves behind its star. The planet's extreme temperature suggests that its heat is not being spread around, so that the day side is very hot, and the night side is much colder. One explanation is that the planet's surface is blacker than charcoal, which is almost unprecedented for planets. In this case, it would reflect almost no starlight, instead absorbing all of the radiation that reaches it. HD 149026b would glow a dull red, like an ember in space.

The spectra of a handful of exoplanets have been measured, and it seems that dark gas giants may not be unusual. The infrared spectrum of HD 189733b, for example, shows no evidence of common molecules, like water, methane, or carbon dioxide. Indeed, the spectrum is flat, with no molecular species that can be detected. Since planet formation presumably works the same way everywhere, and the gases in question should be just as abundant as they are in our Solar System, astronomers speculate that something is hiding the molecules.

One clue comes from the spectrum of one of the most observed hot Jupiters. HD 209458b was the first extrasolar world discovered to transit a star, and the first to have its atmosphere analyzed. The planet is located 150 light years from Earth in the constellation Pegasus. It lies about 7.5 million km from its star and completes one orbit every 3.5 days.

The spectrum shows an unexpected peak at 9.65 microns, which has been attributed to tiny grains of silicate dust. Silicates form rocks on Earth, but on scorching hot worlds they may exist as sand-like grains that occur in high level clouds.

Figure 14.22 The infrared spectra of some hot Jupiters, such as HD 189733b, appear to be missing common molecules like water and methane. Astronomers speculate that these molecules are present, but hidden behind a high layer of dark, silicate clouds. (David A. Aguilar/CfA)

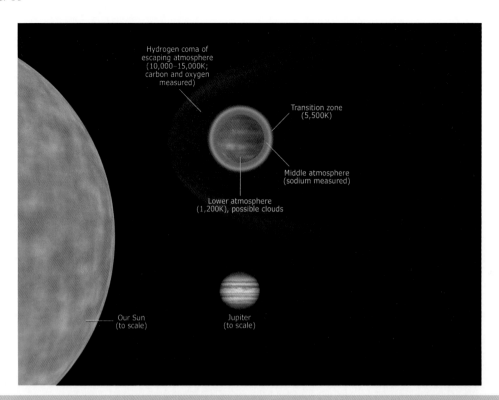

Figure 14.23 Hot exoplanet HD 209458b may be surrounded by an extended envelope of escaping gas. In 2003 the Hubble Space Telescope showed that its hydrogen-rich outer atmosphere appears to be evaporating in the extreme heat from its nearby star. Hubble later observed the first signs of oxygen and carbon in an exoplanet's atmosphere. (ESA, Alfred Vidal-Madjar/Institut d'Astrophysique de Paris, CNRS, France)

Observations with the Hubble Space Telescope have also revealed individual elements, such as sodium, oxygen, carbon and hydrogen, in the upper atmosphere, where molecules like water and methane break apart. This was achieved by measuring changes in the light from the star as the planet passed in front of it. The observations indicated less sodium than predicted, again supporting the presence of high clouds. It seems that some hot Jupiters are cloaked with dark silicate clouds, making them blacker than any planet in our Solar System.

However, the most striking characteristic of HD 209458b is the possible presence of an elongated envelope of escaping gas. If confirmed, it will be the first ever detected. HST observations made in 2003 revealed evidence of a cloud of evaporating hydrogen, resembling a comet's tail. Later studies indicated that oxygen and carbon atoms are also being swept up from the lower atmosphere with the flow of the escaping atomic hydrogen, a process called atmospheric blow off.

This "blow-off" effect is explained by the fact that the planet orbits only seven million km from its yellow, Sun-like star. Its atmosphere is heated to about 1000°C by intense ultraviolet radiation, inflating it like a balloon. Under these scorching conditions, the gas is streaming into space at a speed of more than 35,000 km/h. The amount of hydrogen gas escaping HD 209458b is estimated to be at least 10,000 tonnes per second, and the hydrogen tail is 200,000 km long.[4]

Despite this rapid erosion, HD 209458b is so massive that is expected to survive for more than 5 billion years. However, this remarkably powerful evaporation mechanism may, in some cases, give rise to a strange new class of exoplanets, unlike anything seen in our Solar System: solid, remnant cores of evaporated gas giants, orbiting even closer to their parent star than HD 209458b.

Close-in planets may also be disturbed and disrupted in other ways. WASP-12b orbits a Sun-like star in the constellation Auriga at extremely close range – about two million km or seventy-five times closer than Earth is to the Sun – once every 26 hours. It is also larger than expected. Its mass is estimated to be almost 50% larger than Jupiter's and its diameter 80% larger, giving it six times Jupiter's volume. It is also unusually hot, with a daytime temperature of 2330°C.

Why is it so big? Part of the reason is that the planet is being so strongly heated by its star. However, powerful tidal forces acting on the planet also play a part. These cause the planet's shape to be distorted into something resembling a rugby ball or American football. These tides not only mould the shape of WASP-12b, but they also create friction in its interior. The friction produces heat, which causes the planet to expand. There are also hints that a

[4]The spectral evidence is open to another interpretation. If the exoplanet has a strong magnetic field, it may be able to deflect the stellar wind and trap enough ionized gas to produce the observed spectral signature. In such a case, the planet would exhibit spectacular auroras, where particles spiral down towards its magnetic poles.

Figure 14.24 Hot Jupiter WASP-12b is being destroyed by strong heating and tidal forces which are causing the planet to balloon outward. Material is being stripped away from the outer atmosphere by the star's gravity, creating a disk of material that is spiraling inward toward the star. This rapid rate of erosion will eventually destroy the planet's atmosphere. (Kavli Institute)

second planet may be orbiting nearby. This close neighbor could increase the eccentricity of WASP-12b's orbit, thus enhancing the star's tidal effects.

Observations also show that the planet has ballooned outward so much that it can no longer withstand the star's gravity. Material is being stripped off WASP-12b at a rate of six billion metric tons per second. At this rate, the planet's atmosphere will be completely destroyed by its host star in about ten million years. Instead of falling directly onto the star, the lost material forms a disk around it and slowly spirals inwards.

Wasp-12b is also unique because its spectrum shows the presence of more carbon than oxygen. Not only does its atmosphere contain gases such as methane and carbon monoxide, but the planet itself may consist of pure carbon rocks, like diamond or graphite, as well as compounds such as tar.

Another mysterious, puffed up planet is TrES-4, a Jupiter mass world that was discovered by the Trans-atlantic Exoplanet Survey. TrES-4 is one of a family of planets, which orbits a star about the same age as our Sun. However, its greater mass means that it has evolved much faster. Having exhausted all of the hydrogen fuel in its core, the star has grown into a subgiant and is on its way to becoming a huge, cool, red giant.

TrES-4 orbits the star in 3.5 days at a distance of about 7.2 million km, so its temperature is about 1300°C. It is about 70% bigger than Jupiter, but considerably less massive. Its mean density is only about 0.2 grams per cubic centimeter, or about the density of balsa wood. This makes it the lowest density planet found to date. The planet's relatively weak pull on its upper atmosphere may allow some of the atmosphere to escape.

The reason why TrES-4 is so bloated is unclear. The planet's proximity to its star cannot fully account for the inflation, since

Figure 14.25 An artist's impression of V 391 Pegasi about 100 million years ago, when the star was at its maximum expansion as a red giant and close to explosive helium ignition, the so-called "helium flash." Against all the odds, a giant gas planet, also shown, survived this fiery furnace. (HELAS/Mark Garlick)

it is not as close or as hot as WASP-12b. One possibility is that internal heat may play a part. If the hydrogen and helium separate, with the heavier helium sinking into the core, this contraction could release gravitational energy as heat. However, it is not clear why this would happen to some giant planets and not others. Planets such as TrES-4 are testing current theories of planet formation, since they are larger, relative to their mass, than current models of superheated giant planets can presently explain.

The future of TrES-4 is uncertain. In less than a billion years, the swelling star will engulf the planet. However, despite the odds, there is a chance that it will survive the fiery furnace. The planet V 391 Pegasi b provides a precedent by being the first planet to be discovered around a star that has gone through its red giant phase. This escape act occurred despite an orbital distance of only 1.7 AU. With an age of about ten billion years, it is one of the oldest planets ever discovered.

Another exotic hot Jupiter is the transiting planet WASP-18b, which is about ten times the mass of Jupiter. It orbits its star in 0.94 days, only the second hot Jupiter confirmed to have an orbital period of less than a day. Its large size and close proximity (three stellar radii) should result in powerful tidal interactions between the planet and star. As they rotate, the resulting bulges and torques should cause WASP-18b to spiral inwards, causing it to be engulfed and destroyed by its star in less than a million years.

On the other hand, calculations indicate that that the star is about a billion years old. Since stars and their planets are believed to form at the same time, WASP-18b ought to be the same age. One rather unlikely possibility is that the exoplanet has been discovered just as it is about to be swallowed by the star. Alternatively, perhaps the star is very bad at dissipating the tidal energy between it and the planet. If confirmed, the latter explanation would significantly increase WASP-18b's lifetime, but it would force astronomers to rethink their understanding of tidal interactions in planetary systems and how solar systems evolve.

Another explanation is that WASP-18b may have interacted with a neighboring planet after reaching full size, when it was still relatively far away from the star. Such a process might have caused WASP-18b to migrate closer to the star. This seems quite feasible, since none of the hot Jupiters are likely to have been born in their current locations, and inward migration of many planets almost certainly occurred during the evolution of the young planetary systems (including ours). The mystery may be solved within a few years. If WASP-18b is spiraling rapidly inwards, the effects should be evident within a decade.

Yet another weird exoplanet is WASP-17b, which travels the "wrong" way around its host star.[5] The first sign that WASP-17b was unusual was its large size. Although it has only half the mass of Jupiter, it is bloated to nearly twice Jupiter's size, making it the largest planet known and the planet with the lowest known density. Although it weighs in at 1.6 Saturn masses, its density is only between 6% and 14% that of Jupiter.

Subsequent observations have shown that it has a retrograde orbit. Since planets form out of the same swirling gas cloud that creates a star, they are expected to orbit in the same direction that the star spins. The likely explanation is that WASP-17b was involved in a near collision with another planet early in its history. Ejected into a highly elliptical, retrograde orbit, it would have been subjected to intense tides. Tidal compression and stretching would have heated the gas giant planet so that it became hugely bloated. As a consequence, its density is extremely low, comparable to expanded polystyrene.

From the examples mentioned above, it should also be abundantly clear that many hot Jupiters are undergoing rapid changes, involving factors such extreme heating, stellar winds, mass loss tidal forces and orbital evolution. There is still a great deal to be learned about these mysterious and exotic worlds.

Super-Earths

As instruments and observational techniques have improved, so the sizes and masses of the newly found exoplanets have been falling. The list now includes several dozen with masses similar to Neptune's, some of which lie close to their star, and a handful weighing 5 Earth masses or less. It seems only a matter of time before the first discoveries of exoplanets that are similar in size, or mass, or both, to the Earth. Indeed, as our inventory of alien worlds grows, the smaller members of the family are expected to far outnumber their larger brethren.

Meanwhile, our current instruments are enabling us to find a class of planet that does not exist in our Solar System – super-Earths. For the sake of convenience, super-Earths can be defined as planets with 1.5 to 10 Earth masses. This approximate upper limit is based on the idea that more massive rocky planets would pull in large amounts of gas from the protoplanetary disk in which they were born, so becoming inflated gas giants like Jupiter and Saturn.

There is, however, no reason to believe that all of the exoplanets in this category are alike. At the heavier end, many of the planets may be mini versions of Uranus and Neptune, with small rocky cores and mantles of pressurized ices topped by deep atmospheres of hydrogen and helium. Moving down the scales, a world of 5 Earth masses could have a rock-iron core of 3 Earth masses surrounded by a deep atmosphere, or it could be a solid sphere with no appreciable atmosphere at all.

The best known example of a super-Earth is CoRoT-7b, discovered by the French CoRoT spacecraft in 2009. CoRoT-7b was the first confirmed rocky exoplanet. It is also one of the smallest known exoplanets. Although it has about 5 Earth masses, its diameter is about 70% greater than that of Earth, so its density is quite similar to our planet's, suggesting a solid, rocky world. It travels around a star which is slightly smaller and cooler than our Sun, with an age of about 1.5 billion years.

CoRoT-7b is one of the closest known exoplanets to its star, which also makes it one of the fastest moving. Located only 2.5 million km away, it orbits its star once every 20.4 hours, zipping around at a speed of more than 750,000 km/h. CoRoT-7b is so close that the temperature on its day hemisphere may soar above 2000°C, plunging to −200°C on its night side. Theoretical models suggest that it may have oceans of molten lava on its surface.

With such a high dayside temperature, any rocky surface facing the star must be molten, and the planet would be unable to retain anything more than a tenuous atmosphere, even one of vaporized rock. One possibility is that CoRot-7b weighed in at 100 Earth masses – about the same as Saturn – when it first formed, some 50% further from its star than it is now. After a history of mass loss and migration closer to its star, it could now represent a new class of planet – an evaporated remnant core.

Observations with the High Accuracy Radial velocity Planet Searcher (HARPS) spectrograph attached to the 3.6 m telescope at the La Silla Observatory in Chile, indicate that a second super-Earth orbits slightly further away than CoRoT-7b. Designated CoRoT-7c, it completes one circuit in 3 days 17 hours, and has a mass about eight times that of Earth. Unlike CoRoT-7b, it does not pass in front of its star as seen from Earth, so astronomers cannot measure its radius and determine its density.

GJ 1214b is another transiting super-Earth, but although its mass is similar to CoRoT-7b's (about six times that of Earth), its diameter is 2.7 times greater than Earth's. The larger diameter

---

[5] **WASP-17b and 18b were found by the Wide Area Search for Planets (WASP) consortium of UK universities. The planets were detected using an array of cameras that monitor hundreds of thousands of stars, searching for small dips in their light when a planet transits in front of them.**

Figure 14.26 CoRoT-7b was the first confirmed rocky exoplanet. Transit data show that it is also the smallest and fastest-orbiting exoplanet known. It is larger than Earth but five times more massive, so its density is similar to Earth's. It orbits 2.5 million km from its star, or 23 times closer than Mercury is to the Sun. (ESO)

suggests that the composition of the two super-Earths is very different. Whereas Corot-7b probably has a rocky core and may be covered with lava, 75% of GJ 1214b may be composed of water ice, with the remainder being made of silicon and iron. Since it orbits its star once every 38 hours at a distance of only two million kilometers, GJ 1214b must have a surface temperature of about 200°C, too hot for surface oceans. However, theoretical models suggest that it may support an atmosphere up to 200 km thick. Unfortunately, the high temperature and strong stellar winds may mean that the atmosphere will not survive for long.

At the time of writing, the lowest mass exoplanet to be discovered is probably Gliese 581e, the fourth planet to be found around the star Gliese 581. This relatively lightweight world has a mass of 1.9 to 3 Earth masses, depending on the inclination of its orbit to our line of sight. Unfortunately, since it orbits only 0.03 AU (about 4.5 million km) from its star, there is no possibility of it harboring life. Even the weak output of radiation from the red dwarf heats it to several hundred degrees Celsius.

Brown Dwarfs or Exoplanets?

As mentioned already, many exoplanets are much larger and more massive than Jupiter. This has led to some confusion over whether these gas giants should be regarded as planets or as sub-stellar brown dwarfs. The latter are objects that are not undergoing nuclear fusion in their cores, but display some stellar characteristics. With surface temperatures of between 200 and 2500°C, they appear as faint, dark red objects – just like gas giants.

One of the most extreme exoplanets yet found is CoRoT-3b, which was found during a stellar transit. This strange object is about the size of Jupiter, but more than 20 times as massive. It

takes 4 days 6 hours to orbit its star, which is slightly larger than the Sun.

Whether CoRoT-3b is a giant planet or a brown dwarf depends on how this oddity originally formed. If it was born from a disk-shaped nebula that circled the central star, then it must be classified as a planet. However, if it formed simultaneously with its neighboring star, due to the gravitational collapse of a molecular cloud, then it is a brown dwarf. Unfortunately, this question is extremely difficult to resolve, since there are few tell-tale signs in the object's orbit or physical characteristics.

What is clear is that CoRoT-3b would be the most massive exoplanet yet found, and, with a density more than twice that of lead, the densest exoplanet so far discovered. The surface gravity is over 50 times that of Earth. This density is so high because of the extreme compression of matter in the object's interior.

It is hoped that further studies may throw some light on how this heavyweight object formed and ended up so close to its star. One intriguing possibility is that it might just be the first member of a family of very massive planets that circle stars more massive than our Sun.

Further confusion can arise because isolated planetary mass objects (sometimes dubbed "planemos"), only a few times more massive than Jupiter, have been found far from any stellar companion. Some astronomers believe that these fairly cool objects were planets that were ejected from orbit around their stars. Others believe that they may form on their own through gas cloud collapse, similar to star formation. If they are tiny brown dwarfs (or sub-brown dwarfs), they probably originated as failed stars that formed during the collapse of small molecular clouds. However, like *bona fide* stars, some sub-brown dwarfs and planemos are surrounded by disks of gas and dust. It seems

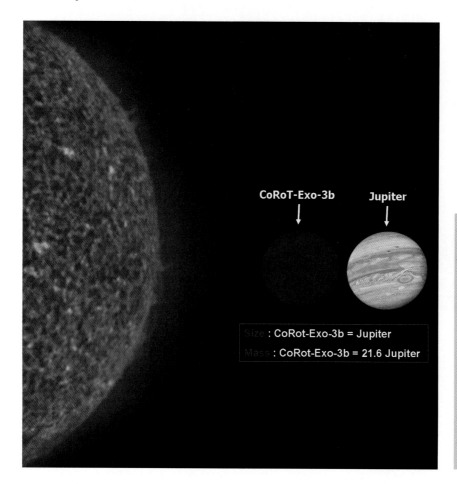

CoRoT-Exo-3b   Jupiter

Size : CoRot-Exo-3b = Jupiter
Mass : CoRot-Exo-3b = 21.6 Jupiter

Figure 14.27 Brown dwarfs are classified as gaseous, sub-stellar objects with masses of between 13 and 80 Jupiter masses. They are similar in size to Jupiter, but their temperatures vary from 200°C to more than 2000°C. Their atmospheres are rich in methane, also a major constituent of Jupiter's atmosphere. With so many characteristics in common, it is very difficult to distinguish a brown dwarf in a binary system from a giant gaseous planet. Here, CoRoT-3b is compared with the Sun and Jupiter. (OAMP)

possible that small planets or asteroids could form in these disks.

Multi-Planet Systems

Individual planets alone cannot give an accurate picture of the planetary systems that are spread throughout the Galaxy. More significant, in many ways, are the multi-planet systems, and the distribution and types of planet within those systems. Only by studying such alien systems is it possible to compare and contrast their characteristics with those of our Solar System, and to discover whether the Sun's family is rare, or even unique.

At present, the limitations of instrumentation and the time it takes for remote exoplanets to orbit their stars make it impossible to state with any certainty whether our Solar System is unusual, since it is still very difficult to detect planets far from a star. However, the data are improving all the time, so that some preliminary conclusions may be drawn.

One piece of evidence comes from a search for exoplanets that are responsible for gravitational microlensing (see above). To date, the Microlensing Follow-Up Network (MicroFUN) survey has discovered only one planetary system like our own, with two gas giants in its outer reaches. The scientists estimate that, if every star had a Solar System like ours, about six comparable systems should have been found.

This lack of discoveries leads to the conclusion that only about 15% of stars in our Galaxy host systems of planets that contain several gas giant planets in fairly remote orbits. Despite this apparently pessimistic outcome, the sheer number of stars in the Milky Way would mean that it could contain a few hundred million systems like our own.

A similar conclusion was drawn in 2007 by an international team who conducted a 3-year survey with telescopes in Arizona and Chile. Their search for giant planets far from their stars used direct detection techniques based on imagers sensitive to methane. The survey looked at 54 young nearby stars that were among the best candidates for having detectable giant Jupiter-like planets at distances beyond 5 AU. Although their technique was capable of detecting super-Jupiters around youthful Sun-like stars at 10 AU and beyond, none were found. The team concluded that "the odds are extremely slight that planets larger than 4 to 5 Jupiter masses exist at distances greater than 20 AU from these stars."

At the time of writing there were 78 known systems with more than one exoplanet. With six exceptions, all of these contained two or three planets. The exceptions were Kepler-11 and HD 10180, both with six known exoplanets, 55 Cancri, with at least five, followed by HD 160691 (mu Ara), HR 8799 and Gliese 581 with at least four.

Kepler-11, located approximately 2000 light years from Earth, is a Sun-like star which possesses the most tightly packed

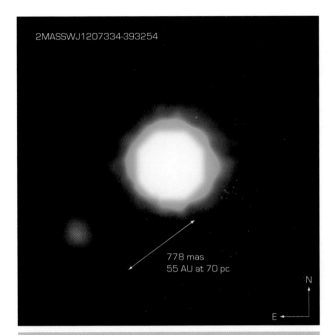

2MASSWJ1207334-393254

778 mas
55 AU at 70 pc

N

E

Figure 14.28 A composite, near-infrared image of the brown dwarf 2M1207 (center) and a planetary mass object known as 2M1207b, which weighs in at about 3–10 Jupiter masses. This was thought to be the first image of an exoplanet, but doubts over its classification as a planet have since arisen. Both objects are thought to be less than 10 million years old. The image was based on three exposures with the NACO adaptive-optics facility at the 8.2 meter VLT Yepun telescope at the ESO Paranal Observatory. (ESO)

planetary system yet discovered. All six of the planets orbit in more or less the same plane. Five of them have orbits smaller than Mercury's, with orbital periods of less than fifty days. They have 2.3 to 13.5 times the mass of the Earth. The sixth planet is larger and farther out, with an orbital period of 118 days and an undetermined mass. All of them are larger than Earth, with the largest of them being comparable in size to Uranus and Neptune.

The red dwarf star, Gliese 581, lies about 20 light years away. All of its planets are much less massive than Jupiter or Saturn. Its most notable member was thought to be Gliese 581g. Initial data indicated that it was a 3–4 Earth mass planet orbiting in the habitable zone of its star, but further studies have suggested that it may not exist. Two other planets probably lie just outside the habitable zone, one on the hot side (planet c) and one on the cold side (planet d).

Gliese 581d follows an eccentric orbit around its star in 66.8 days, at an average distance of 0.22 AU. With almost 8 Earth masses, this super-Earth could be a mixture of ice and rock. There is an outside possibility that it is habitable if it has a thick atmosphere that provides a strong greenhouse effect.

Gliese 581e flies around its star once every 3.15 days. With only 1.9 Earth masses, it is the least massive exoplanet ever detected and probably a rocky world. Also present are the Neptune-sized Gliese 581b (17 Earth masses, orbiting at 0.04 AU), and Gliese 581c, another super-Earth (5 Earth masses, orbiting at 0.07 AU). Gliese 581f (8 Earth masses at 0.76 AU) has yet to be confirmed.

Radial velocity measurements reveal that the planetary system around the Sun-like star HD 10180 contains at least six planets in almost circular orbits. Five of these are between 13 and 25 Earth masses, orbiting at distances between 0.06 and 1.4 AU (closer than Mars to their star), with periods ranging from about 6 to 600 days.

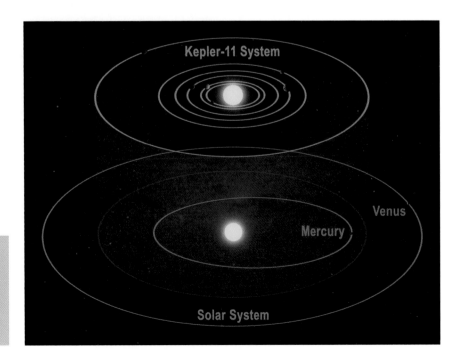

Figure 14.29 A comparison of the Kepler-11 planetary system and our Solar System. Five of the Kepler-11 planets orbit closer to their star than Mercury is to the Sun. (NASA/Tim Pyle)

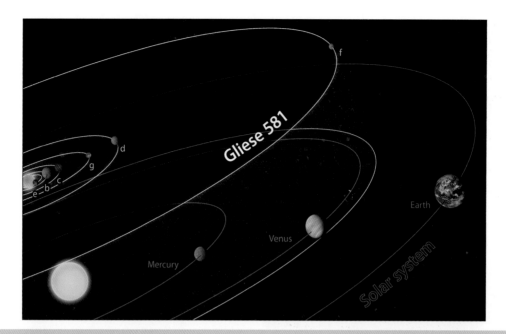

Figure 14.30   Gliese 581 is a red dwarf star with at least four known planets. It has been suggested that a planet known as Gliese 581 g lies well within the habitable zone, where liquid water oceans could exist. However, its existence has not been confirmed. Planets c and d probably lie just outside the habitable zone. The distances of the planets in the Solar System are shown for comparison. Although five of the proposed exoplanets are closer in than Mercury, the weak radiation output from the cool, low mass star means that the habitable zone is much closer than it is to our Sun. (NSF/Zina Deretsky)

There is also a Saturn-like planet (minimum 65 Earth masses) orbiting in 2200 days. An unconfirmed seventh member is very close to its star, at 0.02 AU, completing one orbit in only 1.18 Earth days. It would be the least massive exoplanet ever discovered, with about 1.4 Earth masses.

55 Cancri is located 41 light years away in the constellation Cancer and has nearly the same mass and age as our Sun. Although this system has a dominant gas giant in an orbit similar to Jupiter's, and most of the planets follow nearly circular orbits, it has some significant differences compared with the Sun's family. Three gas giant planets (55 Cancri e, b and c) orbit closer to the star than Mercury does to the Sun (see Fig. 14.31).

55 Cancri f has about 45 times the mass of Earth and may be similar to Saturn in composition and appearance. It completes one orbit every 260 days, at a distance of about 116.7 million km (slightly closer than Earth to our Sun). This places the planet inside the habitable zone, where the temperature would permit liquid oceans to exist.

55 Cancri d is four times the mass of Jupiter and completes one orbit every 14 years at a distance of approximately 867.6 million km, about the same distance as Jupiter from the Sun.

HD 160691, also known as mu Arae, is an aged, Sun-like star approximately 50 light years from Earth. The system consists of an inner, Uranus-mass planet in a tight nine-day orbit, and three massive planets, probably gas giants, at considerable distances from the star. HD 160691d was the first hot Uranus to be discovered, following a 9.5 day orbit at a distance of 0.09 AU.

HD 160691b, a two Jupiter mass object, follows an eccentric orbit similar to an asteroid that swings from near Earth's orbit to beyond the orbit of Mars. HD 160691c, comparable in mass to

Jupiter, is thought to follow an even more eccentric orbit that resembles that of a Jupiter family comet in our Solar System. The eccentric paths of these two giants are likely to cause considerable instability to the entire system. However, despite this potential for chaos, HD 160691d is intermediate in mass between Jupiter and Saturn, and follows a near-circular, Earth-like orbit. Since the star is approaching its subgiant phase and is almost twice as luminous as our Sun, this planet is outside the habitable zone.

Characterization of many exoplanet systems is still uncertain, so it is difficult to make accurate comparisons at this time. However, there is evidence of systems similar to ours that contain debris belts as well as fully grown planets.

One example is the Epsilon Eridani system. Located only 10.5 light years away, it is the closest yet found to Earth. The main sequence star is much younger than the Sun, with an estimated age of about 850 million years. It is also smaller and less massive than the Sun, with a lower metallicity. However, the star has a higher level of magnetic activity than the Sun, with a stellar wind 30 times as strong.

Its close proximity has encouraged various searches for exoplanets, but radial velocity observations over the past twenty years have only yielded tantalizing hints of a gas giant orbiting once every seven years at a distance of about 3.4 AU. The problem of confirmation arises because the star's magnetic activity and surface storms create considerable background noise, from which any potential wobbles caused by nearby planets are difficult to extract.

What is certain is that the system also includes two rocky asteroid belts and an outer icy ring, similar to the Sun's Kuiper Belt. The inner asteroid belt is a virtual twin of the belt in our Solar

Figure 14.31 A comparison between 55 Cancri (top) and our Solar System. The stars are similar in mass and age, and all of their planets have nearly circular orbits. In addition, both systems have giant planets in their outer regions. The furthest exoplanet is 55 Cancri d, four times the mass of Jupiter, which completes one orbit every 14 years, at a similar distance to Jupiter from its star. (NASA/JPL-Caltech)

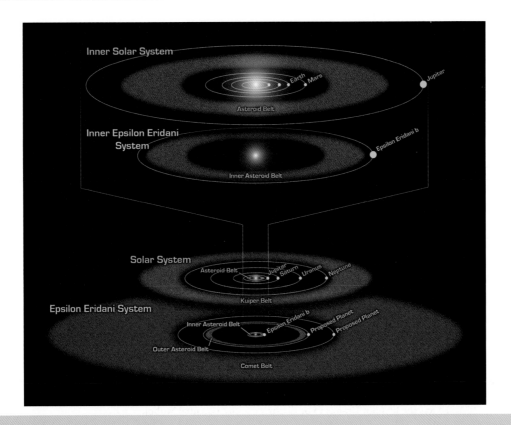

Figure 14.32 Epsilon Eridani is the closest known planetary system, about 10 light years away in the constellation Eridanus. Its central star is a younger, fainter version of the Sun. The system is quite similar to our own, containing asteroids (brown), comets (blue) and at least one planets (white dots). The system hosts two asteroid belts, one at about 3 AU (similar to our asteroid belt) and another at about 20 AU. There is also a dense comet ring at 35–90 AU. Three planets have been suggested as shepherds that shape and confine the ring material. (NASA/JPL-Caltech)

System, located at about 3 AU, while the outer asteroid belt holds 20 times more material and lies at a distance of 20 AU, comparable to the position of Uranus. The outer icy ring extends about 35 to 100 AU from Epsilon Eridani, and contains about one hundred times more material than our Kuiper Belt.

This arrangement bears a close resemblance to the way our Solar System must have appeared in its youth. Moreover, the presence of the three rings of debris implies that unseen planets confine and shape them. This indicates the presence of at least three planets with masses between those of Neptune and Jupiter.

The possible planet near the innermost ring remains a problem. Some radial velocity studies suggested that it follows a highly elliptical path around Epsilon Eridani, but such an orbit would have cleared out the inner asteroid belt long ago through gravitational disruption. If the planet exists, it must follow a much more circular orbit that keeps it just beyond the outer edge of the first asteroid belt. A second planet may lurk near the second asteroid belt, at about 20 AU, while a third could be hiding at about 35 AU, near the inner edge of Epsilon Eridani's Kuiper Belt. Future studies may detect these currently unseen worlds, as well as any terrestrial planets that may orbit inside the innermost asteroid belt.

There is much greater certainty about the multiple planet system around a young, massive star called HR 8799, which lies about 130 light years from Earth. The star has about 1.5 times the mass of the Sun and is five times more luminous, though significantly younger. Infrared observations have shown evidence for a massive disk of cold dust orbiting the star.

Four planets have also been directly imaged by the Gemini North and Keck II ground-based telescopes. Based on the age of the star and their observed luminosities, the objects formed about 60 million years ago, and they are still glowing from heat released as they contracted.

The system is highly unusual, since the planets orbit far from the star, at 14.5, 24, 38 and 68 AU, and their masses are estimated to be 9, 10, 10 and 7 Jupiter masses respectively. The furthest planet orbits just inside a disk of dusty debris, similar to that produced by the comets of our Kuiper Belt, so, in some ways, this planetary system seems to be a scaled-up version of our Solar System in orbit around a larger, brighter star.

However, the fact that such large planets exist so far from the star is hard to explain using the most popular theory of planet formation – steady gravitational coalescence within a disk which leads to the creation of a rocky core capable of drawing in large amounts of gas. Perhaps the more remote planets came about through a different process, such as the gravitational collapse of gaseous clumps. Scattering due to gravitational interactions with other massive objects is another possibility.

Exoplanet Moons

Most of the planets in our Solar System have natural satellites. Some of these moons are as large as terrestrial planets and one of them, Titan, has a dense, nitrogen-rich atmosphere.

The larger satellites, such as the Galilean moons of Jupiter, are thought to have formed from a disk of material surrounding their planet, like a mini-Solar System. A number of giant exoplanets, such as Fomalhaut b, seem to be embedded in such a disk, supporting speculation that they may also be in the process of gaining

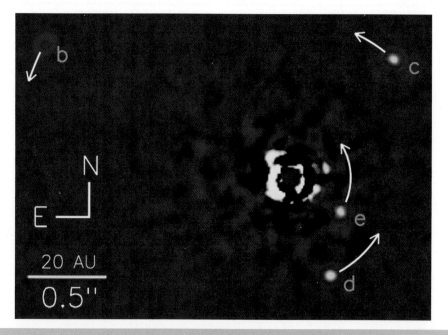

Figure 14.33 An image of the HR 8799 planetary system obtained by using adaptive optics on the Keck II telescope at near-infrared wavelengths. The four planets are seen as small red dots. Their orbits are seen almost pole-on to our line of sight. The planets are at distances of 14.5, 24, 38 and 68 AU, and their masses are 7–10 times that of Jupiter. The noise at the centre of the image is left-over light from the bright central star and the advanced image processing applied. Arrows illustrate possible planet orbital motions for the next 10 years. (NRC-HIA, C. Marois & Keck Observatory)

Box 14.2   The Drake Equation

How can we estimate the number of advanced civilizations that might exist in our Galaxy? In 1961, radio astronomer Frank Drake produced an equation that included all of the terms required to estimate the number of technological civilizations that may exist. The Drake Equation identifies specific factors thought to play a role in the development of such civilizations.

The equation is usually written: $N = R^* \cdot fp \cdot ne \cdot fl \cdot fi \cdot fc \cdot L$, where;

   N = The number of civilizations in the Milky Way galaxy whose electromagnetic
        emissions are detectable.

   R^* = The rate of formation of stars suitable for the development of intelligent life.

   fp = The fraction of those stars with planetary systems.

   ne = The number of planets in each system with an environment suitable for life.

   fl = The fraction of suitable planets on which life actually appears.

   fi = The fraction of life-bearing planets on which intelligent life emerges.

   fc = The fraction of civilizations that develop a technology that releases detectable
        signs of their existence into space.

   L = The length of time such civilizations release detectable signals into space.

Although there is no unique solution to this equation, since many of the numbers are unknown, it is a generally accepted scientific tool that is used to examine the place of humanity in the natural evolution of the Milky Way.

a family of moons. Other satellites, such as the Moon, are more likely to have coalesced from the debris cloud ejected during a collision between two large planetesimals.[6]

There is also evidence that some of the satellites in the Solar System have liquid water beneath their icy surfaces. It would, therefore, be logical to assume that a large proportion of the exoplanet community has satellites, and that some of this multitude of worlds are small and rocky, perhaps with a coating of ice.

Dozens of exoplanets have already been found in the habitable zone of their host star, where liquid water can exist, but all of them are inhospitable giants. However, there is a strong possibility that they may be orbited by large moons, which, of course, will also be in the habitable zone and, therefore, have the potential to harbor life.

Until now, there has been no way to detect such moons, but there is the potential, as measurement techniques are refined, to seek them out by looking for wobbles in the velocity of the planets they orbit.

Precise measurements of variations in the duration of multiple transits would also provide information about the exomoon's mass and its orbital separation from the planet. A moon's gravity would tug on the planet and either speed or slow its transit, depending on whether the moon leads or trails the planet. Once it is possible to calculate a satellite's mass and distance from its host planet, the likely habitability of a moon may be determined.

Meanwhile, there is an outside possibility that the Kepler spacecraft, which has the potential to detect Earth-sized objects, could eventually detect a rocky moon orbiting a gas giant (see above). Looking further ahead, the James Webb Space Telescope may be able to detect an atmosphere of gases such as water, oxygen, carbon dioxide, and methane on a transiting, Earth-like moon.

Figure 14.34   An artist's impression of a rocky exomoon, capable of supporting surface oceans, in orbit around a gas giant. (David Aguilar/CfA)

Small, dim, red dwarf stars are considered to be better targets in the hunt for habitable planets or moons, since the habitable zone is closer to the star and the probability of a transit is markedly higher. In such systems, a planet close enough to be in the habitable zone would also be close enough for the star's gravity

---

[6]Observations from the Spitzer Space Telescope suggest that satellites like our Moon, which formed after gigantic collisions, are uncommon, arising in only 5–10% of planetary systems. A study of 400 stars with ages of about 30 million years – roughly the age of our Sun when the Moon formed – found only one example of a star immersed in the telltale dust.

to slow it until one side always faces the star. One side of the planet would then be baked in constant sunlight, while the other side would freeze in constant darkness.

However, an exomoon circling such a planet would be much more hospitable. The satellite would be tidally locked to its planet, not to the star, so it would have regular – if lengthy – day-night cycles, just like Earth's Moon. Its atmosphere would moderate temperatures, and plant life would have a source of energy moon-wide. Alien moons orbiting gas giant planets may be more likely to be habitable than tidally locked Earth-sized planets or super-Earths.

Alien Life?

The only place in the Universe known to have life is planet Earth, though radio and optical telescopes are searching for signals from intelligent aliens. No one knows how life started on our world. One theory is that comets added "organic" (carbon-based) molecules to the planet's early oceans. Another idea favors chemical reactions near hot springs.

Some scientists believe simple life forms may have evolved long ago on Mars. There is even a faint possibility that life exists in a subsurface ocean on Jupiter's moon Europa. In our Galaxy, there may be billions of planets and moons capable of supporting life.

In the standard view of what makes life possible, the presence of liquid water is the key factor. In the Sun's vicinity this restricts the habitable zone to between 0.8 AU and 2 AU. Hence, our Earth is the "Goldilocks" planet in our Solar System – located where the temperature is (currently) just right – while Venus is too hot and Mars is too cold.

Water-rich worlds may be fairly common across the Galaxy. The best example so far of such an exoplanet is GJ 1214b, which circles a nearby red dwarf that is much smaller than our Sun. It orbits its star once every 1.6 days at a distance of about 15 stellar radii. Earth, by contrast, is located at about 215 solar radii. The planet is about 2.5 times the diameter of Earth and about 6.5 times its mass. Together, these parameters suggest that GJ 1214b has the same density as water. Although its surface temperature is high, between 120–280°C, the higher gravitational field, and hence atmospheric pressure, would ensure that any water is in liquid form. It seems to be very likely that the exoplanet is made entirely of liquid water, although it could have a small rocky core.[7]

The size and mass of an exoplanet are also key factors. Planets such as Mars, which form with less than about 0.5 Earth masses, are unlikely to have enough surface gravity to hold onto a life-sustaining atmosphere for very long. On the other hand, planets with at least 10 Earth masses can hold onto even the lightest and most abundant elements, hydrogen and helium, so they grow into gas giants – highly unsuitable for life as we know it.

Based on these criteria, the search for habitable planets should preferentially concentrate on finding planets with masses between about 0.5 and 10 Earth masses, or, assuming a similar density for terrestrial planets, about 0.8 to 2.2 Earth radii.

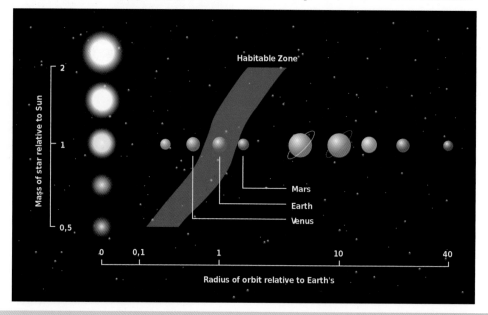

Figure 14.35   The location of the habitable zone varies according to the mass and luminosity of main sequence stars. Regions that could theoretically support life are much closer to cool, red dwarf stars than hot, massive stars. However, the size of the habitable zone around a red dwarf is correspondingly small. On the other hand, the planetary systems around small stars are also scaled down, so rocky planets may exist within their narrow habitable zones. The planets of our Solar System are shown for reference (not to scale) with Venus and Mars just outside the habitable zone around the Sun. (Wikimedia, based on data from James Kasting)

[7]Another possibility is that the planet has a small rocky core, surrounded by a massive water ocean and topped by an atmosphere of hydrogen and helium.

Of course, the habitable zone is not static and unchanging. It depends on the radiation output of the star and on the nature of the planet's atmosphere (e.g. the inventory of greenhouse gases). Stars tend to become more active and luminous as they grow older, so the zone moves outward as a star ages. Some 3.8 billion years ago, when life was first evolving on Earth, the planet would have been too cold without the plentiful presence of greenhouse gases to ameliorate the surface conditions.

Looking into the future, the Sun will continue to grow and Earth's oceans will boil away, leaving our planet a lifeless, arid desert. We have perhaps 500 million to one billion years to move house and find another habitable world. About seven billion years from now, the Sun will swell to become a red giant and Earth may even be swallowed by the bloated star. For a short time, the habitable zone will shift outward, first to Mars and then to the satellites of Jupiter and Saturn. However, this balmy period in the outer Solar System will be short-lived, and as the Sun collapses to a white dwarf, the remaining rocky planets and moons will turn into balls of ice.

The nature of the central star is a key determinant of a planet's potential habitability. Although the habitable zones around cool stars, such as red dwarfs, are very narrow, there is plenty of evidence that they can host rocky planets (*see* Gliese 581 above). On the other hand, although the habitable zones are wider around stars significantly more massive than the Sun, such environments are not very hospitable for life to develop. The stars evolve very quickly (over millions rather than billions of years), so that the habitable zone migrates quickly outward. The intense UV flux may also be a problem.

The habitability of a world may also be influenced by the density and composition of its atmosphere, which, in turn, influences the surface temperature and provides protection against ultraviolet light, X-rays, meteorites and cosmic rays.

Massive neighbors may influence the shape of a rocky planet's orbit, just as Earth's orbit periodically evolves from purely circular to slightly eccentric, giving rise to alternating warm and glacial periods. Giant planets in the outer reaches of the planetary system may also provide protection from comet impacts that can sterilize planetary surfaces or cause mass extinctions. However, worlds such as Jupiter may also lead to the formation of asteroid belts, which turn young planetary systems into shooting galleries. Giant planets in the inner regions may also scatter smaller planets or even eject them altogether from the system.

Despite the devastation caused by asteroid and comet impacts, a planetary system free of these bodies may not be the best thing for the evolution of life. Indeed, ice-rich comets may be necessary for life to emerge on terrestrial planets, since they deliver the organic material and water that are necessary for the formation of oceans and the origin and development of life. In addition, although such impacts cause mass extinctions which wipe out the dominant organisms, they clear the way for other life forms to fill the recently vacated ecological niches.

The transfer of organisms from one world to another may actually be enabled by large impacts. A comet slamming into a terrestrial planet blasts billions of tons of rock into space. A few per cent of this ejected material will escape from the planet and eventually fall onto a neighboring world, possibly seeding a new generation of life forms.

It seems inevitable that, if there were once abundant microbes on Mars, some of them would have reached Earth by hitching a ride on Martian meteorites (and vice versa). Cocooned inside a large rock, they would be screened from the worst effects of radiation and protected from burning up as they plunged into the Earth's atmosphere.

It is interesting to note that the absence of gas giants may open the door to the formation of more small, rocky planets. Hence, if

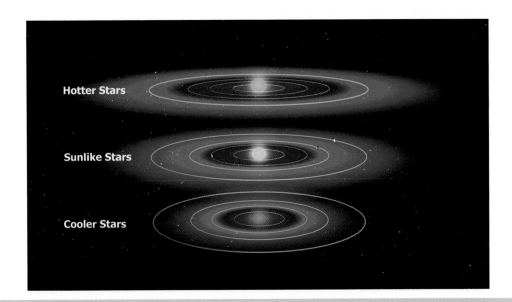

Figure 14.36 The width of the habitable zone (green) and its distance from the central star increases with the size and luminosity of the star. Small rocky planets may be able to form in the habitable zone around cool red dwarfs, despite its limited dimensions. The red area is too hot for life and the blue area is too cold. (NASA)

Jupiter and Saturn had never formed in our Solar System, three Earth-like planets might have developed instead of just one. Lacking the gas giants, icy planets like Uranus and Neptune may have formed at 5–10 AU. Such a system would be fairly asteroid-free, since most of the rocky debris would have coalesced into planets.

The presence of life is further complicated by factors such as tidal locking and internal heating. Tidal locking causes a planet to keep the same face towards the star, so that one hemisphere is boiling hot and the other freezing cold (unless a thick atmosphere and strong winds enable transfer of heat from the dayside to the dark side). Furthermore, the presence of large moons may stabilize the axis of rotation and create tidal effects in the crust and any large, surface water masses.

Internal heating within a rocky planet or satellite, either due to the decay of radioactive elements or tidal flexing, can enable the presence of subsurface water, or even oceans, where alien life could evolve without sunlight. The amino acids and other organic molecules thought to be the building blocks of life are plentiful throughout the Galaxy, as are chemical energy sources such as hydrogen and iron oxides.

It is now clear that life forms on Earth are remarkably resilient, taking hold in extreme environments, where there is no sunlight, temperatures are above the boiling point of water or below freezing, and high acidity or alkalinity occur. Some organisms can even survive exposure to ultraviolet light and high energy particles, or eke out a bare existence inside "solid" rock, deep below the surface.

This apparent hardiness has led to speculation that primitive organisms may exist on worlds where surface conditions are much harsher than those on Earth. Exoplanets with subsurface oceans may also offer potential habitable environments, even though they may orbit well outside their star's traditional habitable zone.

Gliese 581d: A Case Study

Some idea of the complexity of determining the habitability of an exoplanet can be provided by a study of Gliese 581d, one of the most Earth-like planets yet discovered (see also above). This super-Earth orbits a red dwarf, so although it is 14 times closer to its star than Earth is from the Sun, it is cool enough (between 0 and 40°C) to support surface oceans. However, other environmental conditions might be too harsh to allow complex life to appear.

Gliese 581d is tidally locked, like the Moon in our Earth-Moon system, so that one side of the planet is permanently dark. Assuming it has a thick atmosphere, strong winds may be caused by the temperature difference between the day and night hemispheres. If not, the night side would be too frigid for life to survive.

The planet's orbit is short, lasting 83.6 days, and eccentric, so it might leave and re-enter the habitable zone within a short period during each orbit. Since the planet is located at the outer edge of the habitable zone, life forms would have to tolerate reduced stellar irradiation and a very peculiar climate. However, even under these strange conditions, it might still be habitable if its atmosphere is sufficiently dense to cushion their impact.

The carbon dioxide content of the atmosphere and its persistence over long periods may also be an important constraint on habitability. An active planet, such as Earth, maintains a balance between removal of CO_2 in the atmosphere-ocean system and its release through plate tectonics. However, an aged planet that is not tectonically active may not be able to sustain a photosynthetic biosphere. Alternatively, if the atmosphere is dominated by greenhouse gases, the planet may also be uninhabitable, like Venus. Some models suggest that Gliese 581d may possess a thick blanket of CO_2, resulting in a rampant greenhouse effect that cooks the planet and makes it far too hot for life.

The nature of the central star is also significant. Gliese 581d orbits an M-type, red dwarf star. About 75% of all stars in our Galaxy belong to this class. They are very stable and can continue to burn hydrogen over extremely long periods (potentially tens of billions of years). However, as already mentioned, M stars have long been considered as poor candidates for harboring habitable worlds because planets located in their habitable zone are tidally locked, with a permanent dark side, where the frigid atmosphere may condense out irreversibly. Furthermore, in their youth, M stars undergo intense magnetic activity, resulting in violent flares that are associated with high fluxes of X-rays and extreme ultraviolet light that might erode planetary atmospheres.

The climate of exoplanets may also be regulated by the long-term evolution of the planetary orbits in the system. Mutual gravitational perturbations between different planets are present in any planetary system with more than one planet. However, computations of the orbits of the known members of the Gliese 581 system over the next 100 million years show that it appears to be dynamically stable, with only modest periodic orbital changes comparable to those that affect Earth.

If the planet's climate is stable, it may open the door for life to begin – although whether this initial step has actually taken place is another story. We will have to wait until the next generation of space telescopes is launched before the super-Earth can be probed for signs of life.

SETI

A Search for Extraterrestrial Intelligence (SETI) has been taking place for half a century, since Frank Drake began Project OZMA. The most common method is to use large radio telescopes and high speed computers to seek out unusual radio signals or messages from the blizzard of electromagnetic signals that reach our planet from space (and from sources much closer to home!).

NASA began a SETI program in 1992, but it was cancelled soon after, owing to lack of political support and funding. The program was revived in 1995 as a privately funded initiative known as Project Phoenix. Directed by the non-profit SETI Institute in California, the search used some of the most powerful instruments in the world, including the Parkes radio telescope in Australia, the Green Bank telescope in West Virginia, and Arecibo in Puerto Rico. In all, Phoenix observed about 800 stars.

Another effort, called SERENDIP (Search for Extraterrestrial Radio Emissions from Nearby Developed Intelligent Populations), is based at UC Berkeley. The data used in SERENDIP are currently obtained using the giant radio telescope in Puerto Rico. A similar system is being installed for the Southern SETI program in Argentina. The latest spectrometers can track 128 million channels simultaneously and listen instantaneously to a radio band 80 MHz wide.

Most of the SETI programs include large computers that analyze data from a radio telescope in real time. However, a lack of computing power means that none of the programs can scrutinize the data very closely for weak signals or specific signal types. In order to overcome this drawback, the UC Berkeley SETI team came up with the idea of using thousands of home computers to analyze the data streaming from Arecibo. The SETI@home initiative takes advantage of spare computing power whenever a machine switches to screen saver, and then it sends the results back to the project center.

The latest advanced observatory to take part in SETI is the Allen Telescope Array (ATA) in northern California. The ATA currently comprises 42 small dishes and may eventually expand to 350 antennas. It will then have the collecting area equivalent to a single dish 114 meters in diameter and the angular resolution of a dish 700 meters across.[8]

The ATA searches a large area near the galactic center, which contains perhaps 40 billion stars, at frequencies from 1420 to 1720 MHz. This "waterhole" frequency range is defined by the spectral lines for the dissociation products of water, the key to life as we know it. It is also a radio quiet zone, since it features the minimum background noise caused by emission from the Galaxy and by cosmic microwave background radiation. As such, it is where we can best listen for a faint whisper from across the vast expanse of interstellar space.

Radio observations are not the only technique available. In recent years, optical SETI has been introduced. This involves a dedicated search for nanosecond flashes of intense light, using instruments capable of detecting laser beacons from civilizations many light years distant. The latest OSETI program uses a 1.8 meter optical telescope in Massachusetts capable of observing the entire sky simultaneously. One thousand detectors are used to measure the incoming light nearly a billion times a second. Unlike its radio counterpart, optical SETI requires an extraterrestrial civilization to be deliberately signaling in the direction of our Solar System.

A number of efforts have been made to establish contact with extraterrestrials over the years. The most famous of these took place in October 1974, when the Arecibo radio telescope beamed a simple, coded message towards 500,000 mature stars in the globular cluster M13. There was no expectation of a quick response: M13 is so far away that we cannot expect an answer for another 50,000 years.

The message was devised by a group led by Frank Drake of Cornell University, creator of the famous Drake equation. It consisted of seven sections, encoded into 1679 on-off pulses – a multiple of the prime numbers 23 and 73. The message is arranged in a rectangle as 73 rows with 23 columns. The sections were as follows:

1. The numbers 1 to 10 in binary code.
2. The atomic numbers of the five elements that make up deoxyribonucleic acid (DNA): hydrogen, carbon, nitrogen, oxygen, and phosphorus.
3. The chemical formulas for some key biological molecules in DNA, sugar, phosphate and the nucleotides.
4. A graphic of the twisted, double helix structure of DNA.
5. An outline of a human, with the physical height of an average man (equal to 14 wavelengths of signal), flanked by the number of humans on Earth.

Figure 14.37 The Arecibo message was sent from the giant radio telescope in Puerto Rico on November 16, 1974. It was aimed at the globular star cluster M13, some 25,000 light years away. The message forms an image when translated into graphic characters and spaces. Each of the seven sections provides basic information that would inform an alien civilization about human life and the origin of the signal. (Cornell University)

---

[8] The ATA temporarily shut down in 2011, due to lack of funding, but it was reactivated after private donations covered its operational costs.

6.   A graphic of the Solar System, with the main bodies approximately to scale. Earth is displaced to highlight it.
7.   A graphic of the Arecibo radio telescope and how it works.

Some more recent attempts at interstellar communication have been more commercial than scientific. For example on October 9, 2008 a more complex four and a half hour transmission containing 501 photos, drawings and text messages was beamed toward the Gliese 581 planetary system by the RT-70 radar telescope in Evpatoria, Ukraine. It was part of a competition sponsored by a social networking site.

Meanwhile, our TV and radio broadcasts have been spreading out into space for many decades, and, weak as they are, they may one day be picked up by alien civilizations tuning in on a nearby planet.

The chances of ET finding a tiny spacecraft sent from Earth as it travels through the Galaxy are even more remote, but this has not prevented simple attempts at communication involving the four spacecraft that are heading out of the Solar System.

NASA's Pioneer 10 and 11 spacecraft both carry a plaque which shows a man and a woman in front of the spacecraft, and Earth's position in the Solar System. A ray diagram shows the location of the Sun. The two Voyager spacecraft carry a record engraved with Earth's location and playing instructions (see Fig. 14.39). The 30.5 cm, gold-plated, copper disks contain greetings in 60 languages, natural sounds, 90 minutes of music and 118 pictures of Earth and the other planets. They also contain electronic information that an advanced technological civilization could convert into diagrams and photographs.

A number of important questions arise from SETI research. Despite decades of searching, why have no signs of ET ever been found? So far, only false alarms have been detected, the most famous example being a solitary "wow!" signal detected in 1977, the most powerful unidentified transmission ever detected. Is this because such civilizations are very rare in the Galaxy, or is no one using the same technology and listening to the Universe in the same way?[9]

There is also the question of how to react if an alien signal, or message, is ever detected and decoded. How would we know whether the senders were hostile or friendly? Although the vast distances between star systems mean that interstellar travel is highly unlikely, is it wise to advertise our presence, particularly if the receivers may be more technologically advanced?

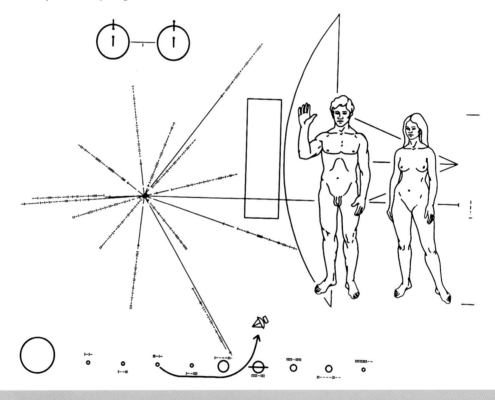

Figure 14.38   Pioneer 10 and 11 carry a 15 × 23 cm, gold-anodized aluminum plaque which carries information about their origin. The radiating lines (left) represent the positions of 14 pulsars, arranged to indicate the Sun's location. The "1-" symbols at the ends of the lines are binary numbers that represent the frequencies of these pulsars relative of that to the hydrogen atom (upper left). The human figures show Pioneer's creators. The man's hand is raised in a gesture of good will. The outline of the Pioneer spacecraft (right) shows the relative size of the humans. At the bottom are the Sun and planets with the spacecraft's trajectory. (NASA)

[9]Radio and TV transmissions, radar and laser beams have only been introduced in the last 100 years. For more than 99.9% of Earth's existence, any advanced aliens would have been unable to detect the presence of intelligent life on Earth. Since technologies may also change rapidly, there is no reason to believe that ET would be at the same stage of technological innovation as ourselves.

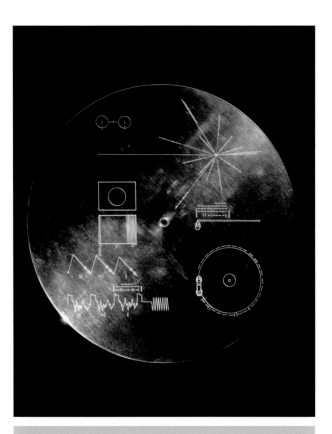

Figure 14.39 This gold-anodized aluminum cover protects the Voyager 1 and 2 "Sounds of Earth" records from micrometeorite bombardment, and includes a guide to playing the record. The explanatory diagram appears on the inner and outer surfaces, as the outside will be eroded in time. (NASA-JPL)

Questions

- Why were no exoplanets discovered until the 1990s?
- Summarize the various methods currently used to find evidence of exoplanets.
- How are exoplanet detection methods likely to improve in the future?
- Why are many of the planets so far detected classed as "hot Jupiters"?
- What is the probability of finding Earth-like planets in the future? Explain your answer.
- Explain the meaning of the term "habitable zone." Describe the factors that may influence the habitability of a planetary system.
- Explain why the Solar System may not be a typical example of a planetary system in our Galaxy.
- Select one of the known multiplanet systems and compare it with our Solar System.
- What are pulsar planets? Why was their discovery a surprise?
- "SETI programs are not a worthwhile use of government funds." Discuss.

Appendices

Appendix 1: Planetary Data (a) Planets

	Mercury	Venus	Earth	Mars	Jupiter	Saturn	Uranus	Neptune
Equatorial diam. (km)	4879	12,104	12,756	6792	142,984	120,536	51,118	49,528
Rotation period (equat.)	58 d 15 h 30 m	243 d 0 h 36 m (R)	23 h 56 m	24 h 37 m	9 h 50 m	10 h 14 m	17 h 14 m (R)	16 h 7 m
Density (water = 1)	5.43	5.24	5.52	3.91	1.33	0.69	1.29	1.64
Mass (Earth = 1)	0.055	0.814	1	0.11	317.8	95.2	14.53	17.14
Surface gravity (Earth = 1)	0.378	0.903	1	0.38	2.69	1.07	0.91	1.14
Equatorial inclination (deg)	0.1	177.3	23.5	25.2	3.1	26.7	97.8	28.3
Orbital period	87.97 d	224.7 d	365.25 d	687 d	11.86 y	29.46 y	84.01 y	164.79 y
Av. distance from Sun (million km)	57.9	108.2	149.6	227.9	778.3	1427	2871	4497.1
Orbital eccentricity	0.2	0.007	0.0167	0.093	0.048	0.056	0.046	0.0097
Av. orbital velocity (km/s)	47.88	35.02	29.8	24.1	13.06	9.65	6.81	5.43
Average temp. (°C)	350 (day), −170 (nt)	467	7	−63	−148	−178	−218	−220
Atmosphere	Potassium, sodium	Carbon dioxide	Nitrogen, oxygen	Carbon dioxide	Hydrogen, helium	Hydrogen, helium	Hydrogen, helium	Hydrogen, helium
Satellites	0	0	1	2	65	62	27	13

R = retrograde; d = days; h = hours; m = minutes

Exploring the Solar System, First Edition. Peter Bond.
© 2012 by Peter Bond. Published 2012 by Blackwell Publishing Ltd.

(b) Dwarf Planets

	Pluto	Eris	Makemake	Haumea	Ceres
Equatorial diameter (km)	2324	2326	Approx. 1500	Approx. 1960 × 1520 × 1000	909 × 975
Rotation period	6d 9h 17m (R)	3h 55m	7h 46m	3h 55m	9h 5m
Density (water = 1)	2.05	2.52	Approx. 2?	Approx. 2.6	2.08
Mass (Earth = 1)	0.0021	0.0027?	0.00067?	0.00067?	0.00015
Surface gravity (Earth = 1)	0.06	0.08?	?	?	0.027
Equatorial inclination (deg)	122	?	?	?	Approx. 3
Orbital period (years)	248.54	557.8	309.88	285.4	4.6
Av. distance from Sun (million km)	5,913.5	10,132	6,850.3	6,484	414.1
Orbital eccentricity	0.249	0.44	0.159	0.189	0.08
Orbital inclination (deg.)	17.14	44.19	28.96	28.19	10.58
Av. orbital velocity (km/s)	4.72	3.44	4.419	4.484	17.882
Average temp. (°C)	−228	−238	−240	−241	−106
Atmosphere	Nitrogen, methane?	Nitrogen, methane?	Nitrogen, methane, ethane?	?	-
Satellites	4	1	0	2	0

R = retrograde; d = days; h = hours; m = minutes

Appendix 2: Satellite Data

Earth

Satellite	Discoverer	Year of discovery	Diameter (km)	Visual magnitude	Distance (km)	Orbital period (d)	Orbital ecc.	Orbital inc. (degrees)
Moon	–	––	3475	−12.7	384,400	27.3	0.055	5.2

Mars

Satellite	Discoverer	Year of discovery	Diameter (km)	Visual magnitude	Distance (km)	Orbital period (d)	Orbital ecc.	Orbital inc. (degrees)
Phobos	A. Hall	1877	13.4 × 11.2 × 9.2	11.4	9380	0.3	0.015	1.1
Deimos	A. Hall	1877	7.5 × 6.1 × 5.2	12.5	23,460	1.3	0.0005	1.8

Jupiter

Satellite	Discoverer	Year of discovery	Diameter (km)	Visual magnitude	Distance (km)	Orbital period (d)	Orbital ecc.	Orbital inc. (degrees)
Metis	S. Synott / Voyager 2	1979	44	17.5	128,100	0.294	0.001	0.0
Adrastea	D. Jewitt, E. Danielson	1979	13 × 10 × 8	18.7	128,900	0.298	0.002	0.0
Amalthea	E. Barnard	1892	270 × 165 × 150	14.1	181,400	0.498	0.003	0.4
Thebe	S. Synott / Voyager 1	1980	116 × 98 × 84	16.0	221,900	0.68	0.018	1.1
Io	Galileo	1610	3643	5.0	421,800	1.77	0.004	0.0
Europa	Galileo	1610	3122	5.3	671,100	3.6	0.009	0.5
Ganymede	Galileo	1610	5262	4.6	1,070,400	7.2	0.001	0.2
Callisto	Galileo	1610	4821	5.7	1,882,700	16.7	0.007	0.3
Themisto	C. Kowal, E. Roemer	1975	8	21.0	7,507,000	130.0	0.242	43.1
Leda	C. Kowal	1974	18	19.5	11,165,000	240.9	0.164	27.5
Himalia	C. Perrine	1904	160	14.6	11,461,000	250.6	0.162	27.5
Lysithea	S. Nicholson	1938	38	18.3	11,717,000	259.2	0.112	28.3
Elara	C. Perrine	1905	78	16.3	11,741,000	259.6	0.217	26.6
S/2000 J11	S. Sheppard & others	2000	4	22.4	12,555,000	287.0	0.248	28.3
Carpo	S. Sheppard & others	2003	3	23.2	16,989,000	456.1	0.430	51.4
S/2003 J3	S. Sheppard & others	2003	2	23.8	18,339,885	583.9R	0.241	143.7
S/2003 J12	S. Sheppard & others	2003	1	23.9	19,002,480	533.3	0.376	145.8
Euporie	S. Sheppard & others	2001	4	23.1	19,302,000	550.7R	0.144	145.8
S/2010 J2	C. Veillet	2010	1	23.9	20,307,150	588.1R	0.307	150.4
S/2003 J18	B. Gladman & others	2003	2	23.4	20,700,000	606.3R	0.119	146.5
Orthosie	S. Sheppard & others	2001	4	23.1	20,721,000	622.6R	0.281	145.9
Euanthe	S. Sheppard & others	2001	6	22.8	20,799,000	620.6R	0.232	148.9
Thyone	S. Sheppard & others	2001	8	22.3	20,940,000	627.3R	0.229	148.5
S/2003 J16	B. Gladman & others	2003	2	23.3	21,000,000	595.4R	0.27	148.6
Mneme	S. Sheppard & others	2003	2	23.3	21,069,000	620.0R	0.227	148.6
Harpalyke	S. Sheppard & others	2000	4	22.2	21,105,000	623.3R	0.226	148.6
Hermippe	S. Sheppard & others	2001	8	22.1	21,131,000	633.9R	0.210	150.7
Praxidike	S. Sheppard & others	2000	7	21.2	21,147,000	625.3R	0.230	149.0
Thelxinoe	S. Sheppard & others	2004	2	23.5	21,162,000	628.1R	0.221	151.4
Helike	S. Sheppard & others	2003	4	22.6	21,263,000	634.8R	0.156	154.8

Satellite	Discoverer	Year of discovery	Diameter (km)	Visual magnitude	Distance (km)	Orbital period (d)	Orbital ecc.	Orbital inc. (degrees)
Iocaste	S. Sheppard & others	2000	5	21.8	21,269,000	631.5R	0.216	149.4
Ananke	S. Nicholson	1951	28	18.8	21,276,000	610.5R	0.244	148.9
S/2003 J15	S. Sheppard & others	2003	2	23.5	22,000,000	668.4R	0.110	140.8
S/2003 J9	S. Sheppard & others	2003	1	23.7	22,441,680	683.0R	0.269	164.5
S/2003 J19	B. Gladman & others	2003	2	23.7	22,800,000	701.3R	0.334	162.9
Eurydome	S. Sheppard & others	2001	6	22.7	22,865,000	717.3R	0.276	150.3
Arche	S. Sheppard & others	2002	3	22.8	22,931,000	723.9R	0.259	165.0
Autonoe	S. Sheppard & others	2001	8	22.0	23,039,000	762.7R	0.334	152.9
Pasithee	S. Sheppard & others	2001	4	23.2	23,096,000	719.5R	0.267	165.1
Herse	B. Gladman & others	2003	2	23.4	23,097,000	715.4R	0.200	164.2
Chaldene	S. Sheppard & others	2000	4	22.5	23,179,000	723.8R	0.251	165.2
Isonoe	S. Sheppard & others	2000	4	22.5	23,217,000	725.5R	0.246	165.2
Kale	S. Sheppard & others	2001	4	23.0	23,217,000	729.5R	0.264	165.1
Aitne	S. Sheppard & others	2001	6	22.7	23,231,000	730.2R	0.264	165.1
S/2003 J4	S. Sheppard & others	2003	2	23.0	23,257,920	723.2R	0.204	144.9
Erinome	S. Sheppard & others	2000	3	22.8	23,279,000	728.3R	0.266	164.9
S/2010 J1	R. Jacobson & others	2010	2	16.7	23,314,335	723.2R	0.320	163.2
Taygete	S. Sheppard & others	2000	5	21.9	23,360,000	732.2R	0.252	165.2
Carme	S. Nicholson	1951	46	17.6	23,404,000	702.3R	0.253	164.9
Sponde	S. Sheppard & others	2001	4	23.0	23,487,000	748.3R	0.312	151.0
Kalyke	S. Sheppard & others	2000	5	21.8	23,583,000	743.0R	0.245	165.2
Pasiphae	P. Melotte	1908	60	17.0	23,624,000	708.0R	0.409	151.4
Eukelade	S. Sheppard & others	2003	4	22.6	23,661,000	746.4R	0.272	165.5
Megaclite	S. Sheppard & others	2000	5	21.7	23,806,000	752.8R	0.421	152.8
Sinope	S. Nicholson	1914	38	18.1	23,939,000	724.5R	0.250	158.1
Hegemone	S. Sheppard & others	2003	3	23.2	23,947,000	739.6R	0.328	155.2
Aoede	S. Sheppard & others	2003	4	22.5	23,981,000	761.5R	0.432	158.3
Kallichore	S. Sheppard & others	2003	2	23.7	24,043,000	764.7R	0.264	165.5

continued

Satellite	Discoverer	Year of discovery	Diameter (km)	Visual magnitude	Distance (km)	Orbital period (d)	Orbital ecc.	Orbital inc. (degrees)
S/2003 J23	S. Sheppard & others	2003	2	23.6	24,055,500	732.4R	0.309	149.2
S/2003 J5	S. Sheppard & others	2003	4	22.4	24,084,180	759.7R	0.210	165.0
Callirrhoe	Spacewatch	1999	9	20.7	24,102,000	758.8R	0.283	147.1
S/2003 J10	S. Sheppard & others	2003	2	23.6	24,249,600	767.0R	0.214	164.1
Cyllene	S. Sheppard & others	2003	2	23.2	24,349,000	737.8R	0.319	149.3
Kore	S. Sheppard & others	2003	2	23.6	24,543,000	779.2R	0.325	145.0
S/2003 J2	S. Sheppard & others	2003	2	23.2	28,570,410	982.5R	0.380	151.8

Saturn

Satellite	Discoverer	Year of discovery	Diameter (km)	Visual magnitude	Distance (km)	Orbital period (d)	Orbital ecc.	Orbital inc. (degrees)
S/2009 S1	C. Porco & others / Cassini	2009	0.3	28	117,000	?	0.000	0.0
Pan	M. Showalter / Voyager 2	1990	20	19.4	133,600	0.575	0.000	0.0
Daphnis	C. Porco & others / Cassini	2005	7	24	136,500	0.59	0.000	0.0
Atlas	R. Terrile / Voyager 1	1980	37 × 34.4 × 27	19.0	137,700	0.60	0.001	0.0
Prometheus	S. Collins / Voyager 1	1980	148 × 100 × 68	15.8	139,400	0.61	0.002	0.0
Pandora	S. Collins / Voyager 1	1980	110 × 88 × 62	16.4	141,700	0.63	0.004	0.1
Epimetheus	J. Fountain & others / Voyager 1	1980	138 × 110 × 110	15.6	151,400	0.69	0.021	0.4
Janus	A. Dollfus	1966	194 × 190 × 154	14.4	151,500	0.7	0.007	0.2
Aegaeon	C. Porco & others / Cassini	2008	0.5	27	167,500	0.808	0.000	0.0
Mimas	W. Herschel	1789	418 × 392 × 383	12.8	185,600	0.9	0.021	1.6
Methone	C. Porco & others / Cassini	2004	3	23	194,300	1.0	0.001	0.0
Anthe	C. Porco & others / Cassini	2007	1	26	197,700	1.04	0.001	0.1
Pallene	C. Porco & others / Cassini	2004	4	23	212,300	1.1	0.004	0.2

Satellite	Discoverer	Year of discovery	Diameter (km)	Visual magnitude	Distance (km)	Orbital period (d)	Orbital ecc.	Orbital inc. (degrees)
Enceladus	W. Herschel	1789	512 × 494 × 489	11.8	238,100	1.4	0.000	0.0
Calypso	D. Pascu & others	1980	30 × 16 × 16	18.7	294,700	1.9	0.001	1.5
Telesto	B. Smith & others/ Voyager 1	1980	30 × 25 × 15	18.5	294,700	1.9	0.001	1.2
Tethys	G. Cassini	1684	1060	10.2	294,700	1.9	0.000	0.17
Dione	G. Cassini	1684	1118	10.4	377,400	2.7	0.000	0.0
Helene	P. Laques, J. Lacacheux	1980	36 × 32 × 30	18.4	377,400	2.7	0.000	0.2
Polydeuces	C. Porco & others / Cassini	2004	13 × 4	23.0	377,400	2.7	0.018	0.2
Rhea	G. Cassini	1672	1528	9.6	527,100	4.5	0.001	0.3
Titan	C. Huygens	1655	5150	8.4	1,221,900	15.95	0.029	1.6
Hyperion	W. Bond, W. Lassell	1848	360 × 280 × 225	14.4	1,464,100	21.3	0.018	0.6
Iapetus	G. Cassini	1671	1436	11.0	3,560,800	79.3	0.028	7.6
Kiviuq	B. Gladman & others	2000	14	22.0	11,111,000	449.2	0.334	46.1
Ijiraq	J. Kavelaars & others	2000	10	22.6	11,124,000	451.4	0.322	46.44
Phoebe	W. Pickering	1898	230 × 220 × 210	16.4	12,944,300	550.2R	0.164	174.8
Paaliaq	B. Gladman & others	2000	19	21.3	15,200,000	686.9	0.363	45.1
Skathi	J. Kavelaars & others	2000	6	23.6	15,541,000	728.2R	0.269	152.6
Albiorix	M. Holman & others	2000	26	20.5	16,182,000	783.5	0.479	34.0
S/2007 S2	S. Sheppard & others	2007	6	24.4	16,560,000	800R	0.107	176.7
Bebhionn	D. Jewitt & others	2004	6	24.1	17,119,000	834.8	0.469	35.01
Erriapus	J. Kavelaars & others	2000	9	23.0	17,343,000	871.2	0.474	34.62
Siarnaq	B. Gladman & others	2000	32	20.1	17,531,000	895.6	0.295	45.56
Skoll	D. Jewitt & others	2006	6	24.5	17,665,000	878.3R	0.464	161.2
Tarvos	J. Kavelaars & others	2000	13	22.1	17,983,000	926.2	0.531	33.82
Tarqeq	S. Sheppard & others	2007	7	23.9	18,009,000	887.5	0.16	46.09
Greip	D. Jewitt & others	2006	6	24.4	18,206,000	921.2R	0.326	179.8
Hyrrokkin	D. Jewitt & others	2004	8	23.5	18,437,000	931.8R	0.333	151.4
S/2004 S13	D. Jewitt & others	2004	6	24.5	18,450,000	906R	0.273	167.4
S/2004 S17	D. Jewitt & others	2004	4	25.2	18,600,000	986R	0.259	166.6

continued

Satellite	Discoverer	Year of discovery	Diameter (km)	Visual magnitude	Distance (km)	Orbital period (d)	Orbital ecc.	Orbital inc. (degrees)
Mundilfari	B. Gladman & others	2000	6	23.8	18,685,000	952.6R	0.21	167.3
Jarnsaxa	D. Jewitt & others	2006	6	24.7	18,811,000	964.7R	0.216	163.3
S/2006 S1	D. Jewitt & others	2006	6	24.6	18,981,135	970R	0.13	154.2
Narvi	S. Sheppard & others	2003	8	24.0	19,007,000	1003.9R	0.431	145.8
Bergelmir	D. Jewitt & others	2004	6	24.2	19,338,000	1005.9R	0.142	158.5
Suttungr	B. Gladman & others	2000	6	23.9	19,459,000	1016.7R	0.114	175.8
S/2004 S12	D. Jewitt & others	2004	5	24.8	19,650,000	1048R	0.401	164.0
S/2004 S7	D. Jewitt & others	2004	6	24.8	19,800,000	1103R	0.58	165.1
Hati	D. Jewitt & others	2004	6	24.4	19,856,000	1038.7R	0.372	165.8
Bestla	D. Jewitt & others	2004	7	23.8	20,129,000	1083.6R	0.521	145.2
Farbauti	D. Jewitt & others	2004	5	24.7	20,390,000	1086.1R	0.206	156.4
Thrymr	B. Gladman & others	2000	6	23.9	20,474,000	1094.3R	0.47	176.0
S/2007 S3	S. Sheppard & others	2007	5	24.9	20,518,500	1100R	0.13	177.2
Aegir	D. Jewitt & others	2004	6	24.4	20,735,000	1116.5R	0.252	166.7
S/2006 S3	D. Jewitt & others	2006	6	24.6	21,132,000	1142R	0.471	150.8
Kari	D. Jewitt & others	2006	7	23.9	22,118,000	1233.6R	0.478	156.3
Fenrir	D. Jewitt & others	2004	4	25.0	22,453,000	1260.3R	0.136	164.9
Surtur	D. Jewitt & others	2006	6	24.8	22,707,000	1297.7R	0.451	177.5
Ymir	B. Gladman & others	2000	16	21.7	23,040,000	1315.4R	0.334	173.1
Loge	D. Jewitt & others	2006	6	24.6	23,065,000	1312R	0.187	167.9
Fornjot	D. Jewitt & others	2004	6	24.6	25,108,000	1491R	0.206	170.4

Uranus

Satellite	Discoverer	Year of discovery	Diameter (km)	Visual magnitude	Distance (km)	Orbital period (d)	Orbital ecc.	Orbital inc. (degrees)
Cordelia	R. Terrile / Voyager 2	1986	40	23.6	49,800	0.4	0.000	0.1
Ophelia	R. Terrile / Voyager 2	1986	42	23.3	53,800	0.4	0.01	0.1
Bianca	B. A. Smith / Voyager 2	1986	51	22.5	59,200	0.4	0.001	0.2
Cressida	S. Synott / Voyager 2	1986	80	21.6	61,800	0.5	0.000	0.0
Desdemona	S. Synott / Voyager 2	1986	64	22.0	62,700	0.5	0.000	0.1
Juliet	S. Synott / Voyager 2	1986	94	21.1	64,400	0.5	0.001	0.1
Portia	S. Synott / Voyager 2	1986	135	20.4	66,100	0.5	0.000	0.1
Rosalind	S. Synott / Voyager 2	1986	72	21.8	69,900	0.6	0.000	0.3
Cupid	M. Showalter, J. Lissauer	2003	24	26.0	74,800	0.6	0.000	0.0
Belinda	S. Synott / Voyager 2	1986	81	21.5	75,300	0.6	0.000	0.0
Perdita	E. Karkoschka / Voyager 2	1999	20	23.6	76,400	0.6	0.000	0.0
Puck	S. Synott / Voyager 2	1985	162	19.8	86,000	0.8	0.000	0.3
Mab	M. Showalter, J. Lissauer	2003	32	26.0	97,734	0.9	0.000	0.0
Miranda	G. Kuiper	1948	472	15.8	129,900	1.4	0.001	4.3
Ariel	W. Lassell	1851	1158	13.7	190,900	2.5	0.001	0.0
Umbriel	W. Lassell	1851	1169	14.5	266,000	4.1	0.004	0.1
Titania	W. Herschel	1787	1578	13.5	436,300	8.7	0.001	0.1
Oberon	W. Herschel	1787	1523	13.7	583,500	13.5	0.001	0.1
Francisco	M. Holman, J. Kavelaars	2001	22	25.0	4,276,000	266.6R	0.146	145.2
Caliban	B. Gladman & others	1997	72	22.4	7,231,000	579.7R	0.159	140.9
Stephano	B. Gladman & others	1999	32	24.1	8,004,000	677.4R	0.23	144.1
Trinculo	M. Holman & others	2001	18	25.4	8,504,000	759.0R	0.208	167.0
Sycorax	B. Gladman & others	1997	150	20.8	12,179,000	1288.3R	0.522	159.4
Margaret	D. Jewitt, S. Sheppard	2001	20	25.2	14,345,000	1694.8	0.783	50.7
Prospero	M. Holman & others	1999	50	23.2	16,256,000	1977.3R	0.445	152.0
Setebos	J. Kavelaars & others	1999	47	23.3	17,418,000	2234.8R	0.591	158.2
Ferdinand	M. Holman, B. Gladman	2001	21	25.1	20,901,000	2823.4R	0.368	169.8

Neptune

Satellite	Discoverer	Year of discovery	Diameter (km)	Visual magnitude	Distance (km)	Orbital period (d)	Orbital ecc.	Orbital inc. (degrees)
Naiad	R. Terrile / Voyager 2	1989	58	24.1	48,227	0.29	0.000	4.7
Thalassa	R. Terrile / Voyager 2	1989	80	23.4	50,075	0.31	0.000	0.2
Despina	S. Synott / Voyager 2	1989	148	22.0	52,526	0.33	0.000	0.1
Galatea	S. Synott / Voyager 2	1989	158	22.0	61,953	0.43	0.000	0.1
Larissa	H. Reitsema & others	1989	192	21.5	73,548	0.55	0.001	0.2
Proteus	S. Synott / Voyager 2	1989	416	20.0	117,647	1.12	0.000	0.0
Triton	W. Lassell	1846	2707	13.0	354,760	5.87R	0.000	156.8
Nereid	G. Kuiper	1949	340	19.2	5,513,400	360.1	0.751	7.2
Halimede	M. Holman & others	2002	61	24.5	15,728,000	1879.7R	0.571	134.1
Sao	M. Holman & others	2002	40	25.4	22,422,000	2914.1	0.293	48.5
Laomedeia	M. Holman & others	2002	40	25.4	23,571,000	3167.9	0.424	34.7
Psamathe	S. Sheppard & others	2003	38	25.6	46,695,000	9115.9R	0.45	137.4
Neso	M. Holman & others	2002	60	24.6	48,387,000	9374.0R	0.495	132.6

Pluto

Satellite	Discoverer	Year of discovery	Diameter (km)	Visual magnitude	Distance (km)	Orbital period (d)	Orbital ecc.	Orbital inc. (degrees)
Charon	J. Christy	1978	1207	17.3	17,536	6.39	0.002	0.001
Nix	H. Weaver, S. Alan Stern & others	2005	46?	23.4	48,708	24.86	0.003	0.195
S/2011 P1 (P4)	M. Showalter & others	2011	?	26.1	56,000	?	?	?
Hydra	H. Weaver, S. Alan Stern & others	2005	61?	24.39	64,749	38.2	0.005	0.212

Eris

Satellite	Discoverer	Year of discovery	Diameter (km)	Visual magnitude	Distance (km)	Orbital period (d)	Orbital ecc.	Orbital inc. (degrees)
Dysnomia	M. Brown & others	2005	150?	18.8	37,350 ± 140	15.774	<0.013	142±3°

Haumea

Satellite	Discoverer	Year of discovery	Diameter (km)	Visual magnitude	Distance (km)	Orbital period (d)	Orbital ecc.	Orbital inc. (degrees)
Namaka	M. Brown & others	2005	170?	Approx. 24	25,657±91	18.28?	0.249 (variable)	113?
Hi'iaka	M. Brown & others	2005	310?	Approx. 24	49,880±198	49.4?	0.25	126.3?

R = retrograde

Appendix 3: Planetary Rings

Rings of Jupiter

Name	Distance From Center of Planet (planet radii)	Distance From Center of Planet (km)	Width (km)	Thickness (km)
Halo	1.40–1.71	100,000–122,000	22,000	(20,000)
Main	1.71–1.81	122,000–129,000	7000	30–300
Gossamer (inner)	1.81–2.55	129,200–182,000	52,800	2600?
Gossamer (outer)	2.55–3.15	182,000–224,900	42,900	8800?

m = meters

Rings of Saturn

Name	Distance from Center of Planet (planet radii)	Distance from Center of Planet (km)	Width (km)	Thickness
D	1.11–1.236	66,900–74,510	7610	?
C	1.239–1.527	74,658–92,000	17,342	5 m
B	1.527–1.951	92,000–117,580	25,580	5–10 m
A	2.027–2.269	122,170–136,775	14,605	10 -30 m
F	2.326	140,180	30–500	1 km
Janus-Epimetheus ring	2.47–2.55	149,000–154,000	5.000	?
G	2.82–2.90	170,000–175,000	5000	100 km
Methone ring arc	3.22	194,230	?	?
Anthe ring	3.28	197,665		
Pallene ring	3.5–3.54	211,000–213,500	2500	
E	3–8	181,000–483,000	302,000	10,000 km
Phoebe ring	59–300?	~4,000,000 ->18,000,000	>14,000,000 ?	2,400,000 km ?

m = meters

Rings of Uranus

Name	Distance From Center of Planet (planet radii)	Distance From Center of Planet (km)	Width (km)	Thickness (m)
zeta (ζ)	1.448–1.545	37,000–39,500	Approx. 2500	(100)
6	1.637	41,837	Approx. 1.5	(100)
5	1.652	42,235	Approx. 2	(100)
4	1.666	42,571	Approx. 2.5	(100)
alpha (α)	1.75	44,718	4–10	(100)
beta (β)	1.786	45,661	5–11	(100)
eta (η)	1.834	47,176	1.6	(100)
gamma (γ)	1.863	47,626	1–4	(100)
delta (δ)	1.900	48,303	3–7	(100)
lambda (λ)	1.958	50,024	Approx. 2	(100)
epsilon (ε)	2.00	51,149	20–96	150
nu (ν)	2.586–2.735	66,100–69,900	3800	?
mu (μ)	3.265–4.03	86,000–103,000	17,000	?

Rings of Neptune

Name	Distance From Center of Planet (planet radii)	Distance From Center of Planet (km)	Width (km)	Thickness
Galle	1.692	41,900–43,900	Approx. 2000	?
Le Verrier	2.148	~53,200	Approx. 110	?
Lassell	2.148–2.31	~53,200–57,200	Approx. 4000	?
Arago	2.31	52,200	Less than 100	?
Unnamed (v. faint)	2.501	61,950	?	?
Adams	2.541	62,933	15–50	?

Arcs in Adams Ring are called Courage, Liberté, Egalité 1, Egalité 2 and Fraternité.

Appendix 4: The Largest Known Trans-Neptunian Objects

Object	Discoverer	Year of discovery	Diameter (km)	Mean Distance (AU)	Orbital period (y)	Orbital ecc.	Orbital inc. (degrees)	Satellites
Eris	M. Brown, C. Trujillo, D. Rabinowitz	2003	Approx. 2400	67.668	557.8	0.442	44.19	1
Pluto	C. Tombaugh	1930	2320	39.481	248.54	0.249	17.2	4
Makemake	M. Brown, C. Trujillo, D. Rabinowitz	2005	Approx. 1500?	45.791	309.88	0.159	28.96	-
Sedna	M. Brown, C. Trujillo, D. Rabinowitz	2003	Approx. 1400?	518.57	11,809	0.853	11.93	-

continued

Object	Discoverer	Year of discovery	Diameter (km)	Mean Distance (AU)	Orbital period (y)	Orbital ecc.	Orbital inc. (degrees)	Satellites
Haumea	M. Brown et al / J. Ortiz et al	2004	Approx. 1960 × 1520 × 1000?	43.132	285.4	0.195	28.22	2
Charon	J. Christy	1978	1207	Satellite of Pluto				-
2007 OR10	M. E. Schwamb, M. Brown, D. Rabinowitz	2007	Approx. 1200?	67.21	550.98	0.5	30.7	-
84522 (2002 TC302)	NEAT	2002	Approx. 1150?	55.23	409.16	0.293	35.1	-
Orcus	M. Brown, C. Trujillo, D. Rabinowitz	2004	Approx. 950?	39.187	247.94	0.226	20.56	1
Quaoar	C. Trujillo & M. Brown	2002	Approx. 890?	43.574	285.97	0.039	7.98	1
Varuna	R. McMillan (Spacewatch)	2000	Approx. 800?	43.129	283.2	0.051	17.2	

N.B. The sizes of most of these objects are very uncertain.

Appendix 5: Lunar and Planetary Missions

Moon

Name	Country	Launch Date	Purpose	Results
Pioneer 0	US	17 Aug 1958	Orbiter	Launch failure
[Unnamed]	USSR	23 Sept 1958	Impact	Launch failure
Pioneer 1	US	11 Oct 1958	Orbiter	Failed to reach escape velocity
[Unnamed]	USSR	11 Oct 1958	Impact	Failed to reach escape velocity
Pioneer 2	US	8 Nov 1958	Orbiter	Failed to reach escape velocity
[Unnamed]	USSR	4 Dec 1958	Impact	Launch failure
Pioneer 3	US	6 Dec 1958	Flyby	Failed to reach escape velocity
Luna 1	USSR	2 Jan 1959	Impact	Lunar flyby at distance of 6400 km on 4 Jan 1959
Pioneer 4	US	3 March 1959	Flyby	Lunar flyby 4 March 1959 at distance of 59,545 km
[Unnamed]	USSR	18 June 1959	Impact	Launch failure
Luna 2	USSR	12 Sept 1959	Impact	Lunar impact
Luna 3	USSR	4 Oct 1959	Circumlunar	Flyby at distance of 7900 km on 6 Oct
Pioneer P-3	US	26 Nov 1959	Orbiter	Launch failure
[Unnamed]	USSR	15 April 1960	Circumlunar	Failed to reach escape velocity
[Unnamed]	USSR	19 April 1960	Circumlunar	Launch failure
Pioneer P-30	US	25 Sept 1960	Orbiter	Launch failure
Pioneer P-31	US	15 Dec 1960	Orbiter	Launch failure
Ranger 3	US	26 Jan 1962	Impact	Flyby on 28 Jan at distance of 36,793 km
Ranger 4	US	23 April 1962	Impact	Impact on lunar farside 26 April. No data
Ranger 5	US	18 Oct 1962	Impact	Flyby on 21 Oct at distance of 724 km
[Unnamed]	USSR	4 Jan 1963	Lander	Failed to leave Earth orbit
[Unnamed]	USSR	3 Feb 1963	Lander	Launch failure
Luna 4	USSR	2 April 1963	Lander	Flyby on 6 April at distance of 8500 km

Name	Country	Launch Date	Purpose	Results
Kosmos 21	USSR	11 Nov 1963	Flyby	Failed to leave Earth orbit
Ranger 6	US	30 Jan 1964	Impact	Impact 2 Feb. No data returned
[Unnamed]	USSR	21 March 1964	Lander	Failed to reach Earth orbit
[Unnamed]	USSR	20 April 1964	Lander	Launch failure
Ranger 7	US	28 July 1964	Impact	Impact 31 July. Returned 4316 pictures
Ranger 8	US	17 Feb 1965	Impact	Impact 20 Feb. Returned 7137 pictures
Ranger 9	US	21 March 1965	Impact	Impact 24 March. Returned 5814 pictures
[Unnamed]	USSR	10 April 1965	Lander	Failed to reach Earth orbit
Luna 5	USSR	9 May 1965	Lander	Crashed on lunar surface
Luna 6	USSR	8 June 1965	Lander	Flyby on 11 June at distance of 161,000 km
Zond 3	USSR	18 July 1965	Flyby	Flyby 20 July at distance of 9220 km
Luna 7	USSR	4 Oct 1965	Lander	Crashed on lunar surface
Luna 8	USSR	3 Dec 1965	Lander	Crashed on lunar surface
Luna 9	USSR	31 Jan 1966	Lander	Landed on Ocean of Storms 3 Feb. Operated until 6 Feb
Kosmos 111	USSR	1 March 1966	Orbiter	Failed to leave Earth orbit
Luna 10	USSR	31 March 1966	Orbiter	Arrived 3 April. Operated until 30 May
Surveyor 1	US	30 May 1966	Lander	Landed on Ocean of Storms 2 June. Operated until 7 Jan 1967
Explorer 33	US	1 July 1966	Orbiter	Entered eccentric Earth orbit
Lunar Orbiter 1	US	10 Aug 1966	Orbiter	Arrived 14 Aug. Deliberately crashed on Moon on 29 Oct
Luna 11	USSR	24 Aug 1966	Orbiter	Arrived 27 Aug. Operated until 1 Oct
Surveyor 2	US	20 Sept 1966	Lander	Crashed on lunar surface
Luna 12	USSR	22 Oct 1966	Orbiter	Arrived 25 Oct. Operated until 19 Jan 1967
Lunar Orbiter 2	US	6 Nov 1966	Orbiter	Arrived 10 Nov. Deliberately crashed on Moon on 11 Oct 1967
Luna 13	USSR	21 Dec 1966	Lander	Landed on Ocean of Storms 24 Dec. Operated until 28 Dec
Lunar Orbiter 3	US	5 Feb 1967	Orbiter	Arrived 7 Feb. Deliberately crashed on lunar surface 9 Oct
Surveyor 3	US	17 April 1967	Lander	Landed on Ocean of Storms 20 April. Operated until 4 May
Lunar Orbiter 4	US	4 May 1967	Orbiter	Arrived 8 May. Operated until 17 July
Surveyor 4	US	14 July 1967	Lander	Contact lost during descent
Lunar Orbiter 5	US	1 Aug 1967	Orbiter	Arrived 5 Aug. Deliberately crashed on Moon 31 Jan 1968
Surveyor 5	US	8 Sept 1967	Lander	Landed on Sea of Tranquillity 11 Sept. Operated until 16 Dec 1967
Surveyor 6	US	7 Nov 1967	Lander	Landed on Central Bay 10 Nov. Operated until 24 Nov
Surveyor 7	US	7 Jan 1968	Lander	Landed near Tycho 10 Jan. Operated until 21 Feb
Luna 14	USSR	7 April 1968	Orbiter	Arrived 10 April. Operated until June/July
Zond 5	USSR	14 Sept 1968	Circumlunar	Flyby 18 Sep at distance of 1950 km
Zond 6	USSR	10 Nov 1968	Circumlunar	Flyby 14 Nov at distance of 2420 km
Apollo 8	US	21 Dec 1968	Orbiter	Arrived 24 Dec. Returned to Earth 25 Dec. Crew: Borman. Lovell, Anders
[Unnamed]	USSR	19 Feb 1969	Lander / rover	Launch failure
[Unnamed]	USSR	14 June 1969	Sample return	Failed to reach Earth orbit
Apollo 10	US	18 May 1969	Orbiter	Arrived 21 May. Returned to Earth 24 May. Crew: Stafford, Young, Cernan
Luna 15	USSR	13 July 1969	Sample return	Crashed on lunar surface

Name	Country	Launch Date	Purpose	Results
Apollo 11	US	16 July 1969	Orbiter / lander / sample return	Arrived 19 July. Returned to Earth 21 July Landed on Sea of Tranquillity 20 July. Crew: Armstrong, Aldrin, Collins
Zond 7	USSR	7 Aug 1969	Circumlunar	Flyby 10 Aug at distance of 1200 km
Kosmos 300	USSR	23 Sept 1969	Sample return	Failed to leave Earth orbit
Kosmos 305	USSR	22 Oct 1969	Sample return	Failed to reach Earth orbit
Apollo 12	US	14 Nov 1969	Orbiter / lander / sample return	Arrived 18 Nov. Returned to Earth 21 Nov. Landed on Ocean of Storms 19 Nov. Crew: Conrad, Bean, Gordon
[Unnamed]	USSR	6 Feb 1970	Sample return	Failed to reach Earth orbit
Apollo 13	US	11 April 1970	Orbiter / lander / sample return	Flyby 14 April. Crew: Lovell, Haise, Swigert
Luna 16	USSR	12 Sept 1970	Sample return	Landed on Sea of Fertility 20 Sept. Returned to Earth 21 Sept with 101 g (3.5 oz) of soil
Zond 8	USSR	20 Oct 1970	Circumlunar	Flyby 24 Oct at distance of 1200 km
Luna 17 / Lunokhod 1	USSR	10 Nov 1970	Lander / rover	Landed 17 Nov. Rover operated until 14 Sept 1971, travelling 10.54 km
Apollo 14	US	31 Jan 1971	Orbiter / lander / sample return	Arrived 4 Feb. Returned to Earth 6 Feb. Landed at Fra Mauro 5 Feb. Crew: Shepard, Roosa, Mitchell
Apollo 15	US	26 July 1971	Orbiter / subsatellite / lander / rover/ sample return	Arrived 29 July. Returned to Earth 4 Aug Landed at Hadley-Apennine 30 July Released subsatellite 4 Aug. Crew: Scott, Irwin, Worden
Luna 18	USSR	2 Sept 1971	Sample return	Contact lost during descent
Luna 19	USSR	28 Sept 1971	Orbiter	Arrived 2 Oct. Operated until Oct 1972
Luna 20	USSR	14 Feb 1972	Sample return	Landed on Sea of Fertility 21 Feb. Returned to Earth 22 Feb with 55 g (2 oz) of soil
Apollo 16	US	16 April 1972	Orbiter / subsatellite / lander / rover/ sample return	Arrived 19 April. Returned to Earth 24 April. Landed at Descartes 20 April Released subsatellite 24 April. Crew: Young, Duke, Mattingley
Apollo 17	US	7 Dec 1972	Orbiter / lander / rover/ sample return	Arrived 10 Dec. Returned to Earth 14 Dec. Landed at Taurus-Littrow 11 Dec. Crew: Cernan, Schmitt, Evans
Luna 21 / Lunokhod 2	USSR	8 Jan 1973	Lander / rover	Landed 15 Jan. Rover operated until 9 May, travelling 37 km (23 ml)
Explorer 49	US	10 June 1973	Orbiter	Arrived 15 June. No lunar data – conducted studies of radio emissions from distant sources. Operated until June 1975
Luna 22	USSR	29 May 1974	Orbiter	Arrived 2 June. Operated until 2 Sept 1975
Luna 23	USSR	28 Oct 1974	Sample return	Damaged during landing on 6 Nov. Contact lost 9 Nov
[Unnamed]	USSR	16 Oct 1975	Sample return	Failed to reach Earth orbit
Luna 24	USSR	9 Aug 1976	Sample return	Landed on Sea of Crises 18 Aug. Returned to Earth 19 Aug with 170 g of soil
Galileo	US	18 Oct 1989	Venus, Earth and asteroid flybys, Jupiter orbiter / probe	Moon flyby 8 Dec 1990 at distance of 565,000 km; Moon flyby 8 Dec 1992 at distance of 110,000 km
Hiten / Hagomoro	Japan	24 Jan 1990	Circumlunar / orbiter	First flyby 19 March at distance of 16,742 km; orbiter released but contact lost
Clementine	US	25 Jan 1994	Lunar orbiter / asteroid flyby	Arrived at Moon 19 Feb. Operated until 3 May
Cassini-Huygens	US/ESA	15 Oct 1997	Venus, Earth and asteroid flybys, Saturn orbiter, Titan probe / lander	Moon flyby 18 Aug 1999 at distance of 377,000 km
Lunar Prospector	US	7 Jan 1998	Orbiter	Arrived 11 Jan. Deliberately crashed on Moon 31 July 1999
Nozomi	Japan	3 July 1998	Earth-Moon flybys, Mars orbiter	Moon flyby 24 Sept 1998 at distance of 2809 km; 2nd flyby 18 Dec 1998 at distance of 1003 km

Name	Country	Launch Date	Purpose	Results
SMART-1	ESA	27 Sept 2003	Orbiter	Arrived 15 Nov 2004. Deliberately crashed on Moon 3 Sept 2006
Kaguya (SELENE)	Japan	14 Sept 2007	Orbiter / 2 subsatellites	Arrived 3 Oct. Released subsatellites on 9 and 12 Oct. Deliberately crashed on Moon 11 June 2009
Chandrayaan-1	India	22 Oct 2008	Orbiter / Moon impactor	Arrived 8 Nov. Moon Impact Probe hit surface on 14 Nov Radio contact lost on 28 Aug 2009
Chang'e-1	China	24 Oct 2007	Orbiter	Arrived 5 Nov. Deliberately crashed on Moon 1 March 2009.
Lunar Reconnaissance Orbiter	US	18 June 2009	Orbiter	Arrived 23 June 2009
LCROSS	US	18 June 2009	Impactor	Impacted Cabeus crater on 9 Oct. 2009
Chang'e-2	China	1 October 2010	Orbiter	Arrived 9 Oct. 2010. Left lunar orbit 9 June 2011.

Mercury

Name	Country	Launch Date	Purpose	Results
Mariner 10	USA	3 Nov 1973	Venus / Mercury flybys	Mercury flybys 29 March 1974; 21 Sep 1974; 16 Mar 1975
MESSENGER	USA	3 Aug 2004	Venus flyby, Mercury flybys / orbiter	Earth flyby 2 Aug 2005; Venus flybys 24 Oct 2006 and 6 June 2007; Mercury flybys 14 Jan 2008, 6 Oct 2008 and 29 Sept 2009. Entered orbit 17 Mar 2011
BepiColombo	ESA/Japan	2014 (?)	2 orbiters (ESA and Japan)	Planned

Venus

Name	Country	Launch Date	Purpose	Results
Sputnik 7	USSR	4 Feb 1961	Impact	Did not reach Earth orbit
Venera 1	USSR	12 Feb 1961	Impact	Contact lost 22 Feb 1961
Mariner 1	US	22 July 1962	Flyby	Did not reach Earth orbit
Sputnik 19	USSR	25 Aug 1962	Impact	Failed to leave Earth orbit
Mariner 2	US	27 Aug 1962	Flyby	Flyby 14 Dec 1962.
Sputnik 20	USSR	1 Sept 1962	Flyby	Failed to leave Earth orbit
Sputnik 21	USSR	12 Sept 1962	Flyby	Destroyed in Earth orbit
[Unnamed]	USSR	19 Feb 1964	Flyby	Did not reach Earth orbit
[Unnamed]	USSR	1 March 1964	Flyby	Did not reach Earth orbit
Kosmos 27	USSR	27 March 1964	Atmospheric probe	Failed to leave Earth orbit
Zond 1	USSR	2 April 1964	Atmospheric probe	Contact lost 25 May
Venera 2	USSR	12 Nov 1965	Flyby	Flyby 27 Feb 1966, no return of data
Venera 3	USSR	16 Nov 1965	Atmospheric probe	Arrived 1 March 1966, no return of data
Kosmos 96	USSR	23 Nov 1965	Flyby	Failed to leave Earth orbit
Venera 4	USSR	12 June 1967	Atmospheric probe	Arrived 18 Oct 1967, sent back data for 93 minutes during descent

Name	Country	Launch Date	Purpose	Results
Mariner 5	US	14 June 1967	Flyby	Flyby 19 Oct
Kosmos 167	USSR	17 June 1967	Atmospheric probe	Failed to leave Earth orbit
Venera 5	USSR	5 Jan1969	Atmospheric probe	Arrived 16 May 1969, sent back data for 53 minutes during descent
Venera 6	USSR	10 Jan 1969	Atmospheric probe	Arrived 17 May 1969, sent back data for 51 minutes during descent
Venera 7	USSR	17 Aug 1970	Atmospheric probe / lander	Landed on 15 Dec 1970, sent back data from surface for 23 minutes
Kosmos 359	USSR	22 Aug 1970	Atmospheric probe / lander	Failed to leave Earth orbit
Venera 8	USSR	27 March 1972	Atmospheric probe / lander	Landed on 22 July 1972, sent back data from surface for 50 minutes
Kosmos 482	USSR	31 March 1972	Atmospheric probe / lander	Failed to leave Earth orbit
Mariner 10	US	4 Nov 1973	Venus / Mercury flybys	Venus flyby 5 Feb 1974
Venera 9	USSR	8 June 1975	Orbiter / lander	Arrived 22 Oct 1975. Lander sent back data from surface for 53 minutes, including first photo
Venera 10	USSR	14 June 1975	Orbiter / lander	Arrived 25 Oct 1975. Lander sent back data from surface for 65 minutes, including a photo
Pioneer Venus 1	US	20 May 1978	Orbiter	Arrived 4 Dec 1978. Operated until 8 Oct 1992
Pioneer Venus 2	US	8 Aug 1978	Main bus / 4 atmospheric probes	Arrived 9 Dec 1978. Day probe sent back data from surface for 67 minutes
Venera 11	USSR	9 Sept 1978	Orbiter / lander	Arrived 25 Dec 1978. Lander sent back data from surface for 95 minutes
Venera 12	USSR	14 Sept 1978	Orbiter / lander	Arrived 9 Dec 1978. Lander sent back data from surface for 110 minutes
Venera 13	USSR	30 Oct 1981	Flyby / lander	Arrived 1 March 1982. Lander sent back data from surface for 127 minutes, including 8 colour photos
Venera 14	USSR	4 Nov 1981	Flyby / lander	Arrived 3 March 1982. Lander sent back data from surface for 57 minutes, including 8 colour photos
Venera 15	USSR	2 June 1983	Orbiter	Arrived 10 Oct 1983. Mapped N. hemisphere with radar until 10 July 1984
Venera 16	USSR	7 June 1983	Orbiter	Arrived 14 Oct 1983. Mapped N. hemisphere with radar until 10 July 1984
Vega 1	USSR	15 Dec 1984	Flyby / lander / balloon; comet Halley flyby	Venus flyby and landing 11 June 1985. Lander sent back 56 minutes of data. Balloon operated until 13 June
Vega 2	USSR	21 Dec 1984	Flyby / lander / balloon; comet Halley flyby	Venus flyby and landing 15 June 1985. Lander sent back 57 minutes of data. Balloon operated until 17 June
Magellan	US	4 May 1989	Orbiter	Arrived 10 Aug 1990. Mapped planet with radar, operational until 12 Oct 1994
Galileo	US	18 Oct 1989	Venus, Earth and asteroid flybys, Jupiter orbiter / probe	Venus flyby 10 Feb 1990
Cassini-Huygens	US/ESA	15 Oct 1997	Venus, Earth and asteroid flybys, Saturn orbiter, Titan probe / lander	Venus flybys 26 April 1998 and 24 June 1999
MESSENGER	US	3 Aug 2004	Venus flyby / Mercury orbiter	Venus flybys Oct 2006 and June 2007
Venus Express	ESA	9 Nov 2005	Orbiter	Arrived 11 April 2006
Akatsuki	Japan	21 May 2010	Orbiter	Failed to enter orbit 7 Dec 2010

Mars

Name	Country	Launch Date	Purpose	Results
[Unnamed]	USSR	10 Oct 1960	Flyby	Did not reach Earth orbit
[Unnamed]	USSR	14 Oct 1960	Flyby	Did not reach Earth orbit
[Unnamed]	USSR	24 Oct 1962	Flyby	Achieved Earth orbit only
Mars 1	USSR	1 Nov 1962	Flyby	Radio failed at 106 million km (65.9 million ml)
[Unnamed]	USSR	4 Nov 1962	Flyby	Achieved Earth orbit only
Mariner 3	US	5 Nov 1964	Flyby	Launcher shroud failed to jettison
Mariner 4	US	28 Nov 1964	Flyby	First successful Mars flyby 14 July 1965. Returned 21 photos
Zond 2	USSR	30 Nov 1964	Flyby	Passed Mars but radio failed. Returned no planetary data
Mariner 6	US	24 Feb 1969	Flyby	Mars flyby 31 July 1969. Returned 75 photos
Mariner 7	US	27 March 1969	Flyby	Mars flyby 5 Aug 1969. Returned 126 photos
[Unnamed]	USSR	27 March 1969	Orbiter	Did not reach Earth orbit
[Unnamed]	USSR	2 April 1969	Orbiter	Did not reach Earth orbit
Mariner 8	US	8 May 1971	Orbiter	Failed during launch
Kosmos 419	USSR	10 May 1971	Orbiter	Achieved Earth orbit only
Mars 2	USSR	19 May 1971	Orbiter / lander	Arrived 27 Nov 1971, no useful data, lander burned up due to steep entry
Mars 3	USSR	28 May 1971	Orbiter / lander	Arrived 3 Dec 1971, lander operated on surface for 20 seconds before failing
Mariner 9	US	30 May 1971	Orbiter	Arrived 13 Nov 1971. End of mission 27 Oct 1972. Returned 7329 photos
Mars 4	USSR	21 July 1973	Orbiter	Flew past Mars 10 Feb 1974
Mars 5	USSR	25 July 1973	Orbiter	Arrived 2/12/74, lasted a few days. Returned 43 photos
Mars 6	USSR	5 Aug 1973	Flyby module / lander	Arrived 12 March 1974, lander failed due to fast impact
Mars 7	USSR	9 Aug 1973	Flyby module / lander	Arrived 9 March 1974, lander missed the planet
Viking 1	US	20 Aug 1975	Orbiter / lander	Arrived 19 June 1976. Orbiter operated until 7 Aug 1980. Lander operated on surface 20 July 1976 to 13 Nov 1982
Viking 2	US	9 Sept 1975	Orbiter / lander	Arrived 7 Aug 1976. Orbiter operated until 24 July 1978. Lander operated on surface 3 Sept 1976 to 11 April 1980. Viking orbiters and landers returned 50,000+ photos
Phobos 1	USSR	7 July 1988	Mars / Phobos orbiter / lander	Contact lost 29 Aug 1988 *en route* to Mars
Phobos 2	USSR	12 July 1988	Mars / Phobos orbiter / lander	Contact lost 27 March 1989 near Phobos. Returned photos of Mars and Phobos
Mars Observer	US	25 Sept 1992	Orbiter	Contact lost just before Mars arrival 21 Aug 1993
Mars Global Surveyor	US	7 Nov 1996	Orbiter	Arrived 12 Sept 1997. Contact lost 2 Nov 2006
Mars 96	Russia	16 Nov 1996	Orbiter and landers	Did not reach Earth orbit
Mars Pathfinder	US	14 Dec 1996	Lander / rover	Landed 4 July 1997, operated until 27 Sept 1997
Nozomi	Japan	4 July 1998	Earth-Moon flybys, Mars orbiter	Mars flyby 14 Dec 2003 at distance of 1000 km. No Mars data returned
Mars Climate Orbiter	US	11 Dec 1998	Orbiter	Contact lost on arrival, 23 Sept 1999
Mars Polar Lander / Deep Space 2	US	3 Jan 1999	Lander / penetrators	Contact lost on arrival 3 Dec 1999
Mars Odyssey	US	7 March 2001	Orbiter	Arrived 24 Oct 2001, still operational
Mars Express / Beagle 2	ESA	2 June 2003	Orbiter / lander	Orbiter arrived 25 Dec 2003, still operational. Lander lost during entry 25 Dec 2003
Mars Exploration Rover A (Spirit)	US	10 June 2003	Lander / rover	Landed 4 Jan 2004, lost contact 22 March 2010

Name	Country	Launch Date	Purpose	Results
Mars Exploration Rover B (Opportunity)	US	7 July 2003	Lander / rover	Landed 25 Jan 2004, still operational
Mars Reconnaissance Orbiter	US	12 Aug 2005	Orbiter	Arrived 10 March 2006, still operational.
Rosetta/Philae	ESA	24 Feb 2004	Earth, Mars and asteroid flybys; comet orbiter / lander	Flyby 25 Feb 2007
Phoenix	US	4 Aug 2007	Lander	Arrived 25 May 2008, contact lost 2 Nov 2008
Dawn	US	27 Sept 2007	Dual asteroid orbiter	Flyby 17 Feb 2009
Mars Science Laboratory	US	Nov 2011 (?)	Lander / rover	Arrival due Aug 2012
Phobos-Grunt / Yinghuo-1	Russia/ China	Nov 2011 (?)	Phobos lander and sample return/Mars orbiter	Arrival due Oct 2012

Asteroids

Name	Country	Launch Date	Purpose	Results
Galileo	US	18 Oct 1989	Venus, Earth and asteroid flybys, Jupiter orbiter / probe	Flyby of Gaspra 29 Oct 1991; flyby of Ida and Dactyl 28 Aug 1993
Clementine	US	25 Jan 1994	Lunar orbiter / asteroid flyby	Mission to Geographos cancelled 3 May 1994 after thruster malfunction
NEAR-Shoemaker	US	17 Feb 1996	Earth and asteroid flybys, Eros orbiter	Mathilde flyby 27 June 1997; Eros flyby 23 Dec 1998. Orbit around Eros 14 Feb 2000. Landed on Eros 12 Feb 2001
Cassini-Huygens	US/ESA	15 Oct 1997	Venus, Earth and asteroid flybys, Saturn orbiter, Titan probe / lander	Flyby of Masursky 23 Jan 2000
Deep Space 1	US	24 Oct 1998	Asteroid and comet flybys	Flyby of Braille 29 July 1999
Stardust	US	7 Feb 1999	Earth, asteroid and comet flybys / comet sample return	Flyby of Anne Frank 2 Nov 2002.
Hayabusa	Japan	9 May 2003	Orbiter / sample return	Arrival at Itokowa 12 Sept 2005; sample return to Earth 13 June 2010
Rosetta	ESA	24 Feb 2004	Earth, Mars and asteroid flybys; comet orbiter / lander	Flyby of Steins 5 Sept 2008; flyby of Lutetia 10 July 2010
Dawn	US	27 Sept 2007	Dual orbiter	Arrival at Vesta 15 July 2011. Depart July 2012. Arrival at Ceres Feb 2015

Jupiter

Name	Country	Launch Date	Purpose	Results
Pioneer 10	US	2 March 1972	Flyby	Jupiter flyby 4 Dec 1973
Pioneer 11	US	6 April 1973	Jupiter and Saturn flybys	Jupiter flyby 1 Sept 1979
Voyager 2	US	20 Aug 1977	Jupiter, Saturn, Uranus and Neptune flybys	Jupiter flyby 9 July 1979
Voyager 1	US	5 Sept 1977	Jupiter and Saturn flybys	Jupiter flyby 5 March 1979
Galileo	US	18 Oct 1989	Venus, Earth and asteroid flybys, Jupiter orbiter / atmospheric probe	Orbiter arrived 8 Dec 1995, operational until atmospheric entry 21 Sept 2003. Probe entry 7 Dec 1995.
Ulysses	ESA/US	6 Oct 1990	Solar orbiter / Jupiter flybys	Jupiter flyby 8 Feb 1992 at distance of 378,400 km; distant flyby 5 Feb 2004
Cassini-Huygens	US/ESA	15 Oct 1997	Venus, Earth and asteroid flybys, Saturn orbiter, Titan probe / lander	Jupiter flyby 30 Dec 2000 at distance of 9.7 million km
New Horizons	US	19 Jan 2006	Jupiter, Pluto and KBO flybys	Jupiter flyby 28 February 2007
Juno	US	5 Aug 2011	Jupiter orbiter	Arrival due 4 July 2016

Saturn

Name	Country	Launch Date	Purpose	Results
Pioneer 11	US	6 April 1973	Jupiter and Saturn flybys	Saturn flyby 1 Sept 1979
Voyager 2	US	20 Aug 1977	Jupiter, Saturn, Uranus and Neptune flybys	Saturn flyby 22 Aug 1981
Voyager 1	US	5 Sept 1977	Jupiter and Saturn flybys	Saturn flyby 12 Nov 1980
Cassini-Huygens	US/ESA	15 Oct 1997	Venus, Earth and asteroid flybys, Saturn orbiter, Titan probe / lander	Orbiter arrived 1 July 2004. Still operational. Huygens landing on Titan 14 Jan 2005

Uranus

Name	Country	Launch Date	Purpose	Results
Voyager 2	US	20 Aug 1977	Jupiter, Saturn, Uranus and Neptune flybys	Uranus flyby 24 Jan 1986

Neptune

Name	Country	Launch Date	Purpose	Results
Voyager 2	US	20 Aug 1977	Jupiter, Saturn, Uranus and Neptune flybys	Neptune flyby 25 Aug 1989

Pluto/Kuiper Belt

Name	Country	Launch Date	Purpose	Results
New Horizons	US	19 Jan 2006	Pluto, Kuiper Belt object flybys	Pluto-Charon flyby 14 July 2015; KBO flybys 2016 to 2020

Comets

Name	Country	Launch Date	Purpose	Results
International Cometary Explorer	US	12 Aug 1978	Study of solar wind etc. from L1; comet flyby	Flyby of Giacobini-Zinner 11 Sept 1985; distant flyby of Halley 28 March 1986
Vega 1	USSR	15 Dec 1984	Flyby / lander / balloon; comet Halley flyby	Halley flyby 6 March 1986
Vega 2	USSR	21 Dec 1984	Flyby / lander / balloon; comet Halley flyby	Halley flyby 9 March 1986
Sakigake	Japan	7 Jan 1985	Comet and Earth flybys	Halley flyby 11 March 1986
Giotto	ESA	2 July 1985	Comet and Earth flybys	Halley flyby 14 March 1986; Grigg-Skjellerup flyby 10 July 1992
Suisei	Japan	18 Aug 1985	Comet flyby	Halley flyby 8 March 1986
Deep Space 1	US	24 Oct 1998	Asteroid and comet flybys	Flyby of Borrelly 22 Sept 2001
Stardust / NExT	US	7 Feb 1999	Earth, asteroid and comet flybys / comet sample return	Flyby of Wild 2 on 2 Jan 2004. Returned sample capsule to Earth 15 Jan 2006. Flyby of Tempel 1 on 14 Feb 2011
Rosetta / Philae	ESA	24 Feb 2004	Earth, Mars and asteroid flybys; comet orbiter / lander	Arrival at Churyumov-Gerasimenko Aug 2014; landing Nov 2014
Deep Impact / EPOXI	US	12 Jan 2005	Flyby / impact	Flyby / impact at Tempel 1 on 4 July 2005. Flyby of Hartley 2 on 4 Nov 2010

Glossary

absolute magnitude the brightness a Solar System object would have if placed 1 AU from both the observer and the Sun, and seen with a phase angle of 0° (fully illuminated).

accretion the collection of gas and dust to form large bodies such as stars, planets, and moons.

achondrite a stony meteorite that lacks chondrules.

active region a disturbed area on the Sun where strong magnetic fields emerge from the interior, resulting in prominences, sunspots, flares and other phenomena.

albedo the proportion of light falling on a body that is reflected.

Alfvén waves magnetic waves that can travel through an electrically conducting fluid or plasma in a magnetic field.

Amor a class of asteroids with perihelion distances between 1.017 and 1.3 AU.

anorthosite a granular, igneous rock composed almost wholly of anorthite, a calcium-rich silicate.

anticyclone a region in a planetary atmosphere where pressure increases toward the center.

aphelion the point in the orbit of an object at which it is furthest from the Sun.

apogee the point in the orbit of an Earth-orbiting object at which it is furthest from Earth.

Apollo a class of asteroids with semimajor axes greater than 1.0 AU and perihelion distances less than 1.017 AU.

apparent magnitude a measure of the apparent brightness of a planet or star. Very bright objects, such as Venus and the Sun have negative magnitudes.

asthenosphere the least rigid region of the mantle, located immediately below the lithosphere.

astronomical unit (AU) 149 597 870 km, which is approximately the average Sun-Earth distance.

Aten a class of asteroids with semimajor axes less than 1.0 AU and aphelion distances greater than 0.983 AU.

aurora a colourful, variable glow in a planetary ionosphere, produced by atoms and ions excited by collisions with energetic particles.

bar a unit of pressure equal to 0.987 atmospheres, the average air pressure on Earth's surface.

barchan a crescent-shaped sand dune.

barycenter the center of mass of a system of celestial bodies, such as the Earth-Moon system.

basalt a volcanic igneous rock which consists primarily of silicon, oxygen, iron, aluminum and magnesium.

biosphere the zone occupied by living organisms on a planet.

black dwarf a dead star that is no longer hot or luminous.

bow shock a sharp, curved boundary in the solar wind, upstream of a planetary magnetosphere or a cometary/planetary ionosphere, where a shock wave is created when the solar wind suddenly slows and is diverted; marked by heating, compression and deflection of the solar wind.

breccia a rock composed of broken rock fragments that were fused together by finer-grained material during an impact.

brown dwarf a gaseous object with a mass between 13 and 80 times the mass of Jupiter, which is too low for fusion to occur at its core.

caldera a large volcanic crater, often resulting from the partial collapse of the summit of a shield volcano or an explosive eruption.

Centaur a small body in the outer Solar System that has been scattered inward from the Kuiper Belt by gravitational interactions.

chaotic terrain areas of the Martian surface where the ground has collapsed, possibly due to the removal of subsurface groundwater or ice, to form a surface of jumbled blocks and isolated hills.

chasmata deep, steep-sided trenches on Venus which may be examples of giant rift valleys.

chondrites a very common class of stony meteorites that usually contain chondrules; carbonaceous chondrites also include carbon compounds.

chondrules small spherical grains, usually composed of iron, aluminum or magnesium silicates, found in abundance in primitive stony meteorites.

chromosphere the region of the solar atmosphere lying between the photosphere and corona.

clathrate a solid in which molecules of one compound are trapped in cavities in the crystalline lattice of another compound.

Exploring the Solar System, First Edition. Peter Bond.
© 2012 by Peter Bond. Published 2012 by Blackwell Publishing Ltd.

coma the spherical, gaseous envelope that surrounds the nucleus of a comet.

composite volcano a volcano composed of both solidified lava and ash.

conjunction an alignment of two objects so that they appear to be in approximately the same position in the sky, as seen from Earth. Mercury and Venus can move directly between Earth and the Sun (**inferior conjunction**) or directly opposite Earth on the far side of the Sun (**superior conjunction**). Planets further than Earth can only come to superior conjunction. This term is also used when planets and the Moon come within a few degrees of each other in the sky.

convection transport of heat energy by vertical atmospheric or fluid motions.

convective zone a region inside the Sun where convection is the dominant process of energy transfer.

core the center of a massive object such as the Sun or a planet.

Coriolis effect the modification of wind direction due to a planet's *Coriolis force.*

Coriolis force the force that appears to deflect anything moving freely between the pole of a rotating body and its equator (e.g. winds); produces cyclonic and anticyclonic weather patterns on Earth.

corona (1) the Sun's hot, highly ionized atmosphere; (2) a circular formation of ridges enclosing a central area of jumbled relief on Venus.

coronal hole an extended region of low density and temperature in the solar corona.

coronal mass ejection (CME) an outward eruption of billions of tons of plasma from the Sun's corona.

cosmic rays high energy atomic nuclei (mostly protons) that enter the Solar System from interstellar space.

cosmogony the study of how the Universe and objects within it (e.g. the Solar System) were formed.

crater density a measure of the closeness of impact crater distribution on a planetary surface, usually given as the number of craters of a given size per unit area.

crater ray a streak of ejected material extending radially beyond a crater's rim.

crust the outer solid layer of a planet.

cryovolcanism volcanism that occurs in icy materials rather than rock.

current sheet the two-dimensional surface within a magnetosphere that separates magnetic fields of opposite polarities.

cyclone a large atmospheric system characterized by the rapid inward circulation of air around a low pressure center.

deflation hollow a surface depression caused by removal of loose material by the wind.

depression a mid-latitude storm on Earth associated with a region of low atmospheric pressure.

deuterium a heavy isotope of hydrogen; its nucleus contains one proton and one neutron.

dune an elevated region of sand, usually in a desert or coastal region.

eccentricity (e) One of the six Keplerian orbital elements, it describes the shape of an orbit, – in particular, how much an elliptical orbit deviates from a circle. When e = 0, the orbit is a circle. When e is close to 1, it is a very elongated and narrow ellipse. Eccentricity is calculated by dividing the distance from the center to the focus of an ellipse (c) with the semimajor axis of the ellipse (a). The eccentricity is then given by the ratio c/a. The eccentricity of an ellipse must always be less than unity (1).

eclipse an event in which the light from a celestial object is partly or totally cut off by the passage of another object in front of it, e.g. a solar eclipse, when the Moon passes directly in front of the Sun. An eclipse of a star by another object is called an **occultation.**

ecliptic the mean plane of the Earth's orbit around the Sun; the Sun's apparent path across the celestial sphere.

ejecta material that was blasted out of an impact crater during its formation.

ellipse a closed, symmetrical curve that has two perpendicular axes, one of which is longer than the other.

equinox one of two dates on which the Sun crosses the celestial equator. At the vernal (or spring) equinox, the Sun crosses the equator from south to north on or near 21 March each year. At the autumnal equinox, the Sun crosses the equator from north to south on or near 23 September. On these dates, the duration of day and night is approximately equal all over Earth. (equinox is Latin for "equal nights")

exoplanet / extrasolar planet a planet orbiting a distant star.

exosphere the outermost, tenuous layer of a planet's atmosphere where it merges into interplanetary space, or the entire gaseous envelope of a planetary object where the atmosphere has similar characteristics to an exosphere.

extremophiles microbes that exist in environmentally extreme regions, such as sea ice, deep-sea vents, hot springs, and salt or soda lakes.

facula (plural faculae) (1) a bright region of the Sun's photosphere.

facula (2) a bright spot on a planet's surface, particularly Ganymede.

feldspar an aluminum-rich silicate mineral found in meteorites and other rocks.

Ferrel cell a mid-latitude circulation in Earth's atmosphere in which air near the surface moves poleward, then rises and returns returns toward lower latitudes.

filament a solar prominence projected against the bright disk of the Sun (i.e. seen from above) which appears as a dark streak across the photosphere.

flux the amount of radiation or number of particles arriving over a given period of time.

geocentric theory a theory in which a stationary Earth lies at the centre of the Universe so that the Sun, Moon, stars and all the other planets move around it.

graben a linear surface depression bounded on its long sides by normal faults and caused by stretching of the crust; commonly known as a rift valley.

granulation a mottled effect in the Sun's photosphere caused by columns of gas rising as a result of convection.

greenhouse effect the trapping of infrared radiation by gases in a planet's atmosphere, resulting in an elevated surface temperature.

habitable zone a region around a star where the temperature is thought to be suitable for the evolution of Earth-like life on a planetary body.

Hadley cell circulation in a planet's atmosphere in which gas heated by the Sun rises and moves poleward, then sinks and returns toward the equator.

heliocentric theory a theory in which the Sun lies at the centre of the Solar System, so that all of the planets move around it.

heliopause the outer edge of the heliosphere, where the pressure of the solar wind equals that of the interstellar medium.

helioseismology the study of the Sun's interior by analysis of oscillations in the solar convective zone.

heliosphere the region of space dominated by the Sun's magnetic field and the solar wind.

horst a block of crust that has been uplifted between parallel faults, sometimes known as a block mountain.

hot spot site above a plume of hot, rising material in the mantle which may result in the formation of volcanoes at the surface.

hydrocarbon a chemical compound made up of hydrogen and carbon atoms.

inclination the angle between an object's orbital plane and a reference plane, usually the ecliptic (for heliocentric orbits) or a planet's equator (for satellites).

inferior conjunction when a planet lies directly between Earth and the Sun.

insolation incoming solar radiation.

ion an atom that has gained or lost one or more electrons so that it has become electrically charged.

ionopause the lower boundary of the ionosphere.

ionosphere the upper region of a planet's atmosphere in which many atoms are ionized.

irregular satellite a satellite whose orbit is highly inclined and/ or highly eccentric.

isostatic uplift a very slow local uplift of crust due to the removal of surface material, (e.g. ice sheets).

jet stream a narrow ribbon of high speed wind circulating around a planet at high altitude.

Kirkwood gaps empty regions in the asteroid belt where the orbital periods of the asteroids are simple fractions of the orbital period of Jupiter.

KREEP acronym for lunar basaltic material rich in radioactive elements; K for potassium, REE for rare-earth elements, P for phosphorus.

Kuiper Belt a disk-shaped region beyond the orbit of Neptune that contains millions of icy objects; the source region for short period comets.

Lagrangian points five points in the orbital plane of two massive bodies where a third body can remain in equilibrium, so that all three bodies have a fixed geometrical configuration. Two Lagrangian points (L4 and L5) form equilateral triangles with the two primaries and are stable; the other three are unstable and lie on the line connecting the two primaries.

latent heat heat released or absorbed as the result of a phase change or change of physical state, for example, from water vapor (a gas) to liquid water.

libration a small oscillation, or apparent rocking motion, exhibited by a synchronously rotating satellite in a slightly eccentric orbit (e.g. the Moon).

lithosphere rigid outer layer of a planet, including the crust and part of the upper mantle.

magma hot, fluid rock material beneath the crust from which igneous rock is formed.

magnetic reconnection the breaking and reconnection of magnetic field lines which results in the release of magnetic energy.

magnetic substorm a sudden instability of the plasma in Earth's magnetic field which results from a build up of energy deposited by the solar wind. Such events cause a dramatic increase in the intensity and activity of the aurora.

magnetopause the outer boundary of a planet's magnetosphere.

magnetosheath the region between a planetary bow shock and magnetopause in which the solar wind plasma flows around the magnetosphere.

magnetosphere the region of space surrounding a planet in which the planet's magnetic field dominates that of the solar wind.

magnetotail the portion of a planetary magnetosphere pulled downstream by the solar wind.

mantle the layer inside a planet between its crust and core.

mare an area on the Moon or Mars that appears darker and smoother than its surroundings; plural "maria."

mascon subsurface mass concentration that causes large-scale gravity anomalies on the Moon and other bodies.

mesosphere an atmospheric region between the stratosphere and the thermosphere.

meteor shower a period of enhanced meteor activity caused by a planet's passage (usually at the same time each year) through the particles distributed along a comet's orbit.

meteor a streak of light in the sky caused when a particle of interplanetary dust burns up in the upper atmosphere; commonly known as a "shooting star."

meteorite a natural rocky/metallic object that survives its fall to Earth from space.

meteoroid a small natural body in orbit around the Sun; may be cometary or asteroidal in origin.

micrometeorite a very small meteorite or meteoritic particle less than a millimeter in diameter.

nanoflares small explosions on the Sun.

near-Earth object (NEO) an asteroid or comet whose orbit sometimes brings it close to Earth.

nebula a large region of interstellar gas and dust. (The word "nebula" is Latin for "cloud").

nebular hypothesis the theory that the Solar System formed from a nebula around the Sun.

neutrino a neutral relativistic particle of very small rest mass.

node one of two points on the celestial sphere where an orbit crosses a reference plane, e.g. the equatorial plane.

nucleus the solid core of a comet, consisting of a mixture of ices and solid silicate and carbonaceous grains.

nuée ardente a hot, glowing cloud of gas and pulverized rock which spreads sideways from an erupting volcano.

oblateness the equatorial distortion of an otherwise spherical planet, caused by its rotation; defined as the difference between the equatorial and polar radii, divided by the equatorial radius.

obliquity angle between a planet's equatorial plane and orbital plane; the tilt of the rotational axis. It causes the planet to have seasons.

occultation the passage of an object in front of another object with a smaller apparent angular size (e.g. when an asteroid or planet temporarily blocks a star from view).

olivine a green, metal-rich silicate mineral found in meteorites and other rocks.

Oort Cloud a roughly spherical region, extending more than 100 000 AU from the Sun, occupied by up to a trillion small icy bodies; the source region for long-period comets.

opposition the time when a planet orbiting beyond Earth is directly opposite the Sun in the sky (i.e. when its elongation is 180 degrees).

orbit the path followed by a small object located within the gravitational field of a larger object.

orbital inclination (see **inclination**)

organic a complex compound containing carbon though not necessarily associated with life.

outflow channels large channels that have few if any tributaries and are believed to have been formed by huge floods.

ozone hole a region in the ozone layer of Earth's upper atmosphere where the amount of ozone gas has been depleted by pollution.

ozone a form of oxygen that contains three atoms instead of the normal two.

palimpsest a roughly circular spot on icy satellites thought to identify a former crater and rim.

parallax the angular distance by which a celestial object appears to be displaced with respect to more distant objects when observed from two widely separated locations. The value of its parallax can be used to calculate an object's distance.

perigee the point in the orbit of an Earth-orbiting object at which it is closest to Earth.

perihelion the point in the orbit of an object at which it is nearest to the Sun.

permafrost near-surface zone within which temperatures are always below 0°C. It may or may not contain ground ice.

phase angle the angle between the direction of illumination (usually the Sun) and a planet or other celestial object as seen by an observer.

phase the extent to which the surface of an illuminated object (e.g. the Moon), is visible.

photodissociation the breakdown of molecules due to the absorption of light, especially ultraviolet sunlight.

photometry the branch of astronomy concerned with measuring the brightness of a celestial object at various wavelengths.

photosphere the visible surface of the Sun.

photosynthesis a chemical process in organisms, such as plants and some bacteria, that converts carbon dioxide into organic compounds, especially sugars, using the energy from sunlight.

plagioclase a common rock-forming silicate mineral.

planetesimal a primordial body of intermediate size, up to perhaps 1 km across, which accreted into planets or asteroids.

plasma torus a donut-shaped ring of plasma.

plasma completely ionized gas, consisting of free electrons and atomic nuclei, in which the temperature is too high for neutral atoms to exist.

plate tectonics the movement of segments of the lithosphere under the influence of convection in the mantle.

plate (see **tectonic plate**)

polar layered deposits layers of dust and ice which surround and lie beneath the permanent ice caps on Mars.

polarization the process of affecting radiation, especially light, so that its electromagnetic vibrations are not randomly oriented.

precession of the ellipse (see **precession**)

precession (1) a slow, periodic conical motion of the rotation axis of a spinning body due to external gravitational influences. (2) a gradual rotation of a planet's elliptical orbit, shown by a shift in the alignment of the line joining aphelion and perihelion.

primordial existing at or near the very beginning of the Solar System, about 4.6 billion years ago.

prograde anticlockwise direction of rotation and orbital motion, as seen from above Earth's north pole. The Sun and most of the planets in the Solar System rotate in this direction, and all of the planets have prograde orbits.

prominence name given to large, flame-like structures in the Sun's chromosphere and corona.

protoplanetary disk a disk of gas and dust around a young star where a planetary system may form.

protoplanets the largest planetary embryos which grew from material in the solar nebula and eventually collided to form planets.

radiation pressure the pressure exerted by photons of electromagnetic radiation when they transfer momentum to matter.

radiative zone a region inside the Sun where high energy photons of radiation collide with electrons and ions and are then reradiated in the form of light and heat.

red giant a large, cool, red star that has grown greatly in size after consuming all of the hydrogen fuel needed for nuclear fusion in its core.

refractory an element that vaporizes at high temperatures (e.g. aluminum, calcium and uranium).

regolith fragmented rocky debris, produced by numerous meteorite impacts, which forms the upper surface on planets, satellites, and asteroids.

regular satellite a prograde-moving satellite whose orbit has low eccentricity and inclination.

remote sensing any technique for measuring an object's characteristics from a distance.

resonance a state in which one orbiting object is subject to periodic gravitational perturbations by another.

retrograde clockwise direction of rotation and orbital motion, as seen from above Earth's north pole. Retrograde motion is unusual in the planets of the Solar System, but common among comets and smaller, irregular planetary satellites.

rille a trenchlike valley, up to several hundred km long and 1 to 2 km wide, on the surface of the Moon or other satellites.

Roche limit the imaginary boundary around a planet within which a "rubble pile" satellite will be torn apart by gravitational forces. Particles in this region cannot gravitationally coalesce to form a satellite because of the disruptive effect of tidal forces. Located at 2.46 planetary radii for objects with little internal strength, and 1.44 radii for small, rocky satellites.

rock pedestal a rock pinnacle whose base has been mostly eroded, leaving it perched on a narrow support.

salt flat a deposit of salt left behind when an inland lake has disappeared due to evaporation.

saltation a process of particle transport by wind or water which involves a series of hops or leaps. When a particle completes each "hop" it may dislodge other particles and set them in motion.

scarp a cliff produced by faulting or erosion.

secondary crater an impact crater produced by ejecta from a primary impact.

seif dune a type of longitudinal dune, generally oriented in a direction parallel to the prevailing wind.

seismology the science of the study of earthquakes.

semimajor axis (a) the mean orbital distance of an object from its primary; half of the longest axis (diameter) of an ellipse.

shepherd satellite a satellite that sustains the structure of a planetary ring through its close gravitational influence.

shield volcano a broad volcano with a large summit pit (caldera) formed by collapse and gently sloping flanks, built mainly from numerous overlapping flows of fluid, basaltic lava.

sidereal day Earth's rotation period with respect to the background of "fixed" stars (23 hours, 56 minutes and 4 seconds).

sidereal month the time taken for one revolution of the Moon around Earth, relative to a fixed star.

sidereal period the orbital period of a planet or other celestial body with respect to a background star; the true orbital period.

SNC meteorites group of meteorites (Shergotty-Nakhla-Chassigny) believed to have come from Mars because of their young ages, basaltic composition, and inclusion of gases with the same composition as the Martian atmosphere.

solar constant the total amount of solar energy irradiating a given surface area in a given interval of time; at 1 AU, this value is 1367 watts per square meter.

solar day the period of time Earth takes to rotate through 360 degrees.

solar flare a sudden release of energy in or near the Sun's photosphere that accelerates charged particles into space.

solar nebula the disk-shaped cloud of gas and dust where the Solar System formed (also known as the protosolar nebula or protoplanetary nebula).

solar wind the high-speed outflow of energetic charged particles (mainly protons and electrons) and entrained magnetic field lines from the solar corona.

solstice time when the Sun reaches the most northerly or southerly position in its yearly path along the ecliptic plane. At the solstices the Sun reaches a declination of 23.5°N or 23.5°S. The summer solstice occurs around 21 June and the winter solstice occurs on or about 22 December.

spectroscopy the study of the light emitted from or reflected by a body (its spectrum).

spectrum the distribution of wavelengths of electromagnetic radiation emitted from or reflected by a body.

spicules spike-like structures in the Sun's chromosphere which may be observed above the solar limb.

spin-orbit coupling a mathematical relationship between an object's orbital period and its period of rotation. The most common example is a 1:1 coupling, which results in the same hemisphere of the Moon always facing toward Earth.

spin-orbit resonance (see **spin-orbit coupling**)

stratosphere the layer of a planet's atmosphere above the troposphere and below the ionosphere; on Earth, its temperature increases with altitude.

stratovolcano a volcano composed of layers of solidified lava and ash.

subduction the process by which by one crustal plate is forced under another.

sunspot a relatively cool, dark area of the solar photosphere.

superior conjunction the point in the orbit of an inferior planet (Mercury or Venus) when it lies on the opposite side of the Sun to Earth.

superrotation the characteristic of an atmosphere that rotates faster than its planet (e.g. Venus).

synchronous rotation when the axial rotation of a celestial body has the same duration as the orbital period of that body (e.g. the Moon).

synodic month the period between two identical phases of the Moon.

tectonic plate a large slab of a planet's lithosphere with independent motion.

terminator the boundary between night and day.

tesserae areas of very rough, elevated terrain on Venus which show evidence of a complex deformational history ridges, fractures and rift valley structures.

thermosphere the atmospheric region where temperature rises due to ionospheric heating.

tholin a complex hydrocarbon created by the action of ultraviolet radiation on a mixture of volatile compounds containing carbon, hydrogen, nitrogen, and oxygen.

tidal heating the heating of a satellite's interior due to the tidal friction induced by the strong gravitational field of its planet.

transform fault a type of crustal fault where slabs of crust or tectonic plates slide sideways past each other (e.g the San Andreas fault in California).

transit the passage of Mercury or Venus across the visible disk of the Sun; the passage of an exoplanet across the visible disk of a distant star; the passage of a satellite across the disk of its planet; the passage of a star or celestial object across the observer's meridian as the result of the daily apparent motion of the celestial sphere (see **occultation**).

transverse dune a large, asymmetrical, elongated dune lying at right angles to the prevailing wind direction.

tropopause the boundary between the troposphere and the stratosphere.

troposphere the lowest layer of Earth's atmosphere, where most weather takes place, and the convection-dominated region of other planetary atmospheres.

Van Allen radiation belts ring-shaped regions around Earth, discovered by James van Allen in 1958, where high-energy charged particles have been trapped by the planet's magnetic field.

volatiles elements or molecules with low melting temperatures (e.g. water and ammonia).

weathering the *in situ* breakdown of rock by elements of weather – rain, ice, temperature change etc.

white dwarf a very small, hot stellar remnant created when a star collapses after consuming all of its possible sources of fuel for thermonuclear fusion.

yardang a large area of soft rock which has been eroded by wind action into alternating ridges and furrows, mostly parallel to the dominant wind direction.

zenith the overhead point.

zodiac a belt of 12 constellations through which the Sun's path across the sky – the ecliptic – passes.

zodiacal light a faint glow in the night sky caused by sunlight scattering off interplanetary dust near the plane of the ecliptic.

Further Reading

General

Beatty, J. Kelly, Peterson, C. C., & Chaiken, A. L. (1999). *The new Solar System* (4th ed.). Sky Publishing Corp. / Cambridge University Press.

Greeley, R., & Batson, R. (2001). *The compact NASA atlas of the Solar System*. Cambridge University Press.

Hartmann, W. K. (2005). *Moons & planets* (5th ed.). Thomson.

Jones, B. W. (2004). *Life in the Solar System and beyond*. Springer-Praxis.

Jones, B. W. (2007). *Discovering the Solar System* (2nd ed.). Wiley.

JPL Solar System. Retrieved from http://www.jpl.nasa.gov/solar-system/index.cfm

Kluger, J. (1999). *Journey beyond Selene*. Simon & Schuster.

Kraemer, R. S. (2000). *Beyond the Moon: Golden age of planetary exploration 1971–1978*, Smithsonian.

Logsdon, J. M. (Ed.). (2001). *Exploring the unknown: selected documents in the history of the U.S. civil space program, volume v, exploring the cosmos*. (NASA SP-4407).

Morrison, D., & Owen, T. (2003). *The planetary system* (3rd ed.). Addison Wesley.

Murray, B. W. (1989). *Journey into space: The first thirty years of space exploration*. W. Norton & Co.

Nicks, O. W. (1985). *Far travelers: The exploring machines*. (NASA SP-480).

Planetary Society. Retrieved from www.planetary.org/

Sagan, C. (1980). *Cosmos*. MacDonald & Co.

Sagan, C. (1995). *Pale blue dot*. Headline.

Siddiqi, A. A. (2002). *Deep space chronicle: A chronology of deep space and planetary probes, 1958–2000*. (NASA Monograph in Aerospace History #24).

Solar System Exploration. Retrieved from http://solarsystem.nasa.gov/index.cfm

Stephenson, D. J. (2002, November). Planetary oceans. *Sky & Telescope. 104*: 38–44.

The 8 Planets. Retrieved from http://nineplanets.org/

Views of the Solar System. Retrieved from http://www.solarviews.com/eng/homepage.htm

Wilson, A. (1987). *Solar System Log*. Jane's Publications.

Chapter 1: Beginnings

Alibert, Y., Broeg, C., Benz, W., et al. (2010, January). Origin and formation of planetary systems. *Astrobiology, 10*(1), 19–32.

Bond, P. (2007). *Distant worlds: Milestones in planetary exploration*. Copernicus/Praxis.

Doody, D. (2007). *Basics of interplanetary flight*. Retrieved from http://www.jpl.nasa.gov/basics

Jayawardhana, R. (2003, April). Planets in production: Making new worlds. *Sky & Telescope. 105*, 36.

Krimigis, S. M., Mitchell, D.G., Roelof, E.C., et al. (2009, November). Re-imagining the heliosphere. [Special issue] *Science, 326*(5951), 959–971.

Laughlin, G. P. (2007, June). From here to eternity: The fate of the sun and the earth. *Sky & Telescope. 112*, 32.

Malhotra, R., & Minton, D. A. (2009, February). A record of planet migration in the main asteroid belt. *Nature, 457*, 1109–1111.

Schilling, G. (1999, October). From a swirl of dust, a planet is born. *Science, 286*, 66–68.

Scott, E. R. D. (2006, June). Meteoritical and dynamical constraints on the growth mechanisms and formation times of asteroids and Jupiter. *Icarus*. Retrieved from http://arxiv.org/ftp/astro-ph/papers/0607/0607317.pdf

Soter, S. (2007, January). What is a Planet? *Scientific American. 296*(1), 34–41.

Stevenson, D. J. (1991). Giant planets and their satellites: What are the relationships between their properties and how they formed? *National Academies Press*. Retrieved from http://www.nap.edu/books/0309043336/html/163.html

Strom, R. G., Michels, J., Malhotral, R., et al. (2005, September). The origin of planetary impactors in the inner Solar System. *Science, 309*, 1847–1850.

Taylor, G. J. (2003, May). Triggering the formation of the Solar System. *PSR Discoveries*. Retrieved from http://www.psrd.hawaii.edu/May03/SolarSystemTrigger.html

Taylor, G. J. (2006, January). Cosmochemistry from nanometers to light-years. *PSR Discoveries*. Retrieved from http://www.psrd.hawaii.edu/Jan06/protoplanetary.html

Exploring the Solar System, First Edition. Peter Bond.
© 2012 by Peter Bond. Published 2012 by Blackwell Publishing Ltd.

Taylor, G. J. (2006, November). Hit-and-run as planets formed. *PSR Discoveries*. Retrieved from http://www.psrd.hawaii.edu/Nov06/hit-and-run.html

Thommes, E. W., Matsumura, S., & Rasio, F. A. (2008, August). Gas disks to gas giants: simulating the birth of planetary systems. *Science, 321*, 814–817.

Tsiganis, K., Gomes, R., Morbidelli, A., et al. (2005, May). Origin of the orbital architecture of the giant planets of the Solar System. *Nature, 435*, 459–461.

Vinković, D. (2009, May). Radiation-pressure mixing of large dust grains in protoplanetary disks. *Nature, 459*, 227–229.

Voyager – The interstellar mission. Retrieved from http://voyager.jpl.nasa.gov/

Walsh, K. J. (2009, February). Asteroids: when planets migrate. *Nature, 457*, 1091–1093.

Wetherill, G. W. (1991). *Formation of the terrestrial planets from planetesimals*. National Academies Press. Retrieved from http://www.nap.edu/books/0309043336/html/98.html

Wood, J. A. (1999, January). Forging the planets. *Sky & Telescope*.

Chapter 2: The Sun

A super solar flare. (2008, May). Science@NASA. Retrieved from http://science.nasa.gov/headlines/y2008/06may_carringtonflare.htm

Are sunspots disappearing? (2009, September). Science@NASA. Retrieved from http://science.nasa.gov/headlines/y2009/03sep_sunspots.htm

Bahcall, J. N. (2004, April). *Solving the mystery of the missing neutrinos*. Retrieved from http://nobelprize.org/nobel_prizes/physics/articles/bahcall/

Big Bear solar observatory. Retrieved from http://www.bbso.njit.edu/

De Pontieu, B., Erdélyi, R., & James, S. P. (2004, July). Solar chromospheric spicules from the leakage of photospheric oscillations and flows. *Nature, 430*, 536–539.

Ehrenstein, D. (1997, September). SOHO traces the Sun's hot currents. *Science, 277*, 1438.

Fleck, B., Müller, D., Haugan, S., Duarte, L. S., & Siili, T. (2006, May). 10 Years of SOHO. *ESA Bulletin 126*, 24–32. http://sohowww.nascom.nasa.gov/publications/ESA_Bull126.pdf

Foukal, P., Fröhlich, C., Spruit, H. et al. (2006, September). Variations in solar luminosity and their effect on the Earth's climate. *Nature, 443*, 161–166.

Global Oscillation Network Group (GONG). Retrieved from http://gong.nso.edu/

Golub, L., & Pasachoff, J. M. (2002). *Nearest star: The surprising science of our Sun*. Harvard University Press.

Gough, D. (2000, March). News from the solar interior. *Science, 287*, 2434–2435.

Gough, D. O., Kosovichev, A. G., Toomre, J., et al. (1996, May). The seismic structure of the Sun. *Science, 272*, 1296–1300.

Haisch, B., & Schmitt, J. (1999, October). The solar-stellar connection. *Sky & Telescope. 98*(4), 46–54.

Hanson, B., & Hoss, D. (2008, December). Results from Hinode. [Special Issue]. *Science, 318*, 1572–1599.

Hassler, D. M., Dammasch, I. E., Lemaire , P., et al, (1999). Solar wind outflow and the chromospheric magnetic network. *Science, 283*, 810–813.

Hill, S., & Carlowicz, M. (2006). *The Sun*. Harry N. Abrams, Inc.

Hinode (Solar-B) mission (JAXA). Retrieved from http://www.isas.jaxa.jp/e/enterp/missions/solar-b/index.shtml

Hinode (Solar-B) mission (NASA). Retrieved from http://solarb.msfc.nasa.gov/

Hudson, H. S. & Kosugi, T. (1999, August). How the Sun's corona gets hot. *Science, 285*, 849.

James, C. R. (2007, July). Solar forecast: Storm ahead. *Sky & Telescope. 114*(1), 24.

Jess, D. B., Mathioudakis, M., Erdélyi, R., et al. (2009, March). Alfven waves in the lower solar atmosphere. *Science, 323*, 1582.

Lang, K. R. (1998). *SOHO Reveals the secrets of the Sun*. Magnificent Cosmos – Scientific American Special, Springer.

Lang, K. R. (2001). *The Cambridge encyclopedia of the Sun*. Cambridge University Press.

Lang, K. R. (2006). *Sun, Earth and Sky* (2nd ed.). Springer.

Lang, K. R. (2008). *The Sun from Space* (2nd ed.). Springer.

Laughlin, G. P. (June, 2007). From here to eternity: The fate of the Sun and the Earth. *Sky & Telescope. 112*, 32.

Lindsey C., & Braun, D.C. (2000, March). Seismic images of the far side of the Sun. *Science, 287*, 1799–1801.

Lockheed Martin Solar and Astrophysics Laboratory. Retrieved from http://www.lmsal.com/

McComas, D. J., Elliott, H. A., Gosling, J.T., et al. (2006, May). Ulysses observations of very different heliospheric structure during the declining phase of solar activity cycle 23. *Geophysical Research Letters, 33*, L09102.

Moran, T. G., & Davila, J. M., (2004, July). Three-dimensional polarimetric imaging of coronal mass ejections, *Science, 305*, 66–70.

Mystery of the missing sunspots solved? (2009, June). Science@NASA. Retrieved from http://science.nasa.gov/headlines/y2009/01apr_deepsolarminimum.htm

National Solar Observatory web site. Retrieved from http://www.nso.edu/

NOAA Space Environment Center. Retrieved from http://www.sec.noaa.gov/

Rempel, M., Schüssler, M., Cameron R. H., et al. (2009, July). Penumbral structure and outflows in simulated sunspots. *Science, 325*, 171–174.

Schriver, C. J. (2006, April). The science behind the solar corona. *Sky & Telescope*.

Science from Hinode. (April 2008). [Special Issue]. *Astronomy and Astrophysics, 481*. Retrieved from http://www.aanda.org/index.php?option=toc&url=/articles/aa/abs/2008/13/contents/contents.html

Shepherd, D. (2007, June). Stellar origins: From the cold depths of space. *Sky & Telescope*.

Smith, E. J., Marsden, R. G., & Page, D. E. (1995, May). Ulysses above the Sun's South Pole. *Science, 268*, 1005–1033.

Smith, E. J., Marsden, R.G., Balogh, A., et al. (2003). The Sun and heliosphere at solar maximum, *Science, 302*(1), 1165–1169.

SOHO mission. Retrieved from http://soho.nascom.nasa.gov/

Solar Physics Group, NASA Marshall Space Flight Center: Retrieved from http://solarscience.msfc.nasa.gov/index.html

Solar Wind Loses Power, Hits 50-year Low. (2008, September). Science@NASA. Retrieved from http://science.nasa.gov/headlines/y2008/23sep_solarwind.htm

STEREO mission. Retrieved from http://stereo.jhuapl.edu/

Sudbury Neutrino Observatory. Retrieved from http://www.sno.phy.queensu.ca/

Swedish 1-meter Solar Telescope. Retrieved from http://www.solarphysics.kva.se/

Taylor, G. J. (2007, July). The Sun's crowded delivery room, PSR Discoveries. Retrieved from http://www.psrd.hawaii.edu/July07/iron-60.html

Ulysses mission. Retrieved from http://helio.esa.int/ulysses/

Yohkoh mission. Retrieved from http://www.lmsal.com/YPOP/homepage.html

Chapter 3: Earth

Bintanja, R., & van de Wal, R. S. W. (2008, August). North American ice-sheet dynamics and the onset of 100,000-year glacial cycles. *Nature, 454,* 869–872.

Destination Earth (NASA). Retrieved from http://www.earth.nasa.gov/

Donnadieu, Y., Godderis, Y., & Ramstein, G. (2004). Cracking continents caused 'snowball Earth.' *Nature, 428,* 303–306.

Durda, D. D. (2006, November). The most dangerous asteroid ever found. *Sky & Telescope, 112*(11), 29.

Earth impact database. University of New Brunswick. Retrieved from http://www.passc.net/EarthImpactDatabase/index.html

Earth impact effects program. Retrieved from http://impact.ese.ic.ac.uk/

Earth pages (Blackwell). Retrieved from http://www.earth-pages.com/default.asp

Earth science world. Retrieved from http://www.earthscienceworld.org

Earth's Fidgeting Climate: http://science.nasa.gov/headlines/y2000/ast20oct_1.htm?list153136

Encyclopedia of the Atmospheric Environment. Retrieved from http://www.ace.mmu.ac.uk/eae/english.html

ESA Earth observation. Retrieved from http://www.esa.int/esaEO/

Hays, J. D., Imbrie, J., & Shackleton, N. J., (1976). Variation in the Earth's orbit: Pacemaker of the ice ages. *Science, 194,* 1121–1132.

Intergovernmental panel on climate change. Retrieved from http://www.ipcc.ch/

James, C. R. (2007, July). Solar forecast: storm ahead. *Sky & Telescope.*

Kasting, J. F. (1993). Earth's early atmosphere. *Science, 259,* 920–926.

Kious, W. J., & Tilling, R. I. (1996). *This dynamic Earth: The story of plate tectonics.* USGS. Retrieved from http://pubs.usgs.gov/gip/dynamic/dynamic.html

Kring, D. A. (1999, December). Calamity at meteor crater. *Sky & Telescope, 98*(3), 48–53.

McPhaden, M. J., Zebiak, S.E., Glantz M.H. et al. (2006). ENSO as an integrating concept in Earth science. *Science, 314,* 1740–1745.

NASA Earth observatory. Retrieved from http://earthobservatory.nasa.gov/

NASA EOS IDS volcanology. Retrieved from http://eos.higp.hawaii.edu/

NASA near Earth object program. Retrieved from http://neo.jpl.nasa.gov/

NASA ozone hole watch. Retrieved from http://ozonewatch.gsfc.nasa.gov/index.html

Ocean surface currents (University of Miami). Retrieved from http://oceancurrents.rsmas.miami.edu/index.html

Ocean surface topography from space (NASA). Retrieved from http://sealevel.jpl.nasa.gov/

Pacchioli, D. (2001, April). Reflections from a warm little pond. *Astrobiology Magazine.* Retrieved from http://www.astrobio.net/news/article5.html

Palaeos: The trace of life on earth. Retrieved from http://www.palaeos.com/Default.htm

Paleomap project (Christopher R. Scotese). Retrieved from http://www.scotese.com/Default.htm

Solar views – terrestrial volcanoes. Retrieved from http://www.solarviews.com/eng/tervolc.htm

Tarduno, J. A., Duncan R.A., Scholl, D.W., et al. (2003). The emperor seamounts: Southward motion of the Hawaiian hotspot plume in Earth's mantle, *Science, 301,* 1064–1069.

USGS earthquake hazard program. Retrieved from http://earthquake.usgs.gov/

USGS volcano hazards program. Retrieved from http://volcanoes.usgs.gov/

Volcano World. Retrieved from http://www.volcanoworld.org/

Walker, A. S. (2000). Deserts: Geology and resources. *USGS.* Retrieved from http://pubs.usgs.gov/gip/deserts/contents/

Chapter 4: The Moon

Apollo image atlas. Retrieved from http://www.lpi.usra.edu/resources/apollo/

Apollo lunar surface journal. Retrieved from http://www.hq.nasa.gov/alsj/

Brandon, A. (2007, December). A younger moon. *Nature, 450,* 1169–1170.

Cameron, A. G. W. (1997). *The origin of the moon and the single impact hypothesis.* Retrieved from http://www.xtec.es/recursos/astronom/moon/camerone.htm

Campbell, B.A., Campbell, D.B., Chandler, J.F., et al. (2003). Radar imaging of the lunar poles. *Nature, 426,* 137–138.

Cohen, B. A. (2001, January). Lunar meteorites and the lunar cataclysm. *PSR Discoveries.* Retrieved from http://www.psrd.hawaii.edu/Jan01/lunarCataclysm.html

Consolidated lunar atlas. Retrieved from http://www.lpi.usra.edu/resources/cla/

Digital lunar orbiter photographic atlas of the moon. Retrieved from http://www.lpi.usra.edu/resources/lunar_orbiter/?info1083

Foust, J. A. (1998, September). NASA's New Moon. *Sky & Telescope.*

Geologic history of the moon. Retrieved from cps.earth.northwestern.edu/ GHM/ghm_hplates.html

Harland, D. M. (1999). *Exploring the Moon: The Apollo expeditions,* Springer Praxis.

Introduction to cratering studies, Planetary Science Institute, Tucson, Arizona. Retrieved from http://www.psi.edu/projects/mgs/cratering.html

Investigating the Moon's far side (3 articles). (2009, February). *Science, 323,* 897–912.

LCROSS and Lunar Reconnaissance Orbiter. (2010, October). Retrieved from http://www.nasa.gov/mission_pages/LCROSS/main/oct_21_media_telecon.html

Lunar meteorites (Randy Korotev). Retrieved from http://epsc.wustl.edu/admin/resources/moon_meteorites.html

Lunar photo of the day. Retrieved from http://www.lpod.org/

Lunar Prospector results. (1998, September). *Science, 281,* 1475–1500.

Lunar Reconnaissance Orbiter mission. Retrieved from http://lunar.gsfc.nasa.gov/

Martel, L. M. V. (1999, September). The Moon at its core. *PSR Discoveries.* Retrieved from http://www.psrd.hawaii.edu/Sept99/MoonCore.html

Martel, L. M. V. (2003, June). The Moon's dark, icy poles. *PSR Discoveries.* Retrieved from http://www.psrd.hawaii.edu/June03/lunarShadows.html

Martel, L. M. V. (2004, December). Composition of the Moon's crust. *PSR Discoveries.* Retrieved from http://www.psrd.hawaii.edu/Dec04/LunarCrust.html

Martel, L. M. V. (2004, September). Lunar crater rays point to a new lunar time scale. *PSR Discoveries.* http://www.psrd.hawaii.edu/Sept04/LunarRays.html

Martel, L. M. V. (2009, December). Celebrated Moon rocks. *PSR Discoveries*. Retrieved from http://www.psrd.hawaii.edu/Dec09/Apollo-lunar-samples.html

Masursky, H., Colton, G. W., & El-Baz, F. (Eds.). (1978). *Apollo over the Moon: A view from orbit*. (NASA SP-362). Retrieved from http://www.hq.nasa.gov/office/pao/History/SP-362/contents.htm

Nozette, S., Lichtenberg, C.L., Spudis, P., et al. (1996). The Clementine Bistatic radar experiment. *Science, 274*, 1495–1498.

Rocks and soils from the Moon (JSC-NASA). Retrieved from http://www-curator.jsc.nasa.gov/curator/lunar/lunar.htm

Selene (Kaguya) mission science reports. (2009, February). *Science, 323*, 897–912.

Smart-1 mission. Retrieved from http://sci.esa.int/science-e/www/area/index.cfm?fareaid=10

Spudis, P. D. (1996). *The once and future moon*. Smithsonian Institution Press.

Spudis, P. D. (1996, December). Ice on the bone dry Moon. *PSR Discoveries*. http://www.psrd.hawaii.edu/Dec96/IceonMoon.html

Spudis, P. D. (2003, January). The new Moon. *Scientific American*.

Taylor, G. J. (1997, October). Moonbeams and elements. *PSR Discoveries*. Retrieved from http://www.psrd.hawaii.edu/Oct97/MoonFeO.html

Taylor, G. J. (1998, December). Origin of the Earth and Moon. *PSR Discoveries*. Retrieved from http://www.psrd.hawaii.edu/Dec98/OriginEarthMoon.html

Taylor, G. J. (2000, August). A new Moon for the twenty-first century. *PSR Discoveries*. Retrieved from http://www.psrd.hawaii.edu/Aug00/newMoon.html

Taylor, G. J. (2000, December). Recipe for high-titanium lunar magmas. *PSR Discoveries*. Retrieved from http://www.psrd.hawaii.edu/Dec00/highTi.html

Taylor, G. J. (2000, June). The surprising Lunar Maria. *PSR Discoveries*. Retrieved from http://www.psrd.hawaii.edu/June00/lunarMaria.html

Taylor, G. J. (2002, October). The Moon beyond 2002. *PSR Discoveries*. Retrieved from http://www.psrd.hawaii.edu/Oct02/moon.html

Taylor, G. J. (2004, October). New lunar meteorite provides its lunar address and some clues about early bombardment of the Moon. *PSR Discoveries*. Retrieved from http://www.psrd.hawaii.edu/Oct04/SaU169.html

Taylor, G. J. (2005, November). Gamma rays, meteorites, lunar samples, and the composition of the Moon. *PSR Discoveries*. Retrieved from http://www.psrd.hawaii.edu/Nov05/MoonComposition.html

Taylor, G. J. (2006, April). Finding basalt chips from distant maria. *PSR Discoveries*. http://www.psrd.hawaii.edu/April06/basaltFragments.html

Taylor, G. J. (2007, April). Two views of the Moon's composition. *PSR Discoveries*. Retrieved from http://www.psrd.hawaii.edu/April07/Moon2Views.html

Terada, K., Anand, M., Sokol A.K., et al. (2007, December). Cryptomare magmatism 4.35 Gyr ago recorded in lunar meteorite Kalahari 009. *Nature, 450*, 849–852.

Weitz, C. M. (1997, February). Explosive volcanic eruptions on the Moon. *PSR Discoveries*. Retrieved from http://www.psrd.hawaii.edu/Feb97/MoonVolcanics.html

Wilhelms, D. (1987). *Geologic history of the Moon*. U.S. Geological Survey Professional Paper 1348. Retrieved from http://cps.earth.northwestern.edu/GHM/

Wilhelms, D. E. (1993). *To a rocky Moon, a geologist's history of lunar exploration*. University of Arizona Press. Retrieved from http://www.lpi.usra.edu/publications/books/rockyMoon/

Wood, C. A. (2007). The Moon's far side: Nearly a new world. *Sky & Telescope*.

Chapter 5: Mercury

BepiColombo web site. Retrieved from http://www.sci.esa.int/bepicolombo

Correia, A. C. M., & Laskar, J. (2004, June). Mercury's capture into the 3/2 spin-orbit resonance as a result of its chaotic dynamics. *Nature, 429*, 848–850.

Dunne, J., & Burgess, E. (1978). *The voyage of Mariner 10*. (NASA SP-424).

Harmon, J. K. (2001). Mercury radar imaging at Arecibo. *LPI abstract*. Retrieved from www.lpi.usra.edu/meetings/mercury01/pdf/8001.pdf

Ice on Mercury. NSSDC. Retrieved from http://nssdc.gsfc.nasa.gov/planetary/ice/ice_mercury.html

Kerr, R. A. (2005, January). Mercury's lively core. *Science Now*.

Mariner 10 archive project (Mark Robinson, Northwestern University). Retrieved from http://cps.earth.northwestern.edu/merc.html

Mariner 10 preliminary science report. (1974). *Science, 185*, 141–180.

MESSENGER web site. Retrieved from http://messenger.jhuapl.edu

MESSENGER's first Mercury flyby. (2008, July). [Special issue]. *Science, 321*, 59–94.

MESSENGER's second Mercury flyby. (2009, May). *Science, 324*, 613–621.

MESSENGER's third Mercury flyby. (2010, August). *Science, 329*, 665–675.

Robinson, M. S., & Lucey, P. G. (1997, January). Recalibrated Mariner 10 color mosaics: implications for mercurian volcanism. *Science, 275*, 197–200.

Strom, R. (1987). *Mercury, the elusive planet*. Smithsonian Institution Press.

Strom, R. G., & Sprague, A. L. (2003). *Exploring Mercury: The iron planet*. Springer-Praxis.

Taylor, G. J. (1997, January). Mercury unveiled. *PSR Discoveries*. Retrieved from http://www.psrd.hawaii.edu/Jan97/MercuryUnveiled.html

Taylor, G. J. (2001, October). New data, new ideas, and lively debate about Mercury. *PSR Discoveries*. Retrieved from http://www.psrd.hawaii.edu/Oct01/MercuryMtg.html

The Atlas of Mercury. (NASA SP-423). (NASA History Web site). Retrieved from http://history.nasa.gov/SP-423/mariner.htm

Vilas, F., Chapman, C. R., & Matthews, M. S. (Eds.). (1988). *Mercury*. The University of Arizona Press.

Watters, T. R., Mark S. R., Bina C.R., et al, (2004). Thrusts faults and the global contraction of Mercury. *Geophysical Research Letters, 31*. Retrieved from http://www.earth.northwestern.edu/people/craig/publish/pdf/grl04.pdf

Watters, T. R., Schultz, R.A., Robinson, M.S., et al. (2002, June). The mechanical and thermal structure of Mercury's early lithosphere. *Geophysical research letters, 29*. Retrieved from http://www.mines.unr.edu/geo-eng/schultz/pdf%20reprints/Watters+Merc.GRL02.pdf

Chapter 6: Venus

Bullock, M. A., & Grinspoon, D. H. (2003). Global climate change on Venus, new light on the Solar System. *Scientific American Special*.

Cattermole, P. (1994). *Venus: The geological story*. London: UCL Press.

Dunne, J., & Burgess, E. (1978). *The voyage of Mariner 10*. (NASA SP-424).

Fimmel R., Colin, L., & Burgess, E. (1983). *Pioneer Venus*. (NASA SP-461).

Fimmel R., Colin, L., & Burgess, E. (1995). *Pioneering Venus*, (NASA

SP-518). Retrieved from http://ntrs.nasa.gov/archive/nasa/casi.ntrs.nasa.gov/19960026995_1996055001.pdf

Grinspoon, D. (1998). *Venus revealed*. New York: Perseus Publishing.

Hansen, Vicki L., & Phillips, Roger J., (1993). Tectonics and volcanism of eastern Aphrodite Terra, Venus: No subduction, no spreading. *Science, 260*, 526–530.

Hunten, D. M., Colin, L., Donahue, T. M., et al. (Eds.). (1993). *Venus*. University of Arizona Press.

Kaula, W. M. (1990, March). Venus: A contrast in evolution to Earth. *Science, 247*, 1191–1196.

Magellan at Venus, (1992, August). *Journal of Geophysical Research, 97*, 13,062–13,689.

Magellan mission. Retrieved from http://www2.jpl.nasa.gov/magellan/

Malin, M. C. (1992). Mass movements on Venus: Preliminary results from Magellan Cycle I observations. *Journal of Geophysical Research. 97* (E10) 16337–16352. Retrieved from http://www.msss.com/venus/landslides/venus_paper.html

Pettengill, G. H., Campbell, D. B., & Masursky, H. (1980, August). The surface of Venus. *Scientific American, 243*, 54–65.

Prinn, R. (1990). The volcanoes and clouds of Venus. Exploring space. *Scientific American Special.*

Sagdeev, R. Z., Linkin V. M., Kerzhanovich, V.V., et al. (1986). Overview of VEGA Venus balloon in situ meteorological measurements. *Science, 231*, 1411–1414.

Sánchez-Lavega, A., Hueso, R., Piccioni, G., et al. (2008, July). Variable winds on Venus mapped in three dimensions. *Geophysical. Research Letters, 35*, L13204.

Saunders, R. S., Arvidson, R.E., Head, J.W. III., et al. (1991). Magellan: A first overview of Venus geology. *Science, 252*, 249–252.

Schubert, G., & Covey, C. (1981, July). The atmosphere of Venus. *Scientific American.*

Smrekar, Suzanne E., Stofan, E. R., et al. (2010). Recent hot-spot volcanism on Venus from VIRTIS emissivity data. *Science, 328*, 605–608.

Solomon, S. C., Bullock, M. A., & Grinspoon, D. H. (1999). Climate change as a regulator of tectonics on Venus. *Science, 286*, 87–90.

Solomon, S. C., Head, J.W., Kaula, W. M., et al. (1991, April). Venus tectonics: Initial analysis from Magellan. *Science, 252*, 297–312.

Taylor, F. W., Read, P. L., Lewis S. R. *The Venusian environment*. Retrieved from http://www.atm.ox.ac.uk/user/fwt/WebPage/Venus%20Review%204.htm

The Soviet exploration of Venus (Don Mitchell). Retrieved from http://www.mentallandscape.com/V_Venus.htm

Transits of Venus and black drop effect. Retrieved from http://www.transitofvenus.info

Venus Climate Orbiter (Akatsuki). Retrieved from http://www.jaxa.jp/projects/sat/planet_c/index_e.html

Venus Express mission. Retrieved from http://sci.esa.int/venusexpress/

Venus Express special – The long-lost twin. (2007, November). *Nature, 450*, 629–662.

Zuber, M. T., & Parmentier, E. M. (2002). Formation of fold-and-thrust belts on Venus by thick-skinned deformation. *Nature, 377*, 704–707.

Chapter 7: Mars

Andrews-Hanna, J. C., Phillips, R. J., & Zuber, M. T., (2007, March). Meridiani Planum and the global hydrology of Mars. *Nature, 446*, 163–166.

Andrews-Hanna, J. C., Zuber, M.T., & Banerdt, W. B. (2008, June). The Borealis basin and the origin of the martian crustal dichotomy. *Nature, 453*, 1212–1215.

Bandfield, J. L., Hamilton, V. E., & Christensen, P. R. (2000, March). A global view of martian surface compositions from MGS-TES. *Science, 287*, 1626–1630.

Barabash, S., Fedorov, A., Lundin, R., et al. (2007, January). Martian atmospheric erosion rates. *Science, 315*, 501–503.

Beatty, J. Kelly. (1999, November). In search of martian seas. *Sky & Telescope, 4*, 38.

Bell, J. (2003, December). Mineral mysteries and planetary paradoxes, *Sky & Telescope*, 34–40.

Bell, J. (2009, July). Uncovering Mars's secret past. *Sky & Telescope.*

Bibring, J-P. (2007, August). Coupled ferric oxides and sulfates on the martian surface. *Science Express.*

Bibring, J-P., Brissaud, O., Schmitt, B., et al. (2004, April). Perennial water ice identified in the south polar cap of Mars. *Nature, 428*, 627–630.

Bishop, J. L., Eldar, Z. N. D., McKeown, N. K., et al. (2008, August). Phyllosilicate Diversity and past aqueous activity revealed at Mawrth Vallis, Mars. *Science, 321*, 830–833.

Carr, M. H. (1996). *Water on Mars*. New York: Oxford University Press.

Christensen, P. R. (2003, March). Formation of recent martian gullies through melting of extensive water-rich snow deposits. *Nature, 422*, 45–48.

Clifford, S. M. (2003, August). The iceball next door. *Sky & Telescope.*

Committee on an astrobiology strategy for the exploration of Mars. (2007). *An astrobiology strategy for the exploration of Mars*, The National Academies Press. Retrieved from http://www.nap.edu/catalog.php?record_id=11937

Connerney, J. E. P., Acuña, M. H., Ness, N. F., et al. (2005, October). Tectonic implications of Mars crustal magnetism. *PNAS 102*(42). 14970–14975. Retrieved from http://www.pnas.org/content/102/42/14970.full

Corrigan, C. M. (2004, July). Carbonates in ALH84001: Part of the story of water on Mars. *PSR Discoveries*. Retrieved from http://www.psrd.hawaii.edu/July04/carbonatesALH84001.html

Cull, Selby, (2007, January). Reverse Panspermia: Seeding life in the Solar System. *Sky & Telescope.*

DiGregorio, B. E. (2004, February). Life on Mars? *Sky & Telescope*, 40–45.

Ezell, E. C., & Ezell, L. N. (1984). *On Mars – exploration of the red planet 1958-1978*. NASA History Office. (NASA SP-4212). Retrieved from http://www.hq.nasa.gov/office/pao/History/SP-4212/on-mars.html

Fenton, L. K., Geussker, P. E., Harberle, R. M. (2007, April). Global warming and climate forcing by recent albedo changes on Mars. *Nature, 446*, 646–649.

Forget, F., Haberle, R. M., Montmessin, F., et al. (2006, January). Formation of glaciers on Mars by atmospheric precipitation at high obliquity. *Science 311*, 368–371.

Golombek, M. P., Arvidson, R. E., Bell, J. F.III., et al. (2005, July). Assessment of Mars exploration rover landing site predictions. *Nature 436*, 44–48.

Google Mars: Retrieved from http://www.google.com/mars/

Gurnett, D.A., Kirchner, D. L., Huff, R. L., et al., (2005, December). *Radar soundings of the ionosphere of Mars. Science, 310*, 1929–1933.

Halevy, I., Zuber, M. T., Schrag, D. P. (2007, December). A sulfur dioxide climate feedback on early Mars. *Science, 318*, 1903–1907.

Hartmann, W. K. (2003, July). Mysteries of Mars. *Sky & Telescope.*

Herkenhoff, K. E., & Keszthelyi, L. P. (2010, January). MRO/HiRISE studies of Mars. *Icarus, 205*(1), 1–320.

Holt, J. W., & Smith, I. B., (2010, May). Onset and migration of spiral troughs on Mars revealed by orbital radar. *Nature, 465*, 450–453.

Holt, J. W., Fishbaugh, K. E., Byrne, S., et al. (2010, May). The construction of Chasma Boreale on Mars. *Nature 465*, 446–449.

Holt, J. W., Safaeinili, A., Plaut, J. J., et al. (2008, November). Radar sounding evidence for buried glaciers in the southern mid-latitudes of Mars. *Science, 322*, 1235–1238.

Jaeger, W. L., Keszthelyi, L. P., McEwen, A. S., et al. (2007, September). Athabasca Valles. Mars: A lava-draped channel system. *Science, 317*, 1709–1711.

Kraal, E. R., van Dikj, M., Postma, G., et al. (2008, February). Martian stepped-delta formation by rapid water release. *Nature, 451*, 973–977.

Lefevre, Franck, Bertaux, J-L., Clancy, R. T., et al, (2008, August). Heterogeneous chemistry in the atmosphere of Mars. *Nature, 454*, 971–975.

Lundin, R., Winningham, D., Barabash, S., et al. (2006, February). Plasma acceleration above martian magnetic anomalies. *Science, 311*, 980–983.

Malin Space Science Systems – images of Mars and other features from Mars spacecraft. Retrieved from http://www.msss.com/mars_images/index.html

Malin, M. C., & Edgett, K. S. (1999). Oceans or seas in the martian northern lowlands: high resolution imaging tests of proposed coastlines. *Geophysical Research Letters, 26*(19), 3049–3052.

Malin, M. C., & Edgett, K. S., (2000, June). Evidence for recent groundwater seepage and surface runoff on Mars. *Science, 288*, 2330–2335.

Malin, M. C., & Edgett, K. S., (2003, December). Evidence for persistent flow and aqueous sedimentation on early Mars. *Science, 302*, 1931–1934.

Mamma, M. J., Villanueva, G. L., Novak, R.E., et al. (2009, February). Strong release of methane on Mars in northern summer 2003. *Science, 323*, 1041–1045.

Mangold N., Quantin, C., Ansan, V., et al. (2004, July). Evidence for precipitation on Mars from dendritic valleys in the Valles Marineris area. *Science, 305*, 78–81.

Mars Exploration Rovers web site. Retrieved from http://marsrovers.jpl.nasa.gov/home/index.html

Mars Express web site. Retrieved from http://www.esa.int/SPECIALS/Mars_Express/index.html

Mars Global Surveyor web site. Retrieved from http://mars.jpl.nasa.gov/mgs/

Mars Odyssey Themis instrument web site. Retrieved from http://themis.asu.edu/

Mars Odyssey web site. Retrieved from http://mars.jpl.nasa.gov/odyssey/

Mars Pathfinder web site. Retrieved from http://mars.jpl.nasa.gov/MPF/index1.html

Mars Phoenix web site. Retrieved from http://phoenix.lpl.arizona.edu

Mars Reconnaissance Orbiter (2007, September). [Special section]. *Science, 317*, 1705–1719.

Mars Reconnaissance Orbiter HiRise instrument web site. Retrieved from http://hirise.lpl.arizona.edu

Mars Reconnaissance Orbiter web site. Retrieved from http://mars.jpl.nasa.gov/mro/

Martel, L. M. V. (2001, June). Outflow channels may make a case for a bygone ocean on Mars. *PSR Discoveries*. Retrieved from http://www.psrd.hawaii.edu/June01/MarsChryse.html

Martel, L. M. V. (2003, August). Gullied slopes on Mars. *PSR Discoveries*. Retrieved from http://www.psrd.hawaii.edu/Aug03/MartianGullies.html

Martel, L. M. V. (2003, July). Ancient floodwaters and seas on Mars. *PSR Discoveries*. Retrieved from http://www.psrd.hawaii.edu/July03/MartianSea.html

Martel, Linda M. V. (2007, January). Did martian meteorites come from these sources? *PSR Discoveries*. Retrieved from http://www.psrd.hawaii.edu/Jan07/MarsRayedCraters.html

McKay, D. S., Gibson, E. K., Thomas-Keprta, K. L., et al, (1996, August). Search for past life on Mars: Possible relic biogenic activity in martian meteorite ALH84001. *Science, 273*, 924–930.

Murray, John B., et al, (2005, March). Evidence from the Mars express high resolution stereo camera for a frozen sea close to Mars' equator. *Nature, 434*, 352–356.

Mustard, J. F., Murchie, S. L., Pelkey, S. M., et al. (2008, July). Hydrated silicate minerals on Mars observed by the Mars Reconnaissance Orbiter CRISM instrument. *Nature, 454*, 305–330.

Naeye, R. (2004, May). Red-Letter Days. *Sky & Telescope, 107*(5), 44.

Neukum, G., Jaumann, R., Hoffmann, H., et al. (2004, December). Recent and episodic volcanic and glacial activity on Mars revealed by the High Resolution Stereo Camera. *Nature, 432*, 971–979.

Okubo, C. H., & McEwen, A. S. (2007, February). Fracture-controlled paleo-fluid flow in candor chasma, Mars. *Science, 315*, 983–985.

Perron, J. Taylor. (2007, June). Evidence for an ancient martian ocean in the topography of deformed shorelines. *Nature, 447*, 840–843.

Phillips, R. J., Smrekar, S. E., Mellon, M. T., et al. (2008, May). Mars north polar deposits: stratigraphy, age, and geodynamical response, *Science, 320*, 1182–1185.

Plaut, J. J., Picardi, G., Safaeinili, A., et al. (2007, April). Subsurface radar sounding of the south polar layered deposits of Mars. *Science, 316*, 92–95.

Raeburn, P. (1998). Mars: Uncovering the secrets of the Red Planet. *National Geographic Society*.

Safe on Mars: Precursor measurements necessary to support human operations on the martian surface. (2002). National Research Council. Retrieved from http://www.nap.edu/catalog/10360.html

Schofield, J. T., Barnes, J. R., Crisp, D., et al. (1997, December) The Mars pathfinder atmospheric structure investigation/meteorology (ASI/MET) experiment. *Science, 278*, 1752–1758.

Schorghofer, N. (2007, September). Dynamics of ice ages on Mars. *Nature, 449*, 192–195.

Solomon, S. C., Bridges, K., Catling, D.C., et al. (2005, February). New perspectives on ancient Mars. *Science, 37*, 1214–1219.

Space Studies Board / Board on Life Sciences. (2007). An astrobiology strategy for the exploration of Mars. The National Academies Press. Retrieved from http://www.nap.edu/openbook.php?record_id=11937&page=23

Squyres, S. (2005). *Roving Mars: Spirit, Opportunity, and the exploration of the red planet*. Hyperion.

Squyres, S.W., Knoll, A. H., Arvidson, R. E., et al. (2006, September). Two years at Meridiani Planum: Results from the Opportunity Rover. *Science, 313*, 1403–1407.

Taylor, G. J. (1996, October). Life on Mars? *PSR Discoveries*. Retrieved from http://www.psrd.hawaii.edu/Oct96/LifeonMars.html

Taylor, G. J. (2000, May). Liquid water on Mars: The story from meteorites. *PSR Discoveries*. Retrieved from http://www.psrd.hawaii.edu/May00/wetMars.html

Taylor, G. J. (2003, November). Hafnium, Tungsten, and the differentiation of the Moon and Mars. *PSR Discoveries*. Retrieved from http://www.psrd.hawaii.edu/Nov03/Hf-W.html

Taylor, G. J. (2004, June). The multifarious martian mantle. *PSR Discoveries*. Retrieved from http://www.psrd.hawaii.edu/June04/martianMantle.html

Taylor, G. J. (2005, December). Magma and water on Mars. *PSR Discoveries*. Retrieved from http://www.psrd.hawaii.edu/Dec05/Magma-WaterOnMars.html

Taylor, G. J. (2005, January). Recent activity on Mars: Fire and ice. *PSR Discoveries*. Retrieved from http://www.psrd.hawaii.edu/Jan05/MarsRecently.html

Taylor, G. J. (2009, May). Mars crust: Made of basalt. *PSR Discoveries*. Retrieved from http://www.psrd.hawaii.edu/May09/Mars.Basaltic.Crust.html

The Mars journal. Retrieved from http://themarsjournal.org/

The quarantine and certification of martian samples. (2002). Committee on planetary and lunar exploration, National Research Council. Retrieved from http://www.nap.edu/catalog/10138.html

Viking lander imaging team. (1978). *The martian landscape*. NASA History Office. (NASA SP-425). Retrieved from http://history.nasa.gov/SP-425/cover.htm

Viking orbiter imaging team, *Viking Orbiter views of Mars*. NASA History Office. (NASA SP-441). Retrieved from http://history.nasa.gov/SP-441/cover.htm

Viking web site. Retrieved from http://nssdc.gsfc.nasa.gov/planetary/viking.html

Withers, P., & Smith, M. D. (2006, August). Atmospheric entry profiles from the Mars exploration Rovers Spirit and Opportunity. *Icarus*, *185*, 133–142. Published online http://sirius.bu.edu/aeronomy/withersmericarus2006.pdf

Zuber, M. T. (2003, December). Mars: The inside story. *Sky & Telescope*.

Zuber, M. T., Phillips, R.J., Andrews-Hanna, J. C., et al. (2007, September). Density of Mars' south polar layered deposits. *Science*, *317*, 1718–1719.

Chapter 8: Jupiter

Anderson, J. D. (2005, May). Amalthea's density is less than that of water. *Science*, *308*, 1291–1293.

Anderson, J. D., Schubert, G., Jacobson, R. A., et al. (1998, June). Distribution of rock, metals, and ices in Callisto. *Science*, *280*, 1573–1576.

Anderson, J. D., Lau, E. L., Schubert, G., et al. (1997, May). Europa's differentiated internal structure: Inferences from two Galileo encounters. *Science*, *276*, 1236–1239.

Burns, J. A., Showwalter, M. R., Hamilton, D. P., et al. (1999, May). The formation of Jupiter's faint rings. *Science*, *284*, 1146–1150.

Cassini Jupiter millennium flyby. Retrieved from http://www.jpl.nasa.gov/jupiterflyby/

Galileo mission (NASA). Retrieved from http://www2.jpl.nasa.gov/galileo/

Greeley, R. (1997, December). The partially watery world of Europa, one of Jupiter's Moons. *Earth in Space*, *10*, 5–8. Retrieved from http://www.agu.org/sci_soc/greeley.html

Harland, D. M. (2000). *Jupiter odyssey: The story of NASA's Galileo mission*. Springer-Praxis.

Ingersoll, A. P. (1990). Jupiter and Saturn. Exploring space. *Scientific American Special*.

Jewitt, D., Sheppard, S. S., & Kleyna, J. (2006, August). The strangest satellites in the Solar System. *Scientific American*. Retrieved from http://www.ifa.hawaii.edu/~jewitt/papers/2006/JSK06.pdf

Johnson, T.V. (2003). The Galileo Mission to Jupiter and its moons. New light on the Solar System. [Special edition] *Scientific American Special*, 54–63.

Kivelson, M. G., Khurana, K. K., Russell, C. T., et al. (2000, August). Galileo magnetometer measurements: A stronger case for a subsurface ocean at Europa. *Science*, *289*, 1340–1343.

Malhotra R. (1999, September). Migrating planets. *Scientific American*.

Marcus, P. S. (2004, April). Prediction of a global climate change on Jupiter. *Nature*, *428*, 828–831.

Martel, L. M. V. (1998, April). Big mountain, big landslide on Jupiter's moon, Io. *PSR Discoveries*. Retrieved from http://www.psrd.hawaii.edu/April98/io.html

Martel, L. M. V. (2001, February). The Europa scene in the Voyager-Galileo Era. *PSR Discoveries*. Retrieved from http://www.psrd.hawaii.edu/Feb01/EuropaGeology.html

Martel, L. M. V. (2002, November). Bands on Europa. *PSR Discoveries*. Retrieved from http://www.psrd.hawaii.edu/Nov02/EuropanBands.html

McCord, T. B., Hansen, G. B., & Hibbitts, C. A. (2001, May). Hydrated salt minerals on Ganymede's surface: Evidence of an ocean below. *Science*, *292*, 1523–1525.

McEwen A. S., Keszthelyi, L., Spencer, J. R., et al, (1998, July). High-temperature silicate volcanism on Jupiter's moon Io. *Science*, *281*, 87–90.

McEwen, A. S., Belton, M. J. S., Breneman, H. H., et al. (2000, May). Galileo: Io reports. *Science*, *288*, 1193–1220.

Morrison, D., & Samz, J. (1980). *Voyage to Jupiter*. (NASA SP-439).

New Horizons at Jupiter. (2007, October). [Special section]. *Science*, *318*, 215-243.

Pappalardo, R. T., Head, J. W., Greeley, R. (2003, September). The hidden ocean of Europa. New light on the Solar System. [Special edition] *Scientific American Special*, 64–73.

Porco, C. C., West, R. A., McEwen, A., et al. (2003, March). Cassini imaging of Jupiter's atmosphere, satellites and rings. *Science*, *299*, 1541–1547.

Reports on Comet Shoemaker-Levy 9 Collision with Jupiter. (1995, March). *Science*, *267*, 1277–1323.

Sánchez-Lavega, A., Orton, S. J., Hueso, R., et al. (2008, January). Depth of a strong Jovian jet from a planetary-scale disturbance driven by storms. *Nature*, *451*, 437–440.

Schenk, P. M., Mc Kinnon, W. B., Gwynn, D., et al. (2001, March). Flooding of Ganymede's bright terrains by low-viscosity water-ice lavas. *Nature*, *410*, 57–60.

Sheppard, S. S. & Jewitt, D. C., (2003, May). An abundant population of small irregular satellites around Jupiter. *Nature*, *423*, 261–263.

Showman, A. P., & Malhotra, R. (1999). The Galilean Satellites. *Science*, *286*, 77–84.

Soderblom, L. A. (1980, January). The Galilean moons of Jupiter. *Scientific American*.

Stevenson, D. (2000, August). Europa's ocean – the case strengthens. *Science*, *289*, 1305–1307.

Taylor, G. J. (1998, September). Europa's salty surface. *PSR Discoveries*. Retrieved from http://www.psrd.hawaii.edu/Sept98/EuropaSalts.html

Taylor, G. J. (2000, February). Jupiter's hot, mushy moon. *PSR Discoveries*. Retrieved from http://www.psrd.hawaii.edu/Feb00/IoMagmaOcean.html

Voyager missions (NASA). Retrieved from http://voyager.jpl.nasa.gov/

Young, R. E., Smith M. A., & Sobeck, C. A. (1996, May). Galileo probes Jupiter's atmosphere. *Science*, *272*, 837–860.

Chapter 9: Saturn

30-day reports from the Voyager 1 and Voyager 2 instrument teams. (1981, April). (1982, January). *Science*.

Ádámkovics, M., Wong, M. H., Laver, C., et al. (2007, October). Widespread morning drizzle on Titan. *Science Express*.

Atreya, S. K. (2009, January). The mystery of methane on Mars and Titan. *Scientific American*.

Bagenal, F. (2005, February). Saturn's mixed magnetosphere. *Nature, 433*, 695–696.

Barnes, J. (2008, December). Titan: Earth in deep freeze. *Sky & Telescope*, 26–32.

Batson, R. M. (1984). Voyager 1 and 2 Atlas of six saturnian satellites. (NASA SP-474) Retrieved from http://ntrs.nasa.gov/archive/nasa/casi.ntrs.nasa.gov/19840027171_1984027171.pdf

Bosh, A. S., & Rivkin, A. S. (1996). Observations of Saturn's inner satellites during the May 1995 ring-plane crossing. *Science, 272*, 518–521.

Brown, R. H., et al. (2008, July). The identification of liquid ethane in Titan's Ontario Lacus. *Nature, 454*, 607–610.

Brown, R., Lebreton, J-P, & Waite, H. (2009). *Titan from Cassini-Huygens*. Springer.

Cassini at Enceladus. (2006, March). *Science, 311*, 1388–1428.

Charnoz, S., Porco, C. C., Déau, E., et al. (2005, November). Cassini discovers a kinematic spiral ring around Saturn. *Science, 310*, 1300–1304.

Coustenis, A., & Taylor, F. W. (2008). Titan: exploring an earthlike world. *World Scientific*.

Cruikshank, D. P., Dalton, J. B., Dalle Ore, et al. (2007, July). Surface composition of Hyperion. *Nature, 448*, 54–56.

Cuzzi, J. N., Burns, J. A., Charnoz, S., et al. (2010, March). An evolving view of Saturn's dynamic rings. *Science, 327*, 1470–1475.

Dyudina, U. A., Ingersoll, P. A., Ewald, P. S., et al. (March, 2008). Dynamics of Saturn's south polar vortex. *Science, 319*, 1801.

Elachi, C., Wall., S., Janssen, M., et al, (2006, June). Titan radar mapper observations from Cassini's T3 fly-by. *Nature, 441*, 709–713.

ESA Huygens web site. Retrieved from http://sci.esa.int/huygens

Gehrels, T., & Matthews, M. S. (Eds.). *Saturn*. University of Arizona Press.

Gombosi, T. I., & Ingersoll, A. P. (2010, March). Saturn: Atmosphere, ionosphere, and magnetosphere. *Science, 327*, 1476–1479.

Greenberg, R., & Brahic, A. (Eds.). (1984). *Planetary rings*. University of Arizona Press.

Griffith, C.A., Penteado, P., & Brown R. (2006, September). Evidence for a polar ethane cloud on Titan. *Science*.

Hansen, C. J., Esposito, L. W., Stewart, A. I. F., et al. (2008, November). Water vapour jets inside the plume of gas leaving Enceladus. *Nature, 456*, 477–479.

Hedman, M. M. (2007, August). The source of Saturn's G Ring. *Science, 317*, 653–656.

Hueso, R., & Sánchez-Lavega, A. (2006, July). Methane storms on Saturn's moon Titan. *Nature, 442*, 428–431.

Ingersoll, A. P. (1990). Jupiter and Saturn. Exploring Space. *Scientific American Special*.

Initial Cassini Results. (2005, February). *Science, 307*, 1226–1251.

Jewitt, D., Sheppard, S. S., & Kleyna, J. (2006, August). The strangest satellites in the Solar System. *Scientific American*. Retrieved from http://www.ifa.hawaii.edu/~jewitt/papers/2006/JSK06.pdf

Jones, G. H., Roussos, E., Krupp, N., et al. (2008, March). The dust halo of Saturn's largest icy moon, Rhea. *Science, 319*, 1380–1384.

Kerr, R. A. (2006, January). How Saturn's icy moons get a (geologic) life. *Science, 311*, 29.

Kieffer, S., & Jakosky, B. M. (2008, June). Enceladus–oasis or ice ball? *Science, 320*, 1432–1433.

Krimigis, S. M., Sergis, N., Mitchell, D. G., et al. (2007, December). A dynamic, rotating ring current around Saturn. *Nature, 450*, 1050–1053.

Lorenz, R. (2000, October). The weather on Titan. *Science, 290*, 467–468.

Lorenz, R. D., Wall, S., Radebaugh, J., et al. (2006, May). The sand seas of Titan: Cassini RADAR observations of longitudinal dunes. *Science, 312*, 724–727.

Lorenz, R., & Mitton, J. (2002). *Lifting Titan's veil: Exploring the giant moon of Saturn*, Cambridge University Press, New York.

Murray, C. D., Beurle, K., Cooper, N. J., et al. (2008, June). The determination of the structure of Saturn's F ring by nearby moonlets. *Nature, 453*, 739–744.

NASA Cassini web site. Retrieved from http://saturn.jpl.nasa.gov

New views of Titan. (2005, December). (Special issue). *Nature, 438*, 538–539 and 765–802.

Nicholson, P. D., Showalter, M. R., Dones., L., et al. (1996). Observations of Saturn's ring-plane crossings in August and November 1995. *Science, 272*, 509–515, 1996

Rannou, P., Montmessin, F., Hourdin, F., et al. (2006, January). The latitudinal distribution of clouds on Titan. *Science, 311*, 201–205.

Schaller, E. L., Roe, H. G., Schneider, T., et al. (2009, August). Storms in the tropics of Titan. *Nature, 460*, 873–875.

Schmidt, J. (2008, February). Slow dust in Enceladus' plume from condensation and wall collisions in tiger stripe fractures. *Nature, 451*, 685–688.

Showalter, M. R. (2005, November). Saturn's strangest ring becomes curiouser and curiouser. *Science, 310*, 1287–1288.

Soderblom, L. A., & Johnson, T. V. (1990). The moons of Saturn. Exploring space. *Scientific American Special*.

Sohl, F. (2010, March). Revealing Titan's interior. *Science, 327*, 1338–1339.

Spitale, J. N., & Porco, C. C. (2007, October). Association of the jets of Enceladus with the warmest regions on its south-polar fractures. *Nature, 449*, 695–697.

Sremčević, M., Schmidt, J., Salo, H., et al. (2007, October). A belt of moonlets in Saturn's A ring. *Nature, 449*, 1019–1021.

Thomas, R. C., Armstrong, J. W., Asmar, S. W., et al. (2007, July). Hyperion's sponge-like appearance. *Nature, 448*, 50–53.

Tiscareno, M. S. (2007, February). Ringworld revelations. *Sky & Telescope*.

Tiscareno, M. S., Burns, J. A., Hedman, M. M., et al. (2006, March). 100-metre-diameter moonlets in Saturn's A ring from observations of 'propeller' structures. *Nature, 440*, 648–650.

Tobie, G., Lunine, J. I., & Sotin, C. (March, 2006). Episodic outgassing as the origin of atmospheric methane on Titan. *Nature, 440*, 61–64.

Tokano, T., McKay, C. P., Neubauer, F. M., et al. (2006, July). Methane drizzle on Titan. *Nature, 442*, 432–435.

Verbiscer, A. J., Skrutskie, M. F., & Hamilton, D. P. (2009, October). Saturn's largest ring. *Nature, 461*, 1098–1100.

Verbiscer, A., French, R., Showalter, M., et al. (2007, February). Enceladus: Cosmic graffiti artist caught in the act. *Science, 315*, 815.

Zarka, P., Lamy, L., Cecconi, B., et al. (2007, November). Modulation of Saturn's radio clock by solar wind speed. *Nature, 450*, 265–267.

Chapter 10: Uranus

30-day reports from the Voyager 2 instrument teams. (1989, December). *Science, 233*.

Benedetti, L. R., Nguyen, J. H., Caldwell, W. A., et al. (1999, October). Dissociation of CH4 at high pressures and temperatures: Diamond formation in giant planet interiors? *Science, 286*, 100–102.

Bergstralh J. T., Miner E. D., & Matthews M. S. (Eds.). (1991). *Uranus*. University of Arizona Press.

Cavazzoni, C., Chiarotti, G. L., Scandolo, S., et al. (1999, January). Superionic and metallic states of water and ammonia at giant planet conditions. *Science, 283*, 44–46.

de Pater, I., Hammel, H. B., Gibbard, S. G., et al. (1996, April). New dust belts of Uranus: One ring, two ring, red ring, blue ring. *Science 312*, 92–94.

de Pater, I., Hammel, H. B., Showalter, M. R., et al. (2007, September). The dark side of the rings of Uranus, *Science, 317,* 1888–1890.

Desch, S. (2007, December). Mass distribution and planet formation in the solar nebula. *Astrophysical Journal, 671,* 878–893.

Guillot, T. (1999, October). Interiors of giant planets inside and outside the Solar System. *Science, 286,* 72–77.

Hubbard, W. B., Nellis, W. J., Mitchell, et al. (1991, August). Interior structure of Neptune: Comparison with Uranus. *Science 253,* 648–651.

Infrared observations of Uranus and its rings. (Imke de Pater). Retrieved from http://astron.berkeley.edu/~newstar/Infrared/UranusAo/UranusAO.htm

Ingersoll, A. (1987). Uranus. *Scientific American.*

Johnson T. V., Brown R. H. & Soderblom L. A. (1987, April). The Moons of Uranus. *Scientific American.*

Laeser, R., McLaughlin, W., & Wolff, D. (1990). Engineering Voyager 2's encounter with Uranus. Exploring Science. *Scientific American Special.*

Littmann, M. (1990). (Revised edition). *Planets beyond: Discovering the outer Solar System.* John Wiley.

Miner, E. D. (1998). *Uranus.* (2nd ed.). Wiley-Praxis.

O'Donel, A. (1965). *The planet Uranus: A history of observation, theory and discovery.* Faber & Faber.

Showalter, M. R., & Lissauer, J. J. (2006, February). The second ring-moon system of Uranus: Discovery and dynamics. *Science, 311,* 973–977.

Stanley, S., & Bloxham, J. (2004, March). Convective-region geometry as the cause of Uranus' and Neptune's unusual magnetic fields. *Nature, 428,* 151–153.

Taylor, G. J. (2001, August). Uranus, Neptune, and the mountains of the Moon. *PSR Discoveries.* Retrieved from http://www.psrd.hawaii.edu/Aug01/bombardment.html

Thommes, E. W., Duncan, M. J., & Levison H. F. (1999, December). The formation of Uranus and Neptune in the Jupiter-Saturn region of the Solar System. *Nature, 402,* 635–638.

Thommes, E. W., Duncan, M. J., & Levison H. F. (2002, May). The formation of Uranus and Neptune among Jupiter and Saturn. *The Astronomical Journal, 123,* 2862–2883.

Voyager missions web site: http://voyager.jpl.nasa.gov/

Chapter 11: Neptune

30-day reports from the Voyager 2 instrument teams. (1989, December). *Science, 246.*

Agnor, C. B., & Hamilton, D. P. (2006, May). Neptune's capture of its moon Triton in a binary–planet gravitational encounter. *Nature, 441,* 192–194.

Cruikshank D. P. (Ed.). (1994). *Neptune and Triton,* University of Arizona Press.

Cruikshank, D. P., Roush, T. L., Owen, T. C., et al. (1993, August). Ices on the surface of Triton. *Science, 261,* 742–745.

de Pater, I., Gibbard, S., Chiang, E., et al. (2004, September). The dynamic Neptunian ring arcs: Evidence for a gradual disappearance of liberte and resonant jump of courage. *Icarus.*

Desch, S. (2007, December). Distribution and planet formation in the solar nebula. *Astrophysical Journal, 671,* 878–893.

Dumas, C., Terrile, R. J., Smith, B. A., et al. (1999, August). Stability of Neptune's ring arcs in question. *Nature, 400,* 733–735.

Dumas, C., Terrile, R., Smith, B. A., et al. (2002, March). Astrometry and near-infrared photometry of Neptune's inner satellites and ring arcs. *The Astronomical Journal, 123,* 1776–1783.

Elliot, J. L., Hammel, H. B., Wasserman, L. H., et al. (1998, June). Global warming on Triton. *Nature, 293,* 785–787.

Holman, M. J., Kavelaars., J. J., Grav, T., et al, (2004, August). Discovery of five irregular moons of Neptune. *Nature, 430,* 865–867.

Hubbard W. B. (1997, February). Neptune's deep chemistry. *Science, 275,* 1279–1280.

Hubbard, W. B., Brahic, A., Sicardy, B., et al. (1986, February). Occultation detection of a Neptunian ring-like arc. *Nature, 319,* 636–640.

Hubbard, W. B.; Nellis, W. J., Mitchell, A. C., et al. (1991, August). Interior structure of Neptune: Comparison with Uranus. *Science, 253,* 648–651.

Kinoshita, J. (1990). *Neptune.* Exploring science. *Scientific American Special.*

Lellouch, E., de Bergh, C., Sicardy, B., et al. (2010, March-April). Detection of CO in Triton's atmosphere and the nature of surface-atmosphere interactions. *Astronomy & Astrophysics, 512,* L8

Miner, Ellis, & Wessen, R. R. (2002). *Neptune: The planet, rings and satellites.* Springer-Praxis.

Moore, P. (1996). *The planet Neptune: An historical survey before Voyager.* (2nd ed.). Wiley-Praxis.

Porco, C.C. (1991, August). An explanation for Neptune's ring arcs. *Science, 253,* 995–1001.

Roddier F., Roddier C., Graves J. E., et al. (1998). Neptune's cloud structure and activity: Ground-based monitoring with adaptive optics. *Icarus,* 168–172, 1998

Sheehan, W., Kollerstrom, N., & Waff, C. B. (2004, December). The Case of the Pilfered Planet. *Scientific American.*

Sicardy B., Roddier F., Roddier C., et al. (1999, August). Images of Neptune's ring arcs obtained by a ground-based telescope. *Nature, 400,* 731–733.

Stanley, S., & Bloxham, J. (2004, March). Convective-region geometry as the cause of Uranus' and Neptune's unusual magnetic fields. *Nature 428,* 151–153.

Thommes, E. W., Duncan, M. J., & Levison H. F. (1999, December). The formation of Uranus and Neptune in the Jupiter-Saturn region of the Solar System. *Nature, 402,* 635–638.

Thommes, E. W., Duncan, M. J., & Levison H. F. (2002, May). The formation of Uranus and Neptune Among Jupiter and Saturn. *The Astronomical Journal, 123,* 2862–2883.

Triton – Reports from Voyager instrument teams. (1990, October). *Science, 250,* 410–443.

Chapter 12: Pluto & Kuiper Belt

Bertoldi, F., Altenhoff, W., Menten, K. M., et al. (2006, April). Comment on the recent Hubble Space Telescope size measurement of 2003 UB313 by Brown et al. Retrieved from http://www.astro.uni-bonn.de/~bertoldi/ub313/

Bertoldi, F., Altenhoff, W., Weiss, A., et al. (2006, February). The trans-Neptunian object UB313 is larger than Pluto. *Nature, 439,* 563–564.

Binzel, R., Tholen, D. J., Tedesco, E. F., et al. (1985, June). The detection of eclipses in the Pluto-Charon system. *Science, 228,* 1193–1195.

Brown M. E. & Schaller E. L. (2007, June). The mass of dwarf planet Eris. *Science, 316,* 1585.

Brown, M. E. (2004, August). Discovery of a candidate inner Oort Cloud planetoids. *Astrophysical Journal Letters.* Retrieved from http://www.gps.caltech.edu/%7Embrown/papers/ps/sedna.pdf

Brown, M. E., & Calvin, W. M. (2000, January). Evidence for crystalline water and ammonia ices on Pluto's satellite Charon. *Science, 287,* 107–109.

Brown, M. E., van Dam M. A., Bouchez, A. H., et al. (2006, March). Satellites of the largest Kuiper Belt objects. *The Astrophysical Journal, 639,* L43–L46.

Brown, M.E., Schaller E. L., Roe, H. G., et al. (2006, May). Direct measurement of the size of 2003 UB313 from the Hubble Space

Telescope (preprint). *The Astrophysical Journal*, *643*, L61–L63. Retrieved from http://www.gps.caltech.edu/%7Embrown/papers/ps/xsize.pdf

Buie, M. W., Grundy, W. M., Young, E. F., et al. (2010). Pluto and Charon with the Hubble Space Telescope: II. Resolving changes on Pluto's surface and a map for Charon. *The Astronomical Journal*, *139*, 1128–1143. Retrieved from http://www.boulder.swri.edu/~buie/biblio/pub073.html

Buie, M.W., Grundy, W. M., Young, E. F., et al. (2005, December). Orbits and photometry of Pluto's satellites: Charon, S/2005 P1 and S/2005 P2. Retrieved from http://arxiv.org/PS_cache/astro-ph/pdf/0512/0512491.pdf

Canup, R. M. (2005, January). A giant impact origin of Pluto-Charon. *Science*, *307*, 546–550.

Clyde Tombaugh web site. Retrieved from http://www.klx.com/clyde/index.html

Cook, J. C., Desch, S. J., Roush, T. L., et al. (2007, July). Near-infrared spectroscopy of Charon: possible evidence for cryovolcanism on Kuiper Belt objects. *Astrophysical Journal*, *663*, 1406–1419.

Davies, J. (2001). *Beyond Pluto: Exploring the outer limits of the Solar System*. Cambridge University Press.

Distant EKOs – The Kuiper Belt electronic newsletters. Retrieved from http://www.boulder.swri.edu/ekonews

Elliot, J. L., Ates, A., Babcock, B. A., et al. (2003, July). The recent expansion of Pluto's atmosphere. *Nature*, *424*, 165–168.

Exploring the trans-Neptunian Solar System. Committee on Planetary and Lunar Exploration, National Research Council (1998). Retrieved from http://www.nap.edu/catalog/6080.html

Gingerich, O. (2006, November). Losing it in Prague: The inside story of Pluto's demotion. *Sky & Telescope*.

Gladman, B. (2005, January). The Kuiper Belt and the Solar System's comet disk. *Science*, *307*, 71–75.

Goldreich, P., Lithwick, Y., & Sari, R. (2002, August). Formation of Kuiper-Belt binaries by dynamical friction and three-body encounters. *Nature*, *420*, 643–646.

Haumea (Mike Brown web site). Retrieved from http://web.gps.caltech.edu/~mbrown/2003EL61/

Jewitt, D.C., & Luu, J. (2004, December). Crystalline water ice on the Kuiper Belt object (50000) Quaoar. *Nature*, *432*, 731–733.

Kenyon, S. J., & Bromley, B.C. (2004, December). Stellar encounters as the origin of distant Solar System objects in highly eccentric orbits. *Nature*, *432*.

Kuiper Belt (David Jewitt). Retrieved from http://www.ifa.hawaii.edu/faculty/jewitt/kb.html

Levison, H. F., & Morbidelli, A. (2003, November). The formation of the Kuiper Belt by the outward transport of bodies during Neptune's migration. *Nature*, *426*, 419–421.

Levy, D. H. (1992). *Clyde Tombaugh: Discoverer of planet Pluto*. University of Arizona Press.

Littmann, M. (2007). Dark beasts of the trans-Neptunian zoo. *Sky & Telescope*.

Littmann, M. (2007, September). From chaos to the Kuiper Belt. *Sky & Telescope*.

Luu, J., & Jewitt, D. C. (2002). Kuiper Belt objects: Relics from the accretion disk of the Sun. *Annual Review of Astronomy & Astrophysics*, *40*, 63–101.

Minor Planet Center. Retrieved from http://www.minorplanetcenter.org/iau/mpc.html

New Horizons mission. Retrieved from http://pluto.jhuapl.edu/

Owen, T., Roush, T. L., Cruikshank, D. P., et al. (1993, August). Surface ices and the atmospheric composition of Pluto. *Science*, *261*, 745–748.

Petit, J.-M., Kavelaars, J. J., Gladman, B. J., et al. (2008, October). The extreme Kuiper Belt binary 2001 QW322. *Science*, *322*, 432–434.

Pluto Today web site. Retrieved from http://www.plutotoday.com/

Rabinowitz, D. L., Barkume, K., Brown, M. E., et al. (2006, March). Photometric observations constraining the size, shape, and albedo of 2003 el61, a rapidly rotating, Pluto-sized object in the Kuiper Belt. *The Astrophysical Journal*, *639*, 1238–1251.

Schaller, E. L., & Brown, M. E. (2007, April). Volatile loss and retention on Kuiper Belt objects. *The Astrophysical Journal*, *659*, L61–L64.

Sicardy B., Widemann, T., Lellouch, E., et al. (2003, July). Large changes in Pluto's atmosphere as revealed by recent stellar occultations. *Nature*, *424*, 168–170.

Stern, A., & Mitton, J. (2005). *Pluto and Charon*. (2nd ed.). Wiley-VCH.

Stern, S. A. (2003, August). The evolution of comets in the Oort Cloud and Kuiper Belt. *Nature*, *424*, 639–642.

Stern, S. A. (2003, November). The 3rd zone: exploring the Kuiper Belt. *Sky & Telescope*.

Stern, S. A. (2003, September). Journey to the farthest planet, new light on the Solar System. *Scientific American Special Edition*.

Stern, S. A., Weaver, H. A., Steffl, A. J., et al. (2006, February). A giant impact origin for Pluto's small moons and satellite multiplicity in the Kuiper Belt. *Nature*, *439*, 946–948.

The Discovery of 2003 UB313 (Mike Brown web site). Retrieved from http://www.gps.caltech.edu/~mbrown/planetlila/

The Pluto portal. Retrieved from http://www.plutoportal.net/

Tombaugh, C., & Moore, P. (1981). Out of the darkness: The planet Pluto. *New American Library*.

Two new satellites of Pluto. Retrieved from http://www.boulder.swri.edu/plutomoons/

Ward, W. R., & Hahn, J. M.. (1998, June). Neptune's eccentricity and the nature of the Kuiper Belt. *Science 280*, 2104–2106.

Ward, W.R., & Canup, R. M. (2006, June). Forced resonant migration of Pluto's outer satellites by Charon. Science Express web site.

Wetterer, M. K. (1996). *Clyde Tombaugh and the search for Planet X*. Carolrhoda Books, 1996.

Young, E. (2000, January). Charon's first detailed spectra hold many surprises. *Science*, *287*, 53–54.

Young, E. F., Binzel, R. P., & Crane, K. (2001, January). A two-color map of Pluto's sub-Charon hemisphere. *The Astronomical Journal*, *121*, 552–561. Retrieved from http://www.boulder.swri.edu/recent/pluto_map.pdf

Chapter 13: Comets, Asteroids and Meteorites

Arctic Asteroid! (Tagish Lake meteorite). Retrieved from http://science.nasa.gov/headlines/y2000/ast01jun_1m.htm

Asteroid and comet impact hazards. Retrieved from http://impact.arc.nasa.gov/

Beatty, J. K. (2006, September). The falcon's wild ride. *Sky & Telescope*.

Bland, P. A., Spurny, P., Towner, M. C., et al. (2009, September). An anomalous basaltic meteorite from the innermost main belt. *Science*, *325*, 1525–1527.

Boehnhardt, H. (2001, May). The death of a comet and the birth of our solar system. *Science*, *292*, 1307–1309.

Bottke, W. F. (2007, March). Spun in the Sun. *Nature*, *446*, 382–383.

Bottke, W. F., & Martel, L. M. V. (2006, July). Iron meteorites as the not-so-distant cousins of Earth. *PSR Discoveries*. Retrieved from http://www.psrd.hawaii.edu/July06/asteroidGatecrashers.html

Bottke, W. F., Bottke, W. F., Vokrouhlicky, D., et al. (2007, September). An asteroid breakup 160 Myr ago as the probable source of the K/T impactor. *Nature*, *449*, 48–53.

Chiron. Retrieved from http://nssdc.gsfc.nasa.gov/planetary/chiron.html

Comet Shoemaker-Levy Collision with Jupiter. (NASA-JPL web site). Retrieved from http://www2.jpl.nasa.gov/sl9/

Comets (David Jewitt). Retrieved from http://www.ifa.hawaii.edu/faculty/jewitt/comet.html

Dawn mission. Retrieved from http://dawn.jpl.nasa.gov/

Deep Impact mission. Retrieved from http://deepimpact.jpl.nasa.gov/home/index.html

Deep Impact special section (2005, October). *Science*, *310*, 258–283.

Deep Space 1 mission. Retrieved from http://nmp.jpl.nasa.gov/ds1/

Festou, M. C., Keller, H. U., & Weaver, H. A. (Eds.). (2005). *Comets II*. University of Arizona Press.

Floss, C. (2003, January). QUE 93148: A part of the mantle of asteroid 4 Vesta? *PSRD Discoveries*. http://www.psrd.hawaii.edu/Jan03/QUE93148.html

Giotto mission. Retrieved from http://sci.esa.int/giotto/

Hayabusa special issue. (August, 2011). Dust from Itokawa. *Science*, *333*, 1113–1131.

Hayabusa special section. (2006, June). *Science*, *312*, 1330–1353.

Jenniskens, P., Shaddad, M. H., Numan, D., et al, (2009, March). The impact and recovery of asteroid 2008 TC3. *Nature*, *458*, 485–488.

Jewitt, D., & Hsieh, H. (2006, October). Physical observations of 2005 UD: A mini-Phaethon. *The Astronomical Journal*, *132*. http://www.ifa.hawaii.edu/faculty/jewitt/papers/2006/JH06.pdf

Kring, D. A. (1998). *Meteorites and their properties*. Lunar and Planetary Laboratory, Department of Planetary Sciences. University of Arizona. Retrieved from http://meteorites.lpl.arizona.edu/index.html

Kuiper Belt (David Jewitt). Retrieved from http://www.ifa.hawaii.edu/faculty/jewitt/kb.html

Küppers, M., Bertini, I., Fornaiser, S., et al. (2005, October). A large dust/ice ratio in the nucleus of comet 9P/Tempel 1. *Nature*, *437*, 987–990.

Li, A. (2009, May). Cosmic crystals caught in the act. *Nature*, *459*, 173–176.

LINEAR NEO program. Retrieved from http://www.ll.mit.edu/mission/space/linear/

Lisse, C. M., VanCleve, J., Adams, A. C., et al, (2006, August). Spitzer spectral observations of the Deep Impact ejecta. *Science*, *313*, 635–640.

Main belt comets. Retrieved from http://www.ifa.hawaii.edu/~hsieh/mbcs.html

Marchis, F. (2006, February). A low density of 0.8 g cm^{-3} for the Trojan binary asteroid 617 Patroclus. *Nature*, *439*, 565–567.

Marcus R., Melosh, H. J., & Collins, G. Earth impact effects program. Retrieved from http://www.lpl.arizona.edu/impacteffects/

Martel, L. M. V. (2001, November). Meteorites on ice. *PSR Discoveries*. Retrieved from http://www.psrd.hawaii.edu/Nov01/metsOnIce.html

Martel, L. M. V. (2002, February). Searching Antarctic ice for meteorites. *PSR Discoveries*, Retrieved from http://www.psrd.hawaii.edu/Feb02/meteoriteSearch.html

Martel, L.M. V. (2002, June). Using chondrites to understand the inside of asteroid 433 Eros. *PSR Discoveries*. http://www.psrd.hawaii.edu/June02/ErosPorosity.html

Martel, L.M. V. (2007, November). Getting to know Vesta. *PSR Discoveries*. Retrieved from http://www.psrd.hawaii.edu/Nov07/HEDs-Vesta.html

Marzari, F. (2006, July). Puzzling Neptune Trojans. *Science*, *313*, 451–452.

Meech, K. J. (1997, February). Apparition of comet Hale-Bopp. *PSR Discoveries*. http://www.psrd.hawaii.edu/Feb97/Hale-Bopp.html

Meteor showers (American Meteor Society). Retrieved from http://www.amsmeteors.org/showers.html

Meteor showers online (Gary W. Kronk). Retrieved from http://meteorshowersonline.com/

Morbidelli, A., Levison, H. F., Tsiganis, K., et al. (2005, May). Chaotic capture of Jupiter's Trojan asteroids in the early Solar System. *Nature*, *435*, 462–465.

Nakamura, T., Noguchi, T., Tsuchiyama, A., Ushikubo, T., Kita, T. N., Valley, J. W., Zolensky, M. E., et al. (2008, September). Chondrule-like objects in short-period comet 81P/Wild 2. *Science*, *321*, 1664–1667.

NASA 2006 Near-Earth object survey and deflection study – final report. Retrieved from http://www.hq.nasa.gov/office/pao/FOIA/NEO_Analysis_Doc.pdf

NASA-JPL Near Earth object program. (2005, May). Retrieved from http://neo.jpl.nasa.gov/neo/

Near-Earth object survey and definition – analysis of alternatives. Report to Congress, NASA, March 2007. Retrieved from http://www.nasa.gov/pdf/171331main_NEO_report_march07.pdf

NEAR-Shoemaker mission. Retrieved from http://near.jhuapl.edu/

NEAR-Shoemaker special edition. (2000, September). *Science*, *289*, 2085–2105.

Paolicchi, P. (2004, March). Solar-powered asteroids. *Nature*, *428*, 400–401.

Radar observations of near-Earth asteroids. Retrieved from http://echo.jpl.nasa.gov/asteroids/

Reports on Comet Shoemaker-Levy 9. (1995, March). *Science*, *267*, 1277–1323.

Rosetta mission. Retrieved from http://sci.esa.int/rosetta/

Russo, N. D., Vervack, R. J., Weaver, H. A., et al. (2007, July). Compositional homogeneity in the fragmented comet 73P/Schwassmann–Wachmann 3. *Nature*, *448*, 172–175.

Ruzicka, A., & Hutson, M. (2005, September). Portales valley: Not just another ordinary chondrite. *PSR Discoveries*. Retrieved from http://www.psrd.hawaii.edu/Sept05/PortalesValley.html

Schmidt, B. E., Thomas, P. C., Bauer, J. M., et al. (2009, October). The Shape and surface variation of 2 Pallas from the Hubble Space Telescope. *Science*, *326*, 275–278.

Schmidt, B.E., Thomas P. C., Bauer, J. M., et al. (2008, March). *Hubble takes a look at Pallas: Shape, size, and surface*. 39th Lunar and Planetary Science Conference (XXXIX, 2008). Retrieved from http://www.lpi.usra.edu/meetings/lpsc2008/pdf/2502.pdf

Scott, E. R. D. (2006, June). Meteoritical and dynamical constraints on the growth mechanisms and formation times of asteroids and Jupiter. *Icarus*. Retrieved from http://arxiv.org/ftp/astro-ph/papers/0607/0607317.pdf

Sheppard, S. S., & Trujillo, C. A. (2006, June). A thick cloud of Neptune Trojans and their colors. Science Express web site.

Soderblom, L. A., Becker T. L., Bennet, G., et al, (2002, May). Observations of comet 19P/Borrelly by the miniature integrated camera and spectrometer aboard Deep Space 1. *Science*, *296*, 1087–1091.

Srinivasan, G. (2007, July). The crystallization age of eucrite zircon. *Science*, *317*, 345–347.

Stardust mission. Retrieved from http://stardust.jpl.nasa.gov/

Stardust special section. (2006, December). *Science*, *314*, 1707–1739.

Stern, A., & Campins, H. (1996, August). Chiron and the Centaurs: Escapees from the Kuiper Belt. *Nature*, *382*, 507–510.

Stern, S. A. (2003, August). The evolution of comets in the Oort Cloud and Kuiper Belt. *Nature*, *424*, 639–642.

Sungrazing comets. Retrieved from http://sungrazer.nrl.navy.mil/

Sunshine, J. M., A'Hearn, M. F., Groussin, O., et al. (2006, March). Exposed water ice deposits on the surface of comet Tempel 1. *Science*, *311*, 1453–1455.

Sunshine, J. M., Connolly, H. C., McCoy, T. J., et al, (2002, May). Observations of comet 19P/Borrelly by the miniature integrated camera and spectrometer aboard Deep Space 1 (2008, April).

Ancient asteroids enriched in refractory inclusions. *Science, 320,* 514–517.

Taylor, G. J. & Martel, L. M. V. (2008, December). Wee rocky droplets in comet dust. *PSR Discoveries.* Retrieved from http://www.psrd.hawaii.edu/Dec08/cometDust.html

Taylor, G. J. (1999, August). Honeycombed asteroids. *PSR Discoveries.* http://www.psrd.hawaii.edu/Aug99/asteroidDensity.html

Taylor, G. J. (2002, February). The composition of asteroid 433 Eros. *PSR Discoveries.* Retrieved from http://www.psrd.hawaii.edu/Feb02/eros.html

Taylor, G. J. (2003, April). Asteroidal lava flows. *PSR Discoveries.*

Taylor, G. J. (2005, May). Making sense of droplets inside droplets. *PSR Discoveries.* http://www.psrd.hawaii.edu/May05/chondrulesCAIs.html

Taylor, G. J. (2006, August). Wandering gas giants and lunar bombardment. *Nature, 435,* 462–465.

Taylor, G. J. (2006, May). Interstellar organic matter in meteorites. *PSR Discoveries.* Retrieved from http://www.psrd.hawaii.edu/May06/meteoriteOrganics.html

Taylor, G. J. (2007, January). Organic globules from the cold far reaches of the proto-solar disk. *PSR Discoveries.* Retrieved from http://www.psrd.hawaii.edu/Jan07/organicGlobules.html

Taylor, G. J. (2009, June). The complicated geologic history of Asteroid 4 vest. *PSR Discoveries.* Retrieved from http://www.psrd.hawaii.edu/June09/Vesta.granite-like.html

The Meteoritical Society. Retrieved from http://www.meteoriticalsociety.org/

Thomas P. C., Parker, J. Wm., McFadden, L. A., et al. (2005, September). Differentiation of the asteroid Ceres as revealed by its shape. *Nature, 437,* 224–226.

Thomas, P. C., Binzel, R. P., Gaffey, M. J., et al. (1997, September). Impact excavation on asteroid 4 Vesta: Hubble Space Telescope results. *Science, 277,* 1492–1495.

Vernazza, P., Binzel, R. P., Rossi, A., et al. (2009, April). Solar wind as the origin of rapid reddening of asteroid surfaces. *Nature, 458,* 993–995.

Vernazza, P., Vernazza, P.,Binzel, R. P., et al, (2008, August). Compositional differences between meteorites and near-Earth asteroids. *Nature, 454,* 858–860.

Vinković, D. (2009, May). Radiation-pressure mixing of large dust grains in protoplanetary disks. *Nature, 459,* 227–229.

Walsh, K.J., Richardson, D.C., & Michel, P. (2008, July). Rotational breakup as the origin of small binary asteroids. *Nature, 454,* 188–191.

Weissman, P. (2003). The Oort Cloud. *Scientific American Special Edition.*

Weissman, P. (2006, February). A comet tale. *Sky & Telescope.*

Chapter 14: Exoplanets

Boss, A. (2009, March). Brave new worlds. *Physics World Online.* Retrieved from http://physicsworld.com/cws/article/print/37980

California & Carnegie Planet Search. Retrieved from http://exoplanets.org/

Carlisle, C. M. (2009, January). The race to find alien Earths. *Sky & Telescope.*

Charbonneau, D., Berta, Z. K., Irwin. J., et al. (2009, December). A super-Earth transiting a nearby low-mass star. *Nature, 462,* 891–894.

Corot mission. Retrieved from http://smsc.cnes.fr/COROT/index.htm

Davies, P. (2004, June). New hope for life beyond Earth. *Sky & Telescope.*

Deming, D., & Seager, S. (2009, November). Light and shadow from distant worlds. *Nature, 462,* 301–306.

DiGregorio, B. E. (2004, April). Life on Mars? *Sky & Telescope.*

ESO Exoplanet media kit. Retrieved from http://www.eso.org/public/products/presskits/exoplanets/

Exoplanet Education Articles. (2010, January). [Special issue]. *Astrobiology, 10*(1), 5–126.

Extrasolar Planets Encyclopedia. Retrieved from http://exoplanet.eu

Gaidos, Eric, Nader Haghighipour, N., Agol. E., et al. (2007, October). New worlds on the horizon: Earth-sized planets close to other stars. *Science, 318,* 210–213.

Gaudi, B. S., Gaudi, B. S., Bennett, D. P., et al. (2008, February). Discovery of Jupiter/Saturn analog with gravitational microlensing. *Science, 319,* 927–930.

Greaves, J. S. (2005, January). Disks around stars and the growth of planetary systems. *Science, 307,* 68–71.

Jakosky, B. M. (1998, Spring). Searching for life in our Solar System. *Magnificent Cosmos – Scientific American Special.*

JPL PlanetQuest. Retrieved from http://planetquest.jpl.nasa.gov/

Kepler mission. Retrieved from http://www.kepler.arc.nasa.gov/

Laughlin, G., Derming D., Langton, J., et al. (2009, January). Rapid heating of the atmosphere of an extrasolar planet. *Nature, 457,* 562–564.

Lissauer, J. J., Fabrycky, D. C., Ford, E. B., et al. (2011, February). A closely packed system of low-mass, low-density planets transiting Kepler-11. *Nature, 470,* 53–58.

Lovis, C., Ségransan, D., Mayor, M., et al. (2010, August). The HARPS search for southern extra-solar planets. XXVII. Up to seven planets orbiting HD 10180: probing the architecture of low-mass planetary systems. *Astronomy & Astrophysics* manuscript no. HD10180. Retrieved from http://www.eso.org/public/archives/releases/sciencepapers/eso1035/eso1035.pdf

Malhotra R. (1999, September). Migrating planets. *Scientific American.*

Marcy, G. W., & Butler, R. P. (1998, Spring). Giant Planets Orbiting Faraway Stars. *Magnificent Cosmos – Scientific American Special.*

Masses and orbital characteristics of extrasolar planets. Retrieved from http://exoplanets.org/almanacframe.html

Naeye, R. (2005, January). Planetary harmony. *Sky & Telescope.*

Naeye, R. (2009, December). Amateur exoplanets. *Sky & Telescope.*

Naeye, R. (2009, March). Exoplanets imaged at last. *Sky & Telescope.*

NASA Astrobiology Institute. Retrieved from http://nai.arc.nasa.gov/

NASA Astrobiology magazine. Retrieved from http://www.astrobio.net/

Planetary Society catalog of exoplanets. Retrieved from http://www.planetary.org/exoplanets/

Planets around pulsars. Retrieved from http://www.astro.psu.edu/users/alex/pulsar_planets_text.htm

Raymond, S. N., Mandel, A. V., Sigurdsson, S. (2006, September). Exotic Earths: Forming habitable worlds with giant planet migration. *Science, 313,* 1413–1416.

Rivera E.J., Lissauer, J. J., Butler, R. P., et al. (2005). A 7.5 Earth-mass planet orbiting the nearby star, GJ 876: Technical figures and table. *Astrophysical Journal, 634,* 625–640.

Schilling, G. (1999, October). From a swirl of dust, a planet is born. *Science, 286,* 66–68.

Seager, S. (2006, February). Unveiling distant worlds. *Sky & Telescope.*

Seager, S. (2008, January). Alien Earths from A to Z. *Sky & Telescope.*

Seager, S. (2010, October). The hunt for super-Earths. *Sky & Telescope.*

SETI Institute. Retrieved from http://www.seti.org/

Setiawan, J., Henning, Th., Launhardt, R., et al. (2008, January). A young massive planet in a star–disk system. *Nature, 451,* 38–41.

Shiga, D. (2004, April). Imaging exoplanets. *Sky & Telescope.*

Silvotti, R., Schuh, S., Janulis, R., et al. (2007, September). A giant planet orbiting the 'extreme horizontal branch' star V391 Pegasi. *Nature, 449,* 189–191.

Snellen, I. A. G. (2009, May). The changing phases of extrasolar planet CoRoT-1b. *Nature, 459,* 543–545.

The search for life. (2011, February). *Astronomy & Geophysics special issue,* Wiley-Blackwell.

Thommes, E. W., Matsumura, S., & Rasio, F. A., (2008, August). Gas disks to gas giants: Simulating the birth of planetary systems. *Science, 321,* 814–817.

Tinetti, Giovanna, Vidal-Madjar, A., Liang, M-C., et al. (2007, July 12). Water vapour in the atmosphere of a transiting extrasolar planet. *Nature, 448,* 169–171.

Tingley, B. (2008, May). The first Earth-size exoplanet. *Sky & Telescope.*

Vogel, G. (1999, October). Expanding the habitable zone. *Science, 286,* 70–71.

Vogt, Steven S., Bulter, R. P., Rivera, E. J., et al. The Lick-Carnegie exoplanet survey: A 3.1M_ Planet in the habitable zone of the nearby M3V Star Gliese 581. *Astrophysical Journal.* Retrieved from http://www.ucolick.org/%7Evogt/ms_press-1.pdf and http://arxiv.org/abs/1009.5733

Weinberger, A. J. (2008, November). Building planets in disks of chaos. *Sky & Telescope.*

Wolszczan, A. (1994, April). Confirmation of earth-mass planets orbiting the millisecond pulsar PSR B1257 + 12, *Science, 264,* 538–542.

Index

Exploring the Solar System, First Edition. Peter Bond.
© 2012 by Peter Bond. Published 2012 by Blackwell Publishing Ltd.